Lecture Notes in Physics

Volume 987

The series Lecture Notes in Physics (LNP), founded in 1969, reports new developments in physics research and teaching - quickly and informally, but with a high quality and the explicit aim to summarize and communicate current knowledge in an accessible way. Books published in this series are conceived as bridging material between advanced graduate textbooks and the forefront of research and to serve three purposes:

- to be a compact and modern up-to-date source of reference on a well-defined topic;
- to serve as an accessible introduction to the field to postgraduate students and non-specialist researchers from related areas;
- to be a source of advanced teaching material for specialized seminars, courses and schools.

Both monographs and multi-author volumes will be considered for publication. Edited volumes should however consist of a very limited number of contributions only. Proceedings will not be considered for LNP.

Volumes published in LNP are disseminated both in print and in electronic formats, the electronic archive being available at springerlink.com. The series content is indexed, abstracted and referenced by many abstracting and information services, bibliographic networks, subscription agencies, library networks, and consortia.

Proposals should be sent to a member of the Editorial Board, or directly to the responsible editor at Springer:

Dr Lisa Scalone
Springer Nature
Physics
Tiergartenstrasse 17
69121 Heidelberg, Germany
lisa.scalone@springernature.com

More information about this series at http://www.springer.com/series/5304

Francesco Becattini • Jinfeng Liao •
Michael Lisa

Editors

Strongly Interacting Matter under Rotation

Springer

Editors
Francesco Becattini
Dipartimento di Fisica e Astronomia
Universita degli Studi Firenze
Sesto Fiorentino, Italy

Jinfeng Liao
Department of Physics
Indiana University
Bloomington, IN, USA

Michael Lisa
Department of Physics
The Ohio State University
Columbus, OH, USA

ISSN 0075-8450   ISSN 1616-6361   (electronic)
Lecture Notes in Physics
ISBN 978-3-030-71426-0   ISBN 978-3-030-71427-7   (eBook)
https://doi.org/10.1007/978-3-030-71427-7

This Springer imprint is published by the registered company Springer Nature Switzerland AG
The registered company address is: Gewerbestrasse 11, 6330 Cham, Switzerland

Contents

About the Editors

Francesco Becattini is a theoretical physicist and Professor of Physics at the University of Florence, Italy. He studied in Florence, Pisa, and CERN, and got his PhD at at the University of Florence. He is an associate member of the INFN and a member of the European science academy Academia Europaea. He is the author of many highly-cited papers in the field of relativistic heavy ion physics. He is a recognized world expert in the subfields of statistical models and of spin and polarization where he was the initiator of the quantum statistical approach and of the first hydrodynamic predictions. Besides heavy ion physics, his main research interests involve relativistic and quantum statistical mechanics, statistical quantum field theory and relativistic hydrodynamics.

Jinfeng Liao is Professor of Physics at the Indiana University Bloomington, USA. He obtained Ph.D. in theoretical nuclear physics from Stony Brook University in the U.S., after studying for B.S. and M.S. at Tsinghua University in China. He then held postdoctoral positions at Lawrence Berkeley National Laboratory and at Brookhaven National Laboratory, before joining the Physics Department of Indiana University Bloomington as a junior faculty in 2011. He was a RBRC Fellow of RIKEN BNL Research Center as well as a recipient of the CAREER Award by the U.S. National Science Foundation. Being the author or co-author of over 110 papers, he has made his own mark in the research field of theoretical nuclear physics,

with notable contributions to the study of novel
phenomena related to topology, chirality, vorticity and
magnetic fields in heavy ion collisions.

Michael Lisa is an experimental nuclear physicist and
Professor of Physics at the Ohio State University. After
undergraduate and master's degrees at the University of
Notre Dame and State University of New York at Stony
Brook, he earned his PhD at Michigan State University
and went on to postdoctoral research at Lawrence
Berkeley National Laboratory. He is a Fellow of the
American Physical Society, an Ohio State University
Distinguished Scholar, and a Fulbright Scholar. He is
co-author of more than 600 publications, has given
more than 150 research presentations, and published an
undergraduate textbook on the Physics of Sports used at
several universities. He is a recognized world leader in
the subfield of femtoscopy, which measures the
space-time substructure of nuclear reactions at the
femtometer scale. He and his graduate student first
discovered global hyperon polarization in relativistic
heavy ion collisions, a measurement that led to the
growing subfield which is the focus of this volume.

Strongly Interacting Matter Under Rotation: An Introduction

Francesco Becattini, Jinfeng Liao and Michael Lisa

Abstract

Ultrarelativistic collisions between heavy nuclei briefly generate the Quark–Gluon Plasma (QGP), a new state of matter characterized by deconfined partons last seen microseconds after the Big Bang. The properties of the QGP are of intense interest, and a large community has developed over several decades, to produce, measure, and understand this primordial plasma. The plasma is now recognized to be a strongly coupled fluid with remarkable properties, and hydrodynamics is commonly used to quantify and model the system. An important feature of any fluid is its vorticity, related to the local angular momentum density; however, this degree of freedom has received relatively little attention because no experimental signals of vorticity had been detected. Thanks to recent high-statistics datasets from experiments with precision tracking and complete kinetic coverage at collider energies, hyperon spin polarization measurements have begun to uncover the vorticity of the QGP created at the Relativistic Heavy Ion Collider. The injection of this new degree of freedom into a relatively mature field of research represents an enormous opportunity to generate new insights into the physics of the QGP. The community has responded with enthusiasm, and this book represents some of the diverse lines of inquiry into aspects of strongly interacting matter under rotation.

F. Becattini
University of Florence, Florence, Italy
e-mail: becattini@fi.infn.it

J. Liao (✉)
Physics Department and Center for Exploration of Energy and Matter, Indiana University, 2401 N Milo B. Sampson Lane, Bloomington, IN 47408, USA
e-mail: liaoji@indiana.edu

M. Lisa (✉)
Department of Physics, The Ohio State University, 191 West Woodruff Avenue, Columbus, OH 43210, USA
e-mail: lisa.1@osu.edu

© Springer Nature Switzerland AG 2021
F. Becattini et al. (eds.), *Strongly Interacting Matter under Rotation*,
Lecture Notes in Physics 987,
https://doi.org/10.1007/978-3-030-71427-7_1

1.1   Milestones

In 2005, Liang and Wang [1] predicted that spin–orbit coupling would polarize strange quarks created in non-central heavy ion collisions, resulting in emitted Λ hyperons globally polarized along the direction of the collision angular momentum. The magnitude and momentum dependence of the predicted polarization depended on details of specific models of quark–quark potentials, small-angle scattering approximations, and details of hadronization mechanisms.

In 2008, Becattini and collaborators [2] noted that in a hydrodynamic picture, local thermodynamic equilibrium implies a relation between the spin polarization and the rotational flow structure (vorticity). In the hydrodynamic model, vorticity can be extracted directly from the evolution, with no need to appeal to specific microscopic processes. In 2013, an equation relating the polarization of Λ hyperons and thermal vorticity was derived [3] and such polarization was predicted to be at the level of a few percent. The first result regarding the systematic dependence of this effect on the collision beam energy, particularly in the range relevant to the beam energy scan program at the Relativistic Heavy Ion Collider (RHIC), was reported in a 2016 paper [4], providing a highly relevant insight for the later experimental measurements.

In 2017, the STAR Collaboration published [5] the first observation of global Λ polarization from non-central heavy ion collisions. As discussed below and throughout this volume, most theoretical interpretations of these observations are based upon this hydrodynamic approach.

While the phenomenon of global polarization was predicted based on particle–particle interaction, the success of quantitative predictions of the hydrodynamic model to reproduce experimental observations (discussed below) seem to confirm that for spin, as for many other observables, microscopic details are less important than bulk thermodynamic properties. Below, we discuss the hydrodynamic approach to vorticity and polarization, followed by experimental observations.

We will briefly discuss the related phenomenon of vector meson spin alignment, also predicted by Liang and Wang [6] in 2005. As of now, measurements of spin alignment at the Large Hadron Collider (LHC) and RHIC are difficult to explain in any theoretical approach.

1.2   Introduction

Twenty years ago, the world's first nuclear collider began producing heavy ion collisions at energies far surpassing those previously achievable in fixed-target experiments. The goal was to produce the Quark–Gluon Plasma (QGP)—a state of matter characterized by partonic (rather than hadronic) degrees of freedom. For decades, production and study of the QGP had long been the focus driving the field of relativistic heavy ion physics, as it holds the promise of shedding light on the non-perturbative region of Quantum Chromodynamics (QCD), the most poorly understood of the fundamental interactions in the Standard Model.

In 2005 [7–10], based on a systematic and comprehensive analysis of available data, the experimental collaborations at the Relativistic Heavy Ion Collider (RHIC) confirmed that QGP is indeed created in ultra-high-energy collisions. Furthermore, the data clearly indicated that the QGP was a strongly coupled fluid, contrary to some expectations that the plasma would be weakly coupled due to the combination of high temperatures and the running of the QCD coupling constant. The evidence driving this conclusion was the collective anisotropic emission distribution of hadrons from the collision—the so-called "elliptic flow." These very strong anisotropies (and the dependence upon mass and momentum) were nearly quantitatively consistent with expectations based on relativistic inviscid (ideal) hydrodynamics.

The discovery of nearly "perfect fluid" behavior had two major outcomes. Firstly, it prompted a re-evaluation of numerical QCD calculations performed on a lattice, the most reliable ab initio calculations of the strong interaction. While numerically correct, lattice calculations could be misinterpreted to suggest that a weakly coupled gas of quarks and gluons was the proper paradigm for modeling collisions at RHIC. It was also realized that the QGP near the pseudo-critical transition temperature is a peculiar system: unlike ordinary matter, its microscopic interaction length is comparable to the thermal de Broglie wavelength, making the kinetic collisional description inappropriate. Nevertheless, even under such unusual conditions, the local thermodynamic equilibrium concept and hydrodynamics are still valid. Hence, the discovery established relativistic fluid dynamics as the new paradigm for the bulk evolution of the system. Confronting increasingly sophisticated hydrodynamic calculations with data has produced valuable estimates of transport coefficients, initial parton distributions, and the QCD equation of state. Triangular and higher-order azimuthal correlations have probed the substructure of the fluid flow fields at ever finer scale.

A relativistic collision between heavy nuclei at finite impact parameter can involve angular momentum of order $10^{3\sim5}\hbar$. In a fluid, angular momentum can manifest as vorticity, rotational gradients of the flow, and temperature fields [2]. Until recently, this aspect of the plasma had been largely ignored, as there had been no experimental observation of its effects.

In 2017, the STAR Collaboration published an observation of global hyperon polarization in Au+Au collisions at RHIC, opening the potential to probe novel substructures of the QGP fluid at the finest possible scale. This is a rare case in which an entirely new direction is introduced to a mature field. It is especially exciting because the natural language for discussing vorticity—three-dimensional relativistic viscous hydrodynamics—has been developed to a high degree of sophistication by a large community of theorists. It is an opportunity for new insights into the physics of deconfined QCD matter, and the heavy ion community has responded with intense focus on the topic. This book represents a broad sampling of directions of inquiry into this new area of research.

1.3   Accessing Subatomic Vorticity

"Lumpy" azimuthal fluid flow patterns (elliptic flow, triangular flow, etc) may be measured by azimuthal correlations between the momenta of emitted particles; this is experimentally straight-forward. Orbital angular momentum in heavy ion collisions, on the other hand, is experimentally inaccessible. Instead, one relies on coupling between the orbital ("mechanical") angular momentum of the fluid and spin of the emitted particles. The first observation of such an effect was reported more than a century ago by Barnett [11], in which an uncharged and un-magnetized solid metal object, when set spinning, spontaneously magnetizes.[1]

The analogous effect in a *fluid*, coupling mechanical vorticity of the bulk fluid and quantum spin polarization, was first reported by Takahashi et al, in 2016 [12]. In their experiment, liquid mercury flowing through a channel acquired local vorticity due to viscous friction with the wall. Spin–vorticity coupling produced a polarization gradient that could then be detected directly through the inverse spin Hall effect. The results could be understood by expanding angular momentum conservation in fluid dynamics, to include angular momentum transfer between the liquid and electron spin [12].

In the Barnett and Takahashi experiments, the macroscopic rotational motion was a controlled variable and the spin polarization straightforward to measure. In high-energy nuclear collisions, the magnitude and direction of the angular momentum fluctuate from one event to the next, and a statistically significant measurement requires combining $\sim 10^7 - 10^8$ events. Furthermore, the particles whose polarization is to be measured are emitted at all angles at speeds approaching that of light.

These challenges are addressed by precision tracking and correlating detector subsystems in different regions of the experiment. In particular, the angular momentum in a collision is given by

$$\mathbf{J} = \mathbf{b} \times \mathbf{p}_{\text{beam}}, \qquad (1.1)$$

where the impact parameter, $\mathbf{b}$, is the transverse (to the beam direction) vector connecting the center of the target nucleus to that of the beam nucleus (where attention to the designation of beam and target is important [13]), and $\mathbf{p}_{\text{beam}}$ is the momentum of the beam in the collision center-of-momentum (c.o.m.) frame. The magnitude of the impact parameter, $|\mathbf{b}|$, is estimated by the total number of charged particles emitted roughly perpendicular to the beam in the collision c.o.m. frame, while its direction, $\hat{b}$, is estimated by the sidewards deflection of particles emitted close to the beam direction. See Fig. 1.1 for an illustration.

The flow pattern of the QGP fluid is complex and any local vorticity may fluctuate as a function of position within each droplet; however, the *average* vorticity must be parallel to $\mathbf{J}$ which is event-specific. For this reason, spin polarization projection along $\hat{J}$ is termed the "global" polarization.

---

[1]That the magnetization arose from spin polarization of the electrons was not known to Barnett and his contemporaries in 1915, as the concept of quantum spin was not introduced until nearly a decade later.

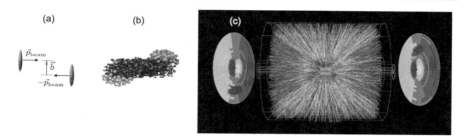

Fig. 1.1 The geometry of a collision. **a** Before collision: the angular momentum is determined by the impact parameter, **b**, an uncontrolled variable that fluctuates from one collision to the next. **b** In a non-central ($|\mathbf{b}| \neq 0$) collision, parts of the nuclei overlap, producing the QGP, while the so-called "spectators" continue to travel forward, experiencing only a slight impulse directed away from the collision. **c** One reconstructed event in two subsystems in STAR experiment. The Time Projection Chamber (TPC) [14] records $\sim 10^3$ charged particles emitted from the QGP created in the collision, while the Event Plane Detector (EPD) [15] measures spectator fragments. The magnitude and direction of **b** are determined, respectively, by the number of charged particles measured in the TPC and the anisotropic hit pattern in the EPD

Having determined the direction[2] of the average vorticity, the second challenge is to measure the spin polarization along that direction.

If the QGP fluid does indeed have non-vanishing vorticity, and if thermalization (complete or partial) of orbital and spin degrees of freedom does occur, then presumably all particles emitted in the collision will have their average spins aligned with $\hat{J}$. Of the zoo of particle types emitted in a heavy ion collision, the spin directions of only a few are easily measurable. In particular, particles undergoing parity-violating weak decay betray their spin direction through asymmetries in the momentum distribution of their daughters. Of this already restricted subset of particles, only a few are created in reasonable numbers to allow a significant measurement. The best candidate is the Λ hyperon, which can be cleanly measured by its $p + \pi^-$ decay in the TPC, as seen in panel (a) of Fig. 1.2. The decay topology is sketched in panel (b) of Fig. 1.2. An ensemble with polarization $\mathbf{P}_\Lambda$ will preferentially emit daughter protons along the direction of polarization according to

$$\frac{dN}{d\cos\theta^*} = \tfrac{1}{2}\left(1 + \alpha_\Lambda \mathbf{P}_\Lambda \cdot \hat{p}_p^*\right), \tag{1.2}$$

where θ^* is the angle between the polarization and daughter proton momentum $\mathbf{p}_p^*$ in the hyperon rest frame. The decay parameter $\alpha_\Lambda = 0.732$ determines the strength of the effect.

---

[2]In principle, the magnitude $|\mathbf{J}|$ of the collision's angular momentum may be estimated as well. However, not all of this angular momentum is transferred to the plasma at midrapidity [4], so usually only the direction $\hat{J}$ is of interest. This quantity is the only important ingredient to estimate vorticity in any event.

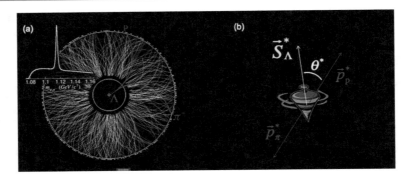

Fig. 1.2 a A Λ hyperon detected in the STAR TPC by combining its charged proton and pion daughters. Inset: the invariant mass of daughter pairs shows a clear peak at the Λ mass. **b** In the parity-violating decay topology, the daughter proton tends to be emitted in the direction of the parent Λ hyperon, in the Λ center of mass frame

The global polarization is then measured by correlating information from both detector subsystems:

$$\left\langle \mathbf{P}_\Lambda \cdot \hat{J} \right\rangle = \frac{8}{\pi \alpha_H R_{\mathrm{EP}}^{(1)}} \left\langle \sin \left(\Psi_{\mathrm{EP},1} - \phi_p^* \right) \right\rangle, \tag{1.3}$$

where ϕ_p^* is the azimuthal angle of the daughter proton in the parent hyperon frame. In Eq. 1.3, $\Psi_{EP,1}$ is the first-order event plane angle, an estimator of the azimuthal angle of the impact parameter $\mathbf{b}$; the resolution of this estimation is $R_{EP}^{(1)}$. Standard

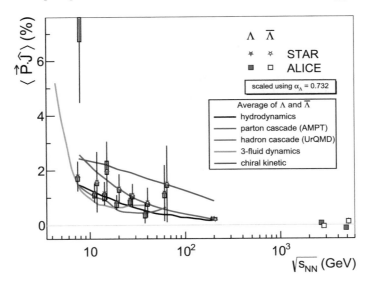

Fig. 1.3 The world dataset of global Λ hyperon polarization in relativistic heavy ion collisions compared to expectations from hydrodynamic and transport simulations. Figure from [16]

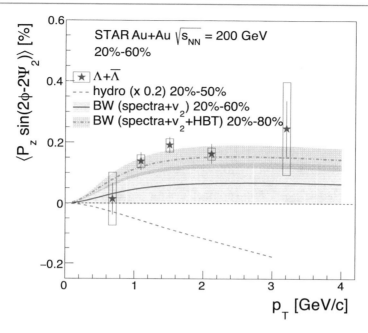

Fig. 1.4 The second-order Fourier coefficient of the azimuthal oscillation of the longitudinal component of the Λ hyperon polarization. Figure from [19]

methods have been developed to extract both the event plane and the resolution from anisotropic particle distributions in the EPD.

The discussion thus far has described the global Λ hyperon polarization measurement in the STAR experiment at RHIC. The ALICE experiment at the LHC performed a similar analysis, tracking charged hyperon daughters with a gas-filled TPC at midrapidity and measuring $\Psi_{EP,1}$ with segmented detectors at forward rapidity. The STAR and ALICE measurements thus far comprise the world's dataset on global polarization and are shown in Fig. 1.3.

We offer some general remarks on Fig. 1.3 in the next section, but at the experimental level, we note that the statistical uncertainties at low $\sqrt{s_{NN}}$ are large. These uncertainties are determined by (1) the number of collision events recorded by the experiment; (2) the per-event hyperon yield; (3) the event plane resolution $R_{EP}^{(1)}$. Measurements by the STAR Collaboration in the second phase of the RHIC Beam Energy Scan (BES-II) [17] will have an order of magnitude better statistics [18] and better event plane resolution [15]; overall, the precision should increase roughly eight-fold, allowing important systematic studies [16] not currently possible.

The average "global" polarization vector must point along the direction of $\hat{J}$. On the other hand, the mean spin polarization vector for particles with specific momentum has three components that can be also measured. The component along the beam (longitudinal component) is expected to show a 2nd-order azimuthal oscillation relative to the event plane. The amplitude and phase of this oscillation has been measured for Au+Au collisions at $\sqrt{s_{NN}} = 200$ GeV by the STAR Collaboration.

Figure 1.4 shows the transverse momentum dependence of the 2nd Fourier component for non-central collisions. Thanks to the excellent tracking, good event plane resolution, and a high-statistics dataset available at RHIC top energy, an oscillating sub-percent polarization signal is easily measured.

1.4   From Signal to Physics

1.4.1   Hydrodynamics as the Basis to Understand Hyperon Polarization

As we discussed in Sect. 1.1, the original idea for global Λ polarization in heavy ion collisions was based on microscopic processes that drove the initial state, the transfer of angular momentum from orbital to spin degrees of freedom, and the subsequent hadronization mechanism. Assumptions and parameters were required to compute each of these components of the calculation. A detailed discussion along this line can be found in Chap. 7.

The tremendous success of hydrodynamics to heavy ion physics suggests that the myriad details of microscopic processes undoubtedly at play in these complex collisions are eventually unimportant, as the system approaches local equilibrium quickly. In the earliest days of RHIC, ideal (inviscid), boost-invariant hydrodynamic calculations with simple initial conditions largely reproduced—nearly "out of the box"—the multiplicity, p_T and mass systematics of measured elliptic flow. This success gave some confidence that equilibrium hydrodynamics was a good paradigm to understand the collective physics of heavy ion collisions.

In the subsequent decades, several important insights have been achieved by working within this framework, using *details* in the data to probe the partonic structure of the initial state, transport coefficients, and hadronization mechanisms. These insights required considerable elaboration of the initial simple models, incorporating viscosity, baryochemical currents, vorticity, three-dimensional dynamics, and event-by-event fluctuations in the initial state. However, the close resemblance of the initial simple calculations with observations set this fruitful enterprise on firm ground.

Figure 1.3 suggests that the same situation exists in the study of global hyperon polarization. Theoretical curves show predictions from hydrodynamic and transport calculations, in which fluid vorticity is assumed to equilibrate with Λ spin degrees of freedom to produce the polarization. Vorticity—more properly, thermal vorticity—is calculated directly from the flow field in the hydrodynamic calculations, as discussed in detail in Chap. 8. On the other hand, in the transport calculations, flow and temperature fields are calculated from the motion of multiple particles in coarse-grained spatial cells; this implicitly assumed local thermalization; see detailed discussions in Chap. 9. Eventually, in both methods, the polarization of a spin 1/2 fermion is obtained from the same formula relating mean spin to the thermal vorticity ϖ at the leading order [3]:

$$S^\mu(p) = -\frac{1}{8m}\epsilon^{\mu\nu\rho\sigma}p_\sigma\frac{\int_\Sigma d\Sigma \cdot p\varpi_{\nu\rho}n_F(1-n_F)}{\int_\Sigma d\Sigma \cdot pn_F}, \qquad (1.4)$$

where $S^\mu(p)$ is the mean spin vector and n_F is the covariant Fermi–Dirac distribution function. The thermal vorticity is defined as the antisymmetric derivative of the four-temperature vector field, that is:

$$\varpi_{\mu\nu} = \frac{1}{2}\left[\partial_\nu\left(\frac{1}{T}u_\mu\right) - \partial_\mu\left(\frac{1}{T}u_\nu\right)\right], \qquad (1.5)$$

where T and u_μ are local temperature field and flow velocity field, respectively. The integration in Eq. (1.4) is performed on the 3-D hadronization hypersurface Σ. The polarization vector P^μ is simply $S^\mu/|S|$ and its global, momentum integrated, value in the particle rest frame turns out to be directed along the angular momentum vector, so that, approximately one has:

$$\langle\mathbf{P}\rangle\cdot\hat{J} \approx \frac{1}{2}\langle\varpi\rangle\cdot\hat{J}, \qquad (1.6)$$

where the $\langle\varpi\rangle$ is the mean thermal vorticity value over the hadronization hypersurface. The above relation is a direct manifestation of the rotational polarization of microscopic spin. The theoretical underpinning of this phenomenon is to be fully elaborated through a variety of approaches such as quantum field theory (in Chaps. 2, 3, and 4) and relativistic kinetic theory (in Chaps. 5 and 6).

For the most part, these models have been used to understand other observations from heavy ion collisions, and the results in Fig. 1.3 are obtained largely "out of the box". The quantitative agreement, as well as the universal decreasing trend of polarization with $\sqrt{s_{NN}}$ (despite the fact that $|\mathbf{J}|$ *increases* with increasing collision energy) is a clear indication that we have at hand a paradigm to understand hyperon polarization.

That said, there are strong tensions with the existing theoretical expectations in certain observables. One is seen in Fig. 1.4; the same hydrodynamic calculation that reproduced $\langle\mathbf{P}_\Lambda\cdot\hat{J}\rangle$ with no special tuning predicts the wrong sign of the longitudinal polarization, $\langle\mathbf{P}_\Lambda\cdot\hat{z}\rangle$. Hence it seems that, similar to the early collective flow studies, the framework is well-grounded, while there is much to learn from the details. Chapters 8, 9, and 10 in this Volume provide an in-depth discussion on the phenomenology study based on this framework. More broadly, the presence of global rotation has opened a new dimension for investigating its nontrivial effects, for example, on the phase structures of matter (see Chap. 11) or on the interplay between orbital and spin angular momentum (c.f. Chap. 12).

1.4.2   Vector Meson Spin Alignment—More Complicated Physics?

In an equilibrium picture, the spins of all emitted particles will be aligned with the total angular momentum of the system. In addition to Λ and $\overline{\Lambda}$ hyperons discussed above, recent results from the STAR collaboration indicate consistent polarization of Ξ, $\overline{\Xi}$ and Ω baryons [21]. Besides baryons, in principle polarization could be detected for vector mesons such as K^* or ϕ.

The spin of a vector meson is quantified by the 3×3 spin-density matrix $\rho_{i,j}$. Becattini discusses the coupling of this quantity to fluid vorticity in Chap. 2. Due to the parity-conserving nature of their strong decay, the elements $\rho_{1,1}$ and $\rho_{-1,-1}$ cannot be separately determined. Because the trace is unity, there is only one independent diagonal element, $\rho_{0,0}$ which quantifies the component of the meson spin perpendicular to the quantization axis. As with the baryons, for the average spin, the axis of interest is $\hat{J}$, perpendicular to the event plane. Random alignment of spins would yield $\rho_{-1,-1} = \rho_{0,0} = \rho_{1,1} = \frac{1}{3}$. Given only experimental access to $\rho_{0,0}$, it is impossible to determine whether the meson spin is parallel or anti-parallel to $\hat{J}$, but in either case, spin alignment would imply $\rho_{0,0} < \frac{1}{3}$ [22].

The two-particle decay topology of a vector meson is related to the alignment according to [23]:

$$\frac{\mathrm{d}N}{\mathrm{d}\cos\theta^*} = \frac{3}{4}\left[1 - \rho_{0,0} + \left(3\rho_{0,0} - 1\right)\cos^2\theta^*\right], \tag{1.7}$$

where θ^* is the angle between the parent spin and a daughter momentum in the parent's rest frame. At local thermodynamic equilibrium, the alignment is quadratic in thermal vorticity to first order [16,24]:

$$\frac{1}{3} - \rho_{0,0} \approx \frac{4}{9}\varpi^2. \tag{1.8}$$

Therefore, consistency with the hyperon results would lead to the expectation $\frac{1}{3} - \rho_{0,0} \approx 10^{-3}$.

Experimental results deviate strongly from that expectation. Figures 1.5 and 1.6 show K^{0*} and ϕ alignment measurements from the STAR and ALICE experiments at RHIC and LHC, respectively. In all cases, $|\frac{1}{3} - \rho_{0,0}| \approx 0.1$, two orders of magnitude larger than expectations based on P_Λ and vorticity considerations. Perhaps more surprisingly, STAR reports $\rho_{0,0} > \frac{1}{3}$ for ϕ mesons.

As discussed above, the hydrodynamic equilibrium ansatz seems a reliable baseline for understanding hyperon polarization, as it is for understanding much else in heavy ion physics. However, there may be many other effects at play. In their original paper [6], Liang and Wang considered different hadronization mechanisms involving polarized quarks. If vector mesons are produced by simple coalescence of a quark and antiquark with polarizations P_q and $P_{\overline{q}}$, respectively, then

$$\rho_{0,0}^{\text{meson}} = \frac{1 - P_q P_{\overline{q}}}{3 + P_q P_{\overline{q}}} \approx \frac{1}{3} - \frac{4}{9}\left(P_q P_{\overline{q}}\right)^2, \tag{1.9}$$

where the approximation holds for small polarizations. This is consistent with the hydrodynamic equilibrium prediction (Eq. 1.8) if $P_q = P_{\overline{q}} = \varpi$. It is not possible to reconcile the ALICE measurements of very small values of hyperon with large values of $|\frac{1}{3} - \rho_{0,0}|$ in a simple recombination picture [25].

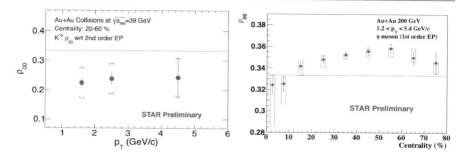

Fig. 1.5 Vector meson alignment in Au+Au collisions at $\sqrt{s_{NN}} = 200$ GeV, measured by the STAR Collaboration at RHIC [20]. Left: $\rho_{0,0}$ for K^{*0} mesons as a function of transverse momentum for mid-central collisions. Right: $\rho_{0,0}$ for ϕ mesons as a function of centrality. Dashed lines indicate $\rho_{0,0} = \frac{1}{3}$, corresponding to no alignment with the normal to the event plane

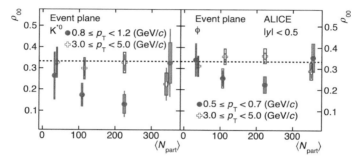

Fig. 1.6 Vector meson alignment, for Pb+Pb collisions at $\sqrt{s_{NN}} = 2.76$ TeV, measured by the ALICE Collaboration at the LHC [25]. Left (right) panel shows $\rho_{0,0}$ for K^{*0} (ϕ) mesons as a function of collision centrality, for two ranges in transverse momentum. Dashed lines indicate $\rho_{0,0} = \frac{1}{3}$, corresponding to no alignment with the normal to the event plane

Perhaps even more surprising are measurements of the STAR Collaboration at RHIC. For K^*, they report [20] values of $\frac{1}{3} - \rho_{0,0}$ similarly large as those seen at the LHC, but for ϕ mesons, $\rho_{0,0} > \frac{1}{3}$; c.f. Fig. 1.5. Liang and Wang pointed out that hadronization via polarized quark fragmentation could result in $\rho_{0,0} > \frac{1}{3}$. This mechanism may be most important at large rapidity or transverse momentum, but could in principle play a role at midrapidity, where these measurements are made. However, naively, if fragmentation is the dominant hadronization mechanism, K^* and ϕ should be affected similarly. Furthermore, it would seem natural that quark hadronization would be more important at LHC energies than at RHIC.

Sheng, Olivia and Wang [26] propose that an entirely new physical effect could be at play, in which a hypothetical mean ϕ field couples to the system angular momentum. Depending on the values of several parameters, $\rho_{0,0}^{\phi}$ could be greater or less than $\frac{1}{3}$. In principle, by fine-tuning [26] the energy dependence of four parameters, this model might accommodate $\rho_{0,0}^{\phi} > \frac{1}{3}$ at RHIC energies and $\rho_{0,0}^{\phi} < \frac{1}{3}$ at the LHC.

Because this model is not expected to apply to K^{*0} mesons [26], it will be important to identify independent measurements that can constrain and verify its assumptions.

In summary, it is clear that the situation with the spin alignment of vector mesons is very different than that for hyperon polarization. In the latter case, the equilibrium hydrodynamic paradigm which works well for other aspects of heavy ion collisions seems a reasonable starting point; polarization then allows a more sensitive probe of the system evolution at the finest scales. For the vector mesons, however, it is clear that observations cannot be explained by this established paradigm. Competing effects from multiple hadronization mechanisms, hadronic effects, and novel mean fields may be at play, differentially affecting the different particle species and different collision energies. See Chap. 7 by Gao, et al. in this volume, for an extensive discussion. The phenomenon of vector meson spin alignment deserves continued intense theoretical focus; at the moment, the situation is too unclear to summarize what might be learned.

1.4.3   Future Experimental Work

Among the most pressing issues in the field of heavy ion physics is the existence and consequences of an intense, long-lived magnetic field. Its presence could allow experimental access to novel effects due to chiral symmetry restoration. Because Λ and $\overline{\Lambda}$ have opposite magnetic moments, a strong **B**-field at hadronization would lead to a polarization "splitting" [24,27,28]. The magnitude of the splitting remains below the statistical sensitivity of existing measurements, but the ongoing BES-II campaign at RHIC is expected to either discover the splitting or set meaningful limits on possible magnetic effects.

While the *average* ("global") polarization must align with the total angular momentum of the collision, hydrodynamic and transport simulations predict a rich flow structure featuring nontrivial local vorticity. The longitudinal polarization results in Fig. 1.4 represent the first observation of such an effect. However, more complicated effects may be present on an event-by-event basis, leading to vorticity "hot spots" that may be revealed by spin–spin correlations [29]. Experimental searches for such a signal are ongoing at RHIC, but two-particle tracking artifacts make them highly challenging.

As discussed above, the physics driving vector meson spin alignment is apparently much more complicated than that behind Λ polarization. The STAR Collaboration at RHIC has presented a study [21] of polarization of Ξ and Ω hyperons which are consistent with the Λ polarizations, with small mass-dependent effects.

While the total system angular momentum decreases with reduced collision energy, the largest global polarization is observed at the lowest energy. It will be important to measure polarization at still lower energies below the energy threshold for QGP formation and at energy densities below the limits of applicability of hydrodynamics. The BES-II program at RHIC includes a fixed-target campaign already producing results [30] in this regime. Much higher statistics datasets at low energy are expected at the NICA and FAIR facilities soon to commence operation.

The energy dependence of the polarization signal may reflect an evolution of rotational flow structure away from midrapidity, where measurements have focused thus far, as the collision energy increases. Experiments with good tracking near beam rapidity may probe strong vorticity resulting from the breakdown of longitudinal boost-invariance, challenging hydrodynamic and transport simulations more stringently than possible previously.

1.5   Summary and Outlook

Over several decades, the field of relativistic heavy ion physics has matured and focused on the creation and study of the quark–gluon plasma. Hydrodynamics and transport theory have provided a useful paradigm in which to interpret a wide diversity of experimental results from high-energy collisions at RHIC and the LHC. Theory and models based on this paradigm have become increasingly sophisticated, simulating the entire evolution of the dynamic system and making quantitative connection to the initial state and fundamental transport coefficients.

The observation of rotational phenomena has opened an exciting new direction into this well-developed and fertile environment, a rare example of a truly new development in a mature field. First measurements of global hyperon polarization are largely consistent with predictions from existing hydrodynamic and transport simulations, indicating that the tools are at hand, to understand the phenomenon. More differential measurements, of the azimuthal dependence of global and longitudinal hyperon polarization, are more difficult to understand; the effects have magnitudes in line with standard expectations, but reproducing the sign of the observed oscillations may require nontrivial revisions to our current understanding. On the other hand, vector meson spin alignment—presumably related to hyperon polarization—is quantitatively and qualitatively impossible to understand solely in terms of the hydrodynamic paradigm that successfully explains other observables; here, there may be numerous competing effects that depend nontrivially on particle species and collision energy, including a newly proposed coherent mesonic mean field.

Thus, it appears that the new phenomena of strongly interacting QCD matter under rotation may be addressed by current theory and models, while at the same time requiring new insights. The contributions to this book represent a broad sample of some of the early theoretical efforts—from fundamental theory to phenomenology—to determine the physics behind these phenomena. New insights are bound to result from continued theoretical focus and upcoming experimental results. The following pages are the first chapters in what will surely be a much longer story.

Acknowledgements This work is supported in part by U.S. Department of Energy grant DE-SC0020651 and by U.S. National Science Foundation grant PHY-1913729.

References

1. Liang, Z.T., Wang, X.N.: Phys. Rev. Lett. **94** (2005). https://doi.org/10.1103/PhysRevLett.94.102301. [Erratum: Phys. Rev. Lett. 96, 039901 (2006)]
2. Becattini, F., Piccinini, F., Rizzo, J.: Phys. Rev. C **77** (2008). https://doi.org/10.1103/PhysRevC.77.024906
3. Becattini, F., Chandra, V., Del Zanna, L., Grossi, E.: Ann. Phys. **338**, 32 (2013). https://doi.org/10.1016/j.aop.2013.07.004
4. Jiang, Y., Lin, Z.W., Liao, J.: Phys. Rev. C **94**(4) (2016). https://doi.org/10.1103/PhysRevC.94.044910, https://doi.org/10.1103/PhysRevC.95.049904. [Erratum: Phys. Rev. C95, no.4,049904(2017)]
5. Adamczyk, L., et al.: Nature **548**, 62 (2017). https://doi.org/10.1038/nature23004
6. Liang, Z.T., Wang, X.N.: Phys. Lett. B **629**, 20 (2005). https://doi.org/10.1016/j.physletb.2005.09.060
7. Adams, J., et al.: Nucl. Phys. A **757**, 102 (2005). https://doi.org/10.1016/j.nuclphysa.2005.03.085
8. Adcox, K., et al.: Nucl. Phys. A **757**, 184 (2005). https://doi.org/10.1016/j.nuclphysa.2005.03.086
9. Back, B., et al.: Nucl. Phys. A **757**, 28 (2005). https://doi.org/10.1016/j.nuclphysa.2005.03.084
10. Arsene, I., et al.: Nucl. Phys. A **757**, 1 (2005). https://doi.org/10.1016/j.nuclphysa.2005.02.130
11. Barnett, S.J.: Phys. Rev. **6**, 239 (1915). https://doi.org/10.1103/PhysRev.6.239. https://link.aps.org/doi/10.1103/PhysRev.6.239
12. Takahashi, M., et al.: Nat. Phys. **12**, 52 (2016). https://doi.org/10.1038/nphys3526
13. Abelev, B., et al.: Phys. Rev. C **95** (2017). https://doi.org/10.1103/PhysRevC.95.039906
14. Anderson, M., et al.: Nucl. Instrum. Meth. **A499**, 659 (2003). https://doi.org/10.1016/S0168-9002(02)01964-2
15. Adams, J., et al.: Nucl. Instrum. Meth. A **968** (2020). https://doi.org/10.1016/j.nima.2020.163970
16. Becattini, F., Lisa, M.A.: Ann. Rev. Nucl. Part. Sci. **70**, 395 (2020). https://doi.org/10.1146/annurev-nucl-021920-095245
17. Bzdak, A., Esumi, S., Koch, V., Liao, J., Stephanov, M., Xu, N.: Phys. Rept. **853**, 1 (2020). https://doi.org/10.1016/j.physrep.2020.01.005
18. Aggarwal, M., et al.: (2010)
19. Adam, J., et al.: Phys. Rev. Lett. **123**(13) (2019). https://doi.org/10.1103/PhysRevLett.123.132301
20. Zhou, C.: Nucl. Phys. A **982**, 559 (2019). https://doi.org/10.1016/j.nuclphysa.2018.09.009
21. Adam, J et al.: arXiv:2012.13601 (2020)
22. Leader, E.: Spin Particle Phys. **15** (2011)
23. Schilling, K., Seyboth, P., Wolf, G.E.: Nucl. Phys. B **15**, 397 (1970). https://doi.org/10.1016/0550-3213(70)90070-2. [Erratum: Nucl. Phys. B 18, 332 (1970)]
24. Becattini, F., Karpenko, I., Lisa, M., Upsal, I., Voloshin, S.: Phys. Rev. C **95**(5) (2017). https://doi.org/10.1103/PhysRevC.95.054902
25. Acharya, S., et al.: Phys. Rev. Lett. **125**(1) (2020). https://doi.org/10.1103/PhysRevLett.125.012301
26. Sheng, X.L., Oliva, L., Wang, Q.: Phys. Rev. D **101**(9) (2020). https://doi.org/10.1103/PhysRevD.101.096005
27. Müller, B., Schäfer, A.: Phys. Rev. D **98**(7) (2018). https://doi.org/10.1103/PhysRevD.98.071902
28. Guo, X., Liao, J., Wang, E.: Sci. Rep. **10**(1), 2196 (2020). https://doi.org/10.1038/s41598-020-59129-6
29. Pang, L.G., Petersen, H., Wang, Q., Wang, X.N.: Phys. Rev. Lett. **117**(19) (2016). https://doi.org/10.1103/PhysRevLett.117.192301
30. Adam, J., et al.: arXiv:2007.14005 (2020)

Polarization in Relativistic Fluids: A Quantum Field Theoretical Derivation

Francesco Becattini

Abstract

We review the calculation of polarization in a relativistic fluid within the framework of statistical quantum field theory. We derive the expressions of the spin density matrix and the mean spin vector both for a single quantum relativistic particle and for a quantum-free field. After introducing the formalism of the covariant Wigner function for the scalar and the Dirac field, the relation between the spin density matrix and the covariant Wigner function is obtained. The formula is applied to the fluid produced in relativistic nuclear collisions by using the local thermodynamic equilibrium density operator and recovering previously known formulae. The dependence of these results on the spin tensor and pseudo-gauge transformations of the stress-energy tensor is addressed.

2.1 Introduction

The discovery of global polarization of Λ hyperons in relativistic nuclear collisions [1–5] has sparked a great interest in the theory of spin and polarization in relativistic fluids and relativistic matter in general. Several approaches have been proposed and several are currently pursued; amongst them, are relativistic kinetic theory [6–11] and a phenomenological treatment of the spin tensor [12–14].

Yet, the most fundamental tool is quantum statistical field theory, which was used to derive the original formula [15] of polarization of quasi-free particles in a relativistic fluid at local thermodynamic equilibrium. All other approaches should, in the first place, reproduce the results obtained with this method, once the state of the system, that is its density operator, is chosen.

F. Becattini (✉)
Università di Firenze and INFN Sezione di Firenze, Via G. Sansone 1, 50019
Sesto Fiorentino (Firenze), Italy
e-mail: becattini@fi.infn.it

© Springer Nature Switzerland AG 2021
F. Becattini et al. (eds.), *Strongly Interacting Matter under Rotation*,
Lecture Notes in Physics 987,
https://doi.org/10.1007/978-3-030-71427-7_2

Carrying out the calculation of the polarization matrix, which for spin 1/2 particles boils down to the mean spin vector, in a quantum field theoretical framework is, however, not an easy task. In the derivation presented in Ref. [15], some approximations were introduced and the final formula admittedly relied on the use of the canonical spin tensor, that is dependent on the specific set of quantum stress-energy and spin tensor (in other words of the *pseudo-gauge* choice [16]). It is the purpose of this paper to review the derivation of the polarization of spin 1/2 particles step by step and fill some of the conceptual gaps. Particularly, the exact formula relating the spin density matrix and the mean spin vector to the covariant Wigner function will be found and it will be thereby conclusively demonstrated that the *expression* of the polarization is independent of the spin tensor. Furthermore, a general formula for particles with any spin S will be derived in the limit of distinguishable quantum particles, that is neglecting quantum statistics. For this purpose, many useful concepts in statistical quantum field theory will be thoroughly reviewed and discussed.

The paper is organized as follows: in Sect. 2.2 the general definitions of spin density matrix and mean spin vector will be given in a quantum relativistic framework. In Sect. 2.3, a formula for the spin density matrix and the mean spin vector will be derived for a single free quantum particle with spin S in a thermal bath with rotation and acceleration, by using only group theory techniques. In Sect. 2.4, the covariant Wigner operator and function will be introduced for the free scalar and Dirac field. In Sect. 2.5, the formula for the mean spin vector of a free Dirac fermion will be obtained as a function of the covariant Wigner function while in Sect. 2.6, the same formula will be obtained with a different method based on total angular momentum, already used in Ref. [15]. In Sect. 2.7, the density operator at local thermodynamic equilibrium will be discussed in detail with emphasis on its application in relativistic heavy ion collisions, and the derivation of the formula of mean spin vector for a fermion in a relativistic fluid at local thermodynamic equilibrium will be outlined.

Notations and Conventions

In this paper, we use the natural units, with $\hbar = c = K = 1$.

We will use the relativistic notation with repeated indices assumed to be saturated, however, contractions of indices will be sometimes denoted with a dot or a colon, e.g. $\beta \cdot p = \beta_\mu p^\mu$ and $\varpi : J = \varpi_{\mu\nu} J^{\mu\nu}$. The Minkowskian metric tensor is diag$(1, -1, -1, -1)$; for the Levi-Civita symbol, we use the convention $\epsilon^{0123} = 1$.

Operators in Hilbert space will be denoted by a large upper hat ($\widehat{T}$) while unit vectors with a small upper hat ($\hat{v}$). Noteworthy exception is, the Dirac field which is expressed by Ψ without an upper hat.

The symbol Tr with a capital T stands for the trace over all states in the Hilbert space, whereas the symbol tr stands for a trace over polarization states or traces of finite-dimensional matrices.

2.2 The Spin Density Matrix and the Definition of Mean Spin

In relativistic quantum mechanics, for a single massive particle, the spin angular momentum vector is defined as

$$\widehat{S}^\mu = -\frac{1}{2m}\epsilon^{\mu\nu\rho\sigma}\widehat{J}_{\nu\rho}\widehat{P}_\sigma, \tag{2.1}$$

where $\widehat{J}_{\nu\rho}$ are the angular momentum-boost operators and $\widehat{P}_\sigma$ the energy-momentum operator. The operator in Eq. (2.1) is also known as Pauli–Lubanski vector and it fulfils the following commutation relations:

$$[\widehat{S}_\mu, \widehat{P}_\nu] = 0 \tag{2.2}$$
$$[\widehat{S}_\mu, \widehat{S}_\nu] = i\epsilon_{\mu\nu\rho\sigma}\widehat{S}^\rho\widehat{P}^\sigma$$
$$\widehat{S}\cdot\widehat{P} = 0.$$

Hence, if the ket $|p\rangle$ is an eigenvector of $\widehat{P}$, so is $\widehat{S}|p\rangle$. The restriction of $\widehat{S}$ to the eigenspace labelled by four-momentum p is defined as $\widehat{S}(p)$. Since $\widehat{S}(p)\cdot p = 0$, it can be decomposed onto three orthonormal space-like four-vectors $n_1(p), n_2(p), n_3(p)$ orthogonal to p, forming a basis of the Minkowski space with the unit vector $\hat{p} = p/\sqrt{p^2}$:

$$\widehat{S}(p) = \sum_{i=1}^{3}\widehat{S}_i(p)n_i(p). \tag{2.3}$$

It can be shown that the operators $\widehat{S}_i(p)$ form a SU(2) algebra and are the generators of the so-called *little group* of massive particles. The third component $\widehat{S}_3(p)$ can be diagonalized along with $\widehat{S}^2$ with corresponding eigenvalues s and $S(S+1)$, S being *the* spin of the particle so that

$$\widehat{P}|p, s\rangle = p|p, s\rangle \quad \text{and} \quad \widehat{S}_3(p)|p, s\rangle = s|p, s\rangle. \tag{2.4}$$

The $\hat{n}_i(p)$, being orthogonal to p, can be written as

$$n_i(p) = [p]\hat{e}_i \tag{2.5}$$

with $\hat{e}_i$ the i-th unit space vector and $[p]$ the so-called *standard Lorentz transformation* bringing the time-like vector $p_0 = (m, 0, 0, 0)$ into the four-momentum p.

The choice of $[p]$ entails a specific physical meaning of the eigenvalue s. For instance, if θ, ϕ are the spherical coordinates of $\mathbf{p}$ and ξ the rapidity of p,

$$[p] \equiv \mathsf{R}_3(\phi)\mathsf{R}_2(\theta)\mathsf{L}_3(\xi),$$

where $\mathsf{R}_k(\psi)$ are rotations around the axis k with angle ψ and $\mathsf{L}_k(\xi)$ a Lorentz boost along the direction k with hyperbolic angle ξ is a typical choice of the standard

Lorentz transformation which makes s the helicity of the particle. Finally, if the states are normalized according to[1]

$$\langle p, r | q, s \rangle = 2\varepsilon\, \delta^3(\mathbf{p} - \mathbf{q})\delta_{rs} \,. \tag{2.6}$$

we have, for the representation of a general Lorentz transformation Λ in the Hilbert space:

$$\widehat{\Lambda}|p, r\rangle = \sum_s |\Lambda p, s\rangle D^S([\Lambda p]^{-1}\Lambda[p])_{sr}, \tag{2.7}$$

where D^S stands for the $(2S+1)$-dimensional irreducible representation of the proper orthocronous Lorentz group SO(1,3) (the so-called $(0, S)$ representation) or—in case—of its universal covering group SL(2,C). The transformation

$$W(\Lambda, p) = [\Lambda p]^{-1}\Lambda[p] \tag{2.8}$$

is the so-called *Wigner rotation*, as it leaves the unit time vector $\hat{t}$ invariant.

The mean value S^μ of the operator (2.1) is the properly called *spin vector* (the polarization vector being the mean spin vector S^μ divided by S so that its maximal magnitude is always 1):

$$S^\mu = \text{Tr}(\widehat{S^\mu}\widehat{\rho}),$$

where $\widehat{\rho}$ is the density operator of the single quantum relativistic particle in the Hilbert space. Its restriction to the subspace of four-momentum p is the *spin density operator* $\widehat{\Theta}(p)$, which is used to express the mean value of the spin vector for a particle with momentum p:

$$S^\mu(p) = \text{Tr}(\widehat{S^\mu}\widehat{\Theta}(p)). \tag{2.9}$$

The matrix:

$$\Theta(p)_{rs} \equiv \langle p, r | \widehat{\rho} | p, s \rangle = \langle p, r | \widehat{\Theta}(p) | p, s \rangle \tag{2.10}$$

is the spin density matrix, in the basis labelled by the eigenvalues of $\widehat{S}_3(p)$. The matrix Θ, which is Hermitian, positive definite and with normalized trace, contains the maximal information about the spin state of the particle. It is dependent on the chosen basis, yet the mean value (2.9) is independent thereof, that is of the standard Lorentz transformation associated with the eigenvalue s.

---

[1]Throughout this paper the symbol ε stands for the on-shell energy, that is $\varepsilon = \sqrt{\mathbf{p}^2 + m^2}$.

Plugging (2.3) into (2.9), we get:

$$
\begin{aligned}
S^\mu(p) &= \sum_r \sum_i \langle p, r | \widehat{S}_i(p) \widehat{\Theta}(p) | p, r \rangle n_i(p) \\
&= \sum_{r,s} \sum_i \langle p, r | \widehat{S}_i(p) | p, s \rangle \langle p, s | \widehat{\Theta}(p) | p, r \rangle n_i(p) \\
&= \sum_{r,s} \sum_i D^S(J^i)_{rs} \Theta(p)_{sr} n_i(p) \\
&= \sum_{i=1}^{3} \mathrm{tr}(D^S(J^i) \Theta(p))[p](\hat{e}_i)^\mu = \sum_{i=1}^{3} [p]_i^\mu \mathrm{tr}(D^S(J^i) \Theta(p)), \quad (2.11)
\end{aligned}
$$

where we have used the fact that $\widehat{S}_i(p)$ are the generators of the little group SU(2) algebra in the subspace spanned by $|p\rangle$; the $D^S(J^i)$ are the familiar matrices of the angular momentum generators for the representation with spin S. The above formula can be made covariant by introducing the definition of angular momenta:

$$
D^S(J^i) = -\frac{1}{2} \epsilon^{i\lambda\nu\rho} D^S(J_{\lambda\nu}) \hat{t}_\rho,
$$

where $\hat{t}$ is the unit time vector and the tensor $J_{\lambda\nu}$ now includes all Lorentz transformation generators, angular momentum and boosts. Since $\hat{t}_\rho = \delta_\rho^0$, we can extend the sum over i from 0 to 4, because of the Levi-Civita tensor and write:

$$
S^\mu(p) = -\frac{1}{2} \epsilon^{\alpha\lambda\nu\rho} \hat{t}_\rho [p]_\alpha^\mu \mathrm{tr}(D^S(J_{\lambda\nu}) \Theta(p)). \quad (2.12)
$$

Now,

$$
\begin{aligned}
[p]_\alpha^\mu \epsilon^{\alpha\lambda\nu\rho} &= [p]_\alpha^\mu ([p]_\phi^\beta [p]_\beta^{-1\lambda})([p]_\chi^\gamma [p]_\gamma^{-1\nu})([p]_\psi^\delta [p]_\delta^{-1\rho}) \epsilon^{\alpha\phi\chi\psi} \\
&= \left([p]_\alpha^\mu [p]_\phi^\beta [p]_\chi^\gamma [p]_\psi^\delta \epsilon^{\alpha\phi\chi\psi} \right) [p]_\beta^{-1\lambda} [p]_\gamma^{-1\nu} [p]_\delta^{-1\rho} \\
&= \det[p] \epsilon^{\mu\beta\gamma\delta} [p]_\beta^{-1\lambda} [p]_\gamma^{-1\nu} [p]_\delta^{-1\rho} = \epsilon^{\mu\beta\gamma\delta} [p]_\beta^{-1\lambda} [p]_\gamma^{-1\nu} [p]_\delta^{-1\rho},
\end{aligned}
$$

and, substituting in (2.12),

$$
S^\mu(p) = -\frac{1}{2} \epsilon^{\mu\beta\gamma\delta} [p]_\delta^{-1\rho} \hat{t}_\rho [p]_\beta^{-1\lambda} [p]_\gamma^{-1\nu} \mathrm{tr}(D^S(J_{\lambda\nu}) \Theta(p)).
$$

By definition of standard Lorentz transformation and taking into account its orthogonality, we have:

$$
[p]_\delta^{-1\rho} \hat{t}_\rho = \frac{p_\delta}{m},
$$

so that the mean spin vector becomes:

$$S^\mu(p) = -\frac{1}{2m}\epsilon^{\mu\beta\gamma\delta} p_\delta [p]_\beta^{-1\lambda}[p]_\gamma^{-1\nu}\mathrm{tr}(D^S(J_{\lambda\nu})\Theta(p)).$$

From group representation theory, we know that

$$[p]_\beta^{-1\lambda}[p]_\gamma^{-1\nu} D^S(J_{\lambda\nu}) = D^S([p])^{-1} D^S(J_{\beta\gamma}) D^S([p]), \tag{2.13}$$

and hence (2.12) can be finally cast as

$$S^\mu(p) = -\frac{1}{2m}\epsilon^{\mu\beta\gamma\delta} p_\delta \mathrm{tr}\left(D^S([p])^{-1} D^S(J_{\beta\gamma}) D^S([p])\Theta(p)\right). \tag{2.14}$$

The "kinematic" part of the spin vector derivation for a single quantum relativistic particle is completed.

In a quantum field theory framework, the single-particle spin density matrix can be still defined with a formula which is a generalization of (2.10):

$$\Theta(p)_{rs} = \frac{\mathrm{Tr}(\widehat\rho\,\widehat a_s^\dagger(p)\widehat a_r(p))}{\sum_t \mathrm{Tr}(\widehat\rho\,\widehat a_t^\dagger(p)\widehat a_t(p))}, \tag{2.15}$$

where $\widehat\rho$ is the density operator for the Hilbert space of the field states. The $\widehat a_r(p)$ are destruction operators of the particle with momentum p and spin state r. The introduction of creation and destruction operators makes it clear that one can define a polarization of particles only when particle is a sensible concept, that is for a non-interacting or a weakly interacting field theory. For instance, defining a spin density matrix of quarks and gluons makes sense only in the perturbative limit of QCD.

The calculation of $\Theta(p)$ is the crucial and hardest part of the procedure. Before tackling the full quantum field theory case, one can obtain a good approximation by using the single quantum relativistic particle formalism, which is the subject of the next section.

2.3   The Single-Particle Limit and Global Equilibrium Factorization

In the fixed-number particle formalism, the Hilbert space of quantum states is the tensor product of single-particle Hilbert spaces. Neglecting symmetrization or anti-symmetrization of the states means disregarding quantum statistics effects and taking the limit of distinguishable particles. Moreover, if the particles are non-interacting, the full density operator can be written as the tensor product of single-particle density operators:

$$\widehat\rho = \otimes_i \widehat\rho_i.$$

We can now set out to get the spin density matrix for the general *global* equilibrium density operator $\widehat{\rho}$ [17, 18]:

$$\widehat{\rho} = \frac{1}{Z} \exp\left[-b \cdot \widehat{P} + \frac{1}{2}\varpi : \widehat{J}\right], \tag{2.16}$$

where b is a constant time-like four-vector and ϖ a constant anti-symmetric tensor. In global equilibrium, the vector field:

$$\beta^{\mu} = b^{\mu} + \varpi^{\mu\nu}x_{\nu} \tag{2.17}$$

is the four-temperature vector [17] fulfilling Killing equation and

$$\varpi_{\mu\nu} = -\frac{1}{2}\left(\partial_{\mu}\beta_{\nu} - \partial_{\nu}\beta_{\mu}\right) \tag{2.18}$$

is called thermal vorticity; the relation (2.18) can be taken as a definition of thermal vorticity in non-equilibrium situation. The operators $\widehat{P}$ and $\widehat{J}$ in (2.16) are the conserved total four-momentum and total angular momentum-boosts, respectively. For a set of non-interacting distinguishable particles, we can write:

$$\widehat{P} = \sum_{i} \widehat{P}_{i}, \qquad \widehat{J} = \sum_{i} \widehat{J}_{i},$$

and consequently,

$$\widehat{\rho}_{i} = \frac{1}{Z_{i}} \exp\left[-b \cdot \widehat{P}_{i} + \frac{1}{2}\varpi : \widehat{J}_{i}\right],$$

so that the single-particle spin density matrix reads:

$$\Theta(p)_{irs} = \frac{\langle p, r|\widehat{\rho}_{i}|p, s\rangle}{\sum_{r}\langle p, t|\widehat{\rho}_{i}|p, t\rangle}. \tag{2.19}$$

In order to calculate the right hand side of (2.19), one can take advantage of a noteworthy factorization:

$$\frac{1}{Z_{i}} \exp\left[-b \cdot \widehat{P}_{i} + \frac{1}{2}\varpi : \widehat{J}_{i}\right] = \frac{1}{Z_{i}} \exp\left[-\tilde{b} \cdot \widehat{P}_{i}\right] \exp\left[\frac{1}{2}\varpi : \widehat{J}_{i}\right], \tag{2.20}$$

where

$$\tilde{b}_{\mu} = \sum_{k=0}^{\infty} \frac{i^{k}}{(k+1)!} \underbrace{\left(\varpi_{\mu\nu_{1}}\varpi^{\nu_{1}\nu_{2}}\ldots\varpi_{\nu_{k-1}\nu_{k}}\right)}_{k \text{ times}} b^{\nu_{k}}. \tag{2.21}$$

Equation (2.20) is a very useful formula, whose derivation is worth being shown in some detail.

Let us start with the following very simple observation concerning the composition of translations and Lorentz transformation in Minkowski space–time. Let x be a four-vector and apply the combination

$$\mathsf{T}(a)\,\Lambda\,\mathsf{T}(a)^{-1},$$

$\mathsf{T}(a)$ being a translation of some four-vector a and Λ a Lorentz transformation. The effect of the above combination on x reads:

$$x \mapsto x - a \mapsto \Lambda(x-a) \mapsto \Lambda(x-a) + a = \Lambda(x) + (\mathrm{I} - \Lambda)(a) = \mathsf{T}((\mathrm{I}-\Lambda)(a))(\Lambda(x)).$$

Since x was arbitrary, we have:

$$\mathsf{T}(a)\,\Lambda\,\mathsf{T}(a)^{-1} = \mathsf{T}((\mathrm{I}-\Lambda)(a)))\Lambda.$$

This relation has a representation of unitary operators in Hilbert space, which can be written in terms of the generators of the Poincaré group:

$$\exp[ia \cdot \widehat{P}]\exp[-i\varphi : \widehat{J}/2]\exp[-ia \cdot \widehat{P}] = \exp[i((\mathrm{I}-\Lambda)(a)) \cdot \widehat{P}]\exp[-i\varphi : \widehat{J}/2], \tag{2.22}$$

where φ are the parameters of the Lorentz transformation.[2] By taking φ infinitesimal, we can obtain a known relation about the effect of translations on angular momentum operators:

$$\exp[ia \cdot \widehat{P}]\widehat{J}_{\mu\nu}\exp[-ia \cdot \widehat{P}] = \widehat{\mathsf{T}}(a)\widehat{J}_{\mu\nu}\widehat{\mathsf{T}}(a)^{-1} = \widehat{J}_{\mu\nu} - a_\mu \widehat{P}_\nu + a_\nu \widehat{P}_\mu.$$

The left hand side of (2.22) can now be worked out by using the above relation:

$$\exp[ia \cdot \widehat{P}]\exp[-i\varphi : \widehat{J}/2]\exp[-ia \cdot \widehat{P}] = \widehat{\mathsf{T}}(a)\exp[-i\varphi : \widehat{J}/2]\widehat{\mathsf{T}}(a)^{-1}$$
$$= \exp[-i\varphi : \widehat{\mathsf{T}}(a)\widehat{J}\widehat{\mathsf{T}}(a)^{-1}/2]$$
$$= \exp[-i\varphi : (\widehat{J} - a \wedge \widehat{P})/2] = \exp[i\varphi_{\mu\nu}a^\mu \widehat{P}^\nu - i\varphi_{\mu\nu}\widehat{J}^{\mu\nu}/2]. \tag{2.23}$$

Hence, combining (2.23) with (2.22), we have obtained the factorization:

$$\exp[i\varphi_{\mu\nu}a^\mu \widehat{P}^\nu - i\varphi_{\mu\nu}\widehat{J}^{\mu\nu}/2] = \exp[i((\mathrm{I}-\Lambda)(a)) \cdot \widehat{P}]\exp[-i\varphi : \widehat{J}/2]. \tag{2.24}$$

Now,

$$i(\mathrm{I}-\Lambda)(a) = ia - i\sum_{k=0}^{\infty} \frac{(-i)^k}{2^k k!}(\varphi : \mathsf{J})^k(a) = -i\sum_{k=1}^{\infty} \frac{(-i)^k}{2^k k!}(\varphi : \mathsf{J})^k(a). \tag{2.25}$$

---

[2]Henceforth, by : we will denote a double contraction of rank 2 tensors, e.g. $\varphi : \widehat{J} = \varphi_{\mu\nu}\widehat{J}^{\mu\nu}$.

Setting

$$V_\mu = i\varphi_{\mu\nu}a^\nu$$

and taking into account that

$$(J_{\mu\nu})^\alpha_\beta = i\left(\delta^\alpha_\mu g_{\nu\beta} - \delta^\alpha_\nu g_{\mu\beta}\right),$$

we have:

$$(\varphi : J)(a)_\alpha = 2i\varphi_{\alpha\beta}a^\beta = 2V_\alpha.$$

Therefore, the right hand side of Eq. (2.25) becomes:

$$-i\sum_{k=1}^\infty \frac{(-i)^k}{2^k k!}(\varphi : J)^k(a) = -i\sum_{k=1}^\infty \frac{(-i)^k}{2^{k-1}k!}(\varphi : J)^{k-1}(V)$$

$$= -\sum_{k=0}^\infty \frac{(-i)^k}{2^k(k+1)!}(\varphi : J)^k(V).$$

Finally, Eq. (2.24) becomes:

$$\exp[-V \cdot \widehat{P} - i\varphi : \widehat{J}/2] = \exp[-\tilde{V}(\varphi) \cdot \widehat{P}]\exp[-i\varphi : \widehat{J}/2], \qquad (2.26)$$

where

$$\tilde{V}(\varphi) \equiv \sum_{k=0}^\infty \frac{(-i)^k}{2^k(k+1)!}(\varphi : J)^k(V) = \sum_{k=0}^\infty \frac{1}{(k+1)!}\underbrace{\left(\varphi_{\mu\nu_1}\varphi^{\nu_1\nu_2}\ldots\varphi_{\nu_{k-1}\nu_k}\right)}_{k \text{ times}} V^{\nu_k}.$$

Equation (2.26) can be read as the factorization of the exponential of a linear combinations of generators of the Poincaré group. For this reason, it must be derivable also by using the known formulae of the factorization of the exponential of the sum of matrices $\exp[A + B]$ in terms of exponentials of commutators of A and B. Indeed, it can be shown, by using the commutation relations of $\widehat{P}$ and $\widehat{J}$, that one precisely gets the Eq. (2.26) for *any* vector V and tensor φ, either real or complex. Hence, the formula (2.26) can be applied to factorize the density operator (2.16) by setting $\varphi = i\varpi$:

$$\widehat{\rho} = \frac{1}{Z}\exp[-b \cdot \widehat{P} + \varpi : \widehat{J}/2] = \frac{1}{Z}\exp[-\tilde{b}(\varpi) \cdot \widehat{P}]\exp[\varpi : \widehat{J}/2] \qquad (2.27)$$

with

$$\tilde{b}(\varpi) = \sum_{k=0}^\infty \frac{1}{2^k(k+1)!}(\varpi : J)^k b = \sum_{k=0}^\infty \frac{i^k}{(k+1)!}\underbrace{\left(\varpi_{\mu\nu_1}\varpi^{\nu_1\nu_2}\ldots\varpi_{\nu_{k-1}\nu_k}\right)}_{k \text{ times}} b^{\nu_k}.$$

We have thus proved the formula (2.20).

The factorization of the density operator in Eq. (2.20) can now be applied to calculate the spin density matrix in Eq. (2.19). The momentum-dependent factor $\exp(-\tilde{b} \cdot p)$ cancels out in the ratio, and one is left with

$$\Theta(p)_{rs} = \frac{\langle p, r| \exp[\varpi : \widehat{J}/2]|p, s\rangle}{\sum_t \langle p, t| \exp[\varpi : \widehat{J}/2]|p, t\rangle}.$$

To derive its explicit form, we use an analytic continuation; namely, we first determine $\Theta(p)$ for imaginary ϖ and then continue the function to real values. If ϖ is imaginary, $\exp[\varpi : \widehat{J}/2] \equiv \widehat{\Lambda}$ is just a unitary representation of a Lorentz transformation, and then one can use known relations of Poincaré group representations [19] to obtain:

$$\Theta(p)_{rs} = \frac{\langle p, r|\widehat{\Lambda}|p, s\rangle}{\sum_t \langle p, t|\widehat{\Lambda}|p, t\rangle} = \frac{2\varepsilon\delta^3(\mathbf{p} - \mathbf{\Lambda}(p))W(\Lambda, p)_{rs}}{2\varepsilon\delta^3(\mathbf{p} - \mathbf{\Lambda}(p)) \sum_t W(\Lambda, p)_{tt}}, \qquad (2.28)$$

where $\mathbf{\Lambda}(p)$ stands for the spacial part of the four-vector $\Lambda(p)$ and $W(\Lambda, p)$ is the Wigner rotation defined in Eq. (2.8). We thus have:

$$\Theta(p)_{rs} = \frac{D^S([p]^{-1}\Lambda[p])_{rs}}{\mathrm{tr}(D^S(\Lambda))},$$

which seems to be an appropriate form to be analytically continued to real ϖ. However, the above form is not satisfactory yet as the continuation to real ϖ, that is:

$$D^S(\Lambda) = \exp\left[-\frac{i}{2}\varpi : \Sigma_S\right] \rightarrow \exp\left[\frac{1}{2}\varpi : \Sigma_S\right],$$

where $\Sigma_S = D^S(J)$ that is the matrix representing the generators does not give rise to a Hermitian matrix for $\Theta(p)$ as it should. This problem can be fixed by taking into account that $W(p)$ is the representation of a rotation, hence unitary. We can thus replace $W(p)$ with $(W(p) + W(p)^{-1\dagger})/2$ in (2.28) and, by using the property of SL(2,C) representations $D^S(A^\dagger) = D^S(A)^\dagger$ [20], we obtain:

$$\Theta(p) = \frac{D^S([p]^{-1}\Lambda[p]) + D^S([p]^\dagger \Lambda^{-1\dagger}[p]^{-1\dagger})}{\mathrm{tr}(D^S(\Lambda) + D^S(\Lambda)^{-1\dagger})},$$

which will give a Hermitian result because the analytic continuation of $\Lambda^{-1\dagger}$ reads[3]:

$$D^S(\Lambda^{-1\dagger}) \rightarrow \exp\left[\frac{1}{2}\varpi : \Sigma_S^\dagger\right].$$

---

[3]Note that the Lorentz transformations in Minkowski space–time and their counterparts of the fundamental $(0, 1/2)$ representation of the SL(2,C) group are henceforth identified. Particularly, the standard Lorentz transformation $[p]$ will indicate either a SO(1,3) transformation or a SL(2,C) transformation.

Altogether, the final expression of the spin density matrix reads:

$$
\Theta(p) = \frac{D^S([p])^{-1} \exp[(1/2)\varpi : \Sigma_S] D^S([p])}{\mathrm{tr}(\exp[(1/2)\varpi : \Sigma_S] + \exp[(1/2)\varpi : \Sigma_S^\dagger])}
$$
$$
+ \frac{D^S([p])^\dagger \exp[(1/2)\varpi : \Sigma_S^\dagger] D^S([p])^{-1\dagger})}{\mathrm{tr}(\exp[(1/2)\varpi : \Sigma_S] + \exp[(1/2)\varpi : \Sigma_S^\dagger])},
\tag{2.29}
$$

which is manifestly Hermitian.

Equation (2.29) can be further developed. By using (2.13), we have:

$$
D^S([p])^{-1} \exp\left[\frac{1}{2}\varpi : \Sigma_S\right] D^S([p]) = \exp\left[\frac{1}{2}\varpi^{\mu\nu} D^S([p])^{-1} \Sigma_{S\,\mu\nu} D^S([p])\right]
$$
$$
= \exp\left[\frac{1}{2}\varpi^{\mu\nu} [p]_\mu^{-1\alpha} [p]_\nu^{-1\beta} \Sigma_{S\,\alpha\beta}\right],
$$

which applies to the original SO(1,3) matrices too. So, if we apply the Lorentz transformation $[p]$ to the tensor ϖ:

$$
\varpi^{\mu\nu} [p]_\mu^{-1\alpha} [p]_\nu^{-1\beta} = \varpi_*^{\alpha\beta}(p),
\tag{2.30}
$$

we realize that $\varpi_*^{\alpha\beta}$ are the components of the thermal vorticity tensor in the rest-frame of the particle with four-momentum p. Note that these components are obtained by back-boosting with $[p]$ (which in fact is not a pure Lorentz boost in the helicity scheme as it includes a rotation). Finally, equation (2.29) becomes:

$$
\Theta(p) = \frac{D^S(\exp[(1/2)\varpi_*(p) : \Sigma_S]) + D^S(\exp[(1/2)\varpi_*(p) : \Sigma_S^\dagger])}{\mathrm{tr}(\exp[(1/2)\varpi : \Sigma_S] + \exp[(1/2)\varpi : \Sigma_S^\dagger])}.
\tag{2.31}
$$

The thermal vorticity ϖ is usually $\ll 1$; in this case, the spin density matrix can be expanded in power series around $\varpi = 0$. Taking into account that $\mathrm{tr}(\Sigma_S) = 0$, we have:

$$
\Theta(p)_{rs} \simeq \frac{\delta_{rs}}{2S+1} + \frac{1}{4(2S+1)}\varpi_*(p)^{\alpha\beta}(\Sigma_{S\alpha\beta} + \Sigma_{S\alpha\beta}^\dagger)_{rs}
$$

to first order in ϖ. Now the Σ_S matrices can be decomposed into representations of angular momentum and boosts:

$$
\Sigma_{S\mu\nu} = D^S(J_{\mu\nu}) = \epsilon_{\mu\nu\rho\sigma} D^S(J^\rho)\hat{t}^\sigma - D^S(K_\mu)\hat{t}_\nu + D^S(K_\nu)\hat{t}_\mu,
$$

and taking into account that the $D^S(J^i)$ are Hermitian while $D^S(K^i)$ are anti-Hermitian, we find:

$$
\Theta(p)_{rs} \simeq \frac{\delta_{rs}}{2S+1} + \frac{1}{2(2S+1)}\varpi_*(p)^{\alpha\beta}\epsilon_{\alpha\beta\rho\nu} D^S(J^\rho)_{r'}^r \hat{t}^\nu.
\tag{2.32}
$$

By plugging (2.32) into (2.12), we get:

$$
\begin{aligned}
S^\mu(p) &= [p]^\mu_\kappa \frac{1}{2(2S+1)} \varpi_*(p)^{\alpha\beta} \epsilon_{\alpha\beta\rho\nu} \operatorname{tr}\left(D^S(J^\rho)D^S(J^\kappa)\right) \hat{t}^\nu \\
&= -\frac{1}{2(2S+1)} \frac{S(S+1)(2S+1)}{3} [p]^\mu_\kappa \varpi_*(p)^{\alpha\beta} \epsilon_{\alpha\beta\rho\nu} g^{\rho\kappa} \hat{t}^\nu \\
&= -\frac{1}{2} \frac{S(S+1)}{3} [p]^\mu_\rho \varpi_*(p)_{\alpha\beta} \epsilon^{\alpha\beta\rho\nu} \hat{t}_\nu = -\frac{1}{2m} \frac{S(S+1)}{3} \varpi_{\alpha\beta} \epsilon^{\alpha\beta\mu\nu} p_\nu,
\end{aligned}
$$

$$(2.33)$$

where, in the last equality, we have boosted the vector to the laboratory frame by using Eq. (2.30).

For a fluid made of distinguishable particles at local thermodynamic equilibrium, the thermal vorticity ϖ is promoted to a function of space and time, so that the expression (2.33) gives rise to the integral average:

$$
S^\mu(p) = -\frac{1}{2m} \frac{S(S+1)}{3} \epsilon^{\mu\alpha\beta\nu} p_\nu \frac{\int_\Sigma d\Sigma_\lambda p^\lambda f(x,p) \varpi_{\alpha\beta}(x)}{\int_\Sigma d\Sigma_\lambda p^\lambda f(x,p)} \tag{2.34}
$$

with $f(x,p)$ the distribution function and Σ a 3D hypersurface from where particles are emitted. The latter is basically the same formula obtained in Refs. [15,21] and should apply to particles with any spin.

2.4   The Covariant Wigner Function

We now turn to the general quantum field formula of the spin density matrix, Eq. (2.15). To develop this expression, we need to introduce an important quantity: the covariant Wigner operator. We will do this first for the scalar field, where the spin plays no role, and later for the Dirac field.

2.4.1   The Scalar Field

The covariant Wigner operator is defined as a Fourier transform of the two-point function of the quantum field:

$$
\widehat{W}(x,k) = \frac{2}{(2\pi)^4} \int d^4y \; : \widehat{\psi}^\dagger(x+y/2)\widehat{\psi}(x-y/2) : e^{-iy\cdot k}, \tag{2.35}
$$

where : stands for the normal ordering of creation and destruction operators; the appearance of a normal ordering implies that this definition is suitable for a free field or a field interacting with an external field. Even if (2.35) is not, strictly speaking, a local operator (it depends on the field in two points), its quasi-locality makes it a suitable tool to deal with local thermodynamic equilibrium in quantum statistical mechanics. Besides, its mean value, namely the covariant Wigner function:

$$W(x, k) = \text{Tr}(\widehat{\rho} \, \widehat{W}(x, k)), \tag{2.36}$$

where $\widehat{\rho}$ is the density operator, is an indispensable tool to reckon quantum corrections to classical kinetic theory [22]. From Eq. (2.36) it turns out that the covariant Wigner function is real, but it does not need to be positive definite. Inserting in Eq. (2.35) the free scalar field expansion in plane waves:

$$\widehat{\psi}(x) = \frac{1}{(2\pi)^{3/2}} \int \frac{d^3 p}{2\varepsilon} \, e^{-ip\cdot x}\widehat{a}(p) + e^{ip\cdot x}\widehat{b}^\dagger(p), \tag{2.37}$$

a_p, b_p being destruction operators of particles with four-momentum p normalized so as to

$$[\widehat{a}(p), \widehat{a}^\dagger(p')] = 2\,\varepsilon\,\delta^3(\mathbf{p} - \mathbf{p}'), \tag{2.38}$$

we get, for the covariant Wigner function:

$$
\begin{aligned}
W(x, k) =& \frac{2}{(2\pi)^7} \int \frac{d^3 p}{2\varepsilon} \frac{d^3 p'}{2\varepsilon'} \int d^4 y \Big[e^{-iy\cdot(k-(p+p')/2)} e^{i(p-p')\cdot x} \langle\widehat{a}^\dagger(p)\widehat{a}(p')\rangle \\
&+ e^{-iy\cdot(k+(p+p')/2)} e^{-i(p-p')\cdot x} \langle\widehat{b}^\dagger(p')\widehat{b}(p)\rangle \\
&+ e^{-iy\cdot(k-(p-p')/2)} e^{i(p+p')\cdot x} \langle\widehat{a}^\dagger(p)\widehat{b}^\dagger(p')\rangle \\
&+ e^{-iy\cdot(k+(p-p')/2)} e^{-i(p+p')\cdot x} \langle\widehat{b}(p)\widehat{a}(p')\rangle \Big] \\
=& \frac{2}{(2\pi)^7} \int \frac{d^3 p}{2\varepsilon} \frac{d^3 p'}{2\varepsilon'} \int d^4 y \, \Big[e^{-iy\cdot(k-(p+p')/2)} e^{i(p-p')\cdot x} \langle\widehat{a}^\dagger(p)\widehat{a}(p')\rangle \\
&+ e^{-iy\cdot(k+(p+p')/2)} e^{i(p-p')\cdot x} \langle\widehat{b}^\dagger(p)\widehat{b}(p')\rangle \\
&+ e^{-iy\cdot(k-(p-p')/2)} e^{i(p+p')\cdot x} \langle\widehat{a}^\dagger(p)\widehat{b}^\dagger(p')\rangle \\
&+ e^{-iy\cdot(k-(p-p')/2)} e^{-i(p+p')\cdot x} \langle\widehat{b}(p')\widehat{a}(p)\rangle \Big] \\
=& \frac{2}{(2\pi)^3} \int \frac{d^3 p}{2\varepsilon} \frac{d^3 p'}{2\varepsilon'} e^{i(p-p')\cdot x} \Big[\delta^4(k - (p+p')/2)\langle\widehat{a}^\dagger(p)\widehat{a}(p')\rangle \\
&+ \delta^4(k + (p+p')/2)\langle\widehat{b}^\dagger(p)\widehat{b}(p')\rangle \Big] \\
&+ \delta^4(k - (p-p')/2) \Big[e^{i(p+p')\cdot x}\langle\widehat{a}^\dagger(p)\widehat{b}^\dagger(p')\rangle + e^{-i(p+p')\cdot x}\langle\widehat{b}(p')\widehat{a}(p)\rangle \Big],
\end{aligned} \tag{2.39}
$$

where $\langle\ \rangle$ stands for the mean $\text{Tr}(\widehat{\rho}\ )$. In the above equalities, we have taken advantage of the symmetric integration in the variables p, p'.

The expression (2.39) makes it apparent that the variable k of the Wigner function $W(x, k)$ is *not* on-shell, i.e. $k^2 \neq m^2$ even in the free case. This makes the definition of a particle distribution function $f(x, p)$ à la Boltzmann not straightforward in a quantum relativistic framework. In the book by De Groot [22], it is shown that a distribution function can be defined in the limit of slowly varying $W(x, k)$ on the microscopic scale of the Compton wavelength. In fact, we will show that in the case of the free scalar field, a distribution function can be defined without introducing such an approximation. From Eq. (2.39), we can infer that W is made up of three terms which can be distinguished for the characteristic of k. For future time-like $k = (p + p')/2$, only the first term involving particles is retained; for past time-like $k = -(p + p')/2$, only the second term involving antiparticles; finally, for space-like $k = (p - p')/2$, the last term with the mean values of two creation/destruction operators is retained. In symbols:

$$
\begin{aligned}
W(x, k) &= W(x, k)\theta(k^2)\theta(k^0) + W(x, k)\theta(k^2)\theta(-k^0) + W(x, k)\theta(-k^2) \\
&\equiv W_+(x, k) + W_-(x, k) + W_S(x, k).
\end{aligned}
\tag{2.40}
$$

Local operators quadratic in the field can be expressed as four-dimensional integrals over k of the covariant Wigner function (2.35). For instance, the mean value of the conserved current of the scalar field is defined as [22]

$$
j^\mu(x) = i \langle : \widehat{\psi}^\dagger(x)\overleftrightarrow{\partial}^\mu \widehat{\psi}(x) : \rangle = \int d^4k \, k^\mu W(x, k).
\tag{2.41}
$$

By using the decomposition (2.40) for $W(x, k)$, the current can be written as the sum of three terms. The particle term, by using (2.39), reads:

$$
\begin{aligned}
j^\mu_+(x) &= \frac{1}{(2\pi)^3} \int \frac{d^3p}{\varepsilon} \frac{d^3p'}{2\varepsilon'} \frac{(p + p')^\mu}{2} \, e^{i(p-p')\cdot x} \langle \widehat{a}^\dagger(p)\widehat{a}(p') \rangle \\
&= \int \frac{d^3p}{\varepsilon} p^\mu \mathrm{Re}\left(\frac{1}{(2\pi)^3} \int \frac{d^3p'}{2\varepsilon'} \, e^{i(p-p')\cdot x} \langle \widehat{a}^\dagger(p)\widehat{a}(p') \rangle \right).
\end{aligned}
\tag{2.42}
$$

To obtain the last expression, we have taken advantage of the hermiticity of the density operator, implying:

$$
\langle \widehat{a}^\dagger(p)\widehat{a}(p') \rangle = \langle \widehat{a}^\dagger(p')\widehat{a}(p) \rangle^*,
$$

which makes it possible to swap the integration variables p and p'. The formula (2.42) brings out a function that we can properly identify as the particle distribution function or phase-space density, as the real part of a *complex distribution function* $f_c(x, p)$:

$$
f(x, p) = \mathrm{Re} f_c(x, p),
\tag{2.43}
$$

where

$$
f_c(x, p) = \frac{1}{(2\pi)^3} \int \frac{d^3p'}{2\varepsilon'} \, e^{i(p-p')\cdot x} \langle \widehat{a}^\dagger(p)\widehat{a}(p') \rangle
\tag{2.44}
$$

Similarly, the antiparticle term in the current leads to a distribution function $\bar{f}(x, p)$ which is obtained from (2.44) replacing $\langle \widehat{a}^\dagger(p)\widehat{a}(p')\rangle$ with $\langle \widehat{b}^\dagger(p)\widehat{b}(p')\rangle$. Finally, the last term in the current, can be written, by using (2.41) and (2.39):

$$
\begin{aligned}
j_S^\mu(x) &= \frac{1}{(2\pi)^3} \int \frac{d^3p\,d^3p'}{\varepsilon\;2\varepsilon'}\,(p - p')^\mu \, \mathrm{Re}\left(e^{i(p+p')\cdot x}\langle \widehat{a}^\dagger(p)\widehat{b}^\dagger(p')\rangle\right) \\
&= \frac{1}{(2\pi)^3} \int \frac{d^3p\,d^3p'}{\varepsilon\;2\varepsilon'}\,p^\mu \, \mathrm{Re}\left[e^{i(p+p')\cdot x}\left(\langle \widehat{a}^\dagger(p)\widehat{b}^\dagger(p')\rangle - \langle \widehat{a}^\dagger(p')\widehat{b}^\dagger(p)\rangle\right)\right],
\end{aligned}
$$

where, again, we have used the hermiticity of the density operator. Thereby, we could define a mixed distribution function g_c:

$$
g_c(x, p) = \frac{1}{(2\pi)^3} \int \frac{d^3p'}{2\varepsilon'}\left[e^{i(p+p')\cdot x}\left(\langle \widehat{a}^\dagger(p)\widehat{b}^\dagger(p')\rangle - \langle \widehat{a}^\dagger(p')\widehat{b}^\dagger(p)\rangle\right)\right]
$$

and write the whole current as

$$
\begin{aligned}
j^\mu(x) &= \int d^4k\, k^\mu\, W(x, k) = \mathrm{Re}\int \frac{d^3p}{2\varepsilon}\,p^\mu\left[f_c(x, p) - \bar{f}_c(x, p) + g_c(x, p)\right] \\
&= \int \frac{d^3p}{2\varepsilon}\,p^\mu\left[f(x, p) - \bar{f}(x, p) + g(x, p)\right].
\end{aligned}
$$

If $g_c = 0$, which is the most common case, the current can be formally written as the familiar relativistic kinetic formula, that is an integral over on-shell four-momenta of the four-momentum vector multiplied by on-shell phase-space densities.

In general, an algebraic relation between, e.g. $W_+(x, k)$ and $f(x, p)$ does not exist. Nevertheless, interesting integral relations between them can be obtained. It is not hard to show that, integrating (2.39) in k, one gets:

$$
\begin{aligned}
\int d^4k\; W(x, k) &= \int \frac{d^3p}{\varepsilon}\,(f_c(x, p) + \bar{f}_c(x, p) + g_c(x, p)) \\
&= \int \frac{d^3p}{\varepsilon}\,(f(x, p) + \bar{f}(x, p) + g(x, p)),
\end{aligned} \tag{2.45}
$$

where the last equality follows from the vanishing of the imaginary part of the integral. Furthermore, if we integrate the time component of the particle current (2.42) over the hypersurface $t = const$, one retrieves the total number of particles:

$$
N = \int d^3x\, j_+^0(x) = \int d^3x \frac{d^3p}{\varepsilon}\varepsilon f(x, p) = \int d^3p \int d^3x\, f(x, p),
$$

and the last expression confirms that $f(x, p)$ is the actual density of particles in phase-space. Furthermore, according to Eq. (2.44),

$$
\frac{dN}{d^3p} = \int d^3x\, f(x, p) = \frac{1}{2\varepsilon}\langle \widehat{a}^\dagger(p)\widehat{a}(p)\rangle,
$$

which is the expected relation between the particle density in momentum space in view of (2.38).

2.4.2 The Dirac Field

A similar connection can be built up for the spin 1/2 particles and the Dirac field. In this case, the covariant Wigner operator is a 4×4 spinorial matrix[4]:

$$
\begin{aligned}
\widehat{W}(x,k)_{AB} &= -\frac{1}{(2\pi)^4} \int d^4y \, e^{-ik\cdot y} : \Psi_A(x-y/2)\overline{\Psi}_B(x+y/2) : \\
&= \frac{1}{(2\pi)^4} \int d^4y \, e^{-ik\cdot y} : \overline{\Psi}_B(x+y/2)\Psi_A(x-y/2) :, \quad (2.46)
\end{aligned}
$$

Ψ being the Dirac field:

$$
\Psi_A(x) = \sum_r \frac{1}{(2\pi)^{3/2}} \int \frac{d^3p}{2\varepsilon} \, \widehat{a}_r(p)u_r(p)_A e^{-ip\cdot x} + \widehat{b}_r^\dagger(p)v_r(p)_A e^{ip\cdot x}. \quad (2.47)
$$

In Eq. (2.47), the creation and destruction operators are normalized according to (2.38) with the anticommutator replacing the commutator, and $u_r(p)$, $v_r(p)$ are the spinors of free particles and antiparticles in their polarization state r (usually a helicity or third spin component) normalized so as to $\bar{u}_r u_s = 2m\delta_{rs}$, $\bar{v}_r v_s = -2m\delta_{rs}$. The covariant Wigner function is, again, the mean value of the Wigner operator in (2.46). The definition (2.46) has to be modified in full spinor electrodynamics [23] to preserve gauge invariance, but this can be neglected for the scope of this work. Because of the Dirac equation, the Wigner operator solves the equation:

$$
\left(m - \not{k} - \frac{i}{2}\not{\partial} \right) \widehat{W}(x,k) = 0.
$$

By plugging (2.47) into (2.46), we obtain:

$$
\begin{aligned}
\widehat{W}(x,k)_{AB} &= \sum_{r,s} \frac{1}{(2\pi)^3} \int \frac{d^3p \, d^3p'}{2\varepsilon \, 2\varepsilon'} e^{-i(p-p')\cdot x} \times \\
&\quad \left[\delta^4(k-(p+p')/2)\widehat{a}_s^\dagger(p')\widehat{a}_r(p)u_r(p)_A \bar{u}_s(p')_B \right. \\
&\quad \left. - \delta^4(k+(p+p')/2)\widehat{b}_r^\dagger(p)\widehat{b}_s(p')v_r(p)_A \bar{v}_s(p')_B \right] \\
&\quad - \delta^4(k-(p-p')/2)\left[e^{-i(p+p')\cdot x}\widehat{a}_r(p)\widehat{b}_s(p')u_r(p)_A \bar{v}_s(p')_B \right. \\
&\quad \left. + e^{i(p+p')\cdot x}v_r(p')_A \bar{u}_s(p)_B \widehat{b}_r^\dagger(p')\widehat{a}_s^\dagger(p) \right]. \quad (2.48)
\end{aligned}
$$

---

[4]It should be reminded that the normal ordering for fermion fields involves a minus sign for each permutation, e.g. $: aa^\dagger := -a^\dagger a$. Therefore, taking into account anticommutation relations, for fields $: \Psi_A(x)\overline{\Psi}_B(y) := - : \overline{\Psi}_B(y)\Psi_A(x) :$

It can be seen that in the fermion case, the covariant Wigner operator can be split into future time-like (particle), past time-like (antiparticle) and space-like parts corresponding to the three terms of the right hand side, just as for the scalar field:

$$\widehat{W}(x, k) = \widehat{W}(x, k)\theta(k^2)\theta(k^0) + \widehat{W}(x, k)_{AB}\theta(k^2)\theta(-k^0) + \widehat{W}(x, k)\theta(-k^2)$$
$$\equiv \widehat{W}_+(x, k) + \widehat{W}_-(x, k) + \widehat{W}_S(x, k).$$

Yet, unlike for the scalar field, we cannot identify a distribution function in phase-space by integrating the covariant Wigner function in k. We can make this clear by calculating the mean current of the Dirac field (without vacuum contribution) which is obtained from the Wigner function through the formula [22]:

$$j^\mu(x) = \langle : \overline{\Psi}(x)\gamma^\mu \Psi(x) : \rangle = \int d^4k \ \text{tr}(\gamma^\mu W(x, k)).$$

We confine ourselves to the particle term of (2.48), which, once fed into the above formula yields:

$$j_+^\mu(x) = \int d^4k \ \text{tr}(\gamma^\mu W(x, k)_+)$$
$$= \sum_{r,s} \frac{1}{(2\pi)^3} \int \frac{d^3 p \ d^3 p'}{2\varepsilon \ 2\varepsilon'} e^{-i(p-p')\cdot x} \langle \widehat{a}_s^\dagger(p')\widehat{a}_r(p)\rangle \bar{u}_s(p')\gamma^\mu u_r(p). \tag{2.49}$$

Unlike in (2.42), we cannot factorize the momentum integration because the spinors u have different momenta as argument. It thus follows that a reasonable definition of a particle distribution function with spin indices $f(x, p)_{rs}$ is precluded, except in the limit of very slow variation of the Wigner function as derived in Ref. [22].

Notwithstanding, it is possible to establish an exact relation between the density of particles in momentum space and the covariant Wigner function. First of all, it can be shown, from (2.49), that

$$\partial_\mu j_+^\mu = 0$$

and likewise for j_-^μ, taking into account that the spinors $u(p)$ and $\bar{u}(p)$ fulfil the equations:

$$(\not{p} - m)u(p) = 0 \qquad \bar{u}(p)(\not{p} - m) = 0.$$

If the divergence of j_+ vanishes, we can integrate the particle current (2.49) over an arbitrary 3D space-like hypersurface to get a constant particle number, provided that boundary fluxes vanish. For instance, we can integrate j_+^0 over the hypersurface $t = const$ to get:

$$
N = \int \mathrm{d}^3x \, j^0_+(x) = \int \mathrm{d}^3x \int \mathrm{d}^4k \, \mathrm{tr}(\gamma^0 W(x,k)_+)
$$

$$
= \sum_{r,s} \int \frac{\mathrm{d}^3p \, \mathrm{d}^3p'}{2\varepsilon \, 2\varepsilon'} \delta^3(\mathbf{p} - \mathbf{p}') \langle \widehat{a}^\dagger_r(p)\widehat{a}_s(p') \rangle \bar{u}_s(p') \gamma^0 u_r(p)
$$

$$
= \sum_{r,s} \int \frac{\mathrm{d}^3p}{4\varepsilon^2} \langle \widehat{a}^\dagger_r(p)\widehat{a}_s(p) \rangle \bar{u}_s(p) \gamma^0 u_r(p) = \sum_{r,s} \int \frac{\mathrm{d}^3p}{4\varepsilon^2} \langle \widehat{a}^\dagger_r(p)\widehat{a}_s(p) \rangle 2\varepsilon \delta_{r,s}
$$

$$
= \int \frac{\mathrm{d}^3p}{2\varepsilon} \sum_r \langle \widehat{a}^\dagger_r(p)\widehat{a}_r(p) \rangle,
$$

whence we obtain the particle density in momentum space, as expected:

$$
\frac{\mathrm{d}N}{\mathrm{d}^3p} = \frac{1}{2\varepsilon} \sum_r \langle \widehat{a}^\dagger_r(p)\widehat{a}_r(p) \rangle \tag{2.50}
$$

An important feature of the Wigner operator of free fields is that integrating it over a 3D hypersurface, k becomes an on-shell vector. Indeed, from (2.48),

$$
k^\mu \partial_\mu \widehat{W}_\pm(x,k) = k^\mu \partial_\mu \widehat{W}_S(x,k) = 0
$$

because in taking the derivative $k \cdot \partial$ the factor $(p - p') \cdot (p + p') = 0$ is generated in all of the terms; the same applies to the Wigner operator of the scalar field. Therefore, provided that suitable boundary conditions are fulfilled, the integral over a space-like 3D hypersurface:

$$
\int_\Sigma \mathrm{d}\Sigma_\mu k^\mu \widehat{W}(x,k)
$$

is independent of the hypersurface Σ. Thus, we can choose Σ at the hyperplane $t = 0$ and obtain, from (2.48):

$$
\int_{t=0} \mathrm{d}\Sigma_\mu k^\mu \widehat{W}(x,k) = k^0 \int \mathrm{d}^3x \, \widehat{W}(x,k)
$$

$$
= \sum_{r,s} k^0 \int \frac{\mathrm{d}^3p \, \mathrm{d}^3p'}{2\varepsilon \, 2\varepsilon'} \delta^3(\mathbf{p} - \mathbf{p}') \left[\delta^4(k - p)\widehat{a}^\dagger_s(p)\widehat{a}_r(p)u_r(p)\bar{u}_s(p) \right.
$$

$$
- \delta^4(k + p)\widehat{b}^\dagger_r(p)\widehat{b}_s(p)v_r(p)_A \bar{v}_s(p)_B \Big]
$$

$$
+ \delta(k^0)\delta^3(\mathbf{k} - \mathbf{p}) \left[\delta^3(\mathbf{p} + \mathbf{p}')\widehat{a}_r(p)\widehat{b}_s(p')u_r(p)\bar{v}_s(p') + v_r(p')\bar{u}_s(p)\widehat{b}^\dagger_r(p')\widehat{a}^\dagger_s(p) \right]
$$

$$
= \sum_{r,s} k^0 \int \frac{\mathrm{d}^3p}{4\varepsilon^2} \left[\delta^4(k - p)\widehat{a}^\dagger_s(p)\widehat{a}_r(p)u_r(p)\bar{u}_s(p) - \delta^4(k + p)\widehat{b}^\dagger_r(p)\widehat{b}_s(p)v_r(p)\bar{v}_s(p) \right]
$$

$$
= \sum_{r,s} \frac{1}{2}\delta(k^2 - m^2) \left[\theta(k^0)\widehat{a}^\dagger_s(k)\widehat{a}_r(k)u_r(k)\bar{u}_s(k) \right.
$$

$$
\left. + \theta(-k^0)\widehat{b}^\dagger_r(-k)\widehat{b}_s(-k)v_r(-k)_A \bar{v}_s(-k)_B \right]; \tag{2.51}
$$

note that the mixed term vanished because of the factor $k^0 \delta(k^0)$. The last obtained result proves the above statement, that is the argument k of the integral over a 3D space-like hypersurface Σ is an on-shell four-vector and it nicely separates particle–antiparticle contribution. Indeed, the on-shell operators $\widehat{w}_\pm$ can be defined such that

$$\frac{1}{2\varepsilon_k}\delta(k^0 - \varepsilon_k)\widehat{w}_+(k) = \int d\Sigma_\mu k^\mu \, \widehat{W}_+(x, k)$$

$$\frac{1}{2\varepsilon_k}\delta(k^0 + \varepsilon_k)\widehat{w}_-(k) = \int d\Sigma_\mu k^\mu \, \widehat{W}_-(x, k),$$

so that, by comparing with (2.51),

$$\widehat{w}_+(k) = \sum_{r,s} \frac{1}{2}\widehat{a}_s^\dagger(k)\widehat{a}_r(k)u_r(k)\bar{u}_s(k) \tag{2.52}$$

with k on-shell. A similar equation can be established for $\widehat{W}_-$ and antiparticles. Note that, from (2.52),

$$(\slashed{k} - m)\widehat{w}_+(k) = \widehat{w}_+(k)(\slashed{k} - m) = 0 \tag{2.53}$$

Now, multiplying (2.52) by $\bar{u}_r(k)$ to the left and $u_s(k)$ to the right, and keeping in mind the normalization of the spinors u, we get:

$$\bar{u}_r(k)\widehat{w}_+(k)u_s(k) = 2m^2\widehat{a}_s^\dagger(k)\widehat{a}_r(k). \tag{2.54}$$

This formula will be used in the next section to express the spin density matrix. Now, setting $r = s$, summing over r and taking the mean value with the suitable density operator, we obtain:

$$\sum_r \bar{u}_r(k)w_+(k)u_r(k) = \mathrm{tr}\left(w_+(k)\sum_r u_r(k)\bar{u}_r(k)\right) = 2m^2\sum_r \langle \widehat{a}_r^\dagger(k)\widehat{a}_r(k)\rangle$$

$$= 4m^2\varepsilon\frac{dN}{d^3k},$$

where we have used (2.50). Since

$$\sum_r u_r(k)\bar{u}_r(k) = \slashed{k} + m,$$

we have:

$$\varepsilon\frac{dN}{d^3k} = \frac{1}{4m^2}\mathrm{tr}\left(w_+(k)(\slashed{k} + m)\right),$$

and, by using (2.53), we finally obtain the sought relation between the momentum spectrum and the covariant Wigner function:

$$\varepsilon\frac{dN}{d^3k} = \frac{1}{2m}\mathrm{tr}\, w_+(k) = \frac{\varepsilon}{m}\int dk^0 \int d\Sigma_\mu \, k^\mu \, \mathrm{tr}\, W_+(x, k). \tag{2.55}$$

2.5 Fermion Polarization and the Covariant Wigner Function

We are now in a position to derive an exact formula connecting the Wigner function to the spin density matrix and the spin vector for spin 1/2 fermions. The derivation of $\Theta(p)$ for particles is now a straightforward consequence of its definition (2.15) and of (2.54):

$$\Theta(p)_{rs} = \frac{\bar{u}_r(p)w_+(p)u_s(p)}{\sum_t \bar{u}_t(p)w_+(p)u_t(p)}.$$

This formula can be also written in an expanded form by using the actual covariant Wigner function by using (2.52) and taking advantage of the cancellation of the Dirac deltas in the ratio:

$$\Theta(p)_{rs} = \frac{\int d\Sigma_\mu p^\mu \bar{u}_r(p) W_+(x, p) u_s(p)}{\sum_t \int d\Sigma_\mu p^\mu \bar{u}_t(p) W_+(x, p) u_t(p)} \tag{2.56}$$

keeping in mind that p is on-shell because of the integration. As we have emphasized in the previous section, as long as one deals with free fields, the integration hypersurface is arbitrary, and this will be important for the use of (2.56) in relativistic heavy ion collisions. The above matrix can be written in a more compact way by introducing 4×2 spinorial matrices U (and corresponding 2×4 $\bar{U}$) such that $U_{A,r}(p) = u_r(p)_A$:

$$\Theta(p) = \frac{\int d\Sigma_\mu p^\mu \bar{U}(p) W_+(x, p) U(p)}{\mathrm{tr}_2 \int d\Sigma_\mu p^\mu \bar{U}(p) W_+(x, p) U(p)} \tag{2.57}$$

where, henceforth, we will make explicit the distinction between the trace over the polarization states tr_2 and the trace over the four spinorial indices tr_4. In Weyl's representation, which is deeply connected to the theory of Lorentz'group representations, these spinors can be written as [20,24]

$$U(p) = \sqrt{m}\begin{pmatrix} D^S([p]) \\ D^S([p]^{\dagger -1}) \end{pmatrix} \qquad V(p) = \sqrt{m}\begin{pmatrix} D^S([p]C^{-1}) \\ D^S([p]^{\dagger -1}C) \end{pmatrix} \tag{2.58}$$

with $C = i\sigma_2$ (σ_i being Pauli matrices).

The mean spin vector can now be calculated from (2.14) by using Eq. (2.56) therein. We first observe that, since $S^\mu(p)$ is a real number, we can also express it as

$$\begin{aligned} S^\mu(p) = &-\frac{1}{4m}\epsilon^{\mu\beta\gamma\delta} p_\delta \Big[\mathrm{tr}_2(D^S([p]^{-1})D^S(J_{\beta\gamma})D^S([p])\Theta(p)) \\ &+ \mathrm{tr}_2(D^S([p]^{-1})D^S(J_{\beta\gamma})D^S([p])\Theta(p))^*\Big] \\ = &-\frac{1}{4m}\epsilon^{\mu\beta\gamma\delta} p_\delta \Big[\mathrm{tr}_2(D^S([p]^{-1})D^S(J_{\beta\gamma})D^S([p])\Theta(p)) \\ &+\mathrm{tr}_2(D^S([p])^\dagger D^S(J_{\beta\gamma})^\dagger D^S([p]^{-1})^\dagger \Theta(p))\Big], \end{aligned} \tag{2.59}$$

where we have taken advantage of the hermiticity of $\Theta(p)$ and the complicity of the trace. We can now use (2.57) and work out the numerator first:

$$\mathrm{tr}_2(D^S([p]^{-1})D^S(J_{\beta\gamma})D^S([p])\bar{U}(p)W_+(x,p)U(p)+$$
$$+\,\mathrm{tr}_2(D^S([p])^\dagger D^S(J_{\beta\gamma})^\dagger D^S([p]^{-1})^\dagger \bar{U}(p)W_+(x,p)U(p)),$$
$$(2.60)$$

where U are the spinors defined in (2.58). This expression can be written in a more compact and familiar form in the Dirac spinorial formalism. We start by defining

$$\Sigma_{\beta\gamma} = \begin{pmatrix} D^S(J_{\beta\gamma}) & 0 \\ 0 & D^S(J_{\beta\gamma})^\dagger \end{pmatrix}, \qquad (2.61)$$

which is just the generator of Lorentz transformations written for the full spinorial representation $(0, 1/2) \oplus (1/2, 0)$ of the Dirac field, equal to $(i/4)[\gamma_\beta, \gamma_\gamma]$. It can be readily seen that (2.60) is equivalent to

$$\frac{1}{m}\mathrm{tr}_2(\bar{U}(p)\Sigma_{\beta\gamma}U(p)\bar{U}(p)W_+(x,p)U(p)).$$

In general, if A is a 2×4 and B is a 4×2 matrix,

$$\mathrm{tr}_2 AB = \mathrm{tr}_4 BA,$$

hence the above trace can be rewritten:

$$\frac{1}{m}\mathrm{tr}_4(\Sigma_{\beta\gamma}U(p)\bar{U}(p)W_+(x,p)U(p)\bar{U}(p)),$$

which can be worked out taking into account that

$$U(p)\bar{U}(p) = \sum_r u_r(p)\bar{u}_r(p) = \slashed{p} + m$$

Likewise, the denominator of (2.56) can be rewritten as

$$\mathrm{tr}_2(\bar{U}(p)W_+(x,p)U(p)) = \mathrm{tr}_4(W_+(x,p)U(p)\bar{U}(p)) = \mathrm{tr}_4((\slashed{p}+m)W_+(x,p)).$$

Putting all together, we can write the mean spin vector as

$$S^\mu(p) = -\frac{1}{4m^2}\epsilon^{\mu\beta\gamma\delta}p_\delta \frac{\int d\Sigma_\lambda p^\lambda \mathrm{tr}_4(\Sigma_{\beta\gamma}(\slashed{p}+m)W_+(x,p)(\slashed{p}+m))}{\int d\Sigma_\lambda p^\lambda \mathrm{tr}_4((\slashed{p}+m)W_+(x,p))}. \qquad (2.62)$$

Likewise, one can also recast Eq. (2.14)—the exact expression of the mean spin vector at global equilibrium in the Boltzmann limit—for spin 1/2 particles in the

Dirac formalism. With the foregoing definitions and notations, Eq. (2.31) can be rewritten as

$$\Theta(p) = \frac{\bar{U}(p) \exp\left[\frac{1}{2}\varpi : \Sigma\right] U(p)}{\text{tr}_2(\bar{U}(p) \exp\left[\frac{1}{2}\varpi : \Sigma\right] U(p))},$$

which is a Hermitian matrix. By using this equation, (2.59) and (2.60), the mean spin vector of (2.14) can be rewritten by simply replacing the integrals of $W_+(x, p)$ with $\exp\left[\frac{1}{2}\varpi : \Sigma\right]$ in Eq. (2.62). We thus obtain:

$$S^\mu(p) = -\frac{1}{4m^2} \epsilon^{\mu\beta\gamma\delta} p_\delta \frac{\text{tr}_4(\Sigma_{\beta\gamma}(\not{p}+m) \exp\left[\frac{1}{2}\varpi : \Sigma\right](\not{p}+m))}{\text{tr}_4((\not{p}+m) \exp\left[\frac{1}{2}\varpi : \Sigma\right])}. \tag{2.63}$$

Taking into account that the trace of an odd number of gamma matrices vanishes and the commutator:

$$[\Sigma_{\beta\gamma}, \gamma_\lambda] = -i g_{\beta\lambda}\gamma_\gamma + i g_{\gamma\lambda}\gamma_\beta, \tag{2.64}$$

it can be readily shown that (2.63) can be written in the simpler form:

$$S^\mu(p) = -\frac{1}{2m} \epsilon^{\mu\beta\gamma\delta} p_\delta \frac{\text{tr}_4(\Sigma_{\beta\gamma} \exp\left[\frac{1}{2}\varpi : \Sigma\right])}{\text{tr}_4(\exp\left[\frac{1}{2}\varpi : \Sigma\right])}, \tag{2.65}$$

which looks certainly more suggestive and compact with respect to the general group-theoretical formula.

Also the more general Eq. (2.62) can be further developed and simplified. According to (2.53),

$$\not{p} \int d\Sigma_\lambda p^\lambda W_+(x, p) = m \int d\Sigma_\lambda p^\lambda W_+(x, p),$$

and (2.62) becomes:

$$S^\mu(p) = -\frac{1}{2m} \epsilon^{\mu\beta\gamma\delta} p_\delta \frac{\int d\Sigma_\lambda p^\lambda \text{tr}_4(\Sigma_{\beta\gamma} W_+(x, p))}{\int d\Sigma_\lambda p^\lambda \text{tr}_4 W_+(x, p)}, \tag{2.66}$$

which is already quite a suggestive formula. Furthermore, because of (2.51), we can write:

$$\int d\Sigma_\lambda p^\lambda \text{tr}_4(\Sigma_{\beta\gamma} W_+(x, p)) = \sum_{r,s} \frac{1}{4\varepsilon_p} \delta(p^0 - \varepsilon_p) \langle \hat{a}_s^\dagger(p)\hat{a}_r(p)\rangle \text{tr}_4(u_r(p)\bar{u}_s(p)\Sigma_{\beta\gamma})$$

$$= \sum_{r,s} \frac{1}{4\varepsilon_p} \delta(p^0 - \varepsilon_p) \langle \hat{a}_s^\dagger(p)\hat{a}_r(p)\rangle \bar{u}_s(p)\Sigma_{\beta\gamma}u_r(p).$$

Now, by using the spinorial relation:

$$m\bar{u}_r(p)\Sigma^{\mu\nu}\gamma_\lambda u_s(p) = \bar{u}_r(p)\Sigma^{\mu\nu}u_s(p)p_\lambda - 2i\bar{u}_r(p)(\gamma^\mu p^\nu - \gamma^\nu p^\mu)\gamma_\lambda u_s(p),$$

the previous equation turns into

$$\int d\Sigma^\lambda \, p_\lambda \text{tr}_4(\Sigma_{\beta\gamma} W_+(x, p)) = \sum_{r,s} \frac{1}{4\varepsilon_p^2} \delta(p^0 - \varepsilon_p) \langle \widehat{a}_s^\dagger(p)\widehat{a}_r(p)\rangle m \, \bar{u}_s(p)\Sigma_{\beta\gamma} \gamma_0 u_r(p)$$

$$+ 2i \sum_{r,s} \frac{1}{4\varepsilon_p^2} \delta(p^0 - \varepsilon_p)\langle \widehat{a}_s^\dagger(p)\widehat{a}_r(p)\rangle \bar{u}_s(p)(\gamma_\beta p_\gamma - \gamma_\gamma p_\beta)\gamma_0 \, u_r(p). \quad (2.67)$$

The second term in the right hand side will not contribute to the mean spin vector (2.66) because of two momenta multiplying the Levi-Civita tensor. By using (2.64), the first term in the right hand side of (2.67) can be rewritten:

$$\sum_{r,s} \frac{m}{4\varepsilon_p^2} \delta(p^0 - \varepsilon_p)\langle \widehat{a}_s^\dagger(p)\widehat{a}_r(p)\rangle \left[\bar{u}_s(p)\frac{1}{2}\{\gamma_0, \Sigma_{\beta\gamma}\} u_r(p) + i g_{0\gamma}\bar{u}_s(p)\gamma_\beta u_r(p) \right.$$

$$\left. - i g_{0\beta}\bar{u}_s(p)\gamma_\gamma u_r(p)\right]$$

$$= \sum_{r,s} \frac{m}{4\varepsilon_p^2} \delta(p^0 - \varepsilon_p)\langle \widehat{a}_s^\dagger(p)\widehat{a}_r(p)\rangle \left[\bar{u}_s(p)\frac{1}{2}\{\gamma_0, \Sigma_{\beta\gamma}\} u_r(p) + 2i\delta_{rs}(p_\beta g_{0\gamma} - p_\gamma g_{0\beta})\right],$$

$$(2.68)$$

where we have used the relation

$$\bar{u}_s(p)\gamma^\lambda u_r(p) = 2p^\lambda \delta_{rs}.$$

Again, the second term on the right hand side of (2.68) does not contribute to the mean spin vector because of the Levi-Civita tensor in (2.66). Now, since

$$\{\gamma_\lambda, \Sigma_{\beta\gamma}\} = \epsilon_{\sigma\lambda\beta\gamma}\gamma^\sigma \gamma^5, \quad (2.69)$$

we can finally rewrite the numerator of the mean spin vector (2.66) as

$$-\frac{1}{4}\epsilon^{\mu\beta\gamma\delta}\epsilon_{\sigma 0\beta\gamma} \, p_\delta \sum_{r,s} \frac{1}{4\varepsilon_p^2}\delta(p^0 - \varepsilon_p)\langle \widehat{a}_s^\dagger(p)\widehat{a}_r(p)\rangle \, \bar{u}_s(p)\gamma^\sigma \gamma^5 u_r(p)$$

$$= p_0 \sum_{r,s} \frac{1}{8\varepsilon_p^2}\delta(p^0 - \varepsilon_p)\langle \widehat{a}_s^\dagger(p)\widehat{a}_r(p)\rangle \, \bar{u}_s(p)\gamma^\mu \gamma^5 u_r(p)$$

$$- \delta_0^\mu \, p_\sigma \sum_{r,s} \frac{1}{8\varepsilon_p^2}\delta(p^0 - \varepsilon_p)\langle \widehat{a}_s^\dagger(p)\widehat{a}_r(p)\rangle \, \bar{u}_s(p)\gamma^\sigma \gamma^5 u_r(p).$$

The second term vanishes because

$$\bar{u}_s(p)\slashed{p}\gamma^5 u_r(p) = m\bar{u}_s(p)\gamma^5 u_r(p) = 0,$$

so we have, for the numerator of (2.66):

$$\sum_{r,s} \frac{1}{8\varepsilon_p}\delta(p^0 - \varepsilon_p)\langle \widehat{a}_s^\dagger(p)\widehat{a}_r(p)\rangle \, \bar{u}_s(p)\gamma^\mu \gamma^5 u_r(p).$$

This expression can be rewritten in form of an integral over an arbitrary hypersurface of a divergence-free integrand, and so we get, by using again (2.51):

$$S^\mu(p) = \frac{1}{2} \frac{\int d\Sigma \cdot p \, \text{tr}_4(\gamma^\mu \gamma^5 W_+(x, p))}{\int d\Sigma \cdot p \, \text{tr}_4 W_+(x, p)}. \tag{2.70}$$

Hence, the mean spin vector is proportional to the integral of the axial vector component of the covariant Wigner function over some arbitrary 3D space-like hypersurface. Note that $S(p)$ is actually orthogonal to the four-momentum p because $\text{tr}_4(\not{p}\gamma^5 W_+) = 0$ [23] (this can be shown also by using the expansion (2.48)). This expression is consistent and extends the relation between $W(x, p)$ and $S^\mu(p)$ at $O(\hbar)$ used in Refs. [9,11,25] to determine the mean spin vector.

Finally, by using the inverse of the (2.69), that is:

$$\gamma^\mu \gamma^5 \delta^\delta_\lambda = \gamma^\delta \gamma^5 \delta^\mu_\lambda - \frac{1}{2} \epsilon^{\mu\delta\alpha\beta} \{\gamma_\lambda, \Sigma_{\alpha\beta}\}$$

and taking into account that $\text{tr}_4(\not{p}\gamma^5 W_+) = 0$, we obtain another form of (2.70):

$$S^\mu(p) = -\frac{1}{4} \epsilon^{\mu\beta\gamma\delta} p_\delta \frac{\int d\Sigma_\lambda \text{tr}_4(\{\gamma^\lambda, \Sigma_{\beta\gamma}\} W_+(x, p))}{\int d\Sigma_\lambda p^\lambda \text{tr}_4 W_+(x, p)}. \tag{2.71}$$

In the numerator, the educated reader shall recognize the matrix defining the canonical spin tensor of the Dirac field. However, it should be pointed out that the appearance of this combination does not imply the need of a particular spin tensor to find the expression (2.71), unlike originally stated in Ref. [15]. The point is that the polarization expression was obtained without any reference whatsoever to the tensors that are used to express the energy-momentum and angular momentum of the fields; this was already implied in the expressions of the mean spin vector quoted in Refs. [13,26] and will be discussed in more detail in Sect. 2.6.

Indeed, one could have chosen to express the relation between mean spin vector and covariant Wigner function through (2.66) or equally well with Eq. (2.70); they are completely equivalent forms of (2.71).

2.6 Polarization From the Angular Momentum Operator

We have seen how to calculate the mean spin vector from the spin density matrix definition in quantum field theory, see Eq. (2.15). It is possible to calculate the same quantity with a different method. Assume that the total angular momentum tensor $J^{\mu\nu}$ can be decomposed on-shell in momentum space, so that we know the total angular momentum tensor of particles with given four-momentum p, say $\tilde{J}^{\mu\nu}(p)$. We could then be able to obtain the mean spin vector $S^\mu(p)$ by simply dividing this quantity by the number of particles with momentum p and multiplying by the usual Levi-Civita like in the definition (2.1):

$$S^\mu(p) = -\frac{1}{2m}\epsilon^{\mu\nu\rho\sigma} p_\sigma \frac{\tilde{J}_{\nu\rho}(p)}{\frac{\mathrm{d}N}{\mathrm{d}^3 p}}.$$ (2.72)

This definition makes perfect sense as all particles with given momentum p have the same rest-frame and was indeed used in Ref. [15] to derive the expression of the mean spin vector at local thermodynamic equilibrium. We will show that (2.72) leads to Eq. (2.70) as well.

The first step is to prove that $\tilde{J}_{\nu\rho}(p)$ exists and to find its form for the free Dirac field. Indeed, it is possible to show, under quite general assumptions, that conserved *charges* of free fields can be written as integrals over on-shell momentum four-vectors. In general, conserved charges can be written as integrals over a space-like 3D hypersurface Σ of a divergenceless current, also at operator level:

$$\widehat{Q}^{\mu_1\cdots\mu_N} = \int_\Sigma \mathrm{d}\Sigma_\lambda \; \widehat{J}^{\lambda\mu_1\cdots\mu_N}.$$

With a suitable choice of the 3D boundaries, the above integral is independent of the hypersurface Σ, and it can then be calculated using any space-like Σ, e.g. $t = 0$. Suppose now that the normal-ordered current : $\widehat{J}$: (normal ordering is necessary to have vanishing currents in vacuum) can be expressed as an integral over k of some tensor functional of the covariant Wigner function $\widehat{W}(x, k)$:

$$: \widehat{J}^{\lambda\mu_1\cdots\mu_N} : = \int \mathrm{d}^4k \; \mathcal{F}[\widehat{W}(x, k)]^{\lambda\mu_1\cdots\mu_N}.$$ (2.73)

This is the case, for instance, for the charge current : $\widehat{j}^\mu$: of the scalar field, for which the functional would simply be $k^\mu \widehat{W}(x, k)$, or the current of the Dirac field for which it would be $\mathrm{tr}_4(\gamma^\mu \widehat{W}(x, k))$, but it also applies to stress-energy tensor and spin tensor, as we will see. Using Eq. (2.73) and taking advantage of the independence of the integration hypersurface, the charge can be written as

$$: \widehat{Q}^{\mu_1\cdots\mu_N} : = \int_\Sigma \mathrm{d}\Sigma_\lambda \; : \widehat{J}^{\lambda\mu_1\cdots\mu_N} : = \int_\Sigma \mathrm{d}\Sigma_\lambda \int \mathrm{d}^4k \; \mathcal{F}[\widehat{W}(x, k)]^{\lambda\mu_1\cdots\mu_N}$$ (2.74)

$$= \int_{t=0} \mathrm{d}^3x \int \mathrm{d}^4k \; \mathcal{F}[\widehat{W}(x, k)]^{0\mu_1\cdots\mu_N}$$

$$= \int \mathrm{d}^4k \left(\int_{t=0} \mathrm{d}^3x \; \mathcal{F}[\widehat{W}(x, k)]^{0\mu_1\cdots\mu_N} \right).$$

For *free fields*, the integration over x generally implies that the four-momentum k is on-shell. In fact, this depends on the specific form of the functional $\mathcal{F}$, but it holds for all cases of interest, and the proof is the same which led to Eq. (2.51); after the last integration in d^3x, a factor $\delta(k^2 - m^2)$ comes in which makes it possible to separate particle and antiparticle contribution and to express the generally conserved normal-ordered charge as

$$\begin{aligned} : \widehat{Q}^{\mu_1 \cdots \mu_N} : &= \int d^4 k \, \delta(k^2 - m^2) \widehat{Q}(k)^{\mu_1 \cdots \mu_N} \\ &= \int d^3 k \left(\int dk^0 \, \delta(k^2 - m^2) \widehat{Q}(k)^{\mu_1 \cdots \mu_N} \right) \\ &\equiv \int \frac{d^3 k}{2\varepsilon_k} \left(\widehat{q}_+(k)^{\mu_1 \cdots \mu_N} + \widehat{q}_-(k)^{\mu_1 \cdots \mu_N} \right). \end{aligned} \tag{2.75}$$

Altogether, the charge can be written as a sum over three-momenta of on-shell particles and antiparticles and a spectral decomposition in momentum space is obtained:

$$\frac{d : \widehat{Q}^{\mu_1 \cdots \mu_N} :}{d^3 k} = \frac{1}{2\varepsilon_k} \widehat{q}_+(k)^{\mu_1 \cdots \mu_N} = \int dk^0 \int d\Sigma_\lambda \, \mathcal{F}[\widehat{W}_+(x,k)]^{\lambda \mu_1 \cdots \mu_N}, \tag{2.76}$$

and likewise for antiparticles. The operators $\widehat{q}_\pm(k)$ are invariant by the addition of a total divergence to the current. For instance, for the vector current:

$$\widehat{j}^\lambda \to \widehat{j}^\lambda + \partial_\alpha \widehat{A}^{\lambda\alpha},$$

where $\widehat{A}^{\lambda\alpha}$ is an anti-symmetric tensor, the corresponding $\widehat{q}_\pm$ get changed by

$$\widehat{q}_\pm \to \widehat{q}_\pm + \int dk^0 \int d\Sigma_\lambda \partial_\alpha \mathcal{A}[\widehat{W}_\pm(x,k)]^{\lambda\alpha},$$

where $\mathcal{A}$ is the suitable functional of the Wigner operator associated to $\widehat{A}^{\lambda\alpha}$. The integral over the 3D hypersurface, for fixed k, can be turned into a boundary surface integral by means of the Stokes theorem and so, provided that the suitable boundary conditions are enforced, vanishes.

We now look for the spectral decomposition of the angular momentum-boost operators, hence the $\widetilde{J}^{\mu\nu}(p)$ of Eq. (2.72). The angular momentum-boost operator is the generator of the Lorentz transformations and can be written as

$$\widehat{J}^{\mu\nu} = \int_\Sigma d\Sigma_\lambda \, \widehat{\mathcal{J}}^{\lambda,\mu\nu} = \int_\Sigma d\Sigma_\lambda \left(x^\mu \widehat{T}^{\lambda\nu} - x^\nu \widehat{T}^{\lambda\mu} + \widehat{S}^{\lambda,\mu\nu} \right) \tag{2.77}$$

with Σ space-like hypersurface. There are two contributing terms: the so-called *orbital* part, depending on the stress-energy tensor, and the *spin tensor* operator $\widehat{S}$. The generator in (2.77) is invariant under a so-called *pseudo-gauge transformation* of the stress-energy and spin tensor [27] which amounts to add a divergence to the angular momentum-boost current $\widehat{\mathcal{J}}^{\lambda,\mu\nu}$. The choice of a stress-energy and a spin tensor is just a matter of convenience, and, for the Dirac field, a convenient choice is the so-called *canonical* stress-energy and spin tensor:

$$: \widehat{T}^{\mu\nu}(x) : = \frac{i}{2} : \overline{\Psi}(x)\gamma^\mu \overset{\leftrightarrow}{\partial}{}^\nu \Psi(x) := \int d^4k \, k^\nu \mathrm{tr}_4(\gamma^\mu \widehat{W}(x,k)) \qquad (2.78)$$

$$: \widehat{S}^{\lambda,\mu\nu}(x) : = \frac{1}{2} : \overline{\Psi}(x)\{\gamma^\lambda, \Sigma^{\mu\nu}\}\Psi(x) := \frac{1}{2} \int d^4k \, \mathrm{tr}_4(\{\gamma^\lambda, \Sigma^{\mu\nu}\}\widehat{W}(x,k)),$$

where their relations with the covariant Wigner function have been written down. Let us start with the *orbital* part of the angular momentum-boost operator:

$$: \widehat{L}^{\mu\nu} : \equiv \int d\Sigma_\lambda \left(x^\mu : \widehat{T}^{\lambda\nu} : -x^\nu : \widehat{T}^{\lambda\mu} :\right)$$

$$= \int d\Sigma_\lambda \int d^4k \left(x^\mu k^\nu \mathrm{tr}_4(\gamma^\lambda \widehat{W}(x,k)) - x^\nu k^\mu \mathrm{tr}_4(\gamma^\lambda \widehat{W}(x,k))\right). \qquad (2.79)$$

The subtlety here is that the functional $\mathcal{F}$ that we can write as

$$\mathcal{F} = x^\mu k^\nu \mathrm{tr}_4(\gamma^\lambda W(x,k)) - (\mu \leftrightarrow \nu)$$

explicitly depends on x and so the proof of the on-shellness of k must be reviewed, what is done in detail in Appendix A. The result of this analysis is that the orbital part of the angular momentum operator, with the canonical stress-energy tensor in (2.78), can be written as

$$: \widehat{L}^{\mu\nu} : = \int \frac{d^3k}{2\varepsilon_k} \left(k^\mu \widehat{G}^\nu_+(k) - k^\nu \widehat{G}^\mu_+(k)\right) + \quad \text{antiparticle term}$$

with k on-shell and with $\widehat{G}$ a vector operator (see Appendix A). Thus, the orbital part of the angular momentum does not contribute to the mean spin vector because of the Levi-Civita tensor which makes the orbital part vanishing.

On the other hand, the canonical spin tensor term in (2.77) has an algebraic dependence on the Wigner function, and according to (2.78) and Eq. (2.76) can be applied with

$$\mathcal{F}[\widehat{W}_+(x,k)] = \frac{1}{2}\mathrm{tr}_4\left(\{\gamma^\lambda, \Sigma^{\mu\nu}\}\widehat{W}_+(x,k)\right),$$

so the spin part of the total angular momentum-boost tensor is:

$$: \widehat{S}^{\mu\nu} : = \int d^4k \int d\Sigma_\lambda \, \mathrm{tr}_4\left(\frac{1}{2}\{\gamma^\lambda, \Sigma^{\mu\nu}\}\widehat{W}_+(x,k)\right) + \quad \text{antiparticle term.} \qquad (2.80)$$

Therefore, its contribution to the function $\tilde{J}^{\mu\nu}(p)$ for particles in (2.72) is (with k renamed p):

$$\tilde{S}^{\mu\nu}_+(p) = \int dp^0 \int d\Sigma_\lambda \, \mathrm{tr}_4\left(\frac{1}{2}\{\gamma^\lambda, \Sigma^{\mu\nu}\}W_+(x,p)\right).$$

Now, in Eq. (2.72), we can replace $\tilde{J}^{\mu\nu}$ with the above expression and use the formula (2.55) for the particle density in momentum space, obtaining:

$$
\begin{aligned}
S^\mu(p) &= -\frac{1}{4m}\epsilon^{\mu\nu\rho\sigma} p_\sigma \frac{\int dp^0 \int d\Sigma_\lambda \ \mathrm{tr}_4\left(\{\gamma^\lambda, \Sigma^{\nu\rho}\}W_+(x,p)\right)}{\frac{1}{2m\varepsilon}\mathrm{tr}_4 w_+(p)} \\
&= -\frac{1}{4}\epsilon^{\mu\nu\rho\sigma} p_\sigma \frac{\int dp^0 \int d\Sigma_\lambda \ \mathrm{tr}_4\left(\{\gamma^\lambda, \Sigma^{\nu\rho}\}W_+(x,p)\right)}{\int dp^0 \int d\Sigma \cdot p \ \mathrm{tr}_4 W_+(x,p)},
\end{aligned}
\tag{2.81}
$$

where we have used Eq. (2.52) integrated in p^0. We can also recast the above formula by taking advantage of the cancellation of $\delta(k^0 - \varepsilon_k)$ in the ratio:

$$
S^\mu(p) = -\frac{1}{4}\epsilon^{\mu\nu\rho\sigma} p_\sigma \frac{\int d\Sigma_\lambda \ \mathrm{tr}_4\left(\{\gamma^\lambda, \Sigma^{\mu\nu}\}W_+(x,p)\right)}{\int d\Sigma_\lambda p^\lambda \ \mathrm{tr}_4 W_+(x,p)},
$$

which is precisely (2.70). Hence, the method described in this section leads to the same result obtained in Sect. 2.5.

It is worth stressing the independence of the expression of $S^\mu(p)$ in Eq. (2.81) of the particular couple of stress-energy and spin tensor chosen to calculate the total angular momentum spectral decomposition $\tilde{J}^{\mu\nu}(p)$. If we had used the Belinfante symmetrized tensor:

$$
\begin{aligned}
: \widehat{T}^{\mu\nu}_B(x) :&= \frac{i}{4} : \overline{\Psi}(x)\gamma^\mu \overset{\leftrightarrow}{\partial^\nu}\Psi(x) + \overline{\Psi}(x)\gamma^\nu \overset{\leftrightarrow}{\partial^\mu}\Psi(x) : \\
&= \frac{1}{2}\int d^4 k \ k^\nu \mathrm{tr}_4(\gamma^\mu \widehat{W}(x,k)) + k^\mu \mathrm{tr}_4(\gamma^\nu \widehat{W}(x,k))
\end{aligned}
\tag{2.82}
$$

with associated vanishing spin tensor, for the derivation of the mean spin vector, we would have obtained the same expression (2.81). This happens because the Belinfante associated "orbital" angular momentum (which is actually the only term as $\widehat{S}_B = 0$) implies more terms in the decomposition with respect to Eq. (2.79) (this is discussed at the end of Appendix A).

To conclude, as it was already discussed at the end of Sect. 2.5, the *expression* of the particle polarization as a function of momentum is independent of the pseudo-gauge transformation of stress-energy and spin tensor. For the Dirac field, the canonical stress-energy and spin tensor are actually the most convenient to obtain it by the method presented in this section, and yet, the same expression could be derived by using the Belinfante pseudo-gauge. The appearance of the canonical spin tensor in Eq. (2.71) does not give it a special physical meaning and, indeed, the equivalent forms (2.70) and (2.66) do not feature the canonical spin tensor. However, the *value* of the mean spin vector, as well as any other quantity, may depend on the spin tensor because the density operator at local thermodynamic equilibrium is sensitive to the pseudo-gauge transformations [16,28]. Particularly, it is the Wigner function itself that acquires a dependence on the pseudo-gauge transformations through the density operator.

2.7   Local Thermodynamic Equilibrium

We have seen in the previous sections how the spin density matrix and the mean spin vector relate to the covariant Wigner function. In turn, the covariant Wigner function depends on the density operator $\widehat{\rho}$, see, e.g. Eq. (2.36) and it is thus necessary to know the density operator to calculate it.

For a relativistic fluid which, at some time, is believed to have achieved local thermodynamic equilibrium, a powerful approach is Zubarev's method of the stationary Non-Equilibrium Density Operator (NEDO) [29]. We refer the reader to the recent paper [30] for a detailed description.

This approach is especially well suited for the physics of relativistic nuclear collisions, where the system supposedly achieves Local Thermodynamic Equilibrium (LTE) at some finite early "time" (in the most used model, a finite hyperbolic time $\tau = \sqrt{t^2 - z^2}$, see Ref. [31]), to form a Quark–Gluon Plasma (QGP) which lives in a finite space–time region before breaking up at some 3D hypersurface Σ_{FO} (see Fig. 2.1). The actual density operator, in the Heisenberg representation, must be a fixed, time and space independent, operator and for a fluid at local thermodynamic equilibrium, it is obtained by maximizing the entropy $S = -\mathrm{tr}(\widehat{\rho}\log\widehat{\rho})$ with the constraints of energy-momentum and charge densities [18]. The result is:

$$\widehat{\rho} = \frac{1}{Z}\exp\left[-\int_{\Sigma_0}\mathrm{d}\Sigma\, n_\mu\left(\widehat{T}^{\mu\nu}(x)\beta_\nu(x) - \zeta(x)\widehat{j}^\mu(x)\right)\right], \qquad (2.83)$$

where β is the four-temperature vector, ζ the ratio between chemical potential and temperature and Σ_0 is the initial 3D hypersurface where LTE is achieved. For relativistic nuclear collisions, this is supposedly the 3D hyperbolic hypersurface $\tau = \tau_0$, the Σ_{eq} in Fig. 2.1. It should be pointed out that the form of the local equilibrium density operator is pseudo-gauge dependent [16]; the above form applies to the Belinfante stress-energy tensor only, so in the rest of the section, it will be understood that $\widehat{T}$ is the Belinfante symmetrized stress-energy tensor.

However, the operator (2.83), as it stands, cannot be used to calculate the polarization of final state particles in practice. The reason is that the operators in the exponent of (2.83) are to be evaluated at the time τ_0, when the system is in the QGP phase and the field operators are those of the fundamental QCD degrees of freedom, quarks and gluons, whereas the creation and destruction operators in a formula such as (2.15) or (2.48) are clearly those of the hadronic asymptotic states, which can be expressed in terms of the effective hadronic fields. Even if we were able to write the effective hadronic fields in terms of the fundamental quark and gluon fields, those should be evaluated at different times, that is the initial "time" τ_0 and the decoupling time, so that the full dynamical problem of the interacting quantum field should be solved. It is indeed convenient to rewrite $\widehat{\rho}_{\mathrm{LE}}(\tau_0)$ in terms of the operators at some present "time" τ by means of Gauss' theorem, taking into account that $\widehat{T}$ and $\widehat{j}$ are conserved currents [30]. Being

$$\mathrm{d}\Sigma_\mu = \mathrm{d}\Sigma\, n_\mu,$$

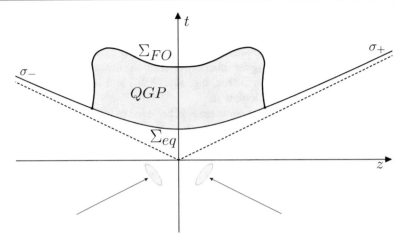

Fig. 2.1 Space–time diagram of a relativistic nuclear collision at very high energy. The hypersurface Σ_{eq} corresponds to the achievement of local thermodynamic equilibrium while Σ_{FO} is the hypersurface where the Quark–Gluon Plasma decouples. The $\sigma_\pm$ are described in the text

where $\hat{n}$ is the unit vector perpendicular to the hypersurface and $d\Omega$ being the measure of a 4D region in space–time, we have

$$-\int_{\Sigma(\tau_0)} d\Sigma_\mu \left(\widehat{T}^{\mu\nu}\beta_\nu - \widehat{j}^\mu\zeta\right) = -\int_{\Sigma(\tau)} d\Sigma_\mu \left(\widehat{T}^{\mu\nu}\beta_\nu - \widehat{j}^\mu\zeta\right) + \int_\Omega d\Omega \left(\widehat{T}^{\mu\nu}\nabla_\mu\beta_\nu - \widehat{j}^\mu\nabla_\mu\zeta\right),$$
(2.84)

where ∇ is the covariant derivative. The region Ω is the portion of space–time enclosed by the two hypersurface $\Sigma(\tau_0)$ and $\Sigma(\tau)$ and the time-like hypersurface at their boundaries, where the flux of $(\widehat{T}^{\mu\nu}\beta_\nu(x) - \widehat{j}^\mu\zeta(x))$ is supposed to vanish [30]. Consequently, the stationary NEDO reads:

$$\begin{aligned}\widehat{\rho} &= \frac{1}{Z}\exp\left[-\int_{\Sigma(\tau_0)} d\Sigma_\mu \left(\widehat{T}^{\mu\nu}\beta_\nu - \widehat{j}^\mu\zeta\right)\right]\\ &= \frac{1}{Z}\exp\left[-\int_{\Sigma(\tau)} d\Sigma_\mu \left(\widehat{T}^{\mu\nu}\beta_\nu - \widehat{j}^\mu\zeta\right) + \int_\Omega d\Omega \left(\widehat{T}^{\mu\nu}\nabla_\mu\beta_\nu - \widehat{j}^\mu\nabla_\mu\zeta\right)\right].\end{aligned}$$
(2.85)

In the case of heavy ion collisions, the $\Sigma(\tau_0)$—looking at Fig. 2.1—is the 3D hypersurface Σ_{eq}, while the hypersurface $\Sigma(\tau)$ is usually the joining of the freeze-out hypersurface Σ_{FO} encompassing the QGP space–time region and the two side branches $\sigma_\pm$ subsets of the $\Sigma(\tau_0)$. A peculiarity of the heavy ion collisions is that the hypersurface of "present" local equilibrium is partly time-like, that is $\hat{n} \cdot \hat{n} = -1$.

The density operator in (2.85) can be expanded perturbatively by identifying the two terms in its exponent:

$$\widehat{A} = -\int_{\Sigma(\tau)} d\Sigma_\mu \left(\widehat{T}^{\mu\nu}\beta_\nu - \widehat{j}^\mu\zeta\right),$$
(2.86)

which is the supposedly, in hydrodynamics, the predominant term, and

$$\widehat{B} = \int_{\Omega} d\Omega \ \left(\widehat{T}^{\mu\nu} \nabla_{\mu} \beta_{\nu} - \widehat{j}^{\mu} \nabla_{\mu} \zeta \right), \tag{2.87}$$

which is supposedly the small term. The $\widehat{A}$ and $\widehat{B}$ terms correspond to the LTE at the current time and the dissipative correction, respectively. Hence, the leading term of the expansion of the mean value of any operator is the local equilibrium one, that is:

$$O \simeq \mathrm{tr}(\widehat{\rho}_{\mathrm{LE}} \widehat{O}) = \frac{\mathrm{tr}(\exp[\widehat{A}] \widehat{O})}{\mathrm{tr}(\exp[\widehat{A}])}.$$

The convenient feature of this approach for the calculation of hydrodynamic constitutive equations is the natural separation between non-dissipative terms—which are obtained by retaining the $\widehat{A}$ term—and the dissipative ones which are obtained by including $\widehat{B}$.

The calculation of $W(x, k)_{\mathrm{LE}}$, that is:

$$W(x, k)_{\mathrm{LE}} = \frac{1}{Z_{\mathrm{LE}}} \mathrm{Tr} \left(\exp \left[-\int_{\Sigma} d\Sigma_{\mu} \ \left(\widehat{T}^{\mu\nu}(y) \beta_{\nu}(y) - \widehat{j}^{\mu}(y) \zeta(y) \right) \right] \widehat{W}(x, k) \right) \tag{2.88}$$

can be tackled by taking advantage of the supposedly slow variation of the fields β and ζ in space–time compared with the variation of the Wigner operator over microscopic scales. Beforehand, it should be pointed out that the point x where the Wigner function is to be evaluated is, to a large extent, arbitrary. For, as we have seen in Sects. 2.4.2 and 2.5, the 3D integration hypersurface of the Wigner function (see, e.g. Eq. (2.56)) can be any hypersurface where the asymptotic hadronic fields are defined, one could choose a hyperplane at a sufficiently large value of the Minkowski time t so as to be completely outside the QGP space–time region (see Fig. 2.1), where hadronic fields cannot be used. However, this is not a convenient choice: at large times, the fields β and ζ are no longer defined because the system is not a fluid anymore and it would then be difficult to estimate the Wigner function therein. A much better choice is an equivalent (from the viewpoint of the Gauss theorem) 3D hypersurface encompassing the QGP and much closer to where the hydrodynamic fields are still defined. This hypersurface Σ can be obtained by joining the break-up hypersurface Σ_{FO} and the two branches $\sigma_{\pm}$, as discussed above. Now one can evaluate $W(x, k)_{\mathrm{LE}}$ in space–time points where β and ζ exist, with the exception of the branches $\sigma_{\pm}$ where the matter is not a fluid. Indeed, those branches involve the cold nuclear matter not participating the QGP formation, and its contribution is usually neglected. Since the hydrodynamic–thermodynamic fields β and ζ are slowly varying, one can expand them in a Taylor series around x, the point where the Wigner operator is evaluated, and, retaining only the first order:

$$\beta_{\nu}(y) \simeq \beta_{\nu}(x) + \partial_{\lambda} \beta_{\nu}(x)(y - x)^{\lambda},$$

and similarly for ζ. Inserting in Eq. (2.88):

$$
\begin{aligned}
W(x,k)_{\mathrm{LE}} &\simeq \frac{1}{Z_{\mathrm{LE}}} \mathrm{Tr} \left(\exp\left[-\int_\Sigma \mathrm{d}\Sigma_\mu \; (\widehat{T}^{\mu\nu}(y)[\beta_\nu(x) + \partial_\lambda\beta_\nu(x)(y-x)^\lambda] \right.\right. \\
&\quad \left.\left. - \widehat{j}^\mu(y)[\zeta(x) + \partial_\lambda\zeta(x)(y-x)^\lambda])\right] \widehat{W}(x,k)\right) \\
&= \frac{1}{Z_{\mathrm{LE}}} \mathrm{Tr} \left(\exp\left[-\beta_\nu(x)\int_\Sigma \mathrm{d}\Sigma_\mu \; \widehat{T}^{\mu\nu}(y) \right.\right. \\
&\quad - \partial_\lambda\beta_\nu(x)\int_\Sigma \mathrm{d}\Sigma_\mu \; (y-x)^\lambda \widehat{T}^{\mu\nu}(y) - \zeta(x)\int_\Sigma \mathrm{d}\Sigma_\mu \; \widehat{j}^\mu \\
&\quad \left.\left. - \partial_\lambda\zeta(x)\int_\Sigma \mathrm{d}\Sigma_\mu \; (y-x)^\lambda \widehat{j}^\mu\right] \widehat{W}(x,k)\right).
\end{aligned}
$$

$$(2.89)$$

This approximation corresponds to the hydrodynamic limit, where the mean value of local operators is determined by the local values of the thermodynamic fields. The gradient of β in the last equation can be split into the symmetric and the anti-symmetric part giving rise to

$$
\begin{aligned}
&-\frac{1}{2}\varpi_{\lambda\nu}\int_\Sigma \mathrm{d}\Sigma_\mu \; (y-x)^\lambda\widehat{T}^{\mu\nu}(y) - (y-x)^\nu\widehat{T}^{\mu\lambda}(y) \\
&+\frac{1}{4}(\partial_\lambda\beta_\nu + \partial_\nu\beta_\lambda)\int_\Sigma \mathrm{d}\Sigma_\mu \; (y-x)^\lambda\widehat{T}^{\mu\nu}(y) + (y-x)^\nu\widehat{T}^{\mu\lambda}(y),
\end{aligned}
$$

$$(2.90)$$

where ϖ is the thermal vorticity (2.18). We can recognize in the first term of the above equation the total angular momentum operator, with a proviso: the above integration is over a 3D hypersurface $\Sigma \supset \Sigma_{FO}$ which is not fully space-like, in fact it has a time-like part. Notwithstanding, as the angular momentum-boost current is divergenceless and being $\Sigma = (\Sigma_{FO} \cup \sigma_\pm)$ as discussed, we can again use the Gauss theorem and write:

$$
\int_\Sigma \mathrm{d}\Sigma_\mu (y-x)^\lambda\widehat{T}^{\mu\nu}(y) - (y-x)^\nu\widehat{T}^{\mu\lambda}(y) = \int_{\Sigma_{eq}} \mathrm{d}\Sigma_\mu \; (y-x)^\lambda\widehat{T}^{\mu\nu}(y) - (y-x)^\nu\widehat{T}^{\mu\lambda}(y),
$$

where Σ_{eq} is the initial, space-like local thermodynamic equilibrium. The latter is, by definition, the conserved total angular momentum-boost generator with centre x, that is $\widehat{J}_x^{\mu\nu}$.

The main contribution to the Wigner function supposedly arises from the terms surviving at the global equilibrium, occurring when $\partial_\mu\beta_\nu + \partial_\nu\beta_\mu = 0$ and $\partial_\mu\zeta = 0$. However, the symmetric term in (2.90) as well as the $\partial\zeta$ term in (2.89) may in principle contribute at LTE (they are non-dissipative, non-equilibrium terms) and it would be interesting to assess their quantitative effect. Assuming that they are negligible, we have:

$$
W(x,k)_{\mathrm{LE}} \simeq \frac{1}{Z}\mathrm{Tr}\left(\exp\left[-\beta_\nu(x)\widehat{P}^\nu + \frac{1}{2}\varpi_{\nu\lambda}(x)\widehat{J}_x^{\nu\lambda} + \zeta(x)\widehat{Q}\right] \widehat{W}(x,k)\right).
$$

$$(2.91)$$

This expression is the *global thermodynamic equilibrium* mean of the Wigner function $W(x, k)_{GE}$ with four-temperature and thermal vorticity values just equal to their values in the point x where the Wigner function is to be evaluated.

2.7.1 Polarization at Local Thermodynamic Equilibrium

Working out (2.91) is, in principle, much easier than a complete local equilibrium calculation and yet, the exact form has not been determined so far. A possible approach is linear response theory, taking as the term $\varpi_{\lambda\nu}(x)\widehat{J}_x^{\lambda\nu}$ in Eq. (2.89) as the small term compared to the main term $-\beta_\nu(x)\widehat{P}^\nu + \zeta(x)\widehat{Q}$. This method, however, involves the calculation of complicated integral correlators between the angular momentum operator and the Wigner operator and, although viable, has never been attempted in literature.

In Ref. [15], an educated *ansatz* was introduced based on the on-shell De Groot's approximation of the general form of the covariant Wigner function:

$$W(x, k) \simeq \frac{1}{2}\sum_{r,s}\int\frac{d^3p}{\varepsilon}\delta^4(k - p)u_r(p)f(x, p)_{rs}\bar{u}_s(p) - \delta^4(k + p)v_r(p)\bar{f}(x, p)_{sr}\bar{v}_s(p),$$

(2.92)

where $f_{rs}(x, p)$ is a 2×2 distribution function and r s label spin states. Now we know that the Boltzmann limit of (2.57) must yield the spin density matrix (2.29), i.e. by using (2.58),

$$\frac{\int d\Sigma_\mu p^\mu \bar{U}(p)W_+(x, k)_{GE}U(p)}{\mathrm{tr}_2 \int d\Sigma_\mu p^\mu \bar{U}(p)W_+(x, k)_{GE}U(p)} \longrightarrow \frac{\bar{U}(p)\exp\left[\frac{1}{2}\varpi : \Sigma\right]U(p)}{\mathrm{tr}_2(\bar{U}(p)\exp\left[\frac{1}{2}\varpi : \Sigma\right]U(p))}. \quad (2.93)$$

Hence, a suitable form of f was assumed giving the correct Boltzmann (as well as the non-relativistic) limit:

$$f_{rs}(x, p) = \bar{u}_r(p)\exp\left[\beta \cdot p - \frac{1}{2}\varpi : \Sigma + I\right]^{-1}u_s(p). \quad (2.94)$$

Equations (2.92) and (2.94) together lead to the following form of the mean spin vector for spin 1/2 particles [15]:

$$S^\mu(p) = -\frac{1}{8m}\epsilon^{\mu\rho\sigma\tau}p_\tau\frac{\int_{\Sigma_{FO}}d\Sigma_\lambda p^\lambda n_F(1 - n_F)\varpi_{\rho\sigma}}{\int_{\Sigma_{FO}}d\Sigma_\lambda p^\lambda n_F}, \quad (2.95)$$

where n_F is the Fermi–Dirac phase space distribution function:

$$n_F = \frac{1}{\exp[\beta \cdot p + \mu q] + 1},$$

q being a charge of the particle and μ the corresponding chemical potential. The formula (2.95) is, in the Boltzmann limit, in full agreement with Eq. (2.34) which is the first-order formula obtained within a single-particle framework. The problem to determine the exact form at global equilibrium including quantum statistics effects is—as mentioned—yet to be solved.

2.8   Summary and Outlook

The calculation of polarization in a relativistic fluid stands out as a fascinating endeavour in quantum field theory. As we have seen, it requires the use of a broad range of concepts and theoretical tools and it involves intriguing fundamental physics problems such as the physical significance of the spin tensor. It should be emphasized that it is not just an academic problem: polarization in the QCD plasma has been observed in experiments, and much of its phenomenological potential as a probe of the hot QCD matter is still to be explored. In this regard, much theoretical and experimental work is ongoing. For a comprehensive review of the status of the subject, we refer the reader to the recent review [32].

The formula (2.95) is the benchmark for most estimates of polarization. While very successful in reproducing the global polarization of Λ hyperons in relativistic heavy ion collisions, a disagreement with the data was found as to the momentum dependence of polarization [32]. These discrepancies could be an effect of incorrect hydrodynamic initial conditions, resulting in a distorted thermal vorticity field at the freeze-out or they could possibly arise from missing theoretical ingredients and major corrections to Eq. (2.34), which is a leading order formula in thermal vorticity. Even though thermal vorticity is apparently a small number in relativistic heavy ion collisions, a quantitative role of the yet unknown exact formula of the Wigner function (2.89) cannot be ruled out for the present. Similarly, dissipative corrections to the mean spin vector are quantitatively unknown thus far. Even the estimate of the first-order correction to the formula (2.95) in the linear approximation with the operator (2.87) is a formidable task as it involves, in heavy ion collisions, the full non-perturbative QCD regime (there is an ongoing effort in this direction [33]). The theory of the polarization in a relativistic fluid is still to be fully developed.

Acknowledgements Stimulating discussions with W. Florkowski and L. Tinti are gratefully acknowledged. I am very grateful to Q. Wang and X. G. Huang for very useful suggestions and to M. Buzzegoli and A. Palermo for a careful review of the manuscript and for making the figure. Special thanks to Enrico Speranza for his very valuable remarks and comments.

Appendix: Angular Momentum Decomposition

We shall prove that the orbital part of the angular momentum operator (2.79) can be written as an integral in momentum space of on-shell functions. We will confine ourselves to the proof for the particle term in (2.48), its extension to the antiparticle term and the proof of the vanishing of the mixed term being alike. By using (2.48) and choosing the hyperplane $t = 0$ as integration hypersurface, for the particle term, we can write:

$$
\int d^3x \, x^\mu k^\nu tr_4(\gamma^0 \widehat{W}(x, k)) =
$$

$$
\int d^3x \, x^\mu k^\nu \sum_{r,s} \frac{1}{(2\pi)^3} \int \frac{d^3p \, d^3p'}{2\varepsilon \, 2\varepsilon'} e^{-i(p-p')\cdot x} \times
$$

$$
\times \, \delta^4\left(k - \frac{p+p'}{2}\right) \widehat{a}_s^\dagger(p')\widehat{a}_r(p)\bar{u}_s(p')\gamma^0 u_r(p) \qquad (2.96)
$$

$$
= 8 \int d^3x \, x^\mu k^\nu \sum_{r,s} \frac{1}{(2\pi)^3} \int \frac{d^3p}{4\varepsilon \, \varepsilon_{k,p}} e^{i(2k-2p)\cdot x} \times
$$

$$
\times \, \delta\left(k^0 - \frac{\varepsilon + \varepsilon_{k,p}'}{2}\right) \widehat{a}_s^\dagger(2k - p)\widehat{a}_r(p)\bar{u}_s(2k - p)\gamma^0 u_r(p),
$$

where

$$
\varepsilon_{k,p} = \sqrt{(2\mathbf{k} - \mathbf{p})^2 + m^2}.
$$

and it is understood that in the arguments of creation and destruction operators, as well as of spinors u, only the spatial part of the four-vector k, that is $\mathbf{k}$, enters.

For $\mu = 0$, the integration is straightforward as x^0 is constant on the hyperplane and we get, after integrating in d^3x:

$$
x^0 k^\nu \sum_{r,s} \int \frac{d^3p}{4\varepsilon \, \varepsilon_{k,p}} \delta^3(\mathbf{p} - \mathbf{k}) \delta\left(k^0 - \frac{\varepsilon + \varepsilon_{k,p}}{2}\right) \widehat{a}_s^\dagger(2k - p)\widehat{a}_r(p)\bar{u}_s(2k - p)\gamma^0 u_r(p)
$$

$$
= x^0 k^\nu \sum_{r,s} \frac{1}{(2\pi)^3} \frac{1}{4\varepsilon_k^2} \delta(k^0 - \varepsilon)\widehat{a}_s^\dagger(k)\widehat{a}_r(k)\bar{u}_s(k)\gamma^0 u_r(k)
$$

$$
= x^0 k^\nu \delta(k^0 - \varepsilon_k) \sum_r \frac{1}{2\varepsilon_k} \widehat{a}_r^\dagger(k)\widehat{a}_r(k)
$$

because $p' = 2k - p$ and $\mathbf{p} = \mathbf{k}$ implies in turn $k = p = p'$, hence k is on-shell; we have also used the known spinor relations.

For $\mu = i \neq 0$, we can replace x^μ with a derivative of the exponential and, integrating by parts,

$$
\int d^3x \, x^i k^\nu tr_4(\gamma^0 \widehat{W}(x, k))
$$

$$
= 4i \int d^3x \, k^\nu \sum_{r,s} \frac{1}{(2\pi)^3} \int \frac{d^3p}{4\varepsilon \, \varepsilon_{k,p}} \frac{\partial}{\partial p^i} e^{i(2k-2p)\cdot x} \times
$$

$$
\times \, \delta\left(k^0 - \frac{\varepsilon + \varepsilon_{k,p}}{2}\right) \widehat{a}_s^\dagger(2k - p)\widehat{a}_r(p)\bar{u}_s(2k - p)\gamma^0 u_r(p)
$$

$$
= 4i \int d^3x \, k^\nu \sum_{r,s} \frac{1}{(2\pi)^3} \int \frac{d^3p}{4\varepsilon \, \varepsilon_{k,p}} \times
$$

$$\times \frac{\partial}{\partial p^i} \left[e^{i(2k-2p)\cdot x} \delta \left(k^0 - \frac{\varepsilon + \varepsilon_{k,p}}{2} \right) \widehat{a}_s^\dagger (2k - p) \widehat{a}_r (p) \bar{u}_s (2k - p) \gamma^0 u_r (p) \right] +$$

$$- 4i \int d^3 x \, k^\nu \sum_{r,s} \frac{1}{(2\pi)^3} \int \frac{d^3 p}{4\varepsilon \, \varepsilon_{k,p}} e^{i(2k-2p)\cdot x} \frac{\partial}{\partial p^i} \times$$

$$\times \left[\delta \left(k^0 - \frac{\varepsilon + \varepsilon_{k,p}}{2} \right) \widehat{a}_s^\dagger (2k - p) \widehat{a}_r (p) \bar{u}_s (2k - p) \gamma^0 u_r (p) \right].$$

The first term gives rise to a boundary integral which vanishes for fixed k and only the second term survives. We can now integrate in $d^3 x$ getting:

$$- \frac{i}{2} k^\nu \sum_{r,s} \int \frac{d^3 p}{4\varepsilon \, \varepsilon_{k,p}} \delta^3 (\mathbf{p} - \mathbf{k}) \frac{\partial}{\partial p^i} \times \tag{2.97}$$

$$\times \left[\delta \left(k^0 - \frac{\varepsilon + \varepsilon_{k,p}}{2} \right) \widehat{a}_s^\dagger (2k - p) \widehat{a}_r (p) \bar{u}_s (2k - p) \gamma^0 u_r (p) \right].$$

There appear two derivative terms in the above expression: the derivative of the δ can be written as

$$\frac{\partial}{\partial p^i} \delta \left(k^0 - \frac{\varepsilon + \varepsilon'_{k,p}}{2} \right) = - \frac{1}{2} \frac{\partial}{\partial k^0} \delta \left(k^0 - \frac{\varepsilon + \varepsilon_{k,p}}{2} \right) \frac{\partial}{\partial p^i} (\varepsilon + \varepsilon_{k,p})$$

$$= - \frac{1}{2} \frac{\partial}{\partial k^0} \delta \left(k^0 - \frac{\varepsilon + \varepsilon_{k,p}}{2} \right) \left(\frac{p^i}{\varepsilon} - \frac{2k^i - p^i}{\varepsilon_{k,p}} \right), \tag{2.98}$$

while the derivative of the factor including creation and destruction operators and spinors yields, taking into account the $\delta^3 (\mathbf{p} - \mathbf{k})$:

$$\delta^3 (\mathbf{p} - \mathbf{k}) \frac{\partial}{\partial p^i} \widehat{a}_s^\dagger (2k - p) \widehat{a}_r (p) \bar{u}_s (2k - p) \gamma^0 u_r (p) \tag{2.99}$$

$$= \delta^3 (\mathbf{p} - \mathbf{k}) \left[\left(\widehat{a}_s^\dagger (p) \frac{\overleftrightarrow{\partial}}{\partial p^i} \widehat{a}_r (p) \right) \bar{u}_s (p) \gamma^0 u_r (p) + \widehat{a}_s^\dagger (p) \widehat{a}_r (p) \left(\bar{u}_s (p) \frac{\overleftrightarrow{\partial}}{\partial p^i} \gamma^0 u_r (p) \right) \right].$$

We can now plug Eqs. (2.98) and (2.99) into (2.97). The term (2.98) vanishes because

$$\delta^3 (\mathbf{p} - \mathbf{k}) \left(\frac{p^i}{\varepsilon} - \frac{2k^i - p^i}{\varepsilon_{k,p}} \right) k^\nu \frac{\partial}{\partial k^0} \delta \left(k^0 - \frac{\varepsilon + \varepsilon_{k,p}}{2} \right)$$

$$= -\delta^3 (\mathbf{p} - \mathbf{k}) \left(\frac{p^i}{\varepsilon} - \frac{2k^i - p^i}{\varepsilon_{k,p}} \right) \delta_0^\nu \delta \left(k^0 - \frac{\varepsilon + \varepsilon_{k,p}}{2} \right)$$

$$= -\delta^3 (\mathbf{p} - \mathbf{k}) \delta_0^\nu \delta \left(k^0 - \varepsilon \right) \left(\frac{k^i}{\varepsilon_k} - \frac{k^i}{\varepsilon_k} \right) = 0,$$

and we are just left with the term from (2.99).

We can now integrate in $d^4 k$ according to Eq. (2.79). For $\mu = i$,

$$\int d^4 k \int d^3 x \, x^i k^\nu \mathrm{tr}_4(\gamma^0 \widehat{W}(x,k)) = -\frac{i}{2} \int d^4 k \, k^\nu \sum_{r,s} \frac{1}{4\varepsilon_k^2} \delta(k_0 - \varepsilon_k) \times$$

$$\times \left[\left(\widehat{a}_s^\dagger(k) \frac{\overset{\leftrightarrow}{\partial}}{\partial k^i} \widehat{a}_r(k) \right) \bar{u}_s(k) \gamma^0 u_r(k) + \widehat{a}_s^\dagger(k) \widehat{a}_r(k) \left(\bar{u}_s(k) \frac{\overset{\leftrightarrow}{\partial}}{\partial k^i} \gamma^0 u_r(k) \right) \right]$$

$$= -\frac{i}{2} \int d^3 k \, k^\nu \sum_{r,s} \frac{1}{4\varepsilon_k^2} \left[\left(\widehat{a}_s^\dagger(k) \frac{\overset{\leftrightarrow}{\partial}}{\partial k^i} \widehat{a}_r(k) \right) \bar{u}_s(k) \gamma^0 u_r(k) \right.$$

$$\left. + \widehat{a}_s^\dagger(k) \widehat{a}_r(k) \left(\bar{u}_s(k) \frac{\overset{\leftrightarrow}{\partial}}{\partial k^i} \gamma^0 u_r(k) \right) \right]$$

(2.100)

with k^ν again on-shell. We can then conclude that

$$\int d^4 k \int_{t=0} d^3 x \, x^\mu k^\nu \mathrm{tr}_4(\gamma^0 \widehat{W}(x,k)) = \int \frac{d^3 k}{2\varepsilon_k} \widehat{G}^\mu(k) k^\nu$$

with k on-shell and $\widehat{G}^0(k) = 0$ if $x^0 = 0$ is chosen.

Finally, we briefly address the calculation of the angular momentum tensor by using the Belinfante stress-energy tensor (2.82) where only the orbital part is involved. The calculation is very similar to the one just described, with the important difference that the second term of (2.82), obtained by swapping the indices of the first term in (2.82), leads to a term akin to the left hand side of Eq. (2.96) with exchanged indices:

$$\int d^3 x \, x^\mu k^0 \mathrm{tr}_4(\gamma^\nu \widehat{W}(x,k)).$$

However, the final result is not proportional to k^ν and a double derivative term appears just like in Eq. (2.100); therefore, this term is not cancelled by the Levi-Civita tensor in the calculation of the mean spin, unlike in the canonical case.

References

1. Adamczyk, L., et al.: Nature **548**, 62 (2017)
2. Voloshin, S.A.: EPJ Web Conf. **171**, 07002 (2018)
3. Adam, J., et al.: Phys. Rev. C **98** (2018)
4. Niida, T.: Nucl. Phys. A **982**, 511 (2019)
5. Adam, J., et al.: Phys. Rev. Lett. **123**(13) (2019)
6. Wang, Z., Guo, X., Shi, S., Zhuang, P.: Phys. Rev. D **100**(1) (2019)
7. Hattori, K., Hidaka, Y., Yang, D.L.: Phys. Rev. D **100**(9) (2019)
8. Gao, J.H., Liang, Z.T.: Phys. Rev. D **100**(5) (2019)

9. Weickgenannt, N., Sheng, X.L., Speranza, E., Wang, Q., Rischke, D.H.: Phys. Rev. D **100**(5) (2019)
10. Yang, D.L., Hattori, K., Hidaka, Y.: JHEP **07**, 070 (2020)
11. Liu, Y.C., Mameda, K., Huang, X.G.: Chin. Phys. C **44**(9), 094101 (2020)
12. Florkowski, W., Friman, B., Jaiswal, A., Speranza, E.: Phys. Rev. C **97**(4) (2018)
13. Florkowski, W., Ryblewski, R., Kumar, A.: Prog. Part. Nucl. Phys. **108** (2019)
14. Florkowski, W., Kumar, A., Ryblewski, R., Singh, R.: Phys. Rev. C **99**(4) (2019)
15. Becattini, F., Chandra, V., Del Zanna, L., Grossi, E.: Ann. Phys. **338**, 32 (2013)
16. Becattini, F., Florkowski, W., Speranza, E.: Phys. Lett. B **789**, 419 (2019)
17. Becattini, F.: Phys. Rev. Lett. **108** (2012)
18. Becattini, F., Bucciantini, L., Grossi, E., Tinti, L.: Eur. Phys. J. C **75**(5), 191 (2015)
19. Weinberg, S.: The Quantum Theory of Fields. Vol. 1: Foundations. Cambridge University Press, Cambridge (2005)
20. Moussa, P., Stora, R.: Angular analysis of elementary particle reactions (1966)
21. Becattini, F., Karpenko, I., Lisa, M., Upsal, I., Voloshin, S.: Phys. Rev. C **95**(5) (2017)
22. De Groot, S.R., Van Leeuwen, W.A., Van Weert, C.G.: Relativistic Kinetic Theory. Principles and Applications. North-Holland, Amsterdam (1980)
23. Vasak, D., Gyulassy, M., Elze, H.T.: Ann. Phys. **173**, 462 (1987)
24. Weinberg, S.: Phys. Rev. **133**, B1318 (1964)
25. Fang, R.H., Pang, L.G., Wang, Q., Wang, X.N.: Phys. Rev. **C94**(2), 024904 (2016)
26. Florkowski, W., Friman, B., Jaiswal, A., Ryblewski, R., Speranza, E.: Phys. Rev. D **97**(11) (2018)
27. Hehl, F.W.: Rept. Math. Phys. **9**, 55 (1976)
28. Becattini, F.: Nucl. Phys. A **1005**, 121833 (2021)
29. Zubarev, D.N., Prozorkevich, A.V., Smolyanskii, S.A.: Theor. Math. Phys. **40**, 821 (1979). https://doi.org/10.1007/BF01032069
30. Becattini, F., Buzzegoli, M., Grossi, E.: Particles **2**(2), 197 (2019)
31. Florkowski, W.: Phenomenology of Ultra-Relativistic Heavy-Ion Collisions (2010)
32. Becattini, F., Lisa, M.A.: (2020). https://doi.org/10.1146/annurev-nucl-021920-095245
33. Hattori, K., Hongo, M., Huang, X.G., Matsuo, M., Taya, H.: Phys. Lett. B **795**, 100 (2019)

Thermodynamic Equilibrium of Massless Fermions with Vorticity, Chirality and Electromagnetic Field

3

Matteo Buzzegoli

Abstract

We present a study of the thermodynamics of the massless free Dirac field at equilibrium with axial charge, angular momentum and external electromagnetic field to assess the interplay between chirality, vorticity and electromagnetic field in relativistic fluids. After discussing the general features of global thermodynamic equilibrium in quantum relativistic statistical mechanics, we calculate the thermal expectation values. Axial imbalance and electromagnetic field are included non-perturbatively by using the exact solutions of the Dirac equation, while a perturbative expansion is carried out in thermal vorticity. It is shown that the chiral vortical effect and the axial vortical effect are not affected by a constant homogeneous electromagnetic field.

3.1 Introduction

The collective macroscopic behaviour of matter in the presence of quantum anomalies and external fields is an increasingly important subject in several fields of physics. Specifically, the experiments of relativistic heavy-ion collisions at RHIC and LHC have posed new and interesting questions about the theoretical foundations of relativistic collective phenomena. The experimental data of heavy-ion collisions indicates the creation of a deconfined quark–gluon plasma in a strongly coupled regime at extreme conditions of temperature, density, thermal vorticity [1] and magnetic fields [2]. Moreover, it was argued [3,4] that the fluctuations of topological configurations of the QCD vacuum in the early stages of a heavy-ion collision generate a chiral imbalance, which is an imbalance between the number of right- and left-handed quarks. Despite the fact that the usual relativistic hydrodynamic has been very

M. Buzzegoli (✉)
Universitá di Firenze and INFN Sezione di Firenze, Via G. Sansone 1, 50019 Sesto Fiorentino (Firenze), Italy
e-mail: buzzegoli@fi.infn.it

© Springer Nature Switzerland AG 2021
F. Becattini et al. (eds.), *Strongly Interacting Matter under Rotation*,
Lecture Notes in Physics 987,
https://doi.org/10.1007/978-3-030-71427-7_3

effective [5] in reproducing the experimental data for collective flow phenomena, it is now essential for the interpretation of heavy-ion collisions to address hydrodynamics in the contemporaneous presence of chiral imbalance, thermal vorticity and external electromagnetic fields.

The first crucial step towards understanding the hydrodynamics of matter subject to external fields is to study its thermodynamic properties. It is the main purpose of this contribution to investigate the effects of an external electromagnetic field on the thermodynamics of a chiral vorticous fluid. The effects of electromagnetic fields on (non-chiral non-vorticous) relativistic quantum fluids were already studied in the past, see for instance [6,7] and reference therein and [8] for the special case of a constant magnetic field. More recently, this topic has been addressed in [9] using Zubarev's non-equilibrium statistical operator, in [10] using the generating functional method and in [11–13] with Wigner function derived from kinetic theory.

This contribution aims to highlight the modifications caused by chiral imbalance and by thermal vorticity. The paper is organized in the following way. In Sect. 3.2, we introduce the global thermal equilibrium of a chiral system with the contemporaneous presence of an external electromagnetic field and a thermal vorticity within Zubarev's non-equilibrium statistical operator formalism. In Sect. 3.2.1, we give a brief overview of the main results for the case of a chiral Dirac field in the absence of the electromagnetic field. In Sect. 3.3, we review the relativistic quantum theory of fermions under the effect of an external magnetic field. Then, we obtain the exact form of the chiral fermionic propagator with an external constant magnetic field and we obtain the exact thermal averages of the axial and electric currents. In Sect. 3.5, we examine the properties of a system at thermal equilibrium with constant vorticity and electromagnetic field. The last part of the paper is concerned with the consequences of an electromagnetic field on the chiral vortical effect and the axial vortical effect.

Notation

In this work, we use the *natural unit* system in which $\hbar = c = G = k_B = 1$. The *Minkowski metric* is defined by the tensor $\eta_{\mu\nu} = \text{diag}(1, -1, -1, -1)$; for the Levi-Civita symbol, we use the convention $\epsilon^{0123} = +1$.

Operators in Hilbert space will be denoted by a large upper hat, e.g. $\widehat{T}$ (with the exception of Dirac field operator that is denoted by Ψ). The stress-energy tensor used to define Poincaré generators is always assumed to be symmetric with an associated vanishing spin tensor.

3.2   General Global Equilibrium with Electromagnetic Field

In this section, we introduce the methods to study the thermodynamic equilibrium of a quantum relativistic system in the presence of a chiral imbalance and of an external electromagnetic field. For that purpose, we review the Zubarev method of stationary non-equilibrium density operator [14,15] (see also [16–19] for recent developments) and we discuss the inclusion of a conserved axial current and of an external electromagnetic field.

When we are dealing with a relativistic system, we must consider local quantities in order to address the appropriate covariant properties. To identify those quantities, we use a Arnowitt–Deser–Misner (ADM) decomposition of space–time [14–17]. Choose then a foliation of space–time and suppose that the system in consideration thermalize faster than the evolution of "time" τ in which we are interested. At each step of evolution $\mathrm{d}\tau$, the system is at local thermal equilibrium and the macroscopic behaviour of the system is described by a stress-energy density $T_{\mu\nu}(x)$, an (electric) current density $j_\mu(x)$ and an axial current $j_{A\mu}(x)$, all lying on a space-like hyper-surface $\Sigma(\tau)$. We can then describe the thermal properties of the system with a density operator which lives on $\Sigma(\tau)$. As in the non-relativistic case, the density operator at local equilibrium $\widehat{\rho}_{LE}$ is obtained as the operator which maximizes the entropy $S = -\mathrm{tr}(\widehat{\rho}\log\widehat{\rho})$. To reproduce the actual thermodynamics on the hyper-surface, we maximize the entropy with the constraints that the mean values of the stress-energy tensor and of the currents on $\Sigma(\tau)$ correspond to the values of the densities $T_{\mu\nu}(x)$, $j_\mu(x)$ and $j_{A\mu}(x)$ [15]. To obtain these densities, we project the stress-energy tensor and the current mean values onto n, i.e. the normalized four-vector perpendicular to Σ:

$$n_\mu(x)\,\mathrm{tr}\left[\widehat{\rho}\;\widehat{T}^{\mu\nu}(x)\right] = n_\mu(x)\,\langle\,\widehat{T}^{\mu\nu}(x)\rangle \equiv n_\mu(x)\,T^{\mu\nu}(x),$$
$$n_\mu(x)\,\mathrm{tr}\left[\widehat{\rho}\;\widehat{j}^{\mu}(x)\right] = n_\mu(x)\,\langle\,\widehat{j}^{\mu}(x)\rangle \equiv n_\mu(x)\,j^{\mu}(x),$$

and similarly for the axial current. We could also impose a constraint on the angular momentum density, but since we are choosing the Belinfante operator as the stress-energy tensor, it turns out that this additional requirement is automatically taken into account [20].

The maximum solution $\widehat{\rho}_{LE}$ gives the *Local Equilibrium Density Operator* (LEDO) [16, 17]:

$$\widehat{\rho}_{LTE} = \frac{1}{Z}\exp\left[-\int_\Sigma \mathrm{d}\Sigma_\mu\left(\widehat{T}^{\mu\nu}(x)\beta_\nu(x) - \zeta(x)\,\widehat{j}^{\mu}(x) - \zeta_A(x)\,\widehat{j}^{\mu}_A(x)\right)\right], \quad (3.1)$$

where β^μ is the four-temperature vector such that $T = 1/\sqrt{\beta^2}$ is the proper comoving temperature, ζ and ζ_A are the ratio of comoving chemical potentials and the temperature (e.g. $\zeta = \mu/T$) and Z is the partition function. In the presence of an external electromagnetic field, we indicate with $A^\mu(x)$ the non-dynamical gauge field and with $F^{\mu\nu} = \partial^\mu A^\nu - \partial^\nu A^\mu$ the electromagnetic strength tensor. Therefore, the operator relations stemming for conservation equations are

$$\partial_\mu \widehat{j}^{\mu} = 0, \quad \partial_\mu \widehat{T}^{\mu\nu} = \widehat{j}_\lambda F^{\nu\lambda}, \quad \partial_\mu \widehat{j}^{\mu}_A = 0. \quad (3.2)$$

Furthermore, the four-momentum operator $\widehat{P}$ and the conserved charges $\widehat{Q}_i$ are obtained by

$$\widehat{P}^\mu = \int_\Sigma \mathrm{d}\Sigma_\lambda\,\widehat{T}^{\lambda\mu}, \quad \widehat{Q}_i = \int_\Sigma \mathrm{d}\Sigma_\lambda\,\widehat{j}^{\lambda}_i,$$

while the angular momentum is

$$\widehat{J}^{\mu\nu} = \int_\Sigma d\Sigma_\lambda \left(x^\mu \widehat{T}^{\lambda\nu} - x^\nu \widehat{T}^{\lambda\mu} \right). \tag{3.3}$$

Notice that, as we will discuss in details in Sect. 3.3, the four-momentum $\widehat{P}$ and the angular momentum $\widehat{J}$ in the presence of external electromagnetic field are neither conserved nor the generators of translations and Lorentz transformations.

In the case of Dirac fermions interacting with an external gauge field, the explicit form of the operators above is

$$\widehat{j}^\mu = q\bar{\Psi}\gamma^\mu\Psi, \quad \widehat{j}_A^\mu = \bar{\Psi}\gamma^\mu\gamma^5\Psi,$$
$$\widehat{T}^{\mu\nu} = \frac{i}{4}\left[\bar{\Psi}\gamma^\mu \overrightarrow{\partial}^\nu \Psi - \bar{\Psi}\gamma^\mu \overleftarrow{\partial}^\nu \Psi + \bar{\Psi}\gamma^\nu \overrightarrow{\partial}^\mu \Psi - \bar{\Psi}\gamma^\nu \overleftarrow{\partial}^\mu \Psi \right] - \frac{1}{2}\left(\widehat{j}^\mu A^\nu + \widehat{j}^\nu A^\mu \right), \tag{3.4}$$

and the stress-energy tensor and the electric current indeed satisfy the relations (3.2). Regarding the axial current $\widehat{j}_A$, we also have to take into account the chiral anomaly. The chiral anomaly affects the axial current divergence as follows:

$$\partial_\mu \widehat{j}_A^\mu = -\frac{1}{8}\epsilon^{\mu\nu\rho\lambda}\frac{q^2}{2\pi^2}F_{\mu\nu}F_{\rho\lambda} = -\frac{q^2}{2\pi^2}(E \cdot B),$$

where q is the electric charge of the fermion, and E and B are comoving electric and magnetic field, defined by

$$F^{\mu\nu} = E^\mu u^\nu - E^\nu u^\mu - \epsilon^{\mu\nu\rho\sigma}B_\rho u_\sigma,$$

with u the fluid velocity. Even when the product $E \cdot B$ is non-vanishing and consequently the axial current is not conserved, we can still define a new conserved "axial" current by means of the Chern–Simons current K, whose divergence gives the chiral anomaly:

$$K^\mu = \epsilon^{\mu\nu\rho\sigma}A_\nu F_{\rho\sigma}, \quad \frac{q^2}{8\pi^2}\partial_\mu K^\mu = \frac{q^2}{2\pi^2}(E \cdot B).$$

The new conserved axial current $\widehat{j}_{CS}$ is then defined as

$$\widehat{j}_{CS}^\mu = \widehat{j}_A^\mu + \frac{q^2}{8\pi^2}K^\mu, \quad \partial_\mu \widehat{j}_{CS}^\mu = 0,$$

and the axial chemical potential μ_A is to be associated with this current. Since the additional current K depends only on external fields, it is not a quantum operator and it does not contribute to thermal averages. Therefore, all the results discussed in the absence of chiral anomaly will also be valid for the case of equilibrium with conserved Chern–Simons current. Because there is no difference in the results, we will continue to denote the current associated to μ_A inside the statistical operator with $\widehat{j}_A$ even when the chiral anomaly is non-vanishing.

Let us now move to describe the system at global thermal equilibrium. The global equilibrium is reached when the statistical operator (3.1) is time independent. This occurs when the integrand inside Eq. (3.1) is divergence-less [21]. Then, it is easily proven using relations (3.2) that global thermal equilibrium is realized when the following relations are satisfied:

$$\partial_\mu \beta_\nu(x) + \partial_\nu \beta_\mu(x) = 0, \qquad \partial^\mu \zeta(x) = F^{\nu\mu}\beta_\nu(x), \qquad \partial^\mu \zeta_A(x) = 0. \qquad (3.5)$$

The inverse four-temperature and the axial chemical potential solves the previous conditions if they are given by [16]:

$$\beta_\mu(x) = b_\mu + \varpi_{\mu\rho}x^\rho, \quad \zeta_A = \text{constant},$$

where b is a constant time-like four-vector and ϖ is a constant anti-symmetric tensor. We refer to ϖ as the *thermal vorticity* because it is the anti-symmetric derivative of inverse four-temperature:

$$\varpi_{\mu\nu} = -\frac{1}{2}\left(\partial_\mu \beta_\nu - \partial_\nu \beta_\mu\right)$$

and because it contains information about the fluid's acceleration and rotation. Indeed, if the β four-vector is a time-like vector, then we can choose the β-frame as hydrodynamic frame [10,21]. The unitary four-vector fluid velocity u is therefore identified with the direction of β:

$$u^\mu(x) = \frac{\beta^\mu(x)}{\sqrt{\beta_\rho(x)\beta^\rho(x)}}.$$

As long as we are considering physical observables in a region where the coordinate x is such that $\beta(x)$ is a time-like vector, this definition provides a proper choice for the fluid velocity. We can decompose the thermal vorticity into two space-like vector fields, each having three independent components, by projecting along the time-like fluid velocity u:

$$\varpi^{\mu\nu} = \epsilon^{\mu\nu\rho\sigma}w_\rho u_\sigma + \alpha^\mu u^\nu - \alpha^\nu u^\mu.$$

The four-vectors α and w are explicitly written inverting the previous relation:

$$\alpha^\mu(x) \equiv \varpi^{\mu\nu}u_\nu, \quad w^\mu(x) \equiv -\frac{1}{2}\epsilon^{\mu\nu\rho\sigma}\varpi_{\nu\rho}u_\sigma.$$

The vectors α and w that depend on the coordinates are space-like and are orthogonal to u. All the quantity u, ϖ, α, w are dimensionless. From their definitions, we can easily show that α and w are given by

$$\alpha^\mu = \sqrt{\beta^2}\,a^\mu, \quad w^\mu = \sqrt{\beta^2}\,\omega^\mu,$$

where a and ω are the local acceleration and rotation of the fluid, which are given by

$$a^\mu = u_\nu \partial^\nu u^\mu, \quad \omega^\mu \equiv \frac{1}{2} \epsilon^{\mu\nu\rho\sigma} u_\sigma \partial_\nu u_\rho.$$

Furthermore, it will prove useful to define the projector into the orthogonal space of fluid velocity:

$$\Delta_{\mu\nu} \equiv g_{\mu\nu} - u_\mu u_\nu,$$

and the four-vector γ orthogonal to the other ones: u, α, w

$$\gamma^\mu = \epsilon^{\mu\nu\rho\sigma} w_\nu \alpha_\rho u_\sigma = (\alpha \cdot \varpi)_\lambda \Delta^{\lambda\mu}.$$

The ϖ decomposition above defines a tetrad $\{u, \alpha, w, \gamma\}$ which can be used as a basis for four-vectors. It must be noticed, however, that the tetrad is neither unitary nor orthonormal, indeed in general we have $\alpha \cdot w \neq 0$.

Returning to the global equilibrium conditions (3.5), notice that in the absence of electromagnetic field also ζ must be a constant. In that case, the global equilibrium statistical operator takes the following form [22,23]:

$$\widehat{\rho} = \frac{1}{Z} \exp \left\{ -b \cdot \widehat{P} + \frac{1}{2}\varpi : \widehat{J} + \zeta \widehat{Q} + \zeta_A \widehat{Q}_A \right\}. \tag{3.6}$$

The thermodynamics of Dirac fermions which follows from this operator is quickly reviewed in Sect. 3.2.1

For the case of a non-vanishing electromagnetic field instead, we need to solve the equation:

$$\partial^\mu \zeta(x) = F^{\sigma\mu} \beta_\sigma. \tag{3.7}$$

To find the solution, we first derive it with respect to ∂^ν:

$$\partial^\nu \partial^\mu \zeta = \partial^\nu (F^{\sigma\mu} \beta_\sigma). \tag{3.8}$$

Since we can exchange the order of the derivatives $\partial^\nu \partial^\mu$ on the l.h.s. of (3.8), it follows that the anti-symmetrization with respect to indices μ and ν of (3.8) must be vanishing:

$$\partial^\nu \partial^\mu \zeta - \partial^\mu \partial^\nu \zeta = 0 = \left[\partial^\nu (F^{\sigma\mu} \beta_\sigma) - \partial^\mu (F^{\sigma\nu} \beta_\sigma) \right]$$
$$= \left[\beta_\sigma (\partial^\nu F^{\sigma\mu} - \partial^\mu F^{\sigma\nu}) + (\partial^\nu \beta_\sigma) F^{\sigma\mu} + (\partial^\mu \beta_\sigma) F^{\nu\sigma} \right].$$

Using the first Bianchi identity $\partial^\nu F^{\sigma\mu} + \partial^\mu F^{\nu\sigma} + \partial^\sigma F^{\mu\nu} = 0$, we obtain

$$\beta_\sigma \partial^\sigma F^{\mu\nu} + (\partial^\nu \beta_\sigma) F^{\mu\sigma} + (\partial^\mu \beta_\sigma) F^{\sigma\nu} = 0.$$

We may recognize the Lie derivative of F along β in the previous equation. This constitutes a first condition for global equilibrium, the system can reach global equilibrium only when

$$\mathcal{L}_\beta(F) = 0, \quad \leftrightarrow \quad \beta_\sigma(x)\partial^\sigma F^{\mu\nu}(x) = \varpi^\mu{}_\sigma F^{\sigma\nu}(x) - \varpi^\nu{}_\sigma F^{\sigma\mu}(x), \qquad (3.9)$$

that is to say when the electromagnetic field follows the field lines of inverse four-temperature.

To actually solve Eq. (3.7), we translate the global equilibrium condition of the strength tensor (3.9) to the four-vector potential A^μ. We see that the constraint (3.9) is satisfied if A solves

$$\beta_\sigma(x)\partial^\sigma A^\mu(x) = \varpi^\mu{}_\sigma A^\sigma(x) + \partial^\mu \Phi(x), \qquad (3.10)$$

where Φ is a smooth function of x. In Ref. [18] was also stated that a gauge potential with vanishing Lie derivative along β gives a stationary statistical operator, which is condition (3.10). It is important to stress that after a gauge transformation, the condition (3.10) still holds true for the new gauge potential because the function Φ is also affected by the gauge transformation. Indeed, let A^μ satisfies Eq. (3.10); after the gauge transformation $A'^\mu = A^\mu + \partial^\mu \Lambda$, we find:

$$\begin{aligned}
\beta_\sigma \partial^\sigma A'^\mu &= \beta_\sigma \partial^\sigma A^\mu + \beta_\sigma \partial^\sigma \partial^\mu \Lambda = \omega^\mu{}_\sigma A^\sigma + \partial^\mu \Phi + \partial^\mu(\beta_\sigma \partial^\sigma \Lambda) - (\partial^\mu \beta_\sigma)\partial^\sigma \Lambda \\
&= \varpi^\mu{}_\sigma(A^\sigma + \partial^\sigma \Lambda) + \partial^\mu(\Phi + \beta_\sigma \partial^\sigma \Lambda) = \varpi^\mu{}_\sigma A'^\sigma + \partial^\mu \Phi',
\end{aligned}$$

which is exactly condition (3.10) for A'^μ and for Φ', that is Φ shifted by the transport of Λ along β.

We can now write Eq. (3.7) by taking advantage of Eq. (3.10):

$$\begin{aligned}
\partial^\mu \zeta &= F^{\sigma\mu}\beta_\sigma = \beta_\sigma(\partial^\sigma A^\mu - \partial^\mu A^\sigma) = \beta_\sigma \partial^\sigma A^\mu - \partial^\mu(\beta_\sigma A^\sigma) + (\partial^\mu \beta_\sigma)A^\sigma \\
&= \varpi^\mu_\sigma A^\sigma + \partial^\mu \Phi - \varpi^\mu_\sigma A^\sigma - \partial^\mu(\beta_\sigma A^\sigma).
\end{aligned}$$

We can then collect all the derivatives together into the equation

$$\partial^\mu\left(\zeta - \Phi + \beta_\sigma A^\sigma\right) = 0,$$

from which we immediately get the solution:

$$\zeta(x) = \zeta_0 - \beta_\sigma(x)A^\sigma(x) + \Phi(x), \qquad (3.11)$$

where ζ_0 is a constant. The parameter Φ is analogous to the parameter which grants gauge invariance to chemical potential in [24]. Even though Eq. (3.11) is given in terms of the gauge potential, it is still gauge invariant. Indeed, we have shown that with a gauge transformation, A^μ and Φ transform as

$$A'^\mu = A^\mu + \partial^\mu \Lambda, \quad \Phi' = \Phi + \beta_\sigma \partial^\sigma \Lambda,$$

therefore, the chemical potential ζ is overall unaffected by gauge transformations:

$$\zeta(x)' = \zeta_0 - \beta_\sigma(x)A'^\sigma + \Phi' = \zeta_0 - \beta_\sigma(x)A^\sigma + \Phi - \beta_\sigma\partial^\sigma\Lambda + \beta_\sigma\partial^\sigma\Lambda = \zeta(x).$$

The global equilibrium statistical operator is then obtained from the local one in Eq. (3.1) by replacing the global equilibrium form of the thermodynamic fields β, ζ, ζ_A:

$$
\begin{aligned}
\widehat{\rho} = \frac{1}{Z}\exp\Big\{ &-\int_\Sigma d\Sigma_\mu \left[\left(\widehat{T}^{\mu\nu}(x) + \widehat{j}^\mu(x)A^\nu(x)\right)\beta_\nu(x) \right. \\
&\left. - (\zeta_0 + \Phi(x))\,\widehat{j}^\mu(x) - \zeta_A\,\widehat{j}_A^\mu(x) \right] \Big\}.
\end{aligned}
\tag{3.12}
$$

This operator, on par with (3.6), is given in terms of global conserved quantities. The difference is that for the general form of the external magnetic field satisfying the constraint (3.9), the integration over the hyper-surface Σ does not give easily recognizable quantities like the four-momentum and the angular momenta in Eq. (3.6). However, the identification of global conserved operators can be carried out in the special case of the constant homogeneous electromagnetic field, and it is discussed in the following sections.

3.2.1 Vanishing Electromagnetic Field

Before proceeding with the effects of electromagnetic fields, we briefly review the thermodynamics properties of a relativistic system in the presence of thermal vorticity but without an external electromagnetic field. Regarding thermal equilibrium in the presence of rotation, exact solutions for the free scalar and Dirac fields are discussed in [25–27]. Instead the effects of acceleration has been recently investigated using the Zubarev method in Ref. [27–32]. Here we want to report the constitutive equations at second order on thermal vorticity discussed in [22,33] and in [23] including an axial current (see also [34] for first order in thermal vorticity and magnetic field). Using linear response theory on thermal vorticity, the thermal expectation value of a local operator $\widehat{O}(x)$ evaluated with statistical operator (3.9) can be written as [22,23]:

$$
\begin{aligned}
\langle\widehat{O}(x)\rangle = &\langle\widehat{O}(0)\rangle_{\beta(x)} - \alpha_\rho\langle\langle\,\widehat{K}^\rho\widehat{O}\,\rangle\rangle - w_\rho\langle\langle\,\widehat{J}^\rho\widehat{O}\,\rangle\rangle + \frac{\alpha_\rho\alpha_\sigma}{2}\langle\langle\,\widehat{K}^\rho\widehat{K}^\sigma\widehat{O}\,\rangle\rangle \\
&+ \frac{w_\rho w_\sigma}{2}\langle\langle\,\widehat{J}^\rho\widehat{J}^\sigma\widehat{O}\,\rangle\rangle + \frac{\alpha_\rho w_\sigma}{2}\langle\langle\,\{\widehat{K}^\rho, \widehat{J}^\sigma\}\widehat{O}\,\rangle\rangle + O(\varpi^3).
\end{aligned}
\tag{3.13}
$$

In the previous expression we indicated with double angular bracket the correlator

$$
\begin{aligned}
\langle\langle\widehat{K}^{\rho_1}\cdots\widehat{K}^{\rho_n}\widehat{J}^{\sigma_1}\cdots\widehat{J}^{\sigma_m}\widehat{O}\rangle\rangle \equiv &\int_0^{|\beta|}\frac{d\tau_1\cdots d\tau_{n+m}}{|\beta|^{n+m}}\times \\
&\times\langle T_\tau\left(\widehat{K}^{\rho_1}_{-i\tau_1 u}\cdots\widehat{K}^{\rho_n}_{-i\tau_n u}\widehat{J}^{\sigma_1}_{-i\tau_{n+1}u}\cdots\widehat{J}^{\sigma_m}_{-i\tau_{n+m}u}\widehat{O}(0)\right)\rangle_{\beta(x),c},
\end{aligned}
$$

where $\widehat{J}$ and $\widehat{K}$ are the comoving rotation and boost generators, identified by

$$\widehat{K}^\mu = u_\lambda \widehat{J}^{\lambda\mu}, \quad \widehat{J}^\mu = \frac{1}{2}\epsilon^{\alpha\beta\gamma\mu} u_\alpha \widehat{J}_{\beta\gamma},$$

and the averages $\langle\cdots\rangle_{\beta(x)}$ are evaluated at a fixed point x with the homogeneous statistical operator

$$\widehat{\rho}_0 = \frac{1}{Z_0}\exp\left\{-\beta(x)\cdot\widehat{P} + \zeta(x)\widehat{Q} + \zeta_A(x)\widehat{Q}_A\right\}.$$

The subscript c on the thermal averages indicates a connected correlator, while the subscript $-i\tau u$ on the operators indicates an imaginary translation along u as follows:

$$J^\mu_{-i\tau u} \equiv e^{-i\tau u\cdot\widehat{P}}\,\widehat{J}^\mu e^{i\tau u\cdot\widehat{P}}.$$

Constitutive equations at second order on thermal vorticity of the stress-energy tensor, the electric current and the axial current can be obtained using the expansion in Eq. (3.13). We obtain [22, 23, 33]

$$
\begin{aligned}
\langle\widehat{T}^{\mu\nu}\rangle = {}& \mathbb{A}\,\epsilon^{\mu\nu\kappa\lambda}\alpha_\kappa u_\lambda + \mathbb{W}_1 w^\mu u^\nu + \mathbb{W}_2 w^\nu u^\mu \\
& + (\rho - \alpha^2 U_\alpha - w^2 U_w)u^\mu u^\nu - (p - \alpha^2 D_\alpha - w^2 D_w)\Delta^{\mu\nu} \\
& + A\,\alpha^\mu\alpha^\nu + W w^\mu w^\nu + G_1 u^\mu\gamma^\nu + G_2 u^\nu\gamma^\mu + O(\varpi^3),
\end{aligned}
\tag{3.14}
$$

$$\langle\widehat{j}^\mu_V\rangle = n_V u^\mu + \left(\alpha^2 N^V_\alpha + w^2 N^V_\omega\right)u^\mu + W^V w^\mu + G^V\gamma^\mu + O(\varpi^3), \tag{3.15}$$

$$\langle\widehat{j}^\mu_A\rangle = n_A u^\mu + \left(\alpha^2 N^A_\alpha + w^2 N^A_\omega\right)u^\mu + W^A w^\mu + G^A\gamma^\mu + O(\varpi^3). \tag{3.16}$$

Not all of these coefficients are independent, indeed conservation equations (3.2) impose the following relations [33] (this is explained in detail in Sect. 3.5.2):

$$U_\alpha = -|\beta|\frac{\partial}{\partial|\beta|}(D_\alpha + A) - (D_\alpha + A),$$

$$U_w = -|\beta|\frac{\partial}{\partial|\beta|}(D_w + W) - D_w + 2A - 3W,$$

$$G_1 + G_2 = 2(D_\alpha + D_w) + A + |\beta|\frac{\partial}{\partial|\beta|}W + 3W,$$

instead, for the first-order coefficients, conservation equations require that

$$-2\mathbb{A} = |\beta|\frac{\partial\mathbb{W}_1}{\partial|\beta|} + 3\mathbb{W}_1 + \mathbb{W}_2.$$

For electric and axial current, we find that only the following equations must be fulfilled:

$$|\beta| \frac{\partial W^V}{\partial |\beta|} + 3W^V = 0, \quad |\beta| \frac{\partial W^A}{\partial |\beta|} + 3W^A = 0. \tag{3.17}$$

We can also take advantage of the Lorentz symmetry to show that the thermal coefficients $\mathbb{S}$ and Γ_w of the canonical spin tensor constitutive equation:

$$\langle \frac{i}{8} \bar{\Psi} \{ \gamma^\lambda, [\gamma^\mu, \gamma^\nu] \} \Psi \rangle = \mathbb{S} \epsilon^{\lambda\mu\nu\rho} u_\rho + \Gamma_w \left(u^\lambda \varpi^{\mu\nu} + u^\nu \varpi^{\lambda\mu} + u^\mu \varpi^{\nu\lambda} \right) + O(\varpi^2)$$

satisfy the following relations:

$$-\left(\frac{\mathbb{S}}{|\beta|} + \frac{\partial \mathbb{S}}{\partial |\beta|} \right) = 2\mathbb{A}^{\text{Can}},$$

$$2\frac{\mathbb{S}}{|\beta|} = \mathbb{W}_1^{\text{Can}} - \mathbb{W}_2^{\text{Can}}, \tag{3.18}$$

$$\frac{\Gamma_w}{|\beta|} - \frac{\partial \Gamma_w}{\partial |\beta|} = 4\frac{\Gamma_w}{|\beta|} = G_1^{\text{Can}} - G_2^{\text{Can}},$$

where $\mathbb{A}^{\text{Can}}$ and $\mathbb{W}_{1,2}^{\text{Can}}$ are the thermal coefficients of Eq. (3.14) related to the mean value of *canonical* stress-energy tensor. Furthermore, because the axial current is dual to the spin tensor, we can show that

$$\mathbb{S} = \frac{1}{2} n_A, \quad \Gamma_w = \frac{1}{2} W^A. \tag{3.19}$$

Then, combining Eq.s (3.19) and (3.18), the coefficients of canonical stress-energy tensor and axial current are related by

$$\mathbb{A}^{\text{Can}} = -\left(\frac{n_A}{|\beta|} + \frac{\partial n_A}{\partial |\beta|} \right),$$

$$\frac{n_A}{|\beta|} = \mathbb{W}_1^{\text{Can}} - \mathbb{W}_2^{\text{Can}},$$

$$\frac{W^A}{|\beta|} = \frac{G_1^{\text{Can}} - G_2^{\text{Can}}}{2},$$

which expose an interesting connection between the Axial Vortical Effect (AVE) conductivity W^A and the second-order thermal coefficients of the canonical stress-energy tensor.

To understand the constraint (3.17) and the relation between axial vortical effect and anomalies, we also consider the case of a free massive field. In that case, the axial current is not conserved, but its divergence is given by

$$\partial_\mu \widehat{j}_A^\mu = 2mi\bar{\Psi}\gamma^5\Psi.$$

It follows that global equilibrium with a conserved axial charge cannot be reached. Then, to still use the previous global equilibrium analysis to the massive field, we simply set the axial chemical potential to zero and we consider global equilibrium with just thermal vorticity and finite electric charge. As a consequence, the symmetries impose that all chiral coefficients (i.e. those which are not parity invariant) must be vanishing. However, the term in W^A of axial current decomposition is not chiral and consequently could be different from zero. Since the conservation equation is changed, we expect that also the condition (3.17) will be modified. We then have to consider the pseudo-scalar operator $i\bar{\Psi}\gamma^5\Psi$ that appears on the divergence of axial current. Pseudo-scalar thermal expectation value can be decomposed at second order in thermal vorticity in the same way as other local operators and we find that it is given by a single term:

$$\langle i\bar{\Psi}\gamma^5\Psi\rangle = (\alpha \cdot w)L^{\alpha \cdot w},$$

where the non-chiral thermal coefficient can be obtained by

$$L^{\alpha \cdot w} = \frac{1}{2}\langle\langle\{\widehat{K}_3, \widehat{J}_3\}i\bar{\Psi}\gamma^5\Psi\rangle\rangle. \tag{3.20}$$

With this definition, we find that the condition on axial vortical effect conductivity W^A becomes:

$$|\beta|\frac{\partial W^A}{\partial|\beta|} + 3W^A = -2mL^{\alpha \cdot w}. \tag{3.21}$$

Differently from (3.17) the constraint (3.21) no longer imposes W^A to be proportional to the third power of temperature and W^A acquires terms which depends on the mass of the fields.

As concluding remarks, we give some results for these coefficients for the free massless Dirac field. In that case, this method reproduces the well-known [2] chiral vortical effect and axial vortical effect conductivities

$$W^V = \frac{\mu \mu_A T}{\pi^2}, \quad W^A = \frac{T^3}{6} + \frac{(\mu^2 + \mu_A^2)T}{2\pi^2}. \tag{3.22}$$

In the case of massive Dirac fields, global thermal equilibrium with thermal vorticity and vanishing axial chemical potential is well defined and the axial currents mean value can be directed along the rotation of the fluid. In that situation, the AVE conductivity for a free massive Dirac field is [33]

$$W^A = \frac{1}{2\pi^2|\beta|}\int_0^\infty dp\left[n_F(E_p - \mu) + n_F(E_p + \mu)\right]\frac{2p^2 + m^2}{E_p}, \tag{3.23}$$

where $E_p = \sqrt{p^2 + m^2}$. This coefficient is related to pseudo-scalar thermal coefficient $L^{\alpha \cdot w}$ via Eq. (3.21) and indeed pseudo-scalar coefficient is given by

$$L^{\alpha \cdot w} = -\frac{m}{4\pi^2\beta^2}\int_0^\infty \frac{dp}{E_p}\left[n'_F(E_p - \mu) + n'_F(E_p + \mu)\right], \tag{3.24}$$

where the prime on distribution functions stands for derivative with respect to E_p. We can give approximate results for integral in Eq. (3.23). For high-temperature regime $(T \gg m)$, if the gas is non-degenerate $(|\mu| < m)$, we extract the AVE conductivity behaviour using the Mellin transformation technique [35]. The result is

$$\frac{W^A}{T} \simeq \frac{T^2}{6} + \frac{\mu^2}{2\pi^2} - \frac{m^2}{4\pi^2} - \frac{7\zeta'(-2)T^2}{8\pi^2}\left(\frac{m}{T}\right)^4 + O\left(\frac{m^6}{T^6}\right). \tag{3.25}$$

The first term in mass was also obtained in [36] where the axial vortical effect was evaluated with the statistical operator (3.6) but in curved space–time. Low temperature behaviour can also be extracted from (3.23), see [33]. For a degenerate gas $(|\mu| > m)$ at zero temperature, we obtain[1]:

$$\frac{W^A}{T} = \frac{\mu^2}{2\pi^2}\frac{\sqrt{\mu^2 - m^2}}{\mu}.$$

Instead for a non-degenerate gas $(|\mu| < m)$ at low temperature $T \ll m$, we have

$$W^A \approx \left(1 + 2\frac{T}{m}\right)\frac{(mT)^{3/2}}{\sqrt{2}\,\pi^{3/2}}e^{|\beta|(\mu-m)}. \tag{3.26}$$

Axial current corrections for rotating and accelerating fluids are also discussed in [37, 38] for both massive and massless fields using an ansatz for Wigner function with thermal vorticity.

3.3   Dirac Field in External Electromagnetic Field

Consider a Dirac field in external electromagnetic field. The Lagrangian of the theory is given by

$$\mathcal{L} = \frac{i}{2}\left[\bar{\Psi}\gamma^\mu\overrightarrow{\partial}_\mu\Psi - \bar{\Psi}\gamma^\mu\overleftarrow{\partial}_\mu\Psi\right] - m\bar{\Psi}\Psi - \hat{j}^\mu A_\mu,$$

where $\hat{j}^\mu = q\bar{\Psi}\gamma^\mu\Psi$, q is the elementary electric charge of the field and the gauge potential A^μ is an external non dynamic field. This Lagrangian is obtained from the free Dirac one with the minimal coupling substitution $\partial_\mu \to \partial_\mu + iqA_\mu$ which ensures gauge invariance to the theory. From Euler Eq.s we obtain the Equations of Motion (EOM) for the Dirac field:

$$\partial\!\!\!/\Psi = -i(q\slashed{A} + m)\Psi, \quad \partial\!\!\!/\bar{\Psi} = \bar{\Psi}i(q\slashed{A} + m).$$

By applying Noether's theorem to this Lagrangian, we obtain the operators in Eq. (3.4).

---

[1]Notice that $W^A w^\mu \to (W^A/T)\omega$, so there is no divergency for $T \to 0$.

3.3.1 Symmetries in Constant Electromagnetic Field

It is worth noticing that the symmetries of the theory of fermions in the external electromagnetic field are different from those of free fermions and from those of quantum electrodynamics. While a system without external forces is symmetric for the full Poincaré group, some of the symmetries are lost when external fields are introduced. Indeed, external fields do not transform together with the rest of the system. In this section, we discuss the symmetries of a system in the presence of an external constant homogeneous electromagnetic field. We will examine the transformations that are still symmetries of the theory, the consequent conserved quantities, and the form of the generators of such transformations.

If the Lagrangian of our theory is invariant under translations, from Noether's theorem, we can identify four operators. Those operators share three properties: they are conserved quantities, they are the generators of translations and they constitute the four-momentum of the system. However, translation invariance by itself does not guarantee that the same quantity must have all the three above properties altogether. Consider again a system under an external electromagnetic field. In this situation, Poincaré symmetry of space–time is broken. Only in the special case of a constant and homogeneous electromagnetic field, translation invariance is restored. However, the Lagrangian is not invariant under space–time translation, but it acquires a term that is a four-divergence. This term, under appropriate boundary conditions, does not affect the action of the system and the overall invariance is preserved. Nevertheless, the consequence of the additional term is that we can distinguish between three different operators, each of them having one of the three properties stated above. This is understood with the Noether–Tassie–Buchdahl theorem [39–41]: *given a Lagrangian $\mathcal{L}(\Psi(x), \partial_\mu \Psi(x), x)$ and the infinitesimal transformation*

$$x'^\mu = x^\mu + \delta x^\mu, \quad \Psi' = \Psi + \delta \Psi$$

such that $\partial_\mu \delta x^\mu = 0$, which transforms the Lagrangian in

$$\mathcal{L}(\Psi'(x'), \partial'_\mu \Psi'(x'), x') = \mathcal{L}(\Psi(x), \partial_\mu \Psi(x), x) + \partial_\mu X^\mu,$$

where X^μ is a functional depending exclusively on $\Psi(x)$ and x, the quantity

$$\Gamma^\mu = \frac{\delta \mathcal{L}}{\delta \partial_\mu \Psi} \delta \Psi - \left(\frac{\delta \mathcal{L}}{\delta \partial_\mu \Psi} \partial_\nu \Psi - \mathcal{L} g^\mu_\nu \right) \delta x^\nu - X^\mu$$

is conserved, i.e. divergence-less.

Consider the Dirac Lagrangian in constant homogeneous electromagnetic field

$$\mathcal{L}(\Psi(x), \partial_\mu \Psi(x), x) = \bar{\Psi}(i\slashed{\partial} - m)\Psi - \widehat{j}^\mu A_\mu.$$

The translation transformation ($\delta \Psi = 0$, $\delta x^\mu = \epsilon^\mu$) acts on the Dirac field but does not act directly on the external gauge field. Therefore, a translation changes the

Lagrangian by

$$\delta\mathcal{L} = \mathcal{L}(\Psi'(x'), \partial'_\mu \Psi'(x'), x') - \mathcal{L}(\Psi(x), \partial_\mu \Psi(x), x)$$
$$= \widehat{j}^\mu \partial_\nu A_\mu \epsilon^\nu = \widehat{j}^\mu (F_{\nu\mu} + \partial_\mu A_\nu)\epsilon^\nu = -\widehat{j}^\mu (F_{\mu\nu} - \partial_\mu A_\nu)\epsilon^\nu.$$

The quantity X of Noether–Tassie–Buchdahl in this case is

$$X^{\mu\nu} \equiv \widehat{j}^\mu (A^\nu + F^{\nu\lambda} x_\lambda).$$

Indeed its divergence is the variation of the Lagrangian

$$\epsilon_\nu \partial_\mu X^{\mu\nu} = (\partial_\mu \widehat{j}^\mu)(A^\nu + F^{\nu\lambda} x_\lambda)\epsilon_\nu + \widehat{j}^\mu (\partial_\mu A_\nu + F_{\nu\mu})\epsilon^\nu = -\widehat{j}^\mu (F_{\mu\nu} - \partial_\mu A_\nu)\epsilon^\nu = \delta\mathcal{L}.$$

Therefore, the theorem implies that the system has a canonical conserved tensor given by

$$\widehat{\pi}^{\mu\nu}_{\text{can}} = \widehat{T}^{\mu\nu}_0 - \widehat{j}^\mu A^\nu - \widehat{j}^\mu F^{\nu\lambda} x_\lambda,$$

where $\widehat{T}^{\mu\nu}_0$ is the free canonical Dirac stress-energy tensor. Using Belinfante procedure, we can transform $\widehat{T}^{\mu\nu}_0 - \widehat{j}^\mu A^\nu$ into the symmetric stress-energy tensor of Dirac field in external magnetic field $\widehat{T}^{\mu\nu}_S$ and the above conserved tensor can be written as

$$\widehat{\pi}^{\mu\nu} \equiv \widehat{T}^{\mu\nu}_S - \widehat{j}^\mu F^{\nu\lambda} x_\lambda. \tag{3.27}$$

From the above equation, we can simply verify that $\partial_\mu \widehat{\pi}^{\mu\nu} = 0$ form Eq. (3.2). Note that $\widehat{\pi}^{\mu\nu}$ is not symmetric and that it is gauge invariant. The conserved quantities are obtained from the previous operators by

$$\widehat{\pi}^\mu = \int d^3x \, \widehat{\pi}^{0\mu},$$

and we can show that this four-vector constitutes the generators of the translation [41]. However, the momentum of the system is still given by

$$\widehat{P}^\mu = \int d^3x \, \widehat{T}^{0\mu}$$

but it is no longer a conserved quantity and it is no longer the generator of translations. Another difference with the four-momenta is that different components of this vectors do not commute, instead they satisfy the commutation relation [41]

$$[\widehat{\pi}^\mu, \widehat{\pi}^\nu] = i\widehat{Q} F^{\mu\nu},$$

where $\widehat{Q}$ is the electric charge operator.

As for Lorentz's transformations, we expect that the variation of the Lagrangian is a full divergence only for specific forms of transformations. For example, with a vanishing electric field and a constant magnetic field, only the rotation along the direction of the magnetic field and the boost along the magnetic field are symmetries of the theory. Therefore, only in these cases, the Lagrangian variation could be vanishing or a full divergence.

Returning to the general case, by repeating the previous argument made for the translations for the Lorentz transformation:

$$\delta x^\mu = \omega^{\mu\nu} x_\nu, \quad \delta \Psi = -\frac{i}{2} \omega_{\mu\nu} \sigma^{\mu\nu} \Psi,$$

we find that the transformed Lagrangian is

$$\delta \mathcal{L} = \omega^{\mu\nu} \widehat{j}^\lambda \left[\partial_\lambda \left(x_\mu A_\nu - x_\nu A_\mu \right) + x_\mu F_{\nu\lambda} - x_\nu F_{\mu\lambda} \right].$$

We can show that the Lagrangian variation can also be written as

$$\delta \mathcal{L} = \frac{1}{2} \omega^{\mu\nu} \partial_\lambda \left[\widehat{j}^\lambda x_\mu \left(A_\nu - \frac{1}{2} x^\sigma F_{\sigma\nu} \right) - \widehat{j}^\lambda x_\nu \left(A_\mu - \frac{1}{2} x^\sigma F_{\sigma\mu} \right) \right]$$
$$- \frac{1}{2} x^\rho \widehat{j}^\lambda (\omega_{\lambda\sigma} F^\sigma_{\ \rho} - \omega_{\rho\sigma} F^\sigma_{\ \lambda}).$$

The first term of the r.h.s. is written as a four-divergence. The remaining part cannot be cast into a four-divergence but it is proportional to the following product:

$$(\omega \wedge F)_{\lambda\rho} = \omega_{\lambda\sigma} F^\sigma_{\ \rho} - \omega_{\rho\sigma} F^\sigma_{\ \lambda}.$$

The product of two non-vanishing anti-symmetric tensor of rank two, $\omega \wedge F$, is zero if and only if ω is a linear combination of F and its dual F^* [41]:

$$(\omega \wedge F)_{\lambda\rho} = 0 \quad \text{iff} \quad \omega_{\mu\nu} = k F_{\mu\nu} + k' F^*_{\mu\nu}, \quad k, k' \in \mathbb{R}. \tag{3.28}$$

Therefore, the part of Lagrangian variation which is not a divergence is vanishing when $\omega^{\mu\nu}$ is a linear combination of electromagnetic stress-energy tensor and its dual:

$$\omega^{\mu\nu} = a F^{\mu\nu} + \frac{b}{2} \epsilon^{\mu\nu\rho\sigma} F_{\rho\sigma}. \tag{3.29}$$

This means, as expected, that the theory is invariant only under a certain type of Lorentz transformations: the ones generated with parameters of the form (3.29). For example, in the case of constant magnetic field, we recover that the system is invariant only for rotation and boost along the magnetic field. Set then ω either as $\omega_{\mu\nu} \propto F_{\mu\nu}$ or $\omega_{\mu\nu} \propto F^*_{\mu\nu}$, so that the Lagrangian variation is a four-divergence. In this case, we can apply Noether–Tassie–Buchdahl theorem and the two following quantities are divergence-less:

$$
\begin{aligned}
\widehat{\Gamma}^\lambda =& \frac{F^{\mu\nu}}{2}\left[x_\mu \left(\widehat{T}^\lambda_{0\,\nu} - \widehat{j}^\lambda A_\nu + \frac{1}{2}\widehat{j}^\lambda x^\rho F_{\rho\nu} \right) \right. \\
& \left. - x_\nu \left(\widehat{T}^\lambda_{0\,\mu} - \widehat{j}^\lambda A_\mu + \frac{1}{2}\widehat{j}^\lambda x^\rho F_{\rho\mu} \right) + \widehat{S}^\lambda_{\mu\nu} \right], \\
\widehat{\Gamma}^{*\lambda} =& \frac{F^{*\mu\nu}}{2}\left[x_\mu \left(\widehat{T}^\lambda_{0\,\nu} - \widehat{j}^\lambda A_\nu + \frac{1}{2}\widehat{j}^\lambda x^\rho F_{\rho\nu} \right) \right. \\
& \left. - x_\nu \left(\widehat{T}^\lambda_{0\,\mu} - \widehat{j}^\lambda A_\mu + \frac{1}{2}\widehat{j}^\lambda x^\rho F_{\rho\mu} \right) + \widehat{S}^\lambda_{\mu\nu} \right],
\end{aligned}
$$

where $\widehat{S}^\lambda_{\mu\nu}$ is the canonical spin tensor of free Dirac field. After Belinfante transformation, the quantities become

$$
\begin{aligned}
\widehat{\Gamma}^\lambda =& \frac{1}{2}F^{\mu\nu}\widehat{M}^\lambda_{\mu\nu}, \quad \widehat{\Gamma}^{*\lambda} = \frac{1}{2}F^{*\mu\nu}\widehat{M}^\lambda_{\mu\nu}, \\
\widehat{M}^\lambda_{\mu\nu} \equiv& x_\mu \left(\widehat{T}^\lambda_{S\,\nu} + \frac{1}{2}\widehat{j}^\lambda x^\rho F_{\rho\nu} \right) - x_\nu \left(\widehat{T}^\lambda_{S\,\mu} + \frac{1}{2}\widehat{j}^\lambda x^\rho F_{\rho\mu} \right) \qquad (3.30) \\
=& x_\mu \left(\widehat{\pi}^\lambda_{\ \nu} - \frac{1}{2}\widehat{j}^\lambda x^\rho F_{\rho\nu} \right) - x_\nu \left(\widehat{\pi}^\lambda_{\ \mu} - \frac{1}{2}\widehat{j}^\lambda x^\rho F_{\rho\mu} \right).
\end{aligned}
$$

We can define the integrals:

$$
\widehat{M}_{\mu\nu} = \int \mathrm{d}^3 x\, \widehat{M}^0_{\mu\nu},
$$

which are conserved quantities only if contracted with $F^{\mu\nu}$ or $F^{*\mu\nu}$. The operators $\widehat{M}_{\mu\nu}$ are the generators of Lorentz transformations if they are also a symmetry for the theory, otherwise the Wigner's theorem does not apply and we cannot say that such transformations admit an unitary and linear (or anti-unitary and anti-linear) representation. For those operators, the following Algebra holds true [41]:

$$
\begin{aligned}
[\widehat{\pi}^\mu, \widehat{\pi}^\nu] =& iF^{\mu\nu}\widehat{Q}, \\
\frac{1}{2}F_{\rho\sigma}[\widehat{\pi}^\mu, \widehat{M}^{\rho\sigma}] =& \frac{i}{2}F_{\rho\sigma}\left(\eta^{\mu\rho}\widehat{\pi}^\sigma - \eta^{\mu\sigma}\widehat{\pi}^\rho \right), \qquad (3.31) \\
\frac{1}{2}F^*_{\rho\sigma}[\widehat{\pi}^\mu, \widehat{M}^{\rho\sigma}] =& \frac{i}{2}F^*_{\rho\sigma}\left(\eta^{\mu\rho}\widehat{\pi}^\sigma - \eta^{\mu\sigma}\widehat{\pi}^\rho \right),
\end{aligned}
$$

where $\widehat{Q}$ is the electric charge operator. In the particular case of vanishing electric field and constant magnetic field along the z axis, the Algebra becomes:

$$
\begin{aligned}
[\widehat{\pi}_x, \widehat{\pi}_y] =& i|\mathbf{B}|\widehat{Q}, \\
[\widehat{J}_z, \widehat{\pi}_x] = i\widehat{\pi}_y, \quad& [\widehat{J}_z, \widehat{\pi}_y] = -i\widehat{\pi}_x, \\
[\widehat{K}_z, \widehat{\pi}_t] = -i\widehat{\pi}_z, \quad& [\widehat{K}_z, \widehat{\pi}_z] = -i\widehat{\pi}_t.
\end{aligned}
$$

3.4   Chiral Fermions in Constant Magnetic Field

Consider now a system consisting of free chiral fermions in an external homogeneous constant magnetic field **B** at global thermal equilibrium with vanishing thermal vorticity. In this configuration, the chiral anomaly is vanishing because there is no electric field ($B \cdot E = 0$). It follows that global equilibrium without vorticity is described by a constant inverse four-temperature β and a constant axial chemical potential ζ_A (see Sect. 3.2). Instead, the condition for the electric chemical potential reads

$$\partial^\mu \zeta(x) = F^{\nu\mu} \beta_\nu = \sqrt{\beta^2} F^{\nu\mu} u_\nu = \sqrt{\beta^2} E^\mu,$$

where u is the fluid velocity directed along β and E is the comoving electric field. Since we are considering the case without electric field, the global equilibrium condition is simply a constant ζ. The global equilibrium statistical operator then becomes

$$\widehat{\rho} = \frac{1}{Z} \exp \left[-\widehat{P}^\mu \beta_\mu + \zeta \widehat{Q} + \zeta_A \widehat{Q}_A \right].$$

Notice that the operators $\widehat{P}^\mu$ are not the generators of translations, which are instead given by $\widehat{\pi}^\mu$ and are obtained by integrating the conserved current in Eq. (3.27). However, in the case of vanishing comoving electric field, the projection of the inverse temperature along the four-momentum is equivalent to the projection along of the generators of translations, that is:

$$\widehat{\pi}^\mu \beta_\mu = \int_\Sigma d\Sigma_\lambda \left(\widehat{T}^{\lambda\nu} - \widehat{j}^\lambda F^{\nu\sigma} x_\sigma \right) \beta_\nu = \widehat{P}^\mu \beta_\mu - \sqrt{\beta^2} E^\sigma \int_\Sigma d\Sigma_\lambda \widehat{j}^\lambda x_\sigma = \widehat{P}^\mu \beta_\mu.$$

The statistical operator can now be written as

$$\widehat{\rho} = \frac{1}{Z} \exp \left[-\widehat{\pi}^\mu \beta_\mu + \zeta \widehat{Q} + \zeta_A \widehat{Q}_A \right].$$

In this form, it is straightforward to use the algebra in Eq. (3.31) and translate the statistical operator of a quantity a^μ. We find

$$\widehat{\mathsf{T}}(a) \widehat{\rho} \widehat{\mathsf{T}}^{-1}(a) = e^{ia \cdot \widehat{\pi}} \widehat{\rho} e^{-ia \cdot \widehat{\pi}} = \frac{1}{Z} \exp \left[-\widehat{\pi}^\mu \beta_\mu + \zeta \widehat{Q} + \zeta_A \widehat{Q}_A + a_\mu F^{\mu\nu} \beta_\nu \widehat{Q} \right].$$

Since $F^{\mu\nu} \beta_\nu$ is the comoving electric field, which is vanishing, the statistical operator is homogeneous:

$$\widehat{\mathsf{T}}(a) \widehat{\rho} \widehat{\mathsf{T}}^{-1}(a) = \widehat{\rho}.$$

3.4.1 Exact Thermal Solutions

Having established the basic quantities of thermal equilibrium with a constant and homogeneous magnetic field, we now move on to introduce the techniques of thermal field theory in order to find exact solutions for thermodynamic equilibrium in a magnetic field. We start by giving a path integral description of the partition function. Since the partition function is a Lorentz invariant, we can choose to evaluate it in the local rest frame where $u = (1, \mathbf{0})$. In this frame, without loss in generality, the magnetic field is chosen along the z axis and we adopt the Landau gauge $A^\mu = (0, 0, Bx_1, 0)$. The path integral formulation of the partition function in local rest frame

$$Z(T, \mu, \mu_5) = \text{tr}\left[e^{-\beta(\hat{H} - \mu\hat{Q} - \mu_A\hat{Q}_A)}\right]$$

is given by[2]

$$Z = C \int_{\Psi(\beta, \mathbf{x}) = -\Psi(0, \mathbf{x})} \mathcal{D}\bar{\Psi}\, \mathcal{D}\Psi\, \exp\left(-S_E(\Psi, \bar{\Psi}, \mu_A)\right),$$

where the Euclidean action of Dirac fermions in external electromagnetic field is

$$S_E(\Psi, \bar{\Psi}, \mu_5) = \int_0^\beta d\tau \int_{\mathbf{x}} \bar{\Psi}(X)\left[i(\gamma \cdot \pi^+) + m - \gamma_0\gamma^5\mu_A\right]\Psi(X)$$

and $\pi_\mu^+ \equiv P_\mu^+ - qA_\mu$, which is not to be confused with the generators of translations.

With regard to the exact solution, instead of solving the Dirac equation directly, we use the Ritus method [42] (see [43] for a brief recap of the method). The core concept of the Ritus method is that we can construct a complete set of orthonormal function, called E_p Ritus functions, such that the Euclidean action is rendered formally identical to the Euclidean action of a free Dirac field in absence of external fields. The E_p functions are constructed such that they are the matrix of the contemporaneous eigenfunctions (eigenvectors) of the maximal set of mutually commuting operators $\{(\gamma \cdot \pi)^2, i\gamma_1\gamma_2, \gamma^5\}$. From gamma algebra, it is straightforward to check that

$$i\gamma_1\gamma_2\Delta(\sigma) = \sigma\Delta(\sigma), \quad \frac{1 + \chi\gamma^5}{2}\gamma^5 = \chi\frac{1 + \chi\gamma^5}{2}$$

with $\sigma = \pm$ and $\chi = \pm$ and we defined

$$\Delta(\sigma) \equiv \frac{1 + i\sigma\gamma_1\gamma_2}{2}.$$

We can then show that [42,43]

$$(\gamma \cdot \pi^+)^2 E_{\widehat{p}\sigma}(X) = \underline{P}^{+2} E_{\widehat{p}\sigma}(X),$$

where $\widehat{p}$ is a label for the quantum numbers $\{l, \omega_n, p_2, p_3\}$, the eigenvalues $\underline{P}^+$ are given by

$$\underline{P}^+ \equiv (\omega_n + i\mu, 0, -\bar{\sigma}\sqrt{2|qB|l}, p_3), \quad \bar{\sigma} \equiv \mathrm{sgn}(qB)$$

and the form of eigenfunction is

$$E_{\widehat{p}\sigma}(X) = N(n)e^{i(P_0\tau + P_2 X_2 + P_3 X_3)} D_n(\rho), \tag{3.32}$$

where $N(n) = (4\pi|qB|)^{1/4}/\sqrt{n!}$ is a normalization factor, and $D_n(\rho)$ denotes the parabolic cylinder functions with argument $\rho = \sqrt{2|qB|}(X_1 - p_2/qB)$ and non-negative integer index $n = 0, 1, 2, \ldots$ given by

$$n = l + \frac{\sigma}{2}\mathrm{sgn}(qB) - \frac{1}{2}.$$

Note that the form of the functions (3.32) strongly depends on the gauge chosen, in our case the Landau gauge. Since the eigenfunction $E_{\widehat{p}\sigma}(X)$ does not depend on chirality, the maximal eigenfunctions of the operators $\{(\gamma \cdot \pi)^2, i\gamma_1\gamma_2, \gamma^5\}$ are given by

$$E_{\widehat{p}}(X) = \sum_{\sigma=\pm}' E_{\widehat{p}\sigma}(X)\Delta(\sigma), \quad \bar{E}_{\widehat{p}}(X) = \gamma_0 E_{\widehat{p}}^\dagger(X)\gamma_0 = \sum_{\sigma'=\pm}' E_{\widehat{p}\sigma'}^*(X)\Delta(\sigma'),$$
$$\tag{3.33}$$

where the prime on the summation symbol denotes that the sum is subject to the constraint

$$\sigma = \begin{cases} \mathrm{sgn}(qB) & l = 0 \\ \pm & l > 0 \end{cases}.$$

Some important properties can be derived from these definitions. Firstly, that the functions E_p commute with γ_0 and with γ^5. Secondly, they satisfy the orthogonality relation

$$\int_X \bar{E}_{\widehat{q}}(X)E_{\widehat{p}}(X) = (2\pi)^4\widehat{\delta}^{(4)}(\widehat{p} - \widehat{q})\Pi(l),$$

where we defined

$$\delta^{(4)}(\widehat{p} - \widehat{p}') \equiv \delta_{l,l'}\beta\delta_{\omega_n,\omega_{n'}}\delta(p_2 - p_2')\delta(p_3 - p_3')$$
$$\Pi(l) \equiv \begin{cases} \frac{1+i\bar{\sigma}\gamma_1\gamma_2}{2} & l = 0 \\ 1 & l > 0 \end{cases}.$$

And lastly, the action of the operator $(\gamma \cdot \pi^+)$ on these function is

$$(\gamma \cdot \pi^+) E_{\widehat{p}}(X) = E_{\widehat{p}}(X)\gamma \cdot \underline{P}.$$

Since we showed that E_p Ritus functions are complete orthonormal functions, we can expand the Dirac fields in these functions:

$$\Psi(X) = T \sum_{\{\omega_n\}} \sum_{l=0}^{\infty} \int dp_2 \int \frac{dp_3}{(2\pi)^3} E_{\widehat{p}}(X)\Psi(\underline{P}) \equiv \oint_{\widehat{P}} E_{\widehat{p}}(X)\Psi(\underline{P}),$$

$$\bar{\Psi}(X) = \oint_{\widehat{Q}} \bar{\Psi}(\underline{Q})\bar{E}_{\widehat{q}}(X).$$

Replacing this expansion on the Euclidean action, we find

$$S_{\mathrm{E}}(\Psi, \bar{\Psi}, \mu_5) = \oint_{\widehat{P}} \oint_{\widehat{Q}} \int_X \bar{\Psi}(\underline{Q})\bar{E}_{\widehat{q}}(X)\left[i(\gamma \cdot \pi^+) + m - \gamma_0\gamma^5\mu_{\mathrm{A}}\right]E_{\widehat{p}}(X)\Psi(\underline{P})$$

and, using the above mentioned properties of E_p functions, we obtain

$$S_{\mathrm{E}}(\Psi, \bar{\Psi}, \mu_5) = \oint_{\widehat{P}} \bar{\Psi}(\underline{P})\Pi(l)\left[i\gamma \cdot \underline{P}^+ + m - \gamma_0\gamma^5\mu_{\mathrm{A}}\right]\Psi(\underline{P}).$$

Notice that this is formally identical to the Euclidean action of the free Dirac field.

We can now proceed to evaluate the partition function. We first change the integration variables in the partition function to the modes of E_p functions $\bar{\Psi}(\underline{P})$ and $\Psi(\underline{P})$. The partition function is then a Gaussian integral of Grassmann variables, whose result is the exponent determinant. Hence the partition function becomes

$$Z = \tilde{C} \int_{\Psi(\beta, \mathbf{x}) = -\Psi(0, \mathbf{x})} \mathcal{D}\bar{\Psi}(\underline{P})\, \mathcal{D}\Psi(\underline{P}) \times$$
$$\times \exp\left\{-\oint_{\widehat{P}} \bar{\Psi}(\underline{P})\Pi(l)\left[i\gamma \cdot \underline{P}^+ + m - \gamma_0\gamma^5\mu_{\mathrm{A}}\right]\Psi(\underline{P})\right\} \qquad (3.34)$$
$$= \tilde{C} \det\left[\Pi(l)\left(i\gamma \cdot \underline{P}^+ + m - \gamma_0\gamma^5\mu_{\mathrm{A}}\right)\right].$$

For the sake of clarity, for now on we will remove the factor $\Pi(l)$:

$$Z = \tilde{C} \det \begin{pmatrix} m\mathbb{I}_{2\times 2} & [i(\omega_n + i\mu) - \mu_{\mathrm{A}}]\mathbb{I}_{2\times 2} + \sigma_i\underline{P}_i \\ (i(\omega_n + i\mu) + \mu_{\mathrm{A}})\mathbb{I}_{2\times 2} - \sigma_i\underline{P}_i & m\mathbb{I}_{2\times 2} \end{pmatrix}.$$

The determinant is evaluated using the standard formula for block matrices

$$\det \begin{pmatrix} A & B \\ C & D \end{pmatrix} = \det\left(AD - BD^{-1}CD\right);$$

replacing that into the partition function, we have

$$
\begin{aligned}
Z &= \tilde{C} \det \left[(\underline{P}^{+2} + m^2 + \mu_A^2)\mathbb{I}_{2\times2} - 2\sigma_i \underline{P}_i \mu_A \right] \\
&= \tilde{C} \det \left[(\underline{P}^{+2} + m^2 + \mu_A^2)^2 - 4|\underline{P}|^2 \mu_A^2 \right] \\
&= \prod_{\omega_n, l, p_3} \tilde{C} \left[(\underline{P}^{+2} + m^2 + \mu_A^2)^2 - 4|\underline{P}|^2 \mu_A^2 \right],
\end{aligned}
$$

where we evaluated the determinant as the product of the eigenvalues of the matrix. To connect this quantity to the thermodynamics of the system, we are actually interested in its logarithm:

$$
\log Z = \sum_{\omega_n, l, p_3} \log \left[(\underline{P}^{+2} + m^2 + \mu_A^2)^2 - 4|\underline{P}|^2 \mu_A^2 \right] + \text{cnst.} \tag{3.35}
$$

In the next subsection, we evaluate the thermodynamic potential of the system from the logarithm of the partition function. Then, by simple derivation, we could obtain other thermodynamic properties. However, starting from partition function in Eq. (3.34), we will obtain the thermal propagator of a chiral fermion in a magnetic field. Once we have the propagator, we can use the point-splitting procedure to evaluate other thermal properties that are not related to the thermodynamic potential. We will use the thermal propagator to evaluate the mean value of the electric current and of the axial current in the following subsections.

3.4.2 Thermodynamic Potential

The thermodynamic potential Ω is derived from the partition function as following

$$
\Omega = \lim_{V \to \infty} -\frac{T}{V} \log Z,
$$

where the logarithm of partition function is given by Eq. (3.35). We can follow the usual techniques used for free fermions to evaluate the thermodynamic potential and to sum the Matsubara frequencies. However, we must first consider that in this case the Landau levels generated by the magnetic field have different degeneracy factors and must be properly taken into account when performing the infinite volume limit. Let be S the area in the $x - y$ plane and $p_{\perp 1}$ and $p_{\perp 2}$ the momenta in that plane. In the limit of infinity area, the sum on modes becomes the following integrals:

$$
\lim_{S \to \infty} \frac{1}{S} \sum_{p_{\perp 1}} \sum_{p_{\perp 2}} = \int_{-\infty}^{\infty} \frac{dp_{\perp 1}}{2\pi} \int_{-\infty}^{\infty} \frac{dp_{\perp 2}}{2\pi}.
$$

Each Landau level has a degeneracy associated with some quantum numbers; this degeneracy is gauge independent and it is given by

$$d_l = \left\lfloor \frac{|qB|S}{2\pi} \right\rfloor.$$

To obtain this degeneracy, we just have to evaluate the quantity

$$\frac{\mathrm{d}p_{\perp 1}}{2\pi} \frac{\mathrm{d}p_{\perp 2}}{2\pi}$$

between two consecutive energy levels:

$$d_l = \left\lfloor \frac{|qB|S}{2\pi} \right\rfloor = \int_l^{l+1} \frac{\mathrm{d}p_{\perp 1}}{2\pi} \frac{\mathrm{d}p_{\perp 2}}{2\pi}.$$

Therefore, removing the floor function, the infinite volume limit of the sum on the states of the system gives:

$$\lim_{V \to \infty} \frac{1}{V} \sum_{\omega_n, l, p_3} = \frac{|qB|}{2\pi} \sum_{l=0}^{\infty} \int_{-\infty}^{\infty} \frac{\mathrm{d}p_3}{2\pi} \sum_{\{\omega_n\}}.$$

Consequently the thermodynamic potential reads:

$$\begin{aligned}
\Omega &= \lim_{V \to \infty} -\frac{T}{V} \log Z \\
&= -\frac{|qB|}{2\pi} \sum_{l=0}^{\infty} \int_{-\infty}^{\infty} \frac{\mathrm{d}p_3}{2\pi} T \sum_{\{\omega_n\}} \log \left[(\underline{P}^{+2} + m^2 + \mu_A^2)^2 - 4|\underline{\mathbf{P}}|^2 \mu_A^2 \right] + \mathrm{cnst}.
\end{aligned}$$

The Matsubara sum can be performed as described in [44]. The final result for thermodynamic potential is

$$\Omega = -\frac{|qB|}{2\pi} \sum_{l=0}^{\infty} {\sum_{s=\pm}}' \int_{-\infty}^{\infty} \frac{\mathrm{d}p_3}{2\pi} \left[E_s + T \sum_{\pm} \log \left(1 + \mathrm{e}^{-\beta(E_s \pm \mu)} \right) \right] + \mathrm{cnst},$$

where $E_s^2 = [(p_3^2 + 2qBl)^{1/2} + s\mu_A]^2 + m^2$ and the constraint of $s = \bar{\sigma}$ for $l = 0$ is caused by the projector $\Pi(l)$. This same thermodynamic potential for chiral fermions in external magnetic field was used in [4] to derive the Chiral Magnetic Effect (CME). This expression can be used to obtain the electric and axial charge density, but instead we are using the point-splitting procedure because the latter can also be used to evaluate other thermodynamic functions related to currents. To do that, we first need the thermal propagator of chiral fermions.

3.4.3 Chiral Fermion Propagator in Magnetic Field

Since we have used the Ritus method, the form of Euclidean action is formally identical to those of a free Dirac field. It is therefore not surprising that the fermionic propagator is obtained in the same way as the free case. The propagator in Fourier modes can be obtained in path integral formulation by [44]

$$\langle \tilde{\Psi}_a(P)\bar{\tilde{\Psi}}_b(Q)\rangle_T = \frac{\int \mathcal{D}\tilde{\Psi}\, \mathcal{D}\bar{\tilde{\Psi}}\ \exp\left(-S_E\right)\tilde{\Psi}_a(P)\bar{\tilde{\Psi}}_b(Q)}{\int \mathcal{D}\tilde{\Psi}\, \mathcal{D}\bar{\tilde{\Psi}}\ \exp\left(-S_E\right)},$$

where in our case the form of the partition function is given in the first line of Eq. (3.34):

$$Z = \tilde{C}\int_{\Psi(\beta,\mathbf{x})=-\Psi(0,\mathbf{x})} \mathcal{D}\bar{\Psi}(\underline{P})\,\mathcal{D}\Psi(\underline{P})\ \times$$
$$\times\, \exp\left\{-\sum_{\hat{P}}\bar{\Psi}(\underline{P})\Pi(l)\left[i\gamma\cdot\underline{P}^+ + m - \gamma_0\gamma^5\mu_A\right]\Psi(\underline{P})\right\}.$$

The Grassmann integrals are straightforward and gives

$$\langle\bar{\Psi}(\underline{Q})_a\Psi(\underline{P})_b\rangle = \delta^{(4)}(\underline{P}-\underline{Q})\mathcal{M}_{ab}^{-1},$$

where a, b denotes spinorial indices and

$$\mathcal{M} = \left[i\gamma\cdot\underline{P}^+ - \gamma_0\gamma^5\mu_A\right].$$

The inverse of $\mathcal{M}$ is easily written in terms of the projector into right and left chirality states, which are defined by

$$\mathbb{P}_\chi = \frac{1+\chi\gamma_5}{2}, \quad \text{i.e.} \quad \mathbb{P}_R = \frac{1+\gamma_5}{2}, \quad \mathbb{P}_L = \frac{1-\gamma_5}{2}.$$

We also introduce right and left chemical potential:

$$\mu_R \equiv \mu + \mu_A, \quad \mu_L \equiv \mu - \mu_A,$$

and we define right and left charged momenta by

$$P_{R/L}^\pm \equiv (\omega_n \pm i\mu_{R/L}, \mathbf{p}).$$

With this notation, after inverting $\mathcal{M}$, the thermal propagator is

$$\langle\bar{\Psi}(\underline{Q})_a\Psi(\underline{P})_b\rangle = \delta^{(4)}(\underline{P}-\underline{Q})\sum_\chi\left(\mathbb{P}_\chi\frac{-i\underline{P}_\chi^+}{\underline{P}_\chi^{+2}}\right)_{ab}.$$

This is the generalization in Euclidean space–time with chemical potentials of the propagator in [42,43]. In the configuration space, the two-point function is

$$\langle \bar{\Psi}(X)_a \Psi(Y)_b \rangle = \sum\!\!\!\!\!\!\!\!\!\!\oint_{\widehat{P}} \sum\!\!\!\!\!\!\!\!\!\!\oint_{\widehat{Q}} \bar{E}_{\widehat{p}}(X)_{a'a} E_{\widehat{q}}(Y)_{bb'} \langle \bar{\Psi}(P)_{a'} \Psi(Q)_{b'} \rangle$$

$$= -\sum\!\!\!\!\!\!\!\!\!\!\oint_{\widehat{P}} \sum\!\!\!\!\!\!\!\!\!\!\oint_{\widehat{Q}} \bar{E}_{\widehat{p}}(X)_{a'a} E_{\widehat{q}}(Y)_{bb'} \langle \Psi(Q)_{b'} \bar{\Psi}(P)_{a'} \rangle$$

$$= -\sum\!\!\!\!\!\!\!\!\!\!\oint_{\widehat{P}} \sum\!\!\!\!\!\!\!\!\!\!\oint_{\widehat{Q}} \bar{E}_{\widehat{p}}(X)_{a'a} E_{\widehat{q}}(Y)_{bb'} \delta^{(4)}(P-Q) \sum_\chi \left(\mathbb{P}_\chi \frac{-i\underline{P}_\chi^+}{\underline{P}_\chi^{+2}} \right)_{b'a'},$$

where to go to second line, we used fermion anti-commutation. Finally, integrating the delta we have

$$\langle \bar{\Psi}(X)_a \Psi(Y)_b \rangle = -\sum\!\!\!\!\!\!\!\!\!\!\oint_{\widehat{P}} \sum_\chi E_{\widehat{p}}(Y)_{bb'} \left(\mathbb{P}_\chi \frac{-i\underline{P}_\chi^+}{\underline{P}_\chi^{+2}} \right)_{b'a'} \bar{E}_{\widehat{p}}(X)_{a'a}. \qquad (3.36)$$

3.4.4 Electric Current Mean Value

Having derived the propagator, we now proceed to evaluate the mean value of electric current. The following method is similar to the one used in [45], as we both use the Ritus method. We take advantage of the point-splitting procedure to compute the thermal expectation value of electric current. First, we write the current in Euclidean space–time and we split the coordinate point in which the fields are evaluated as follows:

$$\langle \widehat{j}_\mu(X) \rangle = (-i)^{1-\delta_{\mu,0}} q \langle \widehat{\bar{\Psi}}(X) \gamma_\mu \widehat{\Psi}(X) \rangle = \lim_{X_1,X_2 \to X} (-i)^{1-\delta_{\mu,0}} q \left(\gamma_\mu \right)_{ab} \langle \widehat{\bar{\Psi}}_a(X_1) \widehat{\Psi}_b(X_2) \rangle.$$

Then we plug the form of fermionic propagator (3.36) and we reconstruct the trace on spinorial indices; eventually we obtain

$$\langle \widehat{j}_\mu(X) \rangle = -(-i)^{1-\delta_{\mu,0}} q \sum\!\!\!\!\!\!\!\!\!\!\oint_{\widehat{P}} \sum_\chi \mathrm{tr} \left[\bar{E}_{\widehat{p}}(X) \gamma_\mu E_{\widehat{p}}(X) \mathbb{P}_\chi \frac{-i\underline{P}_\chi^+}{\underline{P}_\chi^{+2}} \right].$$

It is convenient to indicate the components parallel to the magnetic field, which are the time component and the z component, with the parallel symbol "$\|$". For those components, the following commutator holds true:

$$\left[\gamma_\mu^\|, E_{\widehat{p}}(X) \right] = 0.$$

Therefore, reminding the definitions of the Ritus $E_{\widehat{p}}(X)$ functions (3.33), for the parallel component of electric current, we obtain

$$\langle \widehat{j}_\mu^\parallel (X) \rangle = -(-i)^{1-\delta_{\mu,0}} q \sum_{l=0}^\infty \int \frac{dp_2}{2\pi} \int \frac{dp_3}{(2\pi)^2} \sum_\chi T \sum_{\{\omega_n\}} \sum_{\sigma,\sigma'=\pm}' E_{\widehat{p}\sigma'}^*(X) E_{\widehat{p}\sigma}(X) \times$$

$$\times \mathrm{tr}\left[\Delta(\sigma')\, \Delta(\sigma)\, \gamma_\mu^\parallel\, \mathbb{P}_\chi\, \frac{-i\underline{P}_\chi^+}{\underline{P}_\chi^{+2}} \right].$$

We can simplify the previous expression by taking advantage of the following identity

$$\Delta(\sigma)\Delta(\sigma') = \frac{1 + \sigma\sigma' + i(\sigma + \sigma')\gamma_1\gamma_2}{4} = \delta_{\sigma,\sigma'}\Delta(\sigma).$$

Notice that the dependence on p_2 is only contained inside $E_{\widehat{p}\sigma'}^*(X) E_{\widehat{p}\sigma}(X)$. We can then show that the integration on p_2 gives

$$\int_{-\infty}^\infty \frac{dp_2}{2\pi} E_{\widehat{p}\sigma'}^*(X) E_{\widehat{p}\sigma}(X) = |q\,B|\delta_{n,n'}.$$

Furthermore, it is convenient to split the sum on l between the Lowest Landau Level (LLL) $l = 0$ and the Higher Landau Levels (HLL) $l > 1$. For $l = 0$ the sums on σ are constrained to be equal to $\sigma = \sigma' = \bar\sigma =\mathrm{sgn}(eB)$, and the momenta are given by $\underline{P}_\chi^+ = (\omega_n + i\mu_\chi, 0, 0, p_3)$; then at the lowest Landau level, we have

$$\langle \widehat{j}_\mu^\parallel (X) \rangle_{\mathrm{LLL}} = -(-i)^{1-\delta_{\mu,0}} q |q\,B| \int_{-\infty}^\infty \frac{dp_3}{(2\pi)^2} \sum_\chi T \sum_{\{\omega_n\}} \mathrm{tr}\left[\frac{1 + i\bar\sigma\gamma_1\gamma_2}{2} \gamma_\mu^\parallel \mathbb{P}_\chi \frac{-i\underline{P}_\chi^+}{\underline{P}_\chi^{+2}} \right].$$

After computing the trace, we find that the zero component is

$$\langle \widehat{j}_0(X) \rangle_{\mathrm{LLL}} = -\int_{-\infty}^\infty \frac{q|q\,B|dp_3}{(2\pi)^2} \sum_\chi T \sum_{\{\omega_n\}} \frac{-i\left[(\omega_n + i\mu_\chi) - ip_3\bar\sigma\chi\right]}{(\omega_n + i\mu_\chi)^2 + p_3^2}$$

$$= \int_{-\infty}^\infty \frac{q|q\,B|dp_3}{(2\pi)^2} \sum_\chi T \sum_{\{\omega_n\}} \frac{i\left[(\omega_n + i\mu_\chi)\right]}{(\omega_n + i\mu_\chi)^2 + p_3^2},$$

while the z component is

$$\langle \widehat{j}_3(X) \rangle_{\mathrm{LLL}} = i\int_{-\infty}^\infty \frac{q|q\,B|dp_3}{(2\pi)^2} \sum_\chi T \sum_{\{\omega_n\}} \frac{-i\left[p_3 + i(\omega_n + i\mu_\chi)\bar\sigma\chi\right]}{(\omega_n + i\mu_\chi)^2 + p_3^2}$$

$$= \int_{-\infty}^\infty \frac{\bar\sigma q|q\,B|dp_3}{(2\pi)^2} \sum_\chi T \sum_{\{\omega_n\}} \frac{\chi i(\omega_n + i\mu_\chi)}{(\omega_n + i\mu_\chi)^2 + p_3^2},$$

where the linear terms in p_3 were dropped because they are odd on p_3 and as such they vanish when integrated. After the Matsubara sum, we have

$$\langle \widehat{j}_0(X) \rangle_{\text{LLL}} = q|qB| \sum_\chi \int_{-\infty}^\infty \frac{\mathrm{d}p_3}{(2\pi)^2} \frac{1}{2} \left[n_F(p_3 - \mu_\chi) - n_F(p_3 + \mu_\chi) \right],$$

$$\langle \widehat{j}_3(X) \rangle_{\text{LLL}} = q^2 B \sum_\chi \int_{-\infty}^\infty \frac{\mathrm{d}p_3}{(2\pi)^2} \frac{\chi}{2} \left[n_F(p_3 - \mu_\chi) - n_F(p_3 + \mu_\chi) \right].$$

Finally, taking advantage of the integral in p_3:

$$\int_{-\infty}^\infty \frac{\mathrm{d}p_3}{2} \left[n_F(p_3 - \mu_\chi) - n_F(p_3 + \mu_\chi) \right] = \mu_\chi$$

and summing on chiralities, we obtain

$$\langle \widehat{j}_0(X) \rangle_{\text{LLL}} = \frac{q|qB|}{(2\pi)^2}(\mu_R + \mu_L) = \frac{\mu q|qB|}{2\pi^2},$$

$$\langle \widehat{j}_3(X) \rangle_{\text{LLL}} = \frac{q^2 B}{(2\pi)^2}(\mu_R - \mu_L) = \frac{q^2 \mu_A}{2\pi^2} B.$$

Moving on now to the higher Landau levels, consider

$$\langle \widehat{j}_\mu^\parallel(X) \rangle_{\text{HLL}} = - (-\mathrm{i})^{1-\delta_{\mu,0}} q|qB| \sum_{l=1}^\infty \int_{-\infty}^\infty \frac{\mathrm{d}p_3}{(2\pi)^2} \sum_\chi T \sum_{\{\omega_n\}} \sum_{\sigma,\sigma'=\pm} \delta_{n,n'} \times$$

$$\times \text{tr} \left[\Delta(\sigma') \Delta(\sigma) \gamma_\mu^\parallel \mathbb{P}_\chi \frac{-\mathrm{i} \underline{P}_\chi^+}{\underline{P}_\chi^{+2}} \right].$$

When l is fixed, we can replace the $\delta_{n,n'}$ with the $\delta_{\sigma,\sigma'}$, and the sum on σ' becomes straightforward. The expression is similar to the LLL case, we just have to replace $\bar{\sigma}$ with σ and sum over $\sigma = \pm$. Remind that now $\underline{P}$ has also a y component. After evaluating the trace and removing p_3 odd terms, we obtain

$$\langle \widehat{j}_0(X) \rangle_{\text{HLL}} = \sum_{l=1}^\infty \int_{-\infty}^\infty \frac{q|qB|\mathrm{d}p_3}{(2\pi)^2} \sum_\chi T \sum_{\{\omega_n\}} \frac{2\mathrm{i}\left[(\omega_n + \mathrm{i}\mu_\chi)\right]}{\underline{P}_\chi^{+2}},$$

$$\langle \widehat{j}_3(X) \rangle_{\text{HLL}} = \sum_{l=1}^\infty \int_{-\infty}^\infty \frac{q|qB|\mathrm{d}p_3}{(2\pi)^2} \sum_\chi T \sum_{\{\omega_n\}} \sum_{\sigma=\pm} \sigma \frac{\chi \mathrm{i}(\omega_n + \mathrm{i}\mu_\chi) + p_3}{\underline{P}_\chi^{+2}} = 0.$$

We found that the third component does not get corrections from HLL. Instead for the time component, after the frequency sum, we obtain

$$\langle \widehat{j}_0(X) \rangle_{\text{HLL}} = \sum_{l=1}^\infty \int_{-\infty}^\infty \frac{q|qB|\mathrm{d}p_3}{(2\pi)^2} \sum_\chi \left[n_F(E_{p3,l} - \mu_\chi) - n_F(E_{p3,l} + \mu_\chi) \right],$$

where $E_{p3,l} \equiv \sqrt{p_3^2 + 2|qB|l}$. For the perpendicular components ($\widehat{j}_x$ and $\widehat{j}_y$) we expect a vanishing result because they are not allowed by the symmetries of the system. Indeed the explicit calculations confirmed this expectation.

In summary, restoring covariant expression, we found that the electric current has two thermodynamic function: the electric charge density n_c and the Chiral Magnetic Effect (CME) conductivity σ_B:

$$\langle \widehat{j}_\mu(X) \rangle = n_c u_\mu + \sigma_B B_\mu.$$

The electric charge density is given by the mean value $\langle \widehat{j}_0(X) \rangle$ at the local rest frame:

$$n_c = \frac{q|qB|}{2\pi^2} \left\{ \mu + \sum_{l=1}^{\infty} \int_{-\infty}^{\infty} \frac{dp_3}{2} \left[n_F(E_{p3,l} - \mu_R) - n_F(E_{p3,l} + \mu_R) + \right.\right.$$

$$\left.\left. + n_F(E_{p3,l} - \mu_L) - n_F(E_{p3,l} + \mu_L) \right] \right\},$$

while the CME conductivity is given by $\langle \widehat{j}_3(X) \rangle / B$, that is

$$\sigma_B = \frac{q^2 \mu_A}{2\pi^2}. \tag{3.37}$$

To our knowledge, the equation for the electric charge density of an electron gas in a magnetic medium was first given in [8] and coincides with the expression above. The CME effect evaluated here coincides with the one obtained with many other derivations [2], however we want to point out that this derivation is valid at thermal equilibrium, as the one in [47], and that is non-perturbative in the magnetic field.

3.4.5 Axial Current Mean Value

We can compute the axial current mean value exactly as described above for the electric current. Because of that, we omit all the calculations. The axial current constitutive equation is written in terms of an axial charge density n_A and a Chiral Separation Effect (CSE) conductivity σ_s:

$$\langle \widehat{j}_{A\mu} \rangle = n_A u_\mu + \sigma_s B_\mu.$$

In this case too, we found that only the lowest Landau level contributes to CSE and that the final result is

$$n_A = \frac{|qB|}{2\pi^2} \left\{ \mu_A + \sum_{l=1}^{\infty} \int_{-\infty}^{\infty} \frac{dp_3}{2} \left[n_F(E_{p3,l} - \mu_R) - n_F(E_{p3,l} + \mu_R) + \right.\right.$$

$$\left.\left. - n_F(E_{p3,l} - \mu_L) + n_F(E_{p3,l} + \mu_L) \right] \right\},$$

$$\sigma_s = \frac{q\mu}{2\pi^2}$$

with $E_{p3,l} = \sqrt{p_3^2 + 2|qB|l}$. The last thermal coefficient is exactly the well-known value of Chiral Separation Effect (CSE) conductivity [2].

The same procedure can be followed to evaluate the axial charge density and the CSE conductivity of massive fermions with vanishing axial chemical potential $\mu_A = 0$. In that case, as discussed previously, the thermal equilibrium can be reached and all the quantities discussed in this section are still well defined. The results for the thermodynamic functions related to axial current are:

$$n_A = 0,$$

$$\sigma_s = qB \int_{-\infty}^{\infty} \frac{dp_3}{(2\pi)^2} \left[n_F(E_{p3} - \mu) - n_F(E_{p3} + \mu) \right], \qquad (3.38)$$

with $E_{p3}^2 = p_3^2 + m^2$. The CSE induces an axial current even if the system is not chiral. It is apparent from the result above that CSE has an explicit mass dependence, as it is known that it should have [48].

3.5   Constant Vorticity and Electromagnetic Field

So far this contribution has focussed on the general properties of the statistical operator of global thermodynamic equilibrium with both vorticity and electromagnetic field (3.12). The following section will discuss the special case of a constant homogeneous electromagnetic field ($F^{\mu\nu}$ =constant) for which we already studied the symmetries (Sect. 3.3.1). As discussed in Sect. 3.2, global equilibrium can only be reached if condition (3.9) is satisfied, which in this case becomes

$$\mathcal{L}_\beta(F^{\mu\nu}) = \varpi^\mu_\sigma F^{\sigma\nu} - \varpi^\nu_\sigma F^{\sigma\mu} \equiv (\varpi \wedge F)^{\mu\nu} = 0. \qquad (3.39)$$

We already discussed this wedge product in Eq. (3.28). Equation (3.39) has two independent solutions: $F = k\varpi$ and $F = k'\varpi^*$, with k and k' real numbers. In terms of the gauge potential, the condition (3.10) must be satisfied. From Eq. (3.39), choosing the covariant gauge $A^\mu = \frac{1}{2}F^{\rho\mu}x_\rho$, we find that condition (3.10) is satisfied setting $\Phi = \frac{1}{2}b_\sigma F^{\sigma\lambda}x_\lambda$. The equilibrium chemical potential (3.11) is then written as

$$\zeta(x) = \zeta_0 - \beta_\sigma(x)F^{\lambda\sigma}x_\lambda + \frac{1}{2}\varpi_{\sigma\rho}x^\rho F^{\lambda\sigma}x_\lambda. \qquad (3.40)$$

The same solution can also be obtained by directly solving Eq. (3.7) using Eq. (3.39). This last method to obtain the solution explicitly shows that the chemical potential in Eq. (3.40) is not gauge dependent. For constant magnetic field and vanishing thermal vorticity, the solution (3.40) reduces to ζ =constant, as it was correctly used in Sect. 3.4.

Plugging the form (3.40) inside the operator of Eq. (3.1), we find:

$$\hat{\rho} = \frac{1}{Z} \exp \left\{ -\int d\Sigma_\lambda \left[\left(\hat{T}^{\lambda\nu} - \hat{j}^\lambda F^{\nu\rho} x_\rho \right) \beta_\nu - \frac{1}{2}\varpi_{\sigma\rho}\hat{j}^\lambda x^\rho F^{\tau\sigma} x_\tau - \zeta_0 \hat{j}^\lambda \right] \right\}.$$

Inside the round bracket, we recognize the divergence-less operator $\widehat{\pi}^{\lambda\nu}$ of Eq. (3.27), whose integrals are the generators of translations. Expressing the coordinate dependence of β, we can then write

$$\widehat{\rho} = \frac{1}{Z} \exp\left\{ -\int d\Sigma_\lambda \left[\widehat{\pi}^{\lambda\nu} b_\nu + \varpi_{\nu\tau} x^\tau \widehat{\pi}^{\lambda\nu} - \frac{1}{2} \varpi_{\mu\nu} \widehat{j}^\lambda x^\nu F^{\tau\mu} x_\tau - \zeta_0 \widehat{j}^\lambda \right] \right\}$$

$$= \frac{1}{Z} \exp\left\{ -\int d\Sigma_\lambda \left[\widehat{\pi}^{\lambda\nu} b_\nu + \varpi_{\mu\nu} x^\nu \left(\widehat{\pi}^{\lambda\mu} - \frac{1}{2} \widehat{j}^\lambda F^{\rho\mu} x_\rho \right) - \zeta_0 \widehat{j}^\lambda \right] \right\}$$

$$= \frac{1}{Z} \exp\left\{ -\int d\Sigma_\lambda \left[\widehat{\pi}^{\lambda\nu} b_\nu - \frac{1}{2} \varpi_{\mu\nu} \left[x^\mu \left(\widehat{\pi}^{\lambda\nu} - \frac{1}{2} \widehat{j}^\lambda F^{\rho\nu} x_\rho \right) \right. \right. \right.$$
$$\left. \left. \left. - x^\nu \left(\widehat{\pi}^{\lambda\mu} - \frac{1}{2} \widehat{j}^\lambda F^{\rho\mu} x_\rho \right) \right] - \zeta_0 \widehat{j}^\lambda \right] \right\};$$

this time we have recreated the divergence-less quantity $\varpi_{\mu\nu} \widehat{M}^{\lambda,\mu\nu}$ of Eq. (3.30) that generates the Lorentz transformations and that are symmetries of the system. We can then integrate over the coordinate and we find:

$$\widehat{\rho} = \frac{1}{Z} \exp\left\{ -b \cdot \widehat{\pi} + \frac{1}{2} \varpi : \widehat{M} + \zeta_0 \widehat{Q} \right\}.$$

In the above form, the analogy with statistical operator without electromagnetic field in Eq. (3.6) is evident. In both cases, the statistical operator is written with the sum of conserved operators, each one weighted with a constant Lagrange multiplier. Moreover, starting from a fixed point x, we can write the constants thermal fields as

$$b_\mu = \beta(x)_\mu - \varpi_{\mu\nu} x^\nu, \quad \zeta_0 = \zeta(x) + \beta_\sigma(x) F^{\lambda\sigma} x_\lambda - \frac{1}{2} \varpi_{\sigma\rho} x^\rho F^{\lambda\sigma} x_\lambda,$$

from which the statistical operator becomes

$$\widehat{\rho} = \frac{1}{Z} \exp\left\{ - \beta(x)_\mu \left(\widehat{\pi}^\mu - F^{\lambda\mu} x_\lambda \widehat{Q} \right) + \right.$$
$$\left. + \frac{1}{2} \varpi_{\mu\nu} \left(\widehat{M}^{\mu\nu} + x^\nu \widehat{\pi}^\mu - x^\mu \widehat{\pi}^\nu - x^\nu F^{\lambda\mu} x_\lambda \widehat{Q} \right) + \zeta(x) \widehat{Q} \right\}.$$

It is important to point out that with an external magnetic field, the Poincaré algebra is modified and becomes the Algebra in Eq. (3.31), which we report here for convenience:

$$[\widehat{\pi}^\mu, \widehat{\pi}^\nu] = i F^{\mu\nu} \widehat{Q},$$

$$\frac{1}{2} F_{\rho\sigma} [\widehat{\pi}^\mu, \widehat{M}^{\rho\sigma}] = \frac{i}{2} F_{\rho\sigma} \left(\eta^{\mu\rho} \widehat{\pi}^\sigma - \eta^{\mu\sigma} \widehat{\pi}^\rho \right),$$

$$\frac{1}{2} F^*_{\rho\sigma} [\widehat{\pi}^\mu, \widehat{M}^{\rho\sigma}] = \frac{i}{2} F^*_{\rho\sigma} \left(\eta^{\mu\rho} \widehat{\pi}^\sigma - \eta^{\mu\sigma} \widehat{\pi}^\rho \right).$$

Notice that because F is proportional to ϖ, if we replace F with ϖ and F^* with ϖ^*, the last two algebra identities still hold true. Since the Algebra is known, we can translate the statistical operator. Taking advantage of the unitary of the translation transformation, the translated statistical operator is

$$\widehat{T}(x)\,\widehat{\rho}\,\widehat{T}^{-1}(x) = \frac{1}{Z}\exp\left\{-\widehat{T}(x)\,(b\cdot\widehat{\pi})\,\widehat{T}^{-1}(x) + \right.$$
$$\left. +\,\widehat{T}(x)\left(\frac{\varpi:\widehat{M}}{2}\right)\widehat{T}^{-1}(x) + \zeta_0\widehat{T}(x)\,\widehat{Q}\,\widehat{T}^{-1}(x)\right\}.$$

Therefore we just need to evaluate how the operators $\widehat{\pi}$, $\widehat{M}$ and $\widehat{Q}$ transform under translations. For a unitary transformation, an operator $\widehat{K}$ transforms with

$$e^{i\widehat{A}}\widehat{K}e^{-i\widehat{A}} \simeq \widehat{K} - i\left[\widehat{K},\widehat{A}\right] - \frac{1}{2}\left[\left[\widehat{K},\widehat{A}\right],\widehat{A}\right] + \frac{i}{6}\left[\left[\left[\widehat{K},\widehat{A}\right],\widehat{A}\right],\widehat{A}\right] + \cdots .$$

By applying this formula to our operators, we obtain the complete transformation because after a certain order all the commutators become vanishing. In particular, for the Lorentz transformation generators, we find

$$\frac{1}{2}\varpi_{\mu\nu}\widehat{M}_x^{\mu\nu} \equiv \frac{1}{2}\varpi_{\mu\nu}\widehat{T}(x)\widehat{M}^{\mu\nu}\widehat{T}^{-1}(x) = \frac{1}{2}\varpi_{\mu\nu}\left(\widehat{M}^{\mu\nu} + x^\nu\widehat{\pi}^\mu - x^\mu\widehat{\pi}^\nu - x^\nu F^{\lambda\mu}x_\lambda\widehat{Q}\right).$$

For the other operators instead we find:

$$\widehat{\pi}_x^\mu \equiv \widehat{T}(x)\,\widehat{\pi}^\mu\,\widehat{T}^{-1}(x) = \widehat{\pi}^\mu - x_\rho F^{\rho\mu}\widehat{Q}, \quad \widehat{T}(x)\,\widehat{Q}\,\widehat{T}^{-1}(x) = \widehat{Q}.$$

With these definitions, a translation transformation on the statistical operator acts as following:

$$\widehat{T}(a)\,\widehat{\rho}\,\widehat{T}^{-1}(a) = \frac{1}{Z}\exp\left\{-\beta(x)\cdot\widehat{\pi}_{x+a} + \frac{1}{2}\varpi:\widehat{M}_{x+a} + \zeta(x)\widehat{Q}\right\}$$
$$= \frac{1}{Z}\exp\left\{-\beta(x-a)\cdot\widehat{\pi}_x + \frac{1}{2}\varpi:\widehat{M}_x + \zeta(x-a)\widehat{Q}\right\}.$$

It follows that the statistical operator around a point x can be written as

$$\widehat{\rho} = \frac{1}{Z}\exp\left\{-\beta(x)\cdot\widehat{\pi}_x + \frac{1}{2}\varpi:\widehat{M}_x + \zeta(x)\widehat{Q}\right\}. \tag{3.41}$$

3.5.1 Expansion on Thermal Vorticity

Following Ref. [23], we use linear response theory to evaluate thermal expectation values in the case of the constant electromagnetic field. The purpose of this section is to give the thermal expectation value of an operator $\widehat{O}$ at the point x as a thermal vorticity expansion. Using the properties of the trace, we can transfer the x dependence from the operator $\widehat{O}$ to the statistical operator (3.41) written around the same point x:

$$
\begin{aligned}
\langle \widehat{O}(x) \rangle &= \frac{1}{Z} \mathrm{tr} \left[\exp \left\{ -\beta(x) \cdot \widehat{\pi}_x + \frac{1}{2} \varpi : \widehat{M}_x + \zeta(x) \widehat{Q} \right\} \widehat{O}(x) \right] \\
&= \frac{1}{Z} \mathrm{tr} \left[\widehat{\mathsf{T}}(-x) \exp \left\{ -\beta(x) \cdot \widehat{\pi}_x + \frac{1}{2} \varpi : \widehat{M}_x + \zeta(x) \widehat{Q} \right\} \widehat{\mathsf{T}}^{-1}(-x) \widehat{O}(0) \right] \\
&= \frac{1}{Z} \mathrm{tr} \left[\exp \left\{ -\beta(x) \cdot \widehat{\pi} + \frac{1}{2} \varpi : \widehat{M} + \zeta(x) \widehat{Q} \right\} \widehat{O}(0) \right].
\end{aligned}
$$

To evaluate the mean value, we expand the statistical operator of the last equality around vanishing vorticity. First, we split the exponent of the statistical operator into two parts as follows:

$$
\widehat{\rho} = \frac{1}{Z} \exp \left[\widehat{A} + \widehat{B} \right], \quad \widehat{A} \equiv -\beta_\mu(x) \widehat{\pi}^\mu + \zeta(x) \widehat{Q}, \quad \widehat{B} \equiv \frac{1}{2} \varpi : \widehat{M},
$$

and then we expand on $\widehat{B}$, which is the part containing thermal vorticity. Since $\widehat{B}$ and $\widehat{A}$ satisfy the same algebra of the case discussed in [23], the expansion will lead to the same result, which is

$$
\begin{aligned}
\langle \widehat{O}(x) \rangle =& \langle \widehat{O}(0) \rangle_{\beta(x)} - \alpha_\rho \langle\langle \widehat{K}^\rho \widehat{O} \rangle\rangle - w_\rho \langle\langle \widehat{J}^\rho \widehat{O} \rangle\rangle + \frac{\alpha_\rho \alpha_\sigma}{2} \langle\langle \widehat{K}^\rho \widehat{K}^\sigma \widehat{O} \rangle\rangle \\
&+ \frac{w_\rho w_\sigma}{2} \langle\langle \widehat{J}^\rho \widehat{J}^\sigma \widehat{O} \rangle\rangle + \frac{\alpha_\rho w_\sigma}{2} \langle\langle \{ \widehat{K}^\rho, \widehat{J}^\sigma \} \widehat{O} \rangle\rangle + O(\varpi^3),
\end{aligned}
$$

$$(3.42)$$

where we defined

$$
\begin{aligned}
\langle\langle \widehat{K}^{\rho_1} \cdots \widehat{K}^{\rho_n} \widehat{J}^{\sigma_1} \cdots \widehat{J}^{\sigma_m} \widehat{O} \rangle\rangle &\equiv \int_0^{|\beta|} \frac{\mathrm{d}\tau_1 \cdots \mathrm{d}\tau_{n+m}}{|\beta|^{n+m}} \times \\
&\times \left\langle \mathsf{T}_\tau \left(\widehat{K}^{\rho_1}_{-i\tau_1 u} \cdots \widehat{K}^{\rho_n}_{-i\tau_n u} \widehat{J}^{\sigma_1}_{-i\tau_{n+1} u} \cdots \widehat{J}^{\sigma_m}_{-i\tau_{n+m} u} \widehat{O}(0) \right) \right\rangle_{\beta(x),c}.
\end{aligned}
$$

As discussed in Sect. 3.3.1, the boost and rotation defined starting from $\widehat{M}^{\mu\nu}$, i.e.

$$
\widehat{K}^\mu = u_\lambda \widehat{M}^{\lambda\mu}, \quad \widehat{J}^\mu = \frac{1}{2} \epsilon^{\alpha\beta\gamma\mu} u_\alpha \widehat{M}_{\beta\gamma},
$$

are different from those of a system without external electromagnetic field. The other difference with [23] is that in this case the averages $\langle \cdots \rangle_{\beta(x)}$ are made with the statistical operator

$$
\widehat{\rho}_0 = \frac{1}{Z_0} \exp \left\{ -\beta(x) \cdot \widehat{\pi} + \zeta(x) \widehat{Q} \right\}.
$$

3.5.2 Currents and Chiral Anomaly

In this section, we determine the constitutive equations for the electric and the axial current at first order in thermal vorticity and we investigate the contributions from electric and magnetic fields to the thermal coefficients related to vorticity. The constitutive equations are obtained from the expansion on thermal vorticity given in the previous section. Instead of a direct evaluation, we use the conservation equations to show that indeed no additional corrections from electric and magnetic field occur to first-order vorticous coefficients, such as the conductivity of the Chiral Vortical Effect (CVE) and of the Axial Vortical Effect (AVE). In this way, we obtain several relations between those coefficients and their relation to the chiral anomaly.

Consider the case of global thermal equilibrium with constant vorticity $\varpi_{\mu\nu}$ and an electromagnetic field with strength tensor $F_{\mu\nu} = k\,\varpi_{\mu\nu}$, with k a constant. It then follows that the comoving magnetic and electric fields are parallel respectively to thermal rotation and thermal acceleration:

$$B^\mu(x) = -k\,w^\mu(x), \quad E^\mu(x) = k\,\alpha^\mu(x).$$

For instance, in the case of a constant thermal vorticity caused by a rigid rotation along the z axis and a constant magnetic field along z, we have:

$$\varpi_{\mu\nu} = \frac{\Omega}{T_0}\left(\eta_{\mu 1}\eta_{\nu 2} - \eta_{\nu 1}\eta_{\mu 2}\right), \quad F_{\mu\nu} = B\left(\eta_{\mu 1}\eta_{\nu 2} - \eta_{\nu 1}\eta_{\mu 2}\right), \quad k = \frac{BT_0}{\Omega},$$

where Ω, T_0 and B are constants. In this example, electric and magnetic fields are orthogonal and there is no chiral anomaly. However, in the general case, the product $E \cdot B$ is non-vanishing. In that case, as we showed in Sect. 3.2, we can still discuss global equilibrium with chiral imbalance by defining a conserved Chern–Simons current.

By using the thermal vorticity expansion (3.42), we now proceed to write the thermal expectation value of electric current at first order in thermal vorticity. We want to stress that in the expansion (3.42), no approximations are made on the effects of the external electric and magnetic fields; the expansion (3.42) only approximates the effects of vorticity. At first order on thermal vorticity, the only quantities that can contribute to the mean value of a current are the four-vectors w^μ, α^μ and the scalars $E \cdot \alpha$, $E \cdot w = -B \cdot \alpha$ and $B \cdot w$. We therefore write the thermal expansion in terms of these quantities, which will define several thermal coefficients. Taking into account the symmetries, the thermal vorticity expansion of the electric current is

$$\langle \widehat{j}^\mu(x)\rangle = \left[n_c^0 + n_c^{E\cdot\alpha}(E\cdot\alpha) + n_c^{B\cdot w}(B\cdot w)\right]u^\mu + W^V w^\mu + \sigma_E^{B\cdot\alpha}(B\cdot\alpha)E^\mu$$
$$+ \left[\sigma_B^0 + \sigma_B^{E\cdot\alpha}(E\cdot\alpha) + \sigma_B^{B\cdot w}(B\cdot w)\right]B^\mu + O\left(\varpi^2\right).$$

(3.43)

Since the thermal coefficients n_c^0 and σ_B^0 must be evaluated at vanishing thermal vorticity, they are exactly those computed in Sect. 3.4.4 (for vanishing electric field).

In particular, the Chiral Magnetic Effect (CME) conductivity at vanishing vorticity, Eq. (3.37), is

$$\sigma_B^0(x) = \frac{q\zeta_A}{2\pi^2|\beta(x)|}.$$ (3.44)

All the other coefficients are related to thermal vorticity and they have the following properties under parity, time-reversal and charge conjugation:

	$E \cdot \alpha$	$E \cdot w$	$B \cdot w$	n_c^0	σ_B^0	W^V	σ_E^0	$\sigma_B^{E\cdot\alpha}$	$\sigma_B^{B\cdot w}$	$\sigma_E^{B\cdot\alpha}$
P	+	−	+	+	−	−	+	−	−	−
T	+	−	+	+	+	+	−	+	+	+
C	−	−	−	−	+	−	+	−	−	−

(3.45)

Similarly, the axial current thermal expectation value is

$$\langle \widehat{J}_A^\mu(x) \rangle = \left[n_A^0 + n_A^{E\cdot\alpha}(E \cdot \alpha) + n_A^{B\cdot w}(B \cdot w) \right] u^\mu + W^A w^\mu + \sigma_{sE}^{B\cdot\alpha}(B \cdot \alpha)E^\mu$$
$$+ \left[\sigma_s^0 + \sigma_s^{E\cdot\alpha}(E \cdot \alpha) + \sigma_s^{B\cdot w}(B \cdot w) \right] B^\mu + O\left(\varpi^2\right).$$

Each thermal coefficients is a function depending only on

$$|\beta|, \ \zeta, \ \zeta_A, \ B^2, \ E^2, \ E \cdot B.$$ (3.46)

The coordinate dependence of any thermal coefficients is completely contained inside the Lorentz scalars in (3.46).

With the constitutive equations written down, we are now looking for relations and constraints between those thermodynamic coefficients. The conservation of electric current implies that

$$\partial_\mu \langle \widehat{j}^\mu(x) \rangle = \langle \partial_\mu \widehat{j}^\mu(x) \rangle = 0.$$

The coordinate derivative acts both on thermal coefficients and on thermodynamic fields. We need to establish how the derivative acts on those quantities. For thermodynamic fields, using the equilibrium conditions and the identities in Appendix, we find

$$\partial_\mu u^\mu = 0, \quad \partial_\mu w^\mu = -3\frac{w \cdot \alpha}{|\beta|}, \quad \partial_\mu \alpha^\mu = \frac{2w^2 - \alpha^2}{|\beta|},$$

$$\partial_\mu B^\mu = -3\frac{B \cdot \alpha}{|\beta|}, \quad \partial_\mu E^\mu = -\frac{2(w \cdot B) + (\alpha \cdot E)}{|\beta|}, \quad \partial_\mu(B \cdot \alpha) = 0,$$

$$\partial_\mu(E \cdot \alpha) = -\frac{2}{|\beta|} \left[(w \cdot B)\alpha_\mu + (E \cdot w)w_\mu \right],$$

$$\partial_\mu(B \cdot w) = \frac{2}{|\beta|} \left[(w \cdot B)\alpha_\mu + (E \cdot w)w_\mu \right].$$

Moreover, we can also show that

$$\partial_\mu |\beta| = -\alpha_\mu, \quad \partial_\mu \zeta = \beta^\nu F_{\nu\mu} = -|\beta| E_\mu, \quad \partial_\mu \zeta_A = 0, \quad \partial_\mu (E \cdot B) = 0,$$

$$\partial_\mu |B| = -\frac{(E \cdot B) w_\mu + B^2 \alpha_\mu}{|\beta||B|}, \quad \partial_\mu |E| = -\frac{(E \cdot B) w_\mu + B^2 \alpha_\mu}{|\beta||E|},$$

where

$$|\beta| = \sqrt{\beta^\sigma \beta_\sigma}, \quad |B| = \sqrt{-B^\sigma B_\sigma}, \quad |E| = \sqrt{-E^\sigma E_\sigma}.$$

The derivative with respect to coordinates of a thermodynamic function is

$$\partial_\mu f(|\beta|, \zeta, \zeta_A, |B|, |E|, E \cdot B) = \left(-\partial_\mu |\beta| \frac{\partial}{\partial |\beta|} + \partial_\mu \zeta \frac{\partial}{\partial \zeta} + \partial_\mu \zeta_A \frac{\partial}{\partial \zeta_A} + \partial_\mu |B| \frac{\partial}{\partial |B|} \right.$$

$$\left. + \partial_\mu |E| \frac{\partial}{\partial |E|} + \partial_\mu (E \cdot B) \frac{\partial}{\partial (E \cdot B)} \right) f.$$

Therefore, using the previous identities, the derivative of a thermodynamic function becomes

$$\partial_\mu f = \left[-\alpha_\mu \left(\frac{\partial}{\partial |\beta|} - \frac{|B|^2}{|\beta|} \frac{1}{|B|} \frac{\partial}{\partial |B|} - \frac{|B|^2}{|\beta|} \frac{1}{|E|} \frac{\partial}{\partial |E|} \right) \right.$$

$$\left. - |\beta| E_\mu \frac{\partial}{\partial \zeta} - \frac{(E \cdot B) w_\mu}{|\beta|} \left(\frac{1}{|B|} \frac{\partial}{\partial |B|} + \frac{1}{|E|} \frac{\partial}{\partial |E|} \right) \right] f.$$

We can also define the following short-hand notation:

$$\partial_{\tilde{\beta}} \equiv \frac{\partial}{\partial |\beta|} - \frac{|B|^2}{|\beta|} \partial_{\tilde{B}}, \quad \partial_{\tilde{B}} \equiv \frac{1}{|B|} \frac{\partial}{\partial |B|} + \frac{1}{|E|} \frac{\partial}{\partial |E|},$$

from which the previous derivative is written as

$$\partial_\mu f = \left[-\alpha_\mu \partial_{\tilde{\beta}} - |\beta| E_\mu \partial_\zeta - \frac{(E \cdot B) w_\mu}{|\beta|} \partial_{\tilde{B}} \right] f.$$

We can now use the previous relations to impose electric current conservation by evaluating the divergence of the expansion in Eq. (3.43). For the terms directed along the fluid velocity, we find that no additional constraints are required:

$$\partial_\mu \left(n^0 u^\mu \right) = \partial_\mu \left(n^{E \cdot \alpha} (E \cdot \alpha) u^\mu \right) = \partial_\mu \left(n^{B \cdot w} (B \cdot w) u^\mu \right) = 0.$$

For the terms along the magnetic field, we find:

$$\partial_\mu \left(\sigma_B^0 B^\mu \right) = -(B \cdot \alpha) \left[\frac{3}{|\beta|} + \partial_{\tilde{\beta}} \right] \sigma_B^0 - (E \cdot B)|\beta|\partial_\zeta \sigma_B^0$$
$$- \frac{(E \cdot B)(B \cdot w)}{|\beta|} \partial_{\tilde{B}} \sigma_B^0,$$

$$\partial_\mu \left(\sigma_B^{E \cdot \alpha} (E \cdot \alpha) B^\mu \right) = -(B \cdot \alpha)(E \cdot \alpha) \left[\frac{3}{|\beta|} + \partial_{\tilde{\beta}} \right] \sigma_B^{E \cdot \alpha} - (E \cdot \alpha)(E \cdot B)|\beta|\partial_\zeta \sigma_B^{E \cdot \alpha}$$
$$- \frac{(E \cdot B)(B \cdot w)(E \cdot \alpha)}{|\beta|} \partial_{\tilde{B}} \sigma_B^{E \cdot \alpha}$$
$$- \frac{2}{|\beta|} [(w \cdot B)(B \cdot \alpha) + (E \cdot w)(B \cdot w)] \sigma_B^{E \cdot \alpha},$$

$$\partial_\mu \left(\sigma_B^{B \cdot w} (B \cdot w) B^\mu \right) = -(B \cdot \alpha)(B \cdot w) \left[\frac{3}{|\beta|} + \partial_{\tilde{\beta}} \right] \sigma_B^{B \cdot w} - (B \cdot w)(E \cdot B)|\beta|\partial_\zeta \sigma_B^{B \cdot w}$$
$$- \frac{(E \cdot B)(B \cdot w)^2}{|\beta|} \partial_{\tilde{B}} \sigma_B^{B \cdot w}$$
$$+ \frac{2}{|\beta|} [(w \cdot B)(B \cdot \alpha) + (E \cdot w)(B \cdot w)] \sigma_B^{B \cdot w}.$$

Along electric field, we have

$$\partial_\mu \left(\sigma_E (B \cdot \alpha) E^\mu \right) = -(B \cdot \alpha)(\alpha \cdot E) \left[\frac{1}{|\beta|} + \partial_{\tilde{\beta}} \right] \sigma_E - (B \cdot \alpha) E^2 |\beta|\partial_\zeta \sigma_E$$
$$- \frac{(E \cdot B)(E \cdot w)}{|\beta|} \partial_{\tilde{B}} \sigma_E.$$

Lastly, the divergence of the term along rotation is

$$\partial_\mu \left(W^V w^\mu \right) = -(w \cdot \alpha) \left[\frac{3}{|\beta|} + \partial_{\tilde{\beta}} \right] W^V - (E \cdot w)|\beta|\partial_\zeta W^V - \frac{(E \cdot B)w^2}{|\beta|} \partial_{\tilde{B}} W^V.$$

To impose that $\partial_\mu \langle \widehat{j}^\mu(x) \rangle = 0$, we sum all the previous pieces and we split between the linear independent terms. Those terms must vanish independently of the values of the electromagnetic field and of the thermal vorticity and several equalities are obtained. Among those, we first consider the following identities:

$$\partial_\zeta \sigma_B^0 = 0, \quad \partial_{\tilde{B}} W^V = 0, \quad \partial_\zeta \sigma_B^{E \cdot \alpha} = 0, \quad \partial_{\tilde{B}} \sigma_B^{E \cdot \alpha} = 0, \quad \partial_{\tilde{B}} \sigma_B^{B \cdot w} = 0,$$
$$\partial_\zeta \sigma_E^{B \cdot \alpha} = 0, \quad \partial_{\tilde{B}} \sigma_E^{B \cdot \alpha} = 0.$$

Notice from the table in (3.45) that $\sigma_B^{E \cdot \alpha}$ and $\sigma_E^{B \cdot \alpha}$ are related to C-odd correlator. Therefore they must be odd functions of the electric chemical potential ζ. But the previous constraints require that they do not depend on ζ, therefore they must be vanishing

$$\sigma_B^{E \cdot \alpha} = 0, \quad \sigma_E^{B \cdot \alpha} = 0.$$

The previous constraints also require that W^V and $\sigma_B^{B \cdot w}$ do not depend on $|B|$ and $|E|^3$. That is to say that the CVE conductivity W^V is not affected by electric and magnetic fields. Moreover, electric current conservation also imposes that

$$\left(3 + |\beta|\partial_{\tilde{\beta}}\right)\sigma_B^0 - |\beta|^2\partial_\zeta W^V = 0,$$

$$\partial_{\tilde{\beta}}\sigma_B^0 + |\beta|^2\partial_\zeta\sigma_B^{B \cdot w} = 0,$$

$$\left(3 + |\beta|\partial_{\tilde{\beta}}\right)\sigma_B^{B \cdot w} = 0,$$

$$\left(3 + |\beta|\partial_\beta\right) W^V = 0.$$

We can replace the known result for the CME conductivity σ_B^0 of Eq. (3.44) in the previous constraints to find that

$$\partial_\zeta W^V = \frac{1}{|\beta|^2}\left(3 + |\beta|\partial_{\tilde{\beta}}\right)\sigma_B^0 = \frac{q\zeta_A}{\pi^2|\beta|^3},$$

$$\partial_\zeta\sigma_B^{B \cdot w} = 0,$$

$$\left(3 + |\beta|\frac{\partial}{\partial\beta}\right) W^V = 0.$$

Again, since $\sigma_B^{B \cdot w}$ is C-odd it follows from second equation that it must be vanishing

$$\sigma_B^{B \cdot w} = 0.$$

We want to empathize that we have found a relation between CVE and CME conductivities:

$$\partial_\zeta W^V = \frac{1}{|\beta|^2}\left(3 + |\beta|\partial_{\tilde{\beta}}\right)\sigma_B^0. \tag{3.47}$$

The CVE conductivity W^V in Eq. (3.22) satisfies this relation. It is important to notice that Eq. (3.47) completely determines the CVE conductivity from the CME one. Indeed, since W^V is odd under charge conjugation, by fixing the ζ part of W^V, we obtain the entire coefficient. This also implies that the ζ part of W^V is dictated by the chiral anomaly as found in effective field theories [46]. Therefore, the CVE inherits all the properties proved for the CME. For instance, it is known that the CME conductivity is completely dictated by the chiral anomaly [49] and that it is protected from corrections coming from interactions [50,51]. Since the relation (3.47) holds not only for a free theory but for any microscopic interactions, as long as global thermal equilibrium is concerned, then also the CVE conductivity is dictated by the

---

[3]Note that to reduce the numbers of relations, we have indicated electric field and magnetic field derivative together with one derivative $\partial_{\tilde{\beta}}$. However, electric and magnetic fields are independent and each derivative must be considered independently.

chiral anomaly and it is universal. Despite the CVE can be related to the vector-axial anomalous term of vector current anomaly [52], Eq. (3.47) shows that it can be explained with just the electric charge conservation and the chiral anomaly.

Let us now move to the axial current. Similar steps can be followed to derive the constraint equations between the thermal coefficients of axial current. In this case, we must impose the following identities between thermal expectation values:

$$\partial_\mu \langle \widehat{j}_A^\mu(x) \rangle = 2m \langle i\bar{\Psi}\gamma^5\Psi \rangle - \frac{q^2(E \cdot B)}{2\pi^2},$$

where we also added the naive divergence term $2m\langle i\bar{\Psi}\gamma^5\Psi \rangle$ which is due to the mass of the field. From symmetries, the constitutive equation for the pseudo-scalar is

$$\langle i\bar{\Psi}\gamma^5\Psi \rangle = L^{E \cdot B}(E \cdot B) + L^{\alpha \cdot w}(\alpha \cdot w) + L^{E \cdot w}(E \cdot w)$$
$$+ L^{(E \cdot B)\alpha^2}(E \cdot B)\alpha^2 + L^{(E \cdot B)w^2}(E \cdot B)w^2$$
$$+ L^{(E \cdot w)(E \cdot \alpha)}(E \cdot w)(E \cdot \alpha) + L^{(E \cdot w)(B \cdot w)}(E \cdot w)(B \cdot w) + O(\varpi^3).$$

The value of $L^{\alpha \cdot w}$ for the free Dirac field has been reported in (3.20), and the other coefficients related to the the electromagnetic field can be computed with the Ritus method.

Because the axial current is not conserved, we find different identities compared to the previous case of electric current. For instance, the identities related to the Chiral Separation Effect (CSE) conductivity σ_s^0 and to the AVE conductivity W^A are

$$\left(3 + |\beta|\frac{\partial}{\partial\beta}\right)W^A = -2mL^{\alpha \cdot w},$$

$$\partial_\zeta \sigma_s^0 = \frac{q^2}{2\pi^2|\beta|} - 2mL^{E \cdot B},$$

$$\partial_{\tilde{B}}\sigma_s^0 = -|\beta|^2\partial_\zeta \sigma_s^{B \cdot w},$$

$$\partial_{\tilde{B}}W^A = -2m|\beta|L^{(E \cdot B)w^2},$$

$$\partial_\zeta W^A = \frac{1}{|\beta|^2}\left(3 + |\beta|\partial_{\tilde{B}}\right)\sigma_s^0 - \frac{2m}{|\beta|}L^{E \cdot w}.$$

The first equation has been discussed in [23] and in Sect. 3.2.1. The second equation is similar to the first: the first term on the r.h.s. is coming from the chiral anomaly and the second from the naive anomaly. Therefore for a massive field, as discussed for the AVE, the CSE is not entirely dictated by the anomaly. It is then not surprising to find corrections to the CSE [53] and that it is affected by the mass, see Eq. (3.38). In the massive case, we also expect corrections from the external electromagnetic field both in the AVE and in the CSE conductivities.

On the other hand, for massless field, those constraints become

$$\partial_\zeta \sigma_s^0 = \frac{q^2}{2\pi^2 |\beta|},$$

$$\left(3 + |\beta|\frac{\partial}{\partial\beta}\right) W^A = 0, \quad \partial_{\tilde{B}} W^A = 0,$$

$$\partial_\zeta W^A = \frac{1}{|\beta|^2}\left(3 + |\beta|\partial_\beta\right)\sigma_s^0.$$

In this case, the CSE conductivity is completely fixed by the chiral anomaly as it is clear by the first equation and by the fact that σ_s^0 must be an odd function of ζ. For the symmetries of axial current, W^A has both terms which depend on ζ and terms which depend only on ζ_A and β. All these terms must satisfy the equations in the second line. In particular, we conclude that the AVE is not affected by the electromagnetic field. Moreover, from the third line, we see that the terms related to ζ are fixed by the CSE conductivity and consequently they are dictated by the chiral anomaly. As it is evident from the previous discussion, this only occurs for the massless field.

In summary, by imposing the conservation equation we conclude that, at global thermodynamic equilibrium with thermal vorticity and constant homogeneous electromagnetic field, the chiral vortical effect is dictated by the chiral magnetic effect. Then it is not affected by the mass of the particle, by the external electromagnetic field or by radiative corrections. For the axial current this analysis has showed that we need to distinguish between the massive and the massless case. In the latter case, we found that the chiral anomaly completely fixes the whole Chiral Separation Effect (CSE) but fixes only the part of Axial Vortical Effect (AVE) conductivity which depends on the electric chemical potential. We also found that the AVE is not affected by the external electromagnetic field. For the massive case, despite it exists a relation between the CSE and the AVE, both of them are affected by the mass of the field, the external electromagnetic field and radiative corrections.

Acknowledgements I carried out part of this work while visiting Stony Brook University (New York, U.S.A.). I would like to thank F. Becattini, E. Grossi and D. Kharzeev for stimulating discussions on the subject matter. This research was supported in part by the Florence University with the fellowship "Polarizzazione nei fluidi relativistici".

Appendix: Thermodynamic Relations in Beta Frame

At global thermal equilibrium with thermal vorticity, thermodynamic fields satisfy several equilibrium relations which constraints their coordinate dependence. In the β-frame, we can build several quantities from the four-vector β and thermal vorticity ϖ:

$$u_\mu = \frac{\beta_\mu}{\sqrt{\beta^2}}; \quad \Delta^{\mu\nu} = g^{\mu\nu} - u^\mu u^\nu; \quad \varpi_{\mu\nu} = \partial_\nu \beta_\mu = \epsilon_{\mu\nu\rho\sigma} w^\rho u^\sigma + \alpha_\mu u_\nu - \alpha_\nu u_\mu;$$

$$\alpha_\mu = \varpi_{\mu\nu} u^\nu; \quad w_\mu = -\frac{1}{2}\epsilon_{\mu\nu\rho\sigma}\varpi^{\nu\rho} u^\sigma; \quad \gamma_\mu = (\alpha \cdot \varpi)^\lambda \Delta_{\lambda\mu} = \epsilon_{\mu\nu\rho\sigma} w^\nu \alpha^\rho u^\sigma.$$

Most of these quantities depend on coordinates, and their derivatives are [33]:

$$\partial_\nu \beta_\mu = \varpi_{\mu\nu}; \quad \partial_\nu = -\alpha_\nu \frac{\partial}{\partial \sqrt{\beta^2}}; \quad \varpi : \varpi = 2\left(\alpha^2 - w^2\right)$$

$$\partial_\nu u_\mu = \frac{1}{\sqrt{\beta^2}}\left(\varpi_{\mu\nu} + \alpha_\nu u_\mu\right); \quad \partial^\alpha u_\alpha = 0; \quad u_\alpha \partial^\alpha u_\mu = \frac{\alpha_\mu}{\sqrt{\beta^2}};$$

$$\partial_\mu \alpha_\nu = \frac{1}{\sqrt{\beta^2}}\left(\varpi_{\nu\rho}\varpi^\rho_{\ \mu} + \alpha_\mu \alpha_\nu\right); \quad \partial^\alpha \alpha_\alpha = \frac{1}{\sqrt{\beta^2}}\left(2w^2 - \alpha^2\right); \quad u_\alpha \partial^\alpha \alpha^2 = 0;$$

$$\partial_\mu w_\nu = \frac{1}{\sqrt{\beta^2}}\left(\alpha_\mu w_\nu - \frac{1}{2}\epsilon_{\nu\rho\sigma\lambda}\varpi^{\rho\sigma}\varpi^\lambda_{\ \mu}\right); \quad \partial^\alpha w_\alpha = -3\frac{w \cdot \alpha}{\sqrt{\beta^2}}; \quad u_\alpha \partial^\alpha w^2 = 0;$$

$$\alpha^\sigma \partial_\mu \alpha_\sigma = w^\sigma \partial_\mu w_\sigma = \frac{1}{\sqrt{\beta^2}}\left(w^2 \alpha_\mu - (\alpha \cdot w)w_\mu\right); \quad \partial_\mu(\alpha \cdot w) = 0;$$

$$\partial_\alpha \gamma^\alpha = 0; \quad \partial^\alpha \Delta_{\alpha\beta} = -\frac{\alpha_\beta}{\sqrt{\beta^2}}.$$

References

1. Adamczyk, L., et al.: Nature **548**, 62 (2017). https://doi.org/10.1038/nature23004
2. Kharzeev, D.E., Liao, J., Voloshin, S.A., Wang, G.: Prog. Part. Nucl. Phys. **88**, 1 (2016). https://doi.org/10.1016/j.ppnp.2016.01.001
3. Kharzeev, D.E., McLerran, L.D., Warringa, H.J.: Nucl. Phys. A **803**, 227 (2008). https://doi.org/10.1016/j.nuclphysa.2008.02.298
4. Fukushima, K., Kharzeev, D.E., Warringa, H.J.: Phys. Rev. D **78** (2008). https://doi.org/10.1103/PhysRevD.78.074033
5. Heinz, U., Snellings, R.: Ann. Rev. Nucl. Part. Sci. **63**, 123 (2013). https://doi.org/10.1146/annurev-nucl-102212-170540
6. Hakim, R.: Introduction to Relativistic Statistical Mechanics: Classical and Quantum. World Scientific Publishing Co Inc (2011)
7. Israel, W.: Gen. Rel. Grav. **9**, 451 (1978). https://doi.org/10.1007/BF00759845
8. Canuto, V., Chiu, H.Y.: Phys. Rev. **173**, 1220 (1968). https://doi.org/10.1103/PhysRev.173.1220
9. Huang, X.G., Sedrakian, A., Rischke, D.H.: Ann. Phys. **326**, 3075 (2011). https://doi.org/10.1016/j.aop.2011.08.001
10. Kovtun, P.: JHEP **07**, 028 (2016). https://doi.org/10.1007/JHEP07(2016)028
11. Weickgenannt, N., Sheng, X.L., Speranza, E., Wang, Q., Rischke, D.H.: Phys. Rev. D **100** (2019). https://doi.org/10.1103/PhysRevD.100.056018
12. Gao, J.H., Liang, Z.T.: Phys. Rev. D **100** (2019). https://doi.org/10.1103/PhysRevD.100.056021
13. Chen, J.W., Pu, S., Wang, Q., Wang, X.N.: Phys. Rev. Lett. **110**(26) (2013). https://doi.org/10.1103/PhysRevLett.110.262301
14. Zubarev, D.N., Prozorkevich, A.V., Smolyanskii, S.A.: Theor. Math. Phys. **40**(3), 821 (1979). https://doi.org/10.1007/BF01032069
15. van Weert, C.G.: Ann. Phys. **140**, 133 (1982). https://doi.org/10.1016/0003-4916(82)90338-4
16. Becattini, F., Bucciantini, L., Grossi, E., Tinti, L.: Eur. Phys. J. C **75**(5), 191 (2015). https://doi.org/10.1140/epjc/s10052-015-3384-y
17. Hayata, T., Hidaka, Y., Noumi, T., Hongo, M.: Phys. Rev. D **92**(6) (2015). https://doi.org/10.1103/PhysRevD.92.065008
18. Hongo, M.: Ann. Phys. **383**, 1 (2017). https://doi.org/10.1016/j.aop.2017.04.004

19. Becattini, F., Buzzegoli, M., Grossi, E.: Particles **2**(2), 197 (2019). https://doi.org/10.3390/particles2020014
20. Becattini, F., Florkowski, W., Speranza, E.: Phys. Lett. B **789**, 419 (2019). https://doi.org/10.1016/j.physletb.2018.12.016
21. Becattini, F.: Phys. Rev. Lett. **108** (2012). https://doi.org/10.1103/PhysRevLett.108.244502
22. Becattini, F., Grossi, E.: Phys. Rev. D **92** (2015). https://doi.org/10.1103/PhysRevD.92.045037
23. Buzzegoli, M., Becattini, F.: JHEP **12**, 002 (2018). https://doi.org/10.1007/JHEP12(2018)002
24. Jensen, K., Loganayagam, R., Yarom, A.: JHEP **05**, 134 (2014). https://doi.org/10.1007/JHEP05(2014)134
25. Ambruş, V.E., Winstanley, E.: Phys. Lett. B **734**, 296 (2014). https://doi.org/10.1016/j.physletb.2014.05.031
26. Ambrus, V.E., Winstanley, E.: arXiv: 1908.10244 (2019)
27. Becattini, F., Buzzegoli, M., Palermo, A.: arXiv:2007.08249 (2020)
28. Becattini, F.: Phys. Rev. D **97**(8) (2018). https://doi.org/10.1103/PhysRevD.97.085013
29. Becattini, F., Rindori, D.: Phys. Rev. D **99**(12) (2019). https://doi.org/10.1103/PhysRevD.99.125011
30. Prokhorov, G.Y., Teryaev, O.V., Zakharov, V.I.: Particles **3**(1), 1 (2020). https://doi.org/10.3390/particles3010001
31. Prokhorov, G.Y., Teryaev, O.V., Zakharov, V.I.: Phys. Rev. D **99**(7) (2019). https://doi.org/10.1103/PhysRevD.99.071901
32. Prokhorov, G.Y., Teryaev, O.V., Zakharov, V.I.: JHEP **03**, 137 (2020). https://doi.org/10.1007/JHEP03(2020)137
33. Buzzegoli, M., Grossi, E., Becattini, F.: JHEP **10**, 091 (2017). https://doi.org/10.1007/JHEP07(2018)119, https://doi.org/10.1007/JHEP10(2017)091. [Erratum: JHEP07,119(2018)]
34. Hongo, M., Hidaka, Y.: Particles **2**(2), 261 (2019). https://doi.org/10.3390/particles2020018
35. Landsman, N., van Weert, C.: Phys. Rep. **145**(3), 141 (1987). https://doi.org/10.1016/0370-1573(87)90121-9
36. Flachi, A., Fukushima, K.: Phys. Rev. D **98**(9) (2018). https://doi.org/10.1103/PhysRevD.98.096011
37. Prokhorov, G., Teryaev, O.: Phys. Rev. D **97**(7) (2018). https://doi.org/10.1103/PhysRevD.97.076013
38. Prokhorov, G., Teryaev, O., Zakharov, V.: Phys. Rev. D **98**(7) (2018). https://doi.org/10.1103/PhysRevD.98.071901
39. Tassie, L.J., Buchdahl, H.A.: Austral. J. Phys. **17**, 431 (1964). https://doi.org/10.1071/PH640431
40. Buchdahl, H.A., Tassie, L.J.: Austral. J. Phys. **18**, 109 (1965). https://doi.org/10.1071/PH650109
41. Bacry, H., Combe, P., Richard, J.L.: Nuovo Cim. A **67**, 267 (1970). https://doi.org/10.1007/BF02725178
42. Ritus, V.I.: Ann. Phys. **69**, 555 (1972). https://doi.org/10.1016/0003-4916(72)90191-1
43. Leung, C.N., Wang, S.Y.: Nucl. Phys. B **747**, 266 (2006). https://doi.org/10.1016/j.nuclphysb.2006.04.028
44. Laine, M., Vuorinen, A.: Lect. Notes Phys. **925**, 1 (2016). https://doi.org/10.1007/978-3-319-31933-9
45. Fukushima, K., Kharzeev, D.E., Warringa, H.J.: Nucl. Phys. A **836**, 311 (2010). https://doi.org/10.1016/j.nuclphysa.2010.02.003
46. Sadofyev, A., Shevchenko, V., Zakharov, V.: Phys. Rev. D **83**, 105025 (2011).https://doi.org/10.1103/PhysRevD.83.105025
47. Vilenkin, A.: Phys. Rev. D **22**, 3080 (1980). https://doi.org/10.1103/PhysRevD.22.3080
48. Metlitski, M.A., Zhitnitsky, A.R.: Phys. Rev. D **72** (2005). https://doi.org/10.1103/PhysRevD.72.045011
49. Kharzeev, D.E.: Prog. Part. Nucl. Phys. **75**, 133 (2014). https://doi.org/10.1016/j.ppnp.2014.01.002

50. Zakharov, V.I.: Lect. Notes Phys. **871**, 295 (2013). https://doi.org/10.1007/978-3-642-37305-3_11
51. Feng, B., Hou, D.F., Ren, H.C.: Phys. Rev. D **99**(3) (2019). https://doi.org/10.1103/PhysRevD.99.036010
52. Landsteiner, K., Megias, E., Pena-Benitez, F.: Lect. Notes Phys. **871**, 433 (2013). https://doi.org/10.1007/978-3-642-37305-3_17
53. Gorbar, E.V., Miransky, V.A., Shovkovy, I.A., Wang, X.: Phys. Rev. D **88**(2) (2013). https://doi.org/10.1103/PhysRevD.88.025025

Exact Solutions in Quantum Field Theory Under Rotation

4

Victor E. Ambruş and Elizabeth Winstanley

Abstract

We discuss the construction and properties of rigidly rotating states for free scalar and fermion fields in quantum field theory. On unbounded Minkowski space-time, we explain why such states do not exist for scalars. For the Dirac field, we are able to construct rotating vacuum and thermal states, for which expectation values can be computed exactly in the massless case. We compare these quantum expectation values with the corresponding quantities derived in relativistic kinetic theory.

4.1 Introduction

Rigidly rotating systems are useful toy models for studying the underlying physics of more complex rotating systems in either flat or curved space-times. Consider a rigidly rotating system of classical particles in flat space-time, rotating about a common axis, which we take to be the z-axis in the usual Cartesian coordinates. Assuming that the particles undergo circular motion with constant angular speed Ω about the rotation axis, the linear speed of each particle is then $\rho\Omega$, where ρ is the distance of the particle from the axis of rotation. The speed of the particle therefore increases as the distance from the axis increases, and will become relativistic sufficiently far from the axis. Furthermore, if ρ is sufficiently large, the particle will have a speed greater than the speed of light. Therefore, a simple rigidly rotating system cannot be realized

V. E. Ambruş
Department of Physics, West University of Timişoara, Bd. Vasile Pârvan 4,
300223 Timişoara, Romania
e-mail: Victor.Ambrus@e-uvt.ro

E. Winstanley (✉)
Consortium for Fundamental Physics, School of Mathematics and Statistics,
University of Sheffield, Hicks Building, Hounsfield Road, Sheffield S3 7RH, United Kingdom
e-mail: E.Winstanley@sheffield.ac.uk

© Springer Nature Switzerland AG 2021
F. Becattini et al. (eds.), *Strongly Interacting Matter under Rotation*,
Lecture Notes in Physics 987,
https://doi.org/10.1007/978-3-030-71427-7_4

in nature (at least in flat space-time), and the system must either be bounded in some way to prevent superluminal speeds, or else the system cannot be rigidly rotating.

Although unbounded rigidly rotating systems cannot be realized in flat space-time, nonetheless the study of rigidly rotating systems in both relativistic kinetic theory (RKT) and quantum field theory (QFT) has a long history. The simplicity of the system allows many quantities of physical interest (such as quantum expectation values) to be computed exactly, which enables the extraction of the underlying physics. Many deep physical properties of rotating systems have been revealed by this approach, and in this chapter, we outline some of the most important.

Our motivation for studying rigidly rotating systems in QFT comes from both astrophysics and heavy ion collisions. In astrophysics, rigid rotation can be induced near rapidly-rotating magnetars or in accretion disks around black holes, where the field close to the surface of the star is sufficiently strong to lock charged particles into magnetically dominated accretion flow. The superluminal motion of the plasma constituents can be prohibited by the bending of the magnetic field lines far from the axis of rotation [1]. Particle geodesics on rotating black hole space-times also exhibit rigid rotation close to the event horizon due to the frame-dragging effect [2]. Quantum effects are important for black holes, which emit thermal quantum radiation [3]. Whether or not it is possible to define a quantum state representing a quantum field in thermal equilibrium with a rotating black hole depends on whether one considers a scalar field (in which case such a state does not exist [4,5]) or a fermion field (where a state can be constructed, but is divergent far from the black hole [6]).

In the context of strongly interacting systems, rigid rotation can occur in the quark-gluon plasma (QGP) formed in the early stages following the collision of (ultra-)relativistic heavy ions [7]. Just as a magnetic field can induce a charge current along the magnetic field direction in fermionic matter through the chiral magnetic effect, rigid rotation can induce an axial current through an analogous chiral vortical effect (CVE) [8]. Due to the latter, the rotating fluid becomes polarized along the rotation axis. This polarization was recently demonstrated through measurements of the properties of the decay products of Λ-hyperons [9, 10]. Interest in studying the properties of rigidly rotating quantum systems has surged in the past few years, with recent studies addressing the hydrodynamic description of fluids with spin [11], the role of the spin tensor in nonequilibrium thermodynamics [12] and the properties of thermodynamic equilibrium for the free Dirac field with axial chemical potential [13].

Our focus on this chapter is rigidly rotating systems in flat-space QFT. We consider the simplest types of quantum field, namely a free scalar or Dirac fermion field. By ignoring the self-interactions of the quantum field, and the curvature of space-time, we are able to study in detail the effect of rotation alone. The construction of rotating vacuum and thermal states for these fields is compared with the corresponding construction of nonrotating vacuum and thermal states. Here, the difference between bosonic and fermionic quantum fields plays a major role. Having constructed the rotating states, we then elucidate their physical properties by studying, for the fermion field, the expectation values of the fermion condensate (FC), charge current (CC), axial current (AC) and stress-energy tensor (SET). We compare these with the analogous quantities computed within the framework of RKT, to elucidate the effects of quantum corrections.

This chapter is structured as follows. The problem of rigid rotation at finite temperature is addressed from an RKT perspective in Sect. 4.2. Section 4.3 considers the construction of rigidly rotating states in QFT, showing in particular that these states do not exist for a free quantum scalar field on unbounded flat space-time. The rest of the chapter is therefore devoted to the free Dirac field only. Mode solutions of the Dirac equation are derived with respect to a cylindrical coordinate system in Sect. 4.4. We briefly consider nonrotating thermal expectation values (t.e.v.s) in Sect. 4.5, and demonstrate that there are no quantum corrections for these states. On the other hand, for rotating states, the t.e.v.s constructed in Sect. 4.6 are modified in QFT compared to the RKT results. We examine the physical properties of these quantum corrections for the SET in particular in Sect. 4.7. The above discussion has focussed on unbounded flat space-time, and we briefly review some more general scenarios in Sect. 4.8 before presenting our conclusions in Sect. 4.9.

4.2 Relativistic Kinetic Theory

Before we address the properties of rigidly rotating systems in QFT, we first consider the RKT perspective. We briefly describe the main features of a distribution of Bose–Einstein or Fermi–Dirac particles in global thermal equilibrium (GTE) undergoing rigid rotation.

4.2.1 Rigidly Rotating Thermal Distribution

Consider particles of mass M and four-momentum p^μ in GTE in the absence of external forces. The configuration of particles is described by the distribution function f, which satisfies the relativistic Boltzmann equation [14]

$$p^\mu \partial_\mu f = C[f], \tag{4.1}$$

using Cartesian coordinates on Minkowski space-time so that $x^\mu = (t, x, y, z)^T$. In (4.1), $C[f]$ is the collision operator, which drives the fluid towards local thermal equilibrium and whose properties give the form of the equilibrium distribution function. For neutral scalar particles, the equilibrium is described by the Bose–Einstein distribution function

$$f_S = \frac{g_S}{(2\pi)^3} \left[\exp\left(p_\lambda \beta^\lambda\right) - 1\right]^{-1}, \tag{4.2}$$

where g_S is the number of bosonic degrees of freedom and $\beta^\mu = u^\mu/T$ is the four-temperature, with T the local temperature and u^μ the four-velocity. For simplicity, we do not include a chemical potential in the scalar case. The Fermi–Dirac distribution function, including a local chemical potential μ is

$$f_F = \frac{g_F}{(2\pi)^3} \left[\exp\left(p_\lambda \beta^\lambda - \mu/T\right) + 1 \right]^{-1}, \tag{4.3}$$

where g_F is a degeneracy factor taking into account internal degrees of freedom, such as spin and colour charge.

GTE is achieved when the distribution function (4.2), (4.3) satisfies the Boltzmann equation (4.1). The fluid can be in GTE only when

$$\partial_\lambda (\mu/T) = 0, \qquad \partial_\lambda \beta_\kappa + \partial_\kappa \beta_\lambda = 0. \tag{4.4}$$

The first equality implies that, in the fermion case, the chemical potential is proportional to the temperature. The second equation requires that the four-temperature β^μ is a Killing vector. For Minkowski space-time, the general solution of the Killing equation allows β_μ to be written in the form:

$$\beta_\mu = b_\mu + \varpi_{\mu\nu} x^\nu, \tag{4.5}$$

where the four-vector b^μ and the thermal vorticity tensor $\varpi_{\mu\nu} = -\frac{1}{2}(\partial_\mu \beta_\nu - \partial_\nu \beta_\mu)$ are constants in GTE.

In order to describe a state of rigid rotation with angular velocity $\boldsymbol{\Omega} = \Omega \boldsymbol{k}$ about the z-axis, the constants appearing in (4.5) can be taken to be

$$b^\mu = T_0^{-1} \delta^\mu{}_0, \qquad \varpi_{\mu\nu} = \Omega T_0^{-1} \left(\eta_{\mu x} \eta_{\nu y} - \eta_{\mu y} \eta_{\nu x} \right), \tag{4.6}$$

where $\eta_{\mu\nu} = \mathrm{diag}(1, -1, -1, -1)$ is the usual Minkowski metric. These values correspond to the four-temperature $\beta^\mu = T_0^{-1}(1, -\Omega y, \Omega x, 0)$, where the physical interpretation of the constant T_0 is discussed below. Since the rigidly rotating state is invariant under rotations about the z-axis, it is convenient to employ cylindrical coordinates $x^\mu = (t, \rho, \phi, z)$ to refer to various vector or tensor components. Using the standard transformation formulae for vector components yields:

$$\beta^t = T_0^{-1}, \qquad \beta^\rho = 0, \qquad \beta^\phi = \Omega T_0^{-1}, \qquad \beta^z = 0. \tag{4.7}$$

In our later discussion, it will prove useful to express vector and tensor components of physical quantities relative to an orthonormal (non-holonomic) tetrad $\{e_{\hat{\alpha}}\}$ consisting of four mutually orthogonal vectors of unit norm, $e_{\hat{\alpha}} = e_{\hat{\alpha}}^\mu \partial_\mu$, defined as

$$e_{\hat{t}} = \partial_t, \qquad e_{\hat{\rho}} = \partial_\rho, \qquad e_{\hat{\phi}} = \rho^{-1} \partial_\phi, \qquad e_{\hat{z}} = \partial_z, \tag{4.8}$$

which satisfy the orthogonality relation:

$$g_{\mu\nu} e_{\hat{\alpha}}^\mu e_{\hat{\sigma}}^\nu = \eta_{\hat{\alpha}\hat{\sigma}}, \tag{4.9}$$

where $g_{\mu\nu} = \mathrm{diag}(1, -1, -\rho^2, -1)$ is the metric tensor of Minkowski space-time with respect to the cylindrical coordinates.

Writing the four-temperature (4.7) with respect to the tetrad (4.8) yields the tetrad components:

$$\beta^{\hat{\alpha}} = \eta^{\hat{\alpha}\hat{\sigma}} e^{\mu}_{\hat{\sigma}} \beta_{\mu} = T_0^{-1}(1, 0, \rho\Omega, 0). \tag{4.10}$$

The squared norm of the above expression can be obtained using either the coordinate components β^{μ} or the tetrad components $\beta^{\hat{\alpha}}$, as follows:

$$\beta^2 = g_{\mu\nu}\beta^{\mu}\beta^{\nu} = \eta_{\hat{\alpha}\hat{\sigma}}\beta^{\hat{\alpha}}\beta^{\hat{\sigma}} = T_0^{-2}(1 - \rho^2\Omega^2). \tag{4.11}$$

Since $\beta^{\hat{\alpha}} = u^{\hat{\alpha}}/T$ and the four-velocity $u^{\hat{\alpha}}$ has unit norm by definition, it can be seen that the quantity $\sqrt{\beta^2}$ is the local inverse temperature T^{-1}. For a rigidly rotating system, the four-velocity has the tetrad components:

$$u^{\hat{\alpha}} = \Gamma(1, 0, v^{\hat{\phi}}, 0), \tag{4.12}$$

where we find the following relations:

$$T = T_0\Gamma, \qquad v^{\hat{\phi}} = \rho\Omega, \qquad \Gamma = (1 - \rho^2\Omega^2)^{-1/2}, \tag{4.13}$$

where T is the local temperature. Equation (4.13) shows that T_0 is the temperature on the rotation axis and away from the axis the local temperature increases linearly with the Lorentz factor Γ characterizing the rigid rotation. Furthermore, we can readily identify the speed-of-light surface (SLS), which is the surface where the fluid rotates at the speed of light:

$$\rho_{\mathrm{SLS}} = \Omega^{-1}. \tag{4.14}$$

As expected, the Lorentz factor Γ diverges on the SLS, and so does the local temperature T. Starting from the velocity field in (4.12), it is possible to compute the kinematic vorticity[1], $\omega^{\hat{\alpha}} = \frac{1}{2}\varepsilon^{\hat{\alpha}\hat{\beta}\hat{\gamma}\hat{\sigma}} u_{\hat{\beta}}\nabla_{\hat{\gamma}}u_{\hat{\sigma}}$, acceleration, $a^{\hat{\alpha}} = u^{\hat{\beta}}\nabla_{\hat{\beta}}u^{\hat{\alpha}}$ and circular vector $\tau^{\hat{\alpha}} = -\varepsilon^{\hat{\alpha}\hat{\beta}\hat{\gamma}\hat{\sigma}}\omega_{\hat{\beta}}a_{\hat{\gamma}}u_{\hat{\sigma}}$ [15–18]:

$$\omega^{\hat{\alpha}} = \Gamma^2\Omega(0, 0, 0, 1), \quad a^{\hat{\alpha}} = -\rho\Gamma^2\Omega^2(0, 1, 0, 0), \quad \tau^{\hat{\alpha}} = -\rho\Omega^3\Gamma^5(\rho\Omega, 0, 1, 0). \tag{4.15}$$

---

[1] We use the convention that $\varepsilon^{\hat{0}\hat{1}\hat{2}\hat{3}} = \varepsilon^{\hat{t}\hat{\rho}\hat{\phi}\hat{z}} = 1$.

4.2.2 Macroscopic Quantities

At sufficiently high temperatures, pair production processes can occur. It is thus necessary to account for the presence of both particle and anti-particle species. We consider only the simplest model. For the neutral scalar field in thermal equilibrium, particles and anti-particles have the same distribution function f_S (4.2). Fermions and anti-fermions are distributed according to the Fermi–Dirac distribution (4.3) at the same temperature T and macroscopic velocity $u^{\hat{\alpha}}$, while the chemical potential is taken with the opposite sign for anti-particles:

$$f_{q/\bar{q}} = \frac{g_F}{(2\pi)^3} \left[\exp(p_\lambda \beta^\lambda \mp \mu/T) + 1 \right]^{-1}, \tag{4.16}$$

where f_q is the distribution for fermions and $f_{\bar{q}}$ that for anti-fermions. For the rigidly rotating system, the contraction of the four-temperature β^μ with the particle four-momentum is:

$$p_\lambda \beta^\lambda = T_0^{-1} \left[p^t - \boldsymbol{\Omega} \cdot (\boldsymbol{x} \times \boldsymbol{p}) \right] = T_0^{-1}(p^t - \Omega M^z) = T_0^{-1} \widetilde{p}^t, \tag{4.17}$$

where M^z denotes the z component of the angular momentum, and we have defined the co-rotating energy $\widetilde{p}^t$ by

$$\widetilde{p}^t = p^t - \Omega M^z. \tag{4.18}$$

We first consider the zero-temperature limit. From (4.2), it is clear that the scalar distribution function $f_S \to 0$ as $T_0 \to 0$, as expected. The situation is more complicated for the fermion distribution function (4.16), and depends on the sign of $p_\lambda \beta^\lambda \pm \mu/T$. Noting that $\mu/T = \mu_0/T_0$ (where μ_0 is the chemical potential on the axis of rotation) is a constant from (4.4), the zero-temperature limit of (4.16) is

$$\lim_{T_0 \to 0} f_{q/\bar{q}} = \frac{g_F}{(2\pi)^3} \Theta(\pm \mathcal{E}_F - \widetilde{p}^t), \tag{4.19}$$

where $\mathcal{E}_F = \mu_0$ is the Fermi level and Θ is the Heaviside step function, equal to one when its argument is positive and zero otherwise. Thus, the particle/anti-particle distributions have non-vanishing values only when $\widetilde{p}^t < \mu_0$ for particles and $\widetilde{p}^t < -\mu_0$ for anti-particles.

Starting from the distribution functions, we can define the SET $T_{S/F}^{\hat{\alpha}\hat{\sigma}}$ for either a scalar or fermion field as follows:

$$T_S^{\hat{\alpha}\hat{\sigma}} = \int \frac{d^3 p}{p^{\hat{t}}} p^{\hat{\alpha}} p^{\hat{\sigma}} f_S, \qquad T_F^{\hat{\alpha}\hat{\sigma}} = \int \frac{d^3 p}{p^{\hat{t}}} p^{\hat{\alpha}} p^{\hat{\sigma}} \left[f_q + f_{\bar{q}} \right]. \tag{4.20}$$

For the fermion field, we can also define the macroscopic CC $J^{\hat{\alpha}}$:

$$J^{\hat{\alpha}} = \int \frac{d^3 p}{p^{\hat{t}}} p^{\hat{\alpha}} [f_q - f_{\bar{q}}]. \tag{4.21}$$

By construction, $J^{\hat{\alpha}}$ and $T_{S/F}^{\hat{\alpha}\hat{\sigma}}$ are space-time tensors. Due to the structure of the scalar and Fermi–Dirac distributions, the free indices of these quantities can be carried only by the Minkowski metric tensor $\eta^{\hat{\alpha}\hat{\beta}}$ or the macroscopic velocity $u^{\hat{\alpha}}$. These simple considerations immediately imply the perfect fluid form for the CC and SET:

$$J^{\hat{\alpha}} = Q_F u^{\hat{\alpha}}, \qquad T_{S/F}^{\hat{\alpha}\hat{\sigma}} = (E_{S/F} + P_{S/F})u^{\hat{\alpha}}u^{\hat{\sigma}} - P_{S/F}\eta^{\hat{\alpha}\hat{\sigma}}, \qquad (4.22)$$

where Q_F is the fermion charge density, $E_{S/F}$ is the energy density and $P_{S/F}$ is the pressure. An expression can be obtained for Q_F by contracting $J_F^{\hat{\alpha}}$ with $u_{\hat{\alpha}}$. Similarly, $E_{S/F}$ is obtained by contracting $T_{S/F}^{\hat{\alpha}\hat{\sigma}}$ with $u_{\hat{\alpha}}u_{\hat{\sigma}}$, while a contraction of (4.22) with $\eta_{\hat{\alpha}\hat{\sigma}}$ yields the combination $E_{S/F} - 3P_{S/F}$ on the right-hand side. The above procedure applied to Q_F yields:

$$Q_F = \frac{g_F}{(2\pi)^3} \int \frac{d^3p}{p^{\hat{t}}} \left(u^\lambda p_\lambda\right) \left(\frac{1}{e^{(u^\lambda p_\lambda - \mu)/T} + 1} - \frac{1}{e^{(u^\lambda p_\lambda + \mu)/T} + 1}\right). \quad (4.23)$$

Taking advantage of the Lorentz invariance of the integration measure $d^3p/p^{\hat{t}}$, a Lorentz transformation can be performed on p^λ such that $p^\lambda u_\lambda = p^{\hat{t}}$. Switching to spherical coordinates in momentum space, the integral over the angular coordinates is straightforward and gives

$$Q_F = \frac{g_F}{2\pi^2} \int_0^\infty dp \, p^2 \left(\frac{1}{e^{\left(p^{\hat{t}} - \mu\right)/T} + 1} - \frac{1}{e^{\left(p^{\hat{t}} + \mu\right)/T} + 1}\right), \quad (4.24)$$

where $p = |\boldsymbol{p}|$ is the magnitude of the three-momentum. Similarly, we find, for the scalar field,

$$\begin{pmatrix} E_S \\ E_S - 3P_S \end{pmatrix} = \frac{g_S}{2\pi^2} \int_0^\infty \frac{p^2 dp}{p^{\hat{t}}} \begin{pmatrix} (p^{\hat{t}})^2 \\ M^2 \end{pmatrix} \frac{1}{e^{p^{\hat{t}}/T} - 1}, \quad (4.25)$$

while for the fermion field we have

$$\begin{pmatrix} E_F \\ E_F - 3P_F \end{pmatrix} = \frac{g_F}{2\pi^2} \int_0^\infty \frac{p^2 dp}{p^{\hat{t}}} \begin{pmatrix} (p^{\hat{t}})^2 \\ M^2 \end{pmatrix} \left(\frac{1}{e^{\left(p^{\hat{t}} - \mu\right)/T} + 1} + \frac{1}{e^{\left(p^{\hat{t}} + \mu\right)/T} + 1}\right). \quad (4.26)$$

Since the integrands above exhibit exponential decay at large values of p, they are amenable to numerical integration. The expressions (4.24), (4.25), (4.26) remain valid if the system is stationary rather than rotating, in which case $T = T_0$ and $\mu = \mu_0$ are constants.

In the massless limit, $p^{\hat{t}} = p$, $E_{S/F} = 3P_{S/F}$ and the integrals in (4.24), (4.25), (4.26) can be performed analytically [19], giving the charge density Q_F and pressures P_S and P_F as

$$Q_F = \frac{g_F \mu}{6} \left(T^2 + \frac{\mu^2}{\pi^2} \right), \quad P_S = \frac{\pi^2 g_S T^4}{90}, \quad P_F = \frac{7\pi^2 g_F T^4}{360} + \frac{g_F T^2 \mu^2}{12} + \frac{g_F \mu^4}{24\pi^2},$$

$$(4.27)$$

where $\mu = \mu_0 \Gamma$ and $T = T_0 \Gamma$. We also compute the massless limit of the ratio $T^\mu{}_\mu / M^2 = (E - 3P)/M^2$:

$$\lim_{M \to 0} \frac{E_S - 3P_S}{M^2} = \frac{g_S T^2}{12}, \quad \lim_{M \to 0} \frac{E_F - 3P_F}{M^2} = \frac{g_F T^2}{12} + \frac{g_F \mu^2}{4\pi^2}. \quad (4.28)$$

The Lorentz factor Γ (4.13) and thus μ and T diverge as $\rho \to \Omega^{-1}$ and the SLS is approached. Therefore, for massless particles, all macroscopic quantities are divergent on the SLS. Including the chemical potential does not alter the rate at which the quantities diverge, but does increase their values on the axis of rotation.

To understand the effect of the particle mass, the integrals in (4.24), (4.25), (4.26) are performed numerically. The resulting quantities depend on the angular speed Ω, the temperature on the axis T_0, the chemical potential on the axis μ_0, the particle mass M, the distance from the axis ρ and the numbers of degrees of freedom (dof) g_S, g_F. Here, we consider values of these parameters which are pertinent for the QGP formed in heavy-ion collisions. An analysis of the QGP fluid produced in accelerators indicates that it has the greatest vorticity of any fluid produced in a laboratory [9, 20], with $\hbar\Omega \simeq 6.6$ MeV, where $\hbar$ is the reduced Planck's constant. For this value of Ω, the SLS is located at $c/\Omega \simeq 30$ fm, roughly twice the size of a gold nucleus. For the temperature, we consider a typical value for heavy ion collisions of $k_B T_0 \simeq 0.2$ GeV [9], where k_B is the Boltzmann constant. In the relativistic collision of gold nuclei, a typical value of the chemical potential is $\mu_0 \simeq 0.1$ GeV [21]. For the particle mass Mc^2, we consider the pion mass (0.140 GeV), the ρ meson mass (0.775 GeV), the Λ^0-hyperon mass (1.116 GeV) and the Λ_c^+-charmed hyperon mass (2.286 GeV) [22].[2]

In Fig. 4.1, we plot the radial profile of the energy density E_F (4.26) as a function of ρ (left-hand-plot, linear scale) and as a function of the Lorentz factor Γ (4.13) (right-hand-plot, logarithmic scale). As expected, the energy diverges on the SLS for all values of the particle mass. The results for pions and massless particles are very nearly identical; for larger values of the mass, the energy E_F is lower everywhere. However, close to the SLS, the results for massive particles are indistinguishable from those for massless particles. Similar behaviour is observed for the pressure P_F and charge density Q_F [17, 18]. This is in agreement with the analytic work in the zero chemical potential case [23] (see also [24] for details of relevant techniques), where it was found that the $O(M^2)$ corrections due to the mass make subleading contributions as the SLS is approached.

We now consider more closely the effect of varying the particle mass for both scalars and fermions. To make the comparison relevant, we consider the energy density per particle degree of freedom, which amounts to dividing E_F by $2g_F$ (the

---

[2]Note that, since mesons are bosons, the Fermi–Dirac statistics cannot be strictly applied.

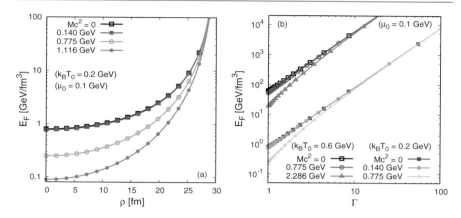

Fig. 4.1 **a** Numerical results for the energy density E_F (4.26), in GeV/fm^3 at $\mu_0 = 0.1$ GeV and $k_B T_0 = 0.2$ GeV, for $Mc^2 = 0$, 0.14 GeV, 0.775 GeV and 1.116 GeV. **b** Log-log plot of E_F, at two temperatures ($k_B T_0 = 0.2$ GeV and 0.6 GeV), for various masses. The number of degrees of freedom was set to $g_F = 6$

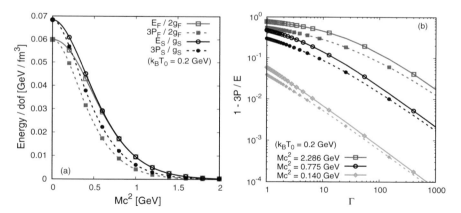

Fig. 4.2 **a** The mass dependence of the energy (continuous lines and empty symbols) and pressure (dashed lines and filled symbols) per dof, computed for the Fermi–Dirac (squares) and Bose–Einstein (circle) statistics. **b** The dependence on the Lorentz factor Γ (4.13) of the quantity $1 - 3P/E$, evaluated for the Fermi–Dirac (continuous lines and empty symbols) and Bose–Einstein (dashed lines and filled symbols) statistics, for various values of the particle mass

factor of two is required since the particle and anti-particle states are explicitly taken into account) and E_S by g_S. Furthermore, we consider the case of vanishing chemical potential, $\mu_0 = 0$, since we have not introduced this quantity for scalars.

Figure 4.2a shows the effect of the particle mass on the energy density and pressure on the rotation axis. For both scalars and fermions, these quantities decrease as the particle mass increases. In Fig. 4.2b, we plot the quantity $1 - 3P/E$, which vanishes in the massless limit. For a constant value of the Lorentz factor Γ, we see that

$1 - 3P/E$ increases as the mass is increased, thus the ratio $3P/E$ decreases. As the SLS is approached, $1 - 3P/E$ decreases, showing that in the vicinity of the SLS, the gas behaves as though its constituents were massless.

4.3   Quantum Rigidly Rotating Thermal States

We now consider the generalization from RKT to QFT, and examine how rigidly rotating quantum states may be defined. In the quantization process, the microscopic momenta are promoted to quantum operators. Thermal states at a temperature T_0 are defined such that the t.e.v. of an operator $\widehat{A}$ takes the form [25]:

$$\langle \widehat{A} \rangle_{T_0} = Z^{-1} \text{Tr}(\widehat{\rho}\widehat{A}), \tag{4.29}$$

where $Z = \text{Tr}\widehat{\rho}$ is the partition function and $\widehat{\rho}$ is the Boltzmann factor, which we define below. The trace is performed over Fock space, that is, the space of all states of the quantum field containing n particles (or anti-particles), for $n = 0, 1, 2, \ldots$.

For a rigidly rotating state with temperature T_0 on the axis of rotation, the Boltzmann factor for a scalar field is given by [26]

$$\widehat{\rho}_{\text{S}} = \exp\left[-(\widehat{H}_{\text{S}} - \Omega\widehat{M}_{\text{S}}^{\widetilde{z}})/T_0\right], \tag{4.30}$$

where $\widehat{H}_{\text{S}}$ is the scalar Hamiltonian operator and $\widehat{M}_{\text{S}}^{z}$ is the z-component of the scalar angular momentum operator. For a fermion field, we include a chemical potential μ_0 on the axis of rotation, which is conjugate to the charge operator. The Boltzmann factor for a fermion field is then given by [26]

$$\widehat{\rho}_{\text{F}} = \exp\left[-(\widehat{H}_{\text{F}} - \Omega\widehat{M}_{\text{F}}^{z} - \mu_0\widehat{Q}_{\text{F}})/T_0\right], \tag{4.31}$$

where $\widehat{H}_{\text{F}}$ is the fermion Hamiltonian operator, $\widehat{M}_{\text{F}}^{z}$ is the z-component of the total fermion angular momentum operator and $\widehat{Q}_{\text{F}}$ is the fermion charge operator.

In order to perform the trace over Fock space in (4.29), we need to define particle creation and annihilation operators acting on the states. For a neutral scalar field, we denote the particle annihilation operators by $\hat{a}_j$, where j labels the quantum properties of the annihilated particle. For a fermion field, the operators $\hat{b}_j$ annihilate fermions, while the $\hat{d}_j$ operators annihilate anti-fermions. In all cases, the adjoint operators are the corresponding particle creation operators. For scalars, the particle creation and annihilation operators satisfy the canonical commutation relations

$$[\hat{a}_j, \hat{a}_{j'}^{\dagger}] = \hat{a}_j\hat{a}_{j'}^{\dagger} - \hat{a}_{j'}^{\dagger}\hat{a}_j = \delta_{j,j'}, \qquad [\hat{a}_j, \hat{a}_{j'}] = 0 = [\hat{a}_j^{\dagger}, \hat{a}_{j'}^{\dagger}], \tag{4.32}$$

where $\delta_{j,j'}$ vanishes unless the labels j and j' are identical. For fermions, canonical anti-commutation relations hold, so that, for the particle operators:

$$\{\hat{b}_j, \hat{b}^\dagger_{j'}\} = \hat{b}_j \hat{b}^\dagger_{j'} + \hat{b}^\dagger_{j'} \hat{b}_j = \delta_{j,j'}, \qquad \{\hat{b}_j, \hat{b}_{j'}\} = 0 = \{\hat{b}^\dagger_j, \hat{b}^\dagger_{j'}\}, \qquad (4.33)$$

and similar relations hold for the anti-particle operators.

Using the particle/anti-particle states corresponding to the above creation and annihilation operators, we consider a quantization which is compatible with the operator $\hat{\rho}$, so that, for scalars:

$$\hat{\rho}_S \hat{a}^\dagger_j (\hat{\rho}_S)^{-1} = e^{-(E_j - \Omega m_j)/T_0} \hat{a}^\dagger_j, \qquad (4.34)$$

where E_j is the energy of the created particle, and $m_j = 0, \pm 1, \pm 2, \ldots$ is the z-component of the angular momentum. Similarly, for fermions we assume that

$$\hat{\rho}_F \hat{b}^\dagger_j (\hat{\rho}_F)^{-1} = e^{-(E_j - \Omega m_j - \mu_0)/T_0} \hat{b}^\dagger_j, \qquad \hat{\rho}_F \hat{d}^\dagger_j (\hat{\rho}_F)^{-1} = e^{-(E_j - \Omega m_j + \mu_0)/T_0} \hat{d}^\dagger_j, \qquad (4.35)$$

where $m_j = \pm\frac{1}{2}, \pm\frac{3}{2}, \ldots$ is the projection of the total fermion angular momentum on the z-axis. The quantities (4.34), (4.35) depend on the energy $\widetilde{E}_j$ of the particle as seen by a co-rotating observer:

$$\widetilde{E}_j = E_j - \Omega m_j. \qquad (4.36)$$

Using the canonical commutation/anti-commutation relations (4.32), (4.33), together with (4.34), (4.35), we find the t.e.v.s of the number operators for scalars to be [26,27]

$$\langle \hat{a}^\dagger_j \hat{a}_{j'} \rangle_{T_0} = \frac{\delta_{j,j'}}{\exp[\widetilde{E}_j/T_0] - 1}, \qquad (4.37)$$

while for fermions we have

$$\langle \hat{b}^\dagger_j \hat{b}_{j'} \rangle_{T_0} = \frac{\delta_{j,j'}}{\exp[(\widetilde{E}_j - \mu_0)/T_0] + 1}, \qquad \langle \hat{d}^\dagger_j \hat{d}_{j'} \rangle_{T_0} = \frac{\delta_{j,j'}}{\exp[(\widetilde{E}_j + \mu_0)/T_0] + 1}. \qquad (4.38)$$

The t.e.v.s (4.37), (4.38) have the expected Bose–Einstein/Fermi–Dirac thermal distributions in terms of the co-rotating energy $\widetilde{E}_j$.

Consider first the scalar field t.e.v. (4.37). This has the correct zero-temperature limit only if $\widetilde{E}_j > 0$. Even with this restriction, it can be seen that (4.37) diverges when $\widetilde{E}_j \to 0$, leading to the divergence of all t.e.v.s (4.29) [26,28]. From this, we deduce that rigidly rotating thermal states cannot be defined for a quantum scalar field on unbounded Minkowski space-time [26,28,29].

To understand this result, we consider how a quantum vacuum state is defined for a scalar field. In the canonical quantization approach to QFT, one starts with an orthonormal basis of scalar field modes φ_j which are solutions of the Klein–Gordon equation for a massive scalar field, $(\partial_\mu \partial^\mu + M^2) \varphi_j = 0$. The scalar field operator $\hat{\Phi}$ is then written as a sum over these field modes and their complex conjugates

$$\hat{\Phi} = \sum_j \left[\hat{a}_j \varphi_j + \hat{a}^\dagger_j \varphi^*_j \right], \qquad (4.39)$$

where the expansion coefficients are the particle creation and annihilation operators. In order that the creation and annihilation operators satisfy the canonical commutation relations (4.32), it must be the case that

$$\langle \varphi_j, \varphi_{j'} \rangle = \delta_{j,j'}, \qquad \langle \varphi_j^*, \varphi_{j'}^* \rangle = -\delta_{j,j'}, \qquad \langle \varphi_j, \varphi_{j'}^* \rangle = 0, \qquad (4.40)$$

where $\langle \, , \, \rangle$ is the Klein–Gordon inner product, defined for two solutions φ_j, $\varphi_{j'}$ of the Klein–Gordon equation by the following integral over a constant-t surface:

$$\langle \varphi_j, \varphi_{j'} \rangle = i \int d^3x \left(\varphi_j^* \partial^t \varphi_{j'} - \varphi_{j'} \partial^t \varphi_j^* \right). \qquad (4.41)$$

In particular, the modes φ_j corresponding to particles must have positive norm $\langle \varphi_j, \varphi_j \rangle$, while those modes φ_j^* corresponding to anti-particle modes must have negative norm. This restricts whether modes can be labelled as "particle" or "anti-particle". Calculating the inner product for a particle mode with energy E_j, we find

$$\langle \varphi_j, \varphi_{j'} \rangle = \frac{E_j}{|E_j|} \delta_{j,j'}, \qquad (4.42)$$

and hence the relations (4.40) hold only if the energy E_j of the mode φ_j is positive, $E_j > 0$ [30]. The vacuum state $|0\rangle$ is then defined as that state which is annihilated by the particle annihilation operators, $\hat{a}_j |0\rangle = 0$, and is simply the (stationary) Minkowski vacuum. For a quantum scalar field, it is not possible to make the choice $\widetilde{E}_j > 0$ because, for fixed $\widetilde{E}_j > 0$, there will be modes with sufficiently large and negative m_j for which $E_j = \widetilde{E}_j + \Omega m_j < 0$, so that (4.40) no longer holds and we do not have a valid quantization [30]. Since there is no rotating vacuum for a quantum scalar field, rotating thermal states for a quantum scalar field are also ill-defined.

One resolution of this difficulty is to insert a reflecting boundary inside the SLS [26,28]. The presence of the boundary means that the energy E_j of the scalar field modes is quantized, and, if the boundary is inside the SLS, it can be shown that $\widetilde{E}_j > 0$ for all m_j [28,31]. In this case, a rotating vacuum state (and also rotating thermal states) can be defined for a quantum scalar field [28].

In view of these difficulties for a quantum scalar field, for the rest of this chapter, we restrict our attention to a quantum fermion field on unbounded Minkowski space-time. First, we consider whether a rotating vacuum state can be defined in canonical quantization. Beginning with an orthonormal basis of particle mode solutions U_j and anti-particle mode solutions V_j of the Dirac equation (which will be discussed in more detail in the next section), the fermion field operator is written as

$$\widehat{\Psi} = \sum_j \left[\hat{b}_j U_j + \hat{d}_j^\dagger V_j \right], \qquad (4.43)$$

where the operators $\hat{b}_j$ and $\hat{d}_j$ satisfy the canonical anti-commutation relations (4.33). In contrast to the scalar field case, all particle and anti-particle modes U_j, V_j have positive Dirac norm, resulting in a greater freedom to label modes as "particle" or "anti-particle". This in turn leads to a greater freedom in how vacuum states (and therefore also thermal states) are defined [29].

One possible quantization is to define "particle" modes as having positive energy E_j [26]. As in the scalar case, the resulting vacuum is simply the usual (nonrotating) Minkowski vacuum state. However, for fermions there is another possibility [32]: particle modes can be defined by setting $\widetilde{E}_j > 0$. This leads to a well-defined quantization and a rotating vacuum state. Furthermore, with this definition the t.e.v.s (4.38) have the correct zero-temperature limit, with contributions only from modes below the Fermi level, for which $\widetilde{E}_j < \mu_0$ for particle modes and $\widetilde{E}_j < -\mu_0$ for anti-particle modes (we remind the reader that $\widetilde{E}_j > 0$ always holds when the rotating vacuum is employed). This is in agreement with the corresponding result (4.19) in the RKT case, and is sufficient to ensure that there are no temperature- and chemical potential-independent contributions to t.e.v.s.

4.4   Mode Solutions in Cylindrical Coordinates

Our purpose for the remainder of this chapter is to compute t.e.v.s of observables for a quantum fermion field of mass M, and compare the results with those for the RKT approach in Sect. 4.2.2. In this section, we lay the groundwork for our computation by considering in more detail the fermion mode solutions discussed schematically in the previous section. Since we are interested in rigidly rotating states, we work in cylindrical coordinates $x^\mu = (t, \rho, \phi, z)$ and follow the approach of [29,33].

The evolution of a free Dirac field with mass M is governed by the least-action principle, starting from the action:

$$S_F = i \int d^4x\, \mathcal{L}, \qquad \mathcal{L} = \frac{i}{2}\left(\overline{\psi}\,\slashed{\partial}\psi - \overline{\slashed{\partial}\psi}\,\psi\right) - M\overline{\psi}\psi, \qquad (4.44)$$

where the Feynman slash denotes contraction with the gamma matrices $\slashed{\partial} = \gamma^\mu \partial_\mu$. The gamma matrices satisfy the canonical anti-commutation relations $\{\gamma^\mu, \gamma^\nu\} = 2\eta^{\mu\nu}$ and in this chapter, we work with the Dirac representation:

$$\gamma^t = \begin{pmatrix} 1 & 0 \\ 0 & -1 \end{pmatrix}, \qquad \gamma^i = \begin{pmatrix} 0 & \sigma^i \\ -\sigma^i & 0 \end{pmatrix}, \qquad (4.45)$$

where the Pauli matrices are given by

$$\sigma^x = \begin{pmatrix} 0 & 1 \\ 1 & 0 \end{pmatrix}, \qquad \sigma^y = \begin{pmatrix} 0 & -i \\ i & 0 \end{pmatrix}, \qquad \sigma^z = \begin{pmatrix} 1 & 0 \\ 0 & -1 \end{pmatrix}. \qquad (4.46)$$

We are considering four-spinors ψ, which have Dirac adjoint $\overline{\psi} = \psi^\dagger \gamma^t$. Demanding that the variation of the action S_F (4.44) with respect to the $\overline{\psi}$ degree of freedom vanishes yields the Dirac equation

$$(i\slashed{\partial} - M)\psi = 0. \qquad (4.47)$$

As outlined in the previous section, in order to construct t.e.v.s, we first require a set of particle modes $\{U_j\}$ and anti-particle modes $\{V_j\}$ satisfying the Dirac equation (4.47). Given a particle mode U_j, the corresponding anti-particle mode V_j is related to U_j by the charge conjugation operation:

$$V_j = i\gamma^y U_j^*. \tag{4.48}$$

In deriving the formal expressions for rigidly rotating t.e.v.s in Sect. 4.3, we have assumed a quantization compatible with $\widehat{\rho}$, see (4.35). This requires that the following commutation relations must hold

$$[\widehat{H}_F, \hat{b}_j^\dagger] = E_j \hat{b}_j^\dagger, \qquad [\widehat{M}_F^z, \hat{b}_j^\dagger] = m_j \hat{b}_j^\dagger, \qquad [\widehat{Q}_F, \hat{b}_j^\dagger] = \hat{b}_j^\dagger,$$

$$[\widehat{H}_F, \hat{d}_j^\dagger] = E_j \hat{d}_j^\dagger, \qquad [\widehat{M}_F^z, \hat{d}_j^\dagger] = m_j \hat{d}_j^\dagger, \qquad [\widehat{Q}_F, \hat{d}_j^\dagger] = -\hat{d}_j^\dagger. \tag{4.49}$$

Taking into account the expression for the conserved operators in the classical Dirac field theory,

$$H_F = i\partial_t, \qquad M_F^z = -i\partial_\phi + S^z, \tag{4.50}$$

where the z-projection of the spin operator S^z is given by

$$S^z = \frac{1}{2}\begin{pmatrix} \sigma^z & 0 \\ 0 & \sigma^z \end{pmatrix}, \tag{4.51}$$

the particle mode solutions U_j must thus be chosen to be simultaneous eigenfunctions of H_F and M_F^z:

$$H_F U_j = E_j U_j, \qquad M_F^z U_j = m_j U_j. \tag{4.52}$$

The above eigenvalue equations are insufficient to specify the particle mode solutions uniquely. The remaining degrees of freedom can be fixed by choosing U_j to be eigenfunctions of the longitudinal momentum operator $P_F^z = -i\partial_z$ and of the helicity operator $W_0 = \boldsymbol{J} \cdot \boldsymbol{P}_F/2p$ (where p is the magnitude of the momentum):

$$P_F^z U_j = k_j U_j, \qquad W_0 U_j = \lambda_j U_j, \tag{4.53}$$

where k_j and λ_j are real constants. The expression for W_0 can be obtained as follows:

$$W_0 = \begin{pmatrix} h & 0 \\ 0 & h \end{pmatrix}, \qquad h = \frac{\boldsymbol{\sigma} \cdot \boldsymbol{P}_F}{2p} = \frac{1}{2p}\begin{pmatrix} P_F^z & P_- \\ P_+ & -P_F^z \end{pmatrix}, \tag{4.54}$$

where $\boldsymbol{P}_F = -i\nabla$, while $P_\pm$ are defined in terms of cylindrical coordinates as

$$P_\pm = P_F^x \pm i P_F^y = -ie^{\pm i\phi}(\partial_\rho \pm i\rho^{-1}\partial_\phi). \tag{4.55}$$

It can be shown that $W_0^2 = \frac{1}{4}$. The eigenvalues $\lambda_j = 1/2$ and $-1/2$ correspond to positive and negative helicity, respectively.

The Dirac equation (4.47) can be written with respect to the above operators as

$$\begin{pmatrix} H_F - M & -2ph \\ 2ph & -H_F - M \end{pmatrix} \psi = 0.$$ (4.56)

The operators H_F and P_F^z are diagonal with respect to the spinor structure, thus the corresponding eigenvalue equations can be solved immediately:

$$U_j = \frac{\mathcal{K}_j}{2\pi} e^{-iE_j t + ik_j z} u_j, \qquad u_j = \begin{pmatrix} C_j^- \varphi_j \\ C_j^+ \varphi_j, \end{pmatrix},$$ (4.57)

where u_j is a four-spinor which depends only on ϕ and ρ and $\mathcal{K}_j$ is a normalization constant. In (4.57), $C_j^{\pm}$ are integration constants and φ_j is a two-spinor satisfying the remaining two eigenvalue equations, namely

$$\left(-i\partial_\phi + \frac{1}{2}\sigma^z\right)\varphi_j = m_j\varphi_j, \qquad h\varphi_j = \lambda_j\varphi_j.$$ (4.58)

Substituting (4.57) into the Dirac equation (4.56) gives

$$\begin{pmatrix} E_j - M & -2p_j\lambda_j \\ 2p_j\lambda_j & -E_j - M \end{pmatrix} u_j = 0,$$ (4.59)

where the magnitude of the momentum is now p_j, and from this the following relation can be established for $C_j^{\pm}$:

$$C_j^- = \frac{2\lambda_j p_j}{E_j - M} C_j^+.$$ (4.60)

Next we consider the angular momentum equation, the first relation in (4.58), which allows φ_j to be written in the form:

$$\varphi_j = \begin{pmatrix} \varphi_j^- e^{i(m_j - \frac{1}{2})\phi} \\ \varphi_j^+ e^{i(m_j + \frac{1}{2})\phi} \end{pmatrix},$$ (4.61)

where $m_j = \pm\frac{1}{2}, \pm\frac{3}{2}, \ldots$ is an odd half-integer, while $\varphi_j^{\pm} \equiv \varphi_j^{\pm}(\rho)$ are functions which depend only on the radial coordinate ρ. Taking into account the result [from (4.55)] $P_+ P_- = P_- P_+ = -\partial_\rho^2 - \rho^{-1}\partial_\rho - \rho^{-2}\partial_\phi^2$, the second relation in (4.58) reduces to

$$\left[\rho^2 \frac{\partial^2}{\partial\rho^2} + \rho\frac{\partial}{\partial\rho} + q_j^2\rho^2 - \left(m_j \pm \frac{1}{2}\right)^2\right]\varphi_j^{\pm} = 0,$$ (4.62)

where the longitudinal momentum q_j is defined by

$$q_j = \sqrt{p_j^2 - k_j^2} = \sqrt{E_j^2 - k_j^2 - M^2}.$$ (4.63)

Equation (4.62) can readily be identified with the Bessel equation [34], having two linearly independent solutions $J_{m\pm1/2}(q\rho)$ and $Y_{m\pm1/2}(q\rho)$. Demanding regularity at the origin discards the Neumann function $Y_{m\pm1/2}(q\rho)$, and therefore

$$\varphi_j^\pm = \mathcal{N}_j^\pm J_{m_j\pm\frac{1}{2}}(q_j\rho). \tag{4.64}$$

The connection between the integration constants $\mathcal{N}_j^+$ and $\mathcal{N}_j^-$ can be established by noting that the operators $P_\pm$ act as ladder operators, in the sense that

$$P_\pm e^{i(m_j\mp\frac{1}{2})\phi} J_{m_j\mp\frac{1}{2}}(q_j\rho) = \pm iq_j e^{i(m_j\pm\frac{1}{2})\phi} J_{m_j\pm\frac{1}{2}}(q_j\rho), \tag{4.65}$$

where the following properties were employed [34]:

$$J'_{m_j+\frac{1}{2}}(q_j\rho) = J_{m_j-\frac{1}{2}}(q_j\rho) - \frac{m_j+\frac{1}{2}}{q_j\rho} J_{m_j+\frac{1}{2}}(q_j\rho),$$

$$J'_{m_j-\frac{1}{2}}(q_j\rho) = -J_{m_j+\frac{1}{2}}(q_j\rho) + \frac{m_j-\frac{1}{2}}{q_j\rho} J_{m_j-\frac{1}{2}}(q_j\rho). \tag{4.66}$$

The helicity equation [the second relation in (4.58)] then yields

$$\mathcal{N}_j^+ = \frac{iq_j}{k_j+2p_j\lambda_j} \mathcal{N}_j^- = 2i\lambda_j\frac{\mathfrak{p}_j^-}{\mathfrak{p}_j^+}\mathcal{N}_j^-, \tag{4.67}$$

with

$$\mathfrak{p}_j^\pm = \left(1 \pm \frac{2\lambda_jk_j}{p_j}\right)^{1/2}. \tag{4.68}$$

Noting that an overall normalization constant, $\mathcal{N}_j^-\sqrt{2}/\mathfrak{p}_j^+$, can be absorbed into $\mathcal{K}_j$ in (4.57), we write φ_j in the form:

$$\varphi_j = \frac{1}{\sqrt{2}}\begin{pmatrix} \mathfrak{p}_j^+ e^{i(m_j-\frac{1}{2})\phi} J_{m_j-\frac{1}{2}}(q_j\rho) \\ 2i\lambda_j\mathfrak{p}_j^- e^{i(m_j+\frac{1}{2})\phi} J_{m_j+\frac{1}{2}}(q_j\rho) \end{pmatrix}. \tag{4.69}$$

Introducing the angle ϑ_j made by the momentum vector with the z-direction, so that $k_j = p_j\cos\vartheta_j$ (with $0 \le \vartheta_j \le \pi$), it can be seen that

$$\frac{1}{\sqrt{2}}\mathfrak{p}_j^\pm = \left(\frac{1}{2} \pm \lambda_j\right)\cos\frac{\vartheta_j}{2} + \left(\frac{1}{2} \mp \lambda_j\right)\sin\frac{\vartheta_j}{2}. \tag{4.70}$$

Thus, the two-spinor φ_j can be written compactly as follows (where we have explicitly written out all the parameters on which this depends):

$$\varphi_{p,k,m}^{1/2} = \begin{pmatrix} \cos\frac{\vartheta}{2}e^{i(m-\frac{1}{2})\phi} J_{m-\frac{1}{2}}(q\rho) \\ i\sin\frac{\vartheta}{2}e^{i(m+\frac{1}{2})\phi} J_{m+\frac{1}{2}}(q\rho) \end{pmatrix}, \quad \varphi_{p,k,m}^{-1/2} = \begin{pmatrix} \sin\frac{\vartheta}{2}e^{i(m-\frac{1}{2})\phi} J_{m-\frac{1}{2}}(q\rho) \\ -i\cos\frac{\vartheta}{2}e^{i(m+\frac{1}{2})\phi} J_{m+\frac{1}{2}}(q\rho) \end{pmatrix}. \tag{4.71}$$

Using the identity

$$\sum_{n=-\infty}^{\infty} J_n^2(q_j\rho) = 1, \tag{4.72}$$

where the sum runs over all integers $n \in \mathbb{Z}$, it can be established that the two-spinors φ_j (4.71) satisfy the normalization condition

$$\sum_{m=-\infty}^{\infty} \varphi_{p,k,m}^{\lambda,\dagger} \varphi_{p,k,m}^{\lambda'} = \delta_{\lambda,\lambda'}, \tag{4.73}$$

We now return to the four-spinors u_j (4.57), for which we impose the normalization condition

$$\sum_{m=-\infty}^{\infty} u_{E,k,m}^{\lambda,\dagger} u_{E,k,m}^{\lambda'} = \delta_{\lambda,\lambda'}. \tag{4.74}$$

This can be achieved by setting $C_j^+ = (2\lambda_j E_j/|E_j|)\mathfrak{E}_j^-/\sqrt{2}$, such that

$$u_j = \frac{1}{\sqrt{2}} \begin{pmatrix} \mathfrak{E}_j^+ \varphi_j \\ \frac{2\lambda_j E_j}{|E_j|} \mathfrak{E}_j^- \varphi_j \end{pmatrix}, \qquad \mathfrak{E}_j^{\pm} = \left(1 \pm \frac{M}{E_j}\right)^{1/2}, \tag{4.75}$$

where $E_j/|E_j|$ is the sign of E_j.

The final piece of the puzzle is to establish unit norm for the modes U_j (4.57). This is achieved using the Dirac inner product, defined for two solutions ψ and χ of the Dirac equation (4.47) by

$$\langle \psi, \chi \rangle = \int d^3x \, \overline{\psi} \gamma^t \chi, \tag{4.76}$$

where the integration is taken over a constant-t surface. Performing the integral with respect to cylindrical coordinates and using the relation

$$\int_0^{\infty} d\rho \, \rho \, J_{m+\frac{1}{2}}(q_j\rho) J_{m+\frac{1}{2}}(q_{j'}\rho) = \frac{\delta(q_j - q_{j'})}{q_j}, \tag{4.77}$$

it can be seen that, with $\mathcal{K}_j = 1$, we have the required normalization condition

$$\begin{aligned} \langle U_j, U_{j'} \rangle &= \delta_{\lambda_j, \lambda_{j'}} \delta_{m_j, m_{j'}} \delta(k_j - k_{j'}) \frac{\delta(q_j - q_{j'})}{q_j} \theta(E_j E_{j'}) \\ &= \delta_{\lambda_j, \lambda_{j'}} \delta_{m_j, m_{j'}} \delta(k_j - k_{j'}) \frac{\delta(E_j - E_{j'})}{|E_j|}. \end{aligned} \tag{4.78}$$

We therefore write the particle modes U_j as

$$U^\lambda_{E,k,m} = \frac{e^{-iEt+ikz}}{2\pi} u^\lambda_{E,k,m}, \qquad u^\lambda_{E,k,m} = \frac{1}{\sqrt{2}} \begin{pmatrix} \mathfrak{C}^+ \varphi^\lambda_{p,k,m} \\ \frac{2\lambda E}{|E|} \mathfrak{C}^- \varphi^\lambda_{p,k,m} \end{pmatrix}. \qquad (4.79)$$

The four-spinors V_j corresponding to the anti-particle modes are then obtained via the charge conjugation operation (4.48):

$$V^\lambda_{E,k,m} = \frac{e^{iEt-ikz}}{2\pi} v^\lambda_{E,k,m}, \qquad v^\lambda_{E,k,m} = \frac{(-1)^{m-\frac{1}{2}}}{\sqrt{2}} \frac{iE}{|E|} \begin{pmatrix} \mathfrak{C}^- \varphi^\lambda_{p,-k,-m} \\ -\frac{2\lambda E}{|E|} \mathfrak{C}^+ \varphi^\lambda_{p,-k,-m} \end{pmatrix}. \qquad (4.80)$$

The two-spinor $\varphi^\lambda_{p,k,m}$ is defined in (4.69), and also in (4.71) in terms of the angle ϑ between the momentum vector and the z-axis. Due to the relationship (4.48) between the particle and anti-particle modes, the anti-particle modes V_j also satisfy the normalization condition (4.78). In particular, anti-particle modes, like particle modes, have positive Dirac norm. As discussed in the previous section, this is crucial for the definition of rigidly rotating quantum states for fermions.

4.5   Quantum Stationary Thermal Expectation Values

With a complete orthonormal basis of fermion modes constructed in the previous section, we are now in a position to compute t.e.v.s of physical quantities. While our primary interest is in rigidly rotating states, we first study the t.e.v.s for stationary, nonrotating states with vanishing angular speed Ω.

At the level of the classical field theory, the CC J^μ and SET $T^{\mu\nu}$ can be constructed using Noether's theorem [27]:

$$J^\mu = \overline{\psi}\gamma^\mu\psi, \qquad T_{\mu\nu} = \frac{i}{2}\left[\overline{\psi}\gamma_{(\mu}\partial_{\nu)}\psi - \partial_{(\mu}\overline{\psi}\gamma_{\nu)}\psi\right]. \qquad (4.81)$$

The trace of the SET is proportional to the FC $\overline{\psi}\psi$:

$$T^\mu{}_\mu = M\overline{\psi}\psi. \qquad (4.82)$$

The generalization to QFT is made by replacing the classical field ψ with the corresponding quantum operator, $\widehat{\Psi}$. Due to the anti-commutation relations (4.33) satisfied by the quantum operators, there is an ambiguity in the ordering of the action of the quantum operators on the Fock space states. For operators which are quadratic in the field operators, such as those arising from (4.81), (4.82), and since we are working on flat space-time, this ambiguity can be overcome by introducing normal ordering, a procedure by which the vacuum expectation value (v.e.v.) is subtracted from the operator itself. For an operator $\widehat{A}$, the normal-ordered operator $: \widehat{A} :$ is therefore defined to be

$$: \widehat{A} := \widehat{A} - \langle 0|\widehat{A}|0\rangle. \qquad (4.83)$$

Inserting the schematic mode expansion (4.43) in (4.81), the following expressions are obtained

$$: \widehat{\overline{\Psi}\Psi} := \sum_{j,j'} \left[\hat{b}_j^\dagger \hat{b}_{j'} \overline{U}_j U_{j'} - \hat{d}_{j'}^\dagger \hat{d}_j \overline{V}_j V_{j'} \right],$$

$$: \widehat{J}^\mu := \sum_{j,j'} \left[\hat{b}_j^\dagger \hat{b}_{j'} \mathfrak{J}^\mu(U_j, U_{j'}) - \hat{d}_{j'}^\dagger \hat{d}_j \mathfrak{J}^\mu(V_j, V_{j'}) \right],$$

$$: \widehat{T}_{\mu\nu} := \sum_{j,j'} \left[\hat{b}_j^\dagger \hat{b}_{j'} \mathcal{T}_{\mu\nu}(U_j, U_{j'}) - \hat{d}_{j'}^\dagger \hat{d}_j \mathcal{T}_{\mu\nu}(V_j, V_{j'}) \right], \qquad (4.84)$$

where we have introduced the sesquilinear forms $\mathfrak{J}^\mu(\psi, \chi)$ and $\mathcal{T}_{\mu\nu}(\psi, \chi)$ for notational brevity, based on the classical quantities (4.81):

$$\mathfrak{J}^\mu(\psi, \chi) = \overline{\psi}\gamma^\mu\chi, \qquad \mathcal{T}_{\mu\nu}(\psi, \chi) = \frac{i}{2}\left[\overline{\psi}\gamma_{(\mu}\partial_{\nu)}\chi - \partial_{(\mu}\overline{\psi}\gamma_{\nu)}\chi \right]. \qquad (4.85)$$

As discussed in Sect. 4.3, the nonrotating Minkowski vacuum is defined by taking all modes corresponding to the positive eigenvalues of the Hamiltonian ($E_j > 0$) as particle modes. This leads to the following decomposition of the field operator:

$$\widehat{\Psi} = \sum_{\lambda=\pm\frac{1}{2}} \sum_{m=-\infty}^{\infty} \int_M^\infty dE\, E \int_{-p}^p dk \left[\hat{b}_{E,k,m}^\lambda U_{E,k,m}^\lambda + \hat{d}_{E,k,m}^\lambda{}^\dagger V_{E,k,m}^\lambda \right], \quad (4.86)$$

where the spinor modes are given by (4.79), (4.80). Substituting the mode expansion (4.86) into (4.84), and using the relations (4.38), we find the following t.e.v.s for a stationary (nonrotating) state at temperature T_0:

$$\langle : \widehat{\overline{\Psi}\Psi} : \rangle_{T_0} = \sum_j \left\{ \frac{\overline{U}_j U_j}{\exp[(E_j - \mu_0)/T_0] + 1} - \frac{\overline{V}_j V_j}{\exp[(E_j + \mu_0)/T_0] + 1} \right\},$$

$$\langle : \widehat{J}^\mu : \rangle_{T_0} = \sum_j \left\{ \frac{\mathfrak{J}^\mu(U_j, U_j)}{\exp[(E_j - \mu_0)/T_0] + 1} - \frac{\mathfrak{J}^\mu(V_j, V_j)}{\exp[(E_j + \mu_0)/T_0] + 1} \right\},$$

$$\langle : \widehat{T}_{\mu\nu} : \rangle_{T_0} = \sum_j \left\{ \frac{\mathcal{T}_{\mu\nu}(U_j, U_j)}{\exp[(E_j - \mu_0)/T_0] + 1} - \frac{\mathcal{T}_{\mu\nu}(V_j, V_j)}{\exp[(E_j + \mu_0)/T_0] + 1} \right\}, \quad (4.87)$$

where $\widetilde{E}_j = E_j$ in the case when $\Omega = 0$.

4.5.1 Fermion Condensate

Using the charge conjugation property (4.48), it can be shown that

$$\overline{V}_j V_j = \overline{U}_j^* \gamma^y \gamma^y U_j^* = -(\overline{U}_j U_j)^*, \tag{4.88}$$

since $(\gamma^y)^2 = -1$. Using the spinor mode (4.79), we have

$$\overline{U}_j U_j = \frac{M}{8\pi^2 E_j} \left[J_{m_j}^+(q_j \rho) + \frac{2\lambda_j k_j}{p_j} J_{m_j}^-(q_j \rho) \right], \tag{4.89}$$

where we define (we will need $J_m^\times(q\rho)$ later)

$$J_m^\pm(q\rho) = J_{m-\frac{1}{2}}^2(q\rho) \pm J_{m+\frac{1}{2}}^2(q\rho), \qquad J_m^\times(q\rho) = 2 J_{m-\frac{1}{2}}(q\rho) J_{m+\frac{1}{2}}(q\rho). \tag{4.90}$$

Since $\overline{U}_j U_j$ is a real scalar, it can be seen that $\overline{V}_j V_j = -\overline{U}_j U_j$. Furthermore, noting that the term proportional to λ_j in (4.89) makes a vanishing contribution under the summation with respect to λ_j, the t.e.v. of the FC, given in the first line of (4.87), is

$$\langle : \widehat{\overline{\Psi}}\widehat{\Psi} : \rangle_{T_0} = \frac{M}{4\pi^2} \sum_{m=-\infty}^{\infty} \int_M^\infty dE \left[\frac{1}{e^{(E-\mu_0)/T_0}+1} + \frac{1}{e^{(E+\mu_0)/T_0}+1} \right] \int_{-p}^{p} dk J_m^+(q\rho). \tag{4.91}$$

Taking into account the identity (4.72), the sum over m can be performed:

$$\sum_{m=-\infty}^{\infty} J_m^+(q\rho) = 2 \sum_{n=-\infty}^{\infty} J_n(q\rho) = 2, \tag{4.92}$$

where $m = \pm\frac{1}{2}, \pm\frac{3}{2}, \ldots$, while $n = 0, \pm 1, \pm 2, \ldots$. After performing the sum over m in (4.91), the integration variable can be changed from E to p, giving

$$\langle : \widehat{\overline{\Psi}}\widehat{\Psi} : \rangle_{T_0} = \frac{M}{\pi^2} \int_0^\infty \frac{dp\, p^2}{E} \left[\frac{1}{e^{(E-\mu_0)/T_0}+1} + \frac{1}{e^{(E+\mu_0)/T_0}+1} \right]. \tag{4.93}$$

The above expression coincides with that for $(E_F - 3P_F)/M$ (4.26), obtained in RKT with $g_F = 2$ (taking into account the fermion helicities) and $\Omega = 0$. Thus, the FC has no corrections in the QFT setting compared to its RKT counterpart.

4.5.2 Charge Current

The charge conjugation property (4.48) can be used to show that

$$\mathfrak{J}^\mu(V_j, V_j) = \overline{U}_j^* \gamma^y \gamma^\mu \gamma^y U_j^* = (\overline{U}_j \gamma^\mu U_j)^* = [\mathfrak{J}^\mu(U_j, U_{j'})]^*, \tag{4.94}$$

where, as well as (4.88), the properties $\gamma^y \gamma^\mu = 2\eta^{y\mu} - \gamma^\mu \gamma^y$ and $(\gamma^y)^* = -\gamma^y$ were used. Thus, it is sufficient to compute $\mathfrak{J}^\mu(U_j, U_j)$. Substituting $\mu = t$ and $\mu = i$ for the index μ, we find

$$\mathfrak{J}^t(U_j, U_j) = \frac{1}{4\pi^2}\varphi_j^\dagger \varphi_j, \qquad \mathfrak{J}^i(U_j, U_j) = \frac{1}{4\pi^2}\frac{2\lambda_j p_j}{E_j}\varphi_j^\dagger \sigma^i \varphi_j. \qquad (4.95)$$

It is convenient to work with components taken with respect to the tetrad introduced in (4.8). The sigma matrices constructed with respect to this tetrad are

$$\sigma^{\hat\rho} = \begin{pmatrix} 0 & e^{-i\phi} \\ e^{i\phi} & 0 \end{pmatrix}, \qquad \sigma^{\hat\phi} = \begin{pmatrix} 0 & -ie^{-i\phi} \\ ie^{i\phi} & 0 \end{pmatrix}. \qquad (4.96)$$

The following relations can be established:

$$\varphi_j^\dagger \varphi_j = \frac{1}{2}J_{m_j}^+(q_j\rho) + \frac{\lambda_j k_j}{p_j}J_{m_j}^-(q_j\rho), \qquad \varphi_j^\dagger \sigma^{\hat\rho}\varphi_j = 0,$$

$$\varphi_j^\dagger \sigma^{\hat z}\varphi_j = \frac{1}{2}J_{m_j}^-(q_j\rho) + \frac{\lambda_j k_j}{p_j}J_{m_j}^+(q_j\rho), \qquad \varphi_j^\dagger \sigma^{\hat\phi}\varphi_j = \frac{\lambda_j q_j}{p_j}J_{m_j}^\times(q_j\rho), \qquad (4.97)$$

where the functions $J_m^\pm(q\rho)$ and $J_m^\times(q\rho)$ were introduced in (4.90).

Noting that the density of states factors $[e^{(E\pm\mu_0)/T_0} + 1]^{-1}$ are invariant under the transformation $k \to -k$, $\lambda \to -\lambda$ and $m \to -m$, it can be seen that the spatial components of J^μ vanish. This is because $J_m^-(q\rho)$ and $J_m^\times(q\rho)$ are odd with respect to $m \to -m$, while $\varphi_j^\dagger \sigma^{\hat z}\varphi_j$ is odd under the transformation $(k, m) \to (-k, -m)$. The time component of the CC can then be written as:

$$\langle : \hat{J}^{\hat t} : \rangle_{T_0} = \frac{1}{4\pi^2}\sum_{m=-\infty}^{\infty}\int_M^\infty dE\, E\left[\frac{1}{e^{(E-\mu_0)/T_0}+1} - \frac{1}{e^{(E+\mu_0)/T_0}+1}\right]\int_{-p}^p dk\, J_m^+(q\rho).$$

$$(4.98)$$

After performing the sum over m using (4.92), an angle ϑ can be introduced such that $k = p\cos\vartheta$ and $q = p\sin\vartheta$. The integration measure $E\,dE\,dk = q\,dq\,dk$ is then changed to $p^2\sin\vartheta\,d\vartheta\,dp$. Since, after the sum over m is performed, the integrand is independent of ϑ, the integration with respect to this variable can be performed automatically, yielding $\int_0^\pi d\vartheta\,\sin\vartheta = 2$. Thus $\langle : J^{\hat t} : \rangle_{T_0}$ reduces to

$$\langle : \hat{J}^{\hat t} : \rangle_{T_0} = \frac{1}{\pi^2}\int_0^\infty dp\, p^2\left[\frac{1}{e^{(E-\mu_0)/T_0}+1} - \frac{1}{e^{(E+\mu_0)/T_0}+1}\right]. \qquad (4.99)$$

As was the case for the FC, the above expression coincides with the fermion charge density Q_F (4.24) obtained using RKT with $g_F = 2$ and $\Omega = 0$, showing that there are no quantum corrections.

4.5.3 Stress-Energy Tensor

In a manner similar to the one employed to derive (4.94), it can be shown that $\mathcal{T}_{\mu\nu}(V_j, V_j)$ can be related to $\mathcal{T}_{\mu\nu}(U_j, U_j)$ via:

$$\mathcal{T}_{\mu\nu}(V_j, V_j) = -\frac{i}{2}[\overline{U}_j \gamma_{(\mu}\partial_{\nu)}U_j - \partial_{(\mu}\overline{U}_j\gamma_{\nu)}U_j]^* = -[\mathcal{T}_{\mu\nu}(U_j, U_j)]^*, \quad (4.100)$$

where the last $-$ sign comes from the complex conjugate of the imaginary unit i prefactor. Using the properties (4.66) of the Bessel functions, we can derive the following relations:

$$\varphi_j^\dagger \sigma^{\hat{\rho}}\partial_\rho\varphi_j = \frac{iq_j^2\lambda_j}{p_j}\left[J_{m_j}^+(q_j\rho) - \frac{m_j}{q_j\rho}J_{m_j}^\times(q_j\rho)\right],$$

$$\varphi_j^\dagger \sigma^{\hat{\phi}}\partial_\phi\varphi_j = \frac{im_j q_j\lambda_j}{p_j}J_{m_j}^\times(q_j\rho). \quad (4.101)$$

For stationary states, all off-diagonal tetrad components of the SET vanish. However, when we consider rigidly rotating states in the next section, the component $T_{\hat{t}\hat{\phi}}$ will be nonzero. We therefore write down the diagonal tetrad components and the $(\hat{t}, \hat{\phi})$ component which we will require later:

$$\mathcal{T}_{\hat{t}\hat{t}}(U_j, U_j) = \frac{E_j}{8\pi^2}\left[J_{m_j}^+(q_j\rho) + \frac{2\lambda_j k_j}{p_j}J_{m_j}^-(q_j\rho)\right],$$

$$\mathcal{T}_{\hat{t}\hat{\phi}}(U_j, U_j) = -\frac{1}{16\pi^2\rho}\left[\left(m_j - \frac{\lambda_j k_j}{p_j}\right)J_{m_j}^+(q_j\rho) - \left(\frac{1}{2} - \frac{2\lambda_j k_j m_j}{p_j}\right)J_{m_j}^-(q_j\rho)\right]$$
$$- \frac{q_j}{16\pi^2}J_{m_j}^\times(q_j\rho),$$

$$\mathcal{T}_{\hat{\rho}\hat{\rho}}(U_j, U_j) = \frac{q_j^2}{8\pi^2 E_j}\left[J_{m_j}^+(q_j\rho) - \frac{m_j}{q_j\rho}J_{m_j}^\times(q_j\rho)\right],$$

$$\mathcal{T}_{\hat{\phi}\hat{\phi}}(U_j, U_j) = \frac{q_j m_j}{8\pi^2 E_j\rho}J_{m_j}^\times(q_j\rho),$$

$$\mathcal{T}_{\hat{z}\hat{z}}(U_j, U_j) = \frac{k_j^2}{8\pi^2 E_j}J_{m_j}^+(q_j\rho) + \frac{\lambda_j k_j p_j}{4\pi^2 E_j}J_{m_j}^-(q_j\rho). \quad (4.102)$$

Using the summation formula,

$$\sum_{m=-\infty}^{\infty} m J_m^\times(q\rho) = \sum_{n=-\infty}^{\infty}(2n+1)J_n(q\rho)J_{n+1}(q\rho) = 1, \quad (4.103)$$

where, as before, $m = \pm\frac{1}{2}, \pm\frac{3}{2}, \ldots$, while $n = 0, \pm 1, \pm 2, \ldots$, it can be shown that the t.e.v. of the SET for nonrotating states has the simple diagonal form

$$\langle : \widehat{T}_{\hat{\alpha}\hat{\sigma}} : \rangle_{T_0} = \text{diag}(E_F, P_F, P_F, P_F), \tag{4.104}$$

where E_F and P_F were obtained in (4.26) using the RKT formulation with $g_F = 2$. Therefore, there are no quantum corrections to t.e.v.s for stationary states.

4.6 Quantum Rigidly Rotating Thermal Expectation Values

In the previous section, the construction of stationary thermal states was based on the nonrotating Minkowski vacuum, defined by setting the energy $E_j > 0$ for particle modes. When the rotation is switched on, as discussed in Sect. 4.3, we can define a rotating vacuum for fermions by instead setting the co-rotating energy $\widetilde{E}_j > 0$ (4.36) to be positive for particle modes [32]. We therefore define the fermion field operator as follows:

$$\widehat{\Psi} = \sum_{\lambda=\pm\frac{1}{2}} \sum_{m=-\infty}^{\infty} \int_{|E|>M} dE\, |E| \int_{-p}^{p} dk\, \Theta(\widetilde{E})$$
$$\times \left[\hat{b}_{E,k,m}^{\lambda} U_{E,k,m}^{\lambda}(x) + \hat{d}_{E,k,m}^{\lambda}{}^{\dagger} V_{E,k,m}^{\lambda}(x) \right], \tag{4.105}$$

where the particle spinors $U_{E,k,m}^{\lambda}$ and anti-particle spinors $V_{E,k,m}^{\lambda}$ can be found in (4.79), (4.80) respectively. The field operator (4.105) should be compared with the corresponding definition (4.86) for the stationary case. In (4.86) the integral over E involves only positive energy $E > 0$, whereas in (4.105) we also take into account negative energy modes, provided that the mass shell condition $|E| > M$ is satisfied. Instead, the requirement that the co-rotating energy is positive, $\widetilde{E} > 0$, is imposed by the presence of the Heaviside step function $\Theta(\widetilde{E})$.

With the decomposition (4.105) of the fermion field operator, we can proceed to construct t.e.v.s using the method employed in Sect. 4.5 in the stationary case. The mode expansion (4.105) is inserted into the FC, CC and SET operators (4.84), to obtain mode sums involving the particle and anti-particle creation and annihilation operators. The t.e.v.s of the particle number operators are then given by (4.38), where the temperature on the axis of rotation is fixed to be T_0. The density of states factor in (4.38) now has a dependence on the angular momentum quantum number m_j as well as the energy E_j. In this section, we study the t.e.v.s of the FC, CC and AC for a rigidly rotating thermal state. We consider the SET separately in Sect. 4.7.

4.6.1   Fermion Condensate

Starting from (4.105), the following expression is obtained for the t.e.v. of the FC:

$$\langle : \widehat{\overline{\Psi}}\widehat{\Psi} : \rangle_{T_0} = \sum_{\lambda=\pm\frac{1}{2}} \sum_{m=-\infty}^{\infty} \int_{|E|>M} dE \, |E| \int_{-p}^{p} dk \, \Theta(\widetilde{E})$$

$$\times \left[\frac{\overline{U}_{E,k,m}^{\lambda} U_{E,k,m}^{\lambda}}{e^{(\widetilde{E}-\mu_0)/T_0} + 1} - \frac{\overline{V}_{E,k,m}^{\lambda} V_{E,k,m}^{\lambda}}{e^{(\widetilde{E}+\mu_0)/T_0} + 1} \right].$$

$$(4.106)$$

Using (4.88), (4.89), the sum over λ can be performed, yielding:

$$\langle : \widehat{\overline{\Psi}}\widehat{\Psi} : \rangle_{T_0} = \frac{M}{4\pi^2} \sum_{m=-\infty}^{\infty} \int_{|E|>M} dE \, \text{sgn}(E) \int_{-p}^{p} dk \, \Theta(\widetilde{E}) \, J_m^+(q\rho)$$

$$\times \left[\frac{1}{e^{(\widetilde{E}-\mu_0)/T_0} + 1} + \frac{1}{e^{(\widetilde{E}+\mu_0)/T_0} + 1} \right],$$

$$(4.107)$$

where $\text{sgn}(E) = |E|/E$ is the sign of the energy of the mode. To simplify the integration above, the integral over E can be split into its positive ($E > M$) and negative ($E < -M$) domains. On the negative branch, the simultaneous sign flip $(E, m) \to (-E, -m)$ can be performed, under which $\widetilde{E} \to -\widetilde{E}$. Noting that $J_{-m}^+(q\rho) = J_m^+(q\rho)$, the following expression is obtained:

$$\langle : \widehat{\overline{\Psi}}\widehat{\Psi} : \rangle_{T_0} = \frac{M}{4\pi^2} \sum_{m=-\infty}^{\infty} \int_{M}^{\infty} dE \int_{-p}^{p} dk \, J_m^+(q\rho) \, \text{sgn}(\widetilde{E})$$

$$\times \left[\frac{1}{e^{(|\widetilde{E}|-\mu_0)/T_0} + 1} + \frac{1}{e^{(|\widetilde{E}|+\mu_0)/T_0} + 1} \right].$$

$$(4.108)$$

In order to study the massless limit of $M^{-1}\langle : \widehat{\overline{\Psi}}\widehat{\Psi} : \rangle_{T_0}$, we now attempt to simplify the integrand, by replacing $\text{sgn}(\widetilde{E}) = 1$ and $|\widetilde{E}| = \widetilde{E}$. To this end, consider the quantity $\mathfrak{F}_1$

$$\mathfrak{F}_1 = \sum_{m=-\infty}^{\infty} \int_{M}^{\infty} dE \left[\frac{\text{sgn}(\widetilde{E})}{e^{(|\widetilde{E}|-\mu_0)/T_0} + 1} + \frac{\text{sgn}(\widetilde{E})}{e^{(|\widetilde{E}|+\mu_0)/T_0} + 1} \right] \mathfrak{f}(m, E), \quad (4.109)$$

where $\mathfrak{f}(m, E)$ is a function depending on m and E, We now write $\mathfrak{F}_1$ as a sum of a term $\{\mathfrak{F}_1\}_{\text{simp}}$ where $|\widetilde{E}|$ is replaced by $\widetilde{E}$ (that is, the modulus is removed) and $\text{sgn}(\widetilde{E})$ is set equal to one, and a remainder $\Delta\mathfrak{F}_1$:

$$\mathfrak{F}_1 = \{\mathfrak{F}_1\}_{\text{simp}} + \Delta\mathfrak{F}_1,$$

$$(4.110)$$

where

$$\{\mathfrak{F}_1\}_{\text{simp}} = \sum_{m=-\infty}^{\infty} \int_M^{\infty} dE \left[\frac{1}{e^{(\widetilde{E}-\mu_0)T_0} + 1} + \frac{1}{e^{(\widetilde{E}+\mu_0)/T_0} + 1} \right] \mathfrak{f}(m, E),$$

$$\Delta\mathfrak{F}_1 = -\sum_{m=m_M}^{\infty} \int_M^{\Omega m} dE \left[\frac{1}{e^{(-\widetilde{E}-\mu_0)/T_0} + 1} + \frac{1}{e^{(-\widetilde{E}+\mu_0)/T_0} + 1} \right.$$

$$\left. + \frac{1}{e^{(\widetilde{E}-\mu_0)/T_0} + 1} + \frac{1}{e^{(\widetilde{E}+\mu_0)/T_0} + 1} \right] \mathfrak{f}(m, E)$$

$$= -\sum_{m=m_M}^{\infty} \int_M^{\Omega m} dE \, 2\mathfrak{f}(m, E), \tag{4.111}$$

where m_M is the minimum value of m for which $\Omega m > M$. The last line follows from the identity $(e^x + 1)^{-1} + (e^{-x} + 1)^{-1} = 1$. The last equality above shows that $\Delta\mathfrak{F}_1$ does not depend on T_0 or μ_0 unless $\mathfrak{f}(m, E)$ explicitly depends on these parameters (which it does not for the FC). The dependence of $\Delta\mathfrak{F}_1$ on Ω is due to the definition of the rotating vacuum, where Ω appears explicitly when restricting the energy spectrum to positive co-rotating energies. We thus find

$$M^{-1} \left\{ \langle : \widehat{\overline{\Psi}\Psi} : \rangle_{T_0} \right\}_{\text{simp}} = \frac{1}{2\pi^2} \sum_{m=-\infty}^{\infty} \int_M^{\infty} dE \left[\frac{1}{e^{(\widetilde{E}-\mu_0)/T_0} + 1} + \frac{1}{e^{(\widetilde{E}+\mu_0)/T_0} + 1} \right]$$

$$\times \int_0^p dk \, J_m^+(q\rho). \tag{4.112}$$

In the massless limit, the following exact result can be obtained (see [17, 18] for further details of the techniques used to perform the integration):

$$M^{-1} \left\{ \langle : \widehat{\overline{\Psi}\Psi} : \rangle_{T_0} \right\}_{\text{simp}} \bigg|_{M=0} = \frac{T^2}{6} + \frac{\mu^2}{2\pi^2} + \frac{3\omega^2 + 2a^2}{24\pi^2}, \tag{4.113}$$

where $\omega^2 = \Omega^2\Gamma^2$ and $a^2 = \rho^2\Omega^2\Gamma^4$ are the squares of the spatial parts of the kinematic vorticity and acceleration introduced in (4.15), while Γ is the Lorentz factor (4.13). The last term is independent of μ and T and hence represents the contribution due to the difference between the rotating and stationary vacua. Subtracting this contribution gives

$$M^{-1} \langle : \widehat{\overline{\Psi}\Psi} : \rangle_{T_0} \bigg|_{M=0} = \frac{T^2}{6} + \frac{\mu^2}{2\pi^2}, \tag{4.114}$$

which agrees with the RKT result (4.28) with $g_F = 2$, diverging as $\Gamma \to \infty$ and the SLS is approached.

4.6.2   Charge Current

Since the density of states factor in (4.38) now has a dependence on the angular momentum quantum number m as well as the energy E, the $\hat{\phi}$ component of the CC no longer vanishes when the state is rigidly rotating. The nonzero components of the t.e.v. of the CC take the form:

$$
\begin{pmatrix} \langle : \widehat{J}^{\hat{t}} : \rangle_{T_0} \\ \langle : \widehat{J}^{\hat{\phi}} : \rangle_{T_0} \end{pmatrix} = \frac{1}{4\pi^2} \sum_{m=-\infty}^{\infty} \int_M^{\infty} dE \left[\frac{1}{e^{(|\widetilde{E}|-\mu_0)/T_0} + 1} - \frac{1}{e^{(|\widetilde{E}|+\mu_0)/T_0} + 1} \right]
$$
$$
\times \int_{-p}^{p} dk \begin{pmatrix} E\, J_m^+(q\rho) \\ q\, J_m^\times(q\rho) \end{pmatrix}. \quad (4.115)
$$

To compute the above integrals in the massless limit, we follow the method employed for the FC and define a quantity

$$
\mathfrak{F}_2 = \sum_{m=-\infty}^{\infty} \int_M^{\infty} dE \left[\frac{1}{e^{(|\widetilde{E}|-\mu_0)/T_0} + 1} - \frac{1}{e^{(|\widetilde{E}|+\mu_0)/T_0} + 1} \right] \mathfrak{f}(m, E). \quad (4.116)
$$

Writing $\mathfrak{F}_2$ as a sum of a term $\{\mathfrak{F}_2\}_{\text{simp}}$ where $|\widetilde{E}|$ is replaced by $\widetilde{E}$ and a remainder $\Delta\mathfrak{F}_2$

$$
\mathfrak{F}_2 = \{\mathfrak{F}_2\}_{\text{simp}} + \Delta\mathfrak{F}_2, \quad (4.117)
$$

we find

$$
\{\mathfrak{F}_2\}_{\text{simp}} = \sum_{m=-\infty}^{\infty} \int_M^{\infty} dE \left[\frac{1}{e^{(\widetilde{E}-\mu_0)T_0} + 1} - \frac{1}{e^{(\widetilde{E}+\mu_0)/T_0} + 1} \right] \mathfrak{f}(m, E),
$$
$$
\Delta\mathfrak{F}_2 = \sum_{m=m_M}^{\infty} \int_M^{\Omega m} dE \left[\frac{1}{e^{(-\widetilde{E}-\mu_0)/T_0} + 1} - \frac{1}{e^{(-\widetilde{E}+\mu_0)/T_0} + 1} \right.
$$
$$
\left. - \frac{1}{e^{(\widetilde{E}-\mu_0)/T_0} + 1} + \frac{1}{e^{(\widetilde{E}+\mu_0)/T_0} + 1} \right] \mathfrak{f}(m, E).
$$
$$
(4.118)
$$

The term inside the square brackets in $\Delta\mathfrak{F}_2$ is identically zero. Thus, it can be concluded that $\mathfrak{F}_2 = \{\mathfrak{F}_2\}_{\text{simp}}$ for any function $\mathfrak{f}(m, E)$, which simplifies the integration. The following expressions are then obtained for massless fermions [17,18]:

$$
\langle : \widehat{J}^{\hat{t}} : \rangle_{T_0} = \frac{\mu_0 \Gamma^4}{3} \left(T_0^2 + \frac{\mu_0^2}{\pi^2} \right) + \frac{\mu_0 \Omega^2 \Gamma^4}{4\pi^2} \left(\frac{4}{3}\Gamma^2 - \frac{1}{3} \right) = \Gamma \left[Q_F + \frac{\mu}{12\pi^2}(3\omega^2 + a^2) \right],
$$
$$
\langle : \widehat{J}^{\hat{\phi}} : \rangle_{T_0} = \rho\Omega\Gamma \left[Q_F + \frac{\mu}{12\pi^2}(\omega^2 + 3a^2) \right]. \quad (4.119)
$$

As expected, the ϕ-component vanishes when $\Omega = 0$ and the state is nonrotating. The first terms appearing on the right-hand-side correspond to the RKT results for $g_F = 2$

(4.27). The second terms are the quantum corrections, and are proportional to Ω^2, vanishing when the rotation is zero. The quantum corrections do not depend on the temperature T, only on the chemical potential, local vorticity and local acceleration. The quantum corrections are therefore present even in the zero-temperature limit. The decomposition of the CC with respect to the kinematic tetrad in (4.15) will be discussed in Sect. 4.7.2.

The t.e.v. of the CC vanishes identically when the chemical potential on the axis μ_0 is zero. This is to be expected since, with vanishing chemical potential, a rigidly rotating thermal state will contain equal numbers of particles and anti-particles. When μ_0 is nonzero, the current diverges as $\Gamma \to \infty$ and the SLS is approached. For both components of the CC, the quantum corrections diverge more rapidly than the RKT contributions as $\rho \to \Omega^{-1}$. Therefore, close to the SLS, the CC is completely dominated by quantum effects and the RKT contributions are subleading.

4.6.3 Axial Current

The classical AC J_5^μ is defined by

$$J_5^\mu = \overline{\psi}\gamma^\mu\gamma_5\psi, \tag{4.120}$$

where we have introduced the chirality matrix

$$\gamma_5 = i\gamma^t\gamma^x\gamma^y\gamma^z = \begin{pmatrix} 0 & 1 \\ 1 & 0 \end{pmatrix}. \tag{4.121}$$

Using the Dirac equation (4.47), and taking into account that γ_5 anti-commutes with all of the other γ matrices, $\{\gamma_5, \gamma^\mu\} = 0$, we find $\partial_\mu J_5^\mu = 2iM\overline{\psi}\gamma_5\psi$, and hence J_5^μ is conserved for massless particles. Non-vanishing values of J_5^μ can be induced through the chiral vortical effect (for a review, see [8]). The expectation values of J_5^μ computed for massless fermions using a perturbative approach were recently reported in [13]. Here we consider the t.e.v. of J_5^μ using QFT techniques.

Using the mode expansion (4.105), the t.e.v. of (4.120) takes the form:

$$\langle : \widehat{J}_5^\mu : \rangle_{T_0} = \sum_j \left\{ \frac{\mathfrak{J}_5^\mu(U_j, U_j)}{\exp[(\widetilde{E}_j - \mu_0)/T_0] + 1} - \frac{\mathfrak{J}_5^\mu(V_j, V_j)}{\exp[(\widetilde{E}_j + \mu_0)/T_0] + 1} \right\}, \tag{4.122}$$

where $\mathfrak{J}_5^\mu(\psi, \chi) = \overline{\psi}\gamma^\mu\gamma_5\chi$. Following the same reasoning applied to obtain (4.94), it is not difficult to show that $\mathfrak{J}_5^\mu(V_j, V_j) = -[\mathfrak{J}_5^\mu(U_j, U_j)]^*$, while

$$\mathfrak{J}_5^t(U_j, U_j) = \frac{p_j}{8\pi^2 E_j} \left[2\lambda_j J_{m_j}^+(q_j\rho) + \frac{k_j}{p_j} J_{m_j}^-(q_j\rho) \right],$$

$$\mathfrak{J}_5^\phi(U_j, U_j) = \frac{\lambda_j q_j}{4\pi^2 p_j} J_{m_j}^\times(q_j\rho),$$

$$\mathfrak{J}_5^z(U_j, U_j) = \frac{1}{8\pi^2} \left[J_{m_j}^-(q_j\rho) + \frac{2\lambda_j k_j}{p_j} J_{m_j}^+(q_j\rho) \right]. \tag{4.123}$$

When considering the sum over j in (4.122), the terms which are odd with respect to λ and k vanish. Thus, the only non-vanishing component of the t.e.v. of the AC is

$$\langle : \widehat{J}_5^z : \rangle_{T_0} = \frac{1}{4\pi^2} \sum_{m=-\infty}^{\infty} \int_M^\infty dE \, E \int_{-p}^p dk \, J_m^-(q\rho) \mathrm{sgn}(\widetilde{E})$$

$$\times \left\{ \frac{1}{\exp[(|\widetilde{E}| - \mu_0)/T_0] + 1} + \frac{1}{\exp[(|\widetilde{E}| + \mu_0)/T_0] + 1} \right\}. \tag{4.124}$$

As in the cases of the FC and CC, the t.e.v. of the axial current can be computed exactly in the massless limit. We simplify as discussed in Sect. 4.6.1, replacing $\mathrm{sgn}(\widetilde{E}) = 1$ and $|\widetilde{E}| = \widetilde{E}$, to find

$$\{\langle : \widehat{J}_5^z : \rangle_{T_0}\}_{\mathrm{simp}} = \frac{1}{2\pi^2} \sum_{m=-\infty}^{\infty} \int_M^\infty dE \, E \left[\frac{1}{e^{(\widetilde{E}-\mu_0)/T_0} + 1} + \frac{1}{e^{(\widetilde{E}+\mu_0)/T_0} + 1} \right]$$

$$\times \int_0^p dk \, J_m^-(q\rho). \tag{4.125}$$

In the massless limit, the following exact result can be obtained [17, 18]:

$$\{\langle : \widehat{J}_5^z : \rangle_{T_0}\}_{\mathrm{simp}}\Big|_{M=0} = \frac{\Omega T_0^2 \Gamma^4}{6} \left(1 + \frac{3\mu_0^2}{\pi^2 T_0^2} \right) + \frac{\Omega^3 \Gamma^4}{24\pi^2}(4\Gamma^2 - 3)$$

$$= \omega^{\hat{z}} \left(\frac{T^2}{6} + \frac{\mu^2}{2\pi^2} + \frac{\omega^2 + 3a^2}{24\pi^2} \right), \tag{4.126}$$

where $\omega^{\hat{\alpha}}$ is the kinematic vorticity introduced in (4.15). The last term is independent of μ_0 and T_0 and hence represents the contribution due to the difference between the rotating and stationary vacua [29]. Eliminating this term allows the t.e.v. of the AC to be obtained as

$$\langle : \widehat{J}_5^{\hat{\alpha}} : \rangle_{T_0}\Big|_{M=0} = \sigma_A^\omega \omega^{\hat{\alpha}}, \qquad \sigma_A^\omega = \frac{T^2}{6} + \frac{\mu^2}{2\pi^2}, \tag{4.127}$$

where σ_A^ω is the axial vortical conductivity, which allows an axial charge flow to develop along the kinematic vorticity vector. As expected, the AC (4.127) vanishes in the stationary case, but, unlike the CC, it is nonzero even when the chemical potential vanishes [29].

The AC vanishes in classical RKT. Restoring the reduced Planck's constant, the AC (4.127) is proportional to $\hbar\Omega$ and is therefore larger than the quantum corrections to the CC, which are $O(\hbar^2\Omega^2)$.

The AC has been studied previously by a number of authors [8,36,37]. Up to possible overall factors due to differences in definitions, (4.127) agrees with the corresponding quantity in [8] only on the rotation axis, where $\Gamma = 1$. The axial current in [36] (derived using the ansatz for the Wigner function proposed in [39]) matches (4.126) only on the axis of rotation, but no distinction is made in [36] between the stationary and rotating vacua. Constructed using a QFT approach and considering the stationary Minkowski vacuum, the AC in [37] agrees with (4.126), again only on the axis of rotation. Finally, the result obtained in [38] using perturbative QFT agrees fully with (4.126).

4.7   Hydrodynamic Analysis of the Quantum Stress-Energy Tensor

In this section, we consider in detail the t.e.v. of the SET for rigidly rotating states. Following the approach of the previous section, we first derive the components of this t.e.v. with respect to the orthonormal tetrad (4.8). For comparison with the RKT results from Sect. 4.2, we then consider quantities defined with respect to the β-frame (or thermometer frame).

4.7.1   Stress-Energy Tensor Expectation Values

The t.e.v. of the SET can be written compactly as

$$\langle : \widehat{T}_{\hat{\alpha}\hat{\sigma}} : \rangle_{T_0} = \frac{1}{4\pi^2} \sum_{m=-\infty}^{\infty} \int_M^{\infty} dE\, E \int_{-p}^{p} dk\, \mathfrak{T}_{\hat{\alpha}\hat{\sigma}}\, \mathrm{sgn}(\widetilde{E})$$

$$\times \left[\frac{1}{e^{(|\widetilde{E}|-\mu_0)/T_0}+1} + \frac{1}{e^{(|\widetilde{E}|+\mu_0)/T_0}+1} \right],$$

(4.128)

where the tensor $\mathfrak{T}_{\hat{\alpha}\hat{\sigma}}$ has the following non-vanishing components:

$$\mathfrak{T}_{\hat{t}\hat{t}} = E\, J_m^+(q\rho), \qquad \mathfrak{T}_{\hat{t}\hat{\phi}} = -\frac{1}{2\rho}\left[m J_m^+(q\rho) - \frac{1}{2} J_m^-(q\rho) \right] - \frac{q}{2} J_m^\times(q\rho),$$

$$\mathfrak{T}_{\hat{\rho}\hat{\rho}} = \frac{q^2}{E}\left[J_m^+(q\rho) - \frac{m}{q\rho} J_m^\times(q\rho) \right], \quad \mathfrak{T}_{\hat{\phi}\hat{\phi}} = \frac{mq}{\rho E} J_m^\times(q\rho), \quad \mathfrak{T}_{\hat{z}\hat{z}} = \frac{k^2}{E} J_m^+(q\rho).$$

(4.129)

As with the t.e.v.s considered in Sect. 4.6, we can obtain closed-form expressions in the massless limit. We first simplify using the approach of Sect. 4.6.1, and then integrate using a procedure whose details can be found in [17,18]. The results are

$$
\langle : \widehat{T}_{\hat{t}\hat{t}} : \rangle_{T_0} = P_{\rm F}(4\Gamma^2 - 1) + \frac{\Omega^2\Gamma^2}{8}\left(T^2 + \frac{3\mu^2}{\pi^2}\right)\left(\frac{8}{3}\Gamma^4 - \frac{16}{9}\Gamma^2 + \frac{1}{9}\right),
$$

$$
\langle : \widehat{T}_{\hat{t}\hat{\phi}} : \rangle_{T_0} = -\rho\Omega\Gamma^2\left[4P_{\rm F} + \frac{2\Omega^2\Gamma^2}{9}\left(T^2 + \frac{3\mu^2}{\pi^2}\right)\left(\frac{3}{2}\Gamma^2 - \frac{1}{2}\right)\right],
$$

$$
\langle : \widehat{T}_{\hat{\rho}\hat{\rho}} : \rangle_{T_0} = P_{\rm F} + \frac{\Omega^2\Gamma^2}{24}\left(T^2 + \frac{3\mu^2}{\pi^2}\right)\left(\frac{4}{3}\Gamma^2 - \frac{1}{3}\right),
$$

$$
\langle : \widehat{T}_{\hat{\phi}\hat{\phi}} : \rangle_{T_0} = P_{\rm F}(4\Gamma^2 - 3) + \frac{\Omega^2\Gamma^2}{24}\left(T^2 + \frac{3\mu^2}{\pi^2}\right)(8\Gamma^4 - 8\Gamma^2 + 1), \quad (4.130)
$$

while $\langle : \widehat{T}_{\hat{z}\hat{z}} : \rangle_{T_0} = \langle : \widehat{T}_{\hat{\rho}\hat{\rho}} : \rangle_{T_0}$ (this relation holds also in the case of massive field quanta [17,29]). The first term in each component of the SET is the contribution from RKT (see Sect. 4.2), while the second term is the quantum correction. As for the CC (see Sect. 4.6.2), the quantum corrections are all proportional to Ω^2 and, as expected from Sect. 4.5, vanish in the stationary case. Unlike the CC, the quantum corrections are now temperature-dependent. All components of the t.e.v. of the SET diverge on the SLS, and, once again, the quantum corrections diverge more quickly as $\Gamma \to \infty$.

4.7.2 Thermometer Frame

Further insight into the effect of quantum corrections can be gleaned from a hydrodynamic analysis of the SET. In relativistic fluid dynamics, the equivalence between mass and energy transfer makes the macroscopic four-velocity u^μ an ambiguous concept. A frame is defined by making a choice for the definition of u^μ. Here we work in the β-frame, also termed the natural frame [40], or thermometer frame [41] (see also [15] for an analysis of the properties of this frame). In the β-frame, the macroscopic four-velocity u^μ is proportional to the temperature four-vector β^μ, that is, $u^\mu = T\beta^\mu$, where T is the local temperature. For rigidly rotating states, the macroscopic four-velocity is then given by (4.12).

With this definition of u^μ, we decompose the CC and SET as follows [42]:

$$
J^\mu = Q_\beta u^\mu + \mathcal{J}_\beta^\mu, \qquad T^{\mu\nu} = E_\beta u^\mu u^\nu - (P_\beta + \varpi)\Delta^{\mu\nu} + \Pi^{\mu\nu} + u^\mu W^\nu + u^\nu W^\mu,
$$
$$
(4.131)
$$

where Q_β, E_β and P_β are the usual equilibrium quantities, $\mathcal{J}^\mu$ and W^μ represent the charge and heat flux in the local rest frame, ϖ is the dynamic pressure and $\Pi^{\mu\nu}$ is the pressure deviator. The tensor $\Delta^{\mu\nu} = g^{\mu\nu} - u^\mu u^\nu$ is a projector on the hypersurface orthogonal to u^μ. The nonequilibrium quantities $\mathcal{J}^\mu$, $\Pi^{\mu\nu}$ and W^μ are also orthogonal to u^μ, by construction. The isotropic pressure $P_\beta + \varpi$ is given as the sum of the hydrostatic pressure P_β, computed using the equation of state of the fluid, and of the dynamic pressure ϖ, which in general depends on the divergence of the velocity. In the case of massless (or ultrarelativistic) particles, the SET is traceless, since the massless Dirac field is conformally coupled and the conformal trace anomaly vanishes on flat space-time [43]. From (4.131), the SET trace is $T^\mu{}_\mu = E_\beta - 3(P_\beta + \varpi)$, and therefore ϖ vanishes for massless particles since $E_\beta = 3P_\beta$. Moreover, since the velocity field is divergenceless ($\nabla_\mu u^\mu = 0$), it is reasonable to assume that $\varpi = 0$ also when $M > 0$. However, below we keep this term for clarity.

For both massive and massless particles, the macroscopic quantities can be extracted from the components of J^μ and $T^{\mu\nu}$ as follows [14]:

$$Q_\beta = u_\mu J^\mu, \qquad E_\beta = u_\mu u_\nu T^{\mu\nu}, \qquad P_\beta + \varpi = -\frac{1}{3}\Delta_{\mu\nu}T^{\mu\nu},$$
$$\mathcal{J}^\mu = \Delta^{\mu\nu}J_\nu, \qquad W^\mu = \Delta^{\mu\nu}u^\lambda T_{\nu\lambda}, \qquad \Pi^{\mu\nu} = T^{\langle\mu\nu\rangle}, \tag{4.132}$$

where the notation $A^{\langle\mu\nu\rangle}$ for a general two-index tensor denotes

$$A^{\langle\mu\nu\rangle} = \left[\frac{1}{2}\left(\Delta^{\mu\lambda}\Delta^{\nu\sigma} + \Delta^{\nu\lambda}\Delta^{\mu\sigma}\right) - \frac{1}{3}\Delta^{\mu\nu}\Delta^{\lambda\sigma}\right]A_{\lambda\sigma}. \tag{4.133}$$

Since in general, $\mathcal{J}^{\hat{\rho}} = \mathcal{J}^{\hat{z}} = 0$ and $\mathcal{J}^{\hat{\alpha}}u_{\hat{\alpha}} = 0$, it can be seen that $\mathcal{J}^{\hat{\alpha}}$ points along the circular vector $\tau^{\hat{\alpha}}$, introduced in (4.15)

$$\mathcal{J}^{\hat{\alpha}} = \sigma_V^\tau \tau^{\hat{\alpha}}, \qquad \sigma_V^\tau = \frac{\rho\Omega J^{\hat{t}} - J^{\hat{\phi}}}{\rho\Omega^3\Gamma^3}, \tag{4.134}$$

where σ_V^τ is the circular vector (electric) charge conductivity. Similarly, the structure of $T^{\mu\nu}$ indicates that $W^{\hat{\rho}} = W^{\hat{z}} = 0$, while the orthogonality between $W^{\hat{\alpha}}$ and $u^{\hat{\alpha}}$ allows $W^{\hat{\alpha}}$ to be written as

$$W^{\hat{\alpha}} = \sigma_\varepsilon^\tau \tau^{\hat{\alpha}}, \qquad \sigma_\varepsilon^\tau = \frac{1}{\Omega^2\Gamma^2}\left(T_{\hat{t}\hat{t}} + T_{\hat{\phi}\hat{\phi}} + \frac{1+\rho^2\Omega^2}{\rho\Omega}T_{\hat{t}\hat{\phi}}\right), \tag{4.135}$$

where σ_ε^τ is the circular heat conductivity. Finally, noting that $\Pi^{\hat{\alpha}\hat{\sigma}}$ is symmetric, traceless and orthogonal to $u^{\hat{\alpha}}$ with respect to both indices, as well as the property $T^{\hat{\rho}\hat{\rho}} = T^{\hat{z}\hat{z}}$, only one degree of freedom is required to characterize $\Pi^{\hat{\alpha}\hat{\sigma}}$, introduced as Π_β below [17, 18]

$$\Pi^{\hat{\alpha}\hat{\sigma}} = \Pi_\beta\left(\tau^{\hat{\alpha}}\tau^{\hat{\sigma}} - \frac{\omega^2}{2}a^{\hat{\alpha}}a^{\hat{\sigma}} - \frac{a^2}{2}\omega^{\hat{\alpha}}\omega^{\hat{\sigma}}\right) = \rho^2\Omega^6\Gamma^8\Pi_\beta\begin{pmatrix}\rho^2\Omega^2\Gamma^2 & 0 & \rho\Omega\Gamma^2 & 0 \\ 0 & -\frac{1}{2} & 0 & 0 \\ \rho\Omega\Gamma^2 & 0 & \Gamma^2 & 0 \\ 0 & 0 & 0 & -\frac{1}{2}\end{pmatrix},$$

$$\Pi_\beta = \frac{2(P_\beta + \varpi - T_{\hat{z}\hat{z}})}{\rho^2\Omega^6\Gamma^8}. \tag{4.136}$$

In the case of massless fermions, substituting the SET components (4.130) into the contractions (4.132) yields the following closed-form results:

$$Q_\beta = Q_{\rm F} + \Delta Q, \quad Q_{\rm F} = \frac{\mu}{3}\left(T^2 + \frac{\mu^2}{\pi^2}\right), \qquad\qquad \Delta Q = \frac{\mu(\omega^2 + a^2)}{4\pi^2},$$

$$P_\beta = P_{\rm F} + \Delta P, \quad P_{\rm F} = \frac{7\pi^2 T^4}{180} + \frac{T^2\mu^2}{6} + \frac{\mu^4}{12\pi^2}, \quad \Delta P = \frac{3\omega^2 + a^2}{72}\left(T^2 + \frac{3\mu^2}{\pi^2}\right),$$

$$\sigma_V^\tau = \frac{\mu}{6\pi^2}, \qquad \sigma_\varepsilon^\tau = -\frac{1}{18}\left(T^2 + \frac{3\mu^2}{\pi^2}\right), \qquad\qquad \Pi_\beta = 0. \tag{4.137}$$

The above results agree with [13,17,18,44]. The first terms in Q_β and P_β coincide with the RKT results in (4.27) with $g_{\rm F} = 2$. On the rotation axis, where $\rho = 0$, equation (4.137) shows that the conductivities σ_V^τ and σ_ε^τ remain finite, while the circular vector $\tau^{\hat\alpha}$ vanishes. This conclusion holds also in the massive case. This can be seen by noting that, according to equations (4.115, 4.128), both $\langle: \widehat{J}^{\hat\phi} :\rangle_{T_0}$ and $\langle: \widehat{T}_{\hat t\hat\phi} :\rangle_{T_0}$ vanish when $\rho = 0$. Furthermore, $E_\beta = \langle: T_{\hat t\hat t} :\rangle_{T_0}$ (since $\rho\Omega = 0$) and it can be shown that $\langle: T_{\hat\rho\hat\rho} :\rangle_{T_0} = \langle: T_{\hat\phi\hat\phi} :\rangle_{T_0} = \langle: T_{\hat z\hat z} :\rangle_{T_0}$ and thus, the SET takes the perfect fluid form at $\rho = 0$.

4.7.3 Quantum Corrections to the SET

We now examine the effect of quantum corrections on the SET, comparing first the exact RKT results (4.27) and QFT results (4.137) in the massless case. There are three features of note.

First, quantum corrections mean that the SET no longer has the perfect fluid form, due to the presence of nonequilibrium terms, except on the axis of rotation, where the circular vector $\tau^{\hat\alpha}$ (4.15) vanishes. Second, the quantum corrections to the equilibrium quantities Q_β, E_β and P_β are proportional to Ω^2. Third, the quantum corrections in (4.137) diverge more quickly than the RKT quantities as $\Gamma \to \infty$ and the SLS is approached. Therefore, there is a neighbourhood of the SLS where quantum corrections become dominant.

In order to assess the relative contribution made by quantum corrections with respect to the RKT results, we first focus on the energy density for massless particles and consider two quantities (we restore the reduced Planck's constant $\hbar$ and Boltzmann constant $k_{\rm B}$):

$$\frac{E_\beta}{E_{\rm F}} - 1 = \frac{15}{14}\left(\frac{\hbar\Omega}{\pi k_{\rm B} T_0}\right)^2\left(\frac{4}{3}\Gamma^2 - \frac{1}{3}\right)\frac{1 + 3(\mu_0/\pi k_{\rm B} T_0)^2}{1 + \frac{30}{7}(\mu_0/\pi k_{\rm B} T_0)^2 + \frac{15}{7}(\mu_0/\pi k_{\rm B} T_0)^4},$$

$$1 - \frac{E_{\rm F}}{E_\beta} = \left[1 + \frac{14}{5(4\Gamma^2 - 1)}\left(\frac{\pi k_{\rm B} T_0}{\hbar\Omega}\right)^2\frac{1 + \frac{30}{7}(\mu_0/\pi k_{\rm B} T_0)^2 + \frac{15}{7}(\mu_0/\pi k_{\rm B} T_0)^4}{1 + 3(\mu_0/\pi k_{\rm B} T_0)^2}\right]^{-1}. \tag{4.138}$$

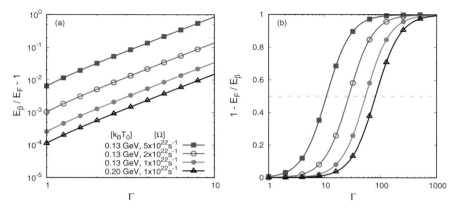

Fig. 4.3 Relative differences **a** $E_\beta/E_F - 1$ and **b** $1 - E_F/E_\beta$ between the β-frame energy density E_β (4.137) and the RKT result E_F (4.27) for massless fermions. The curves correspond to $k_B T_0 = 0.13$ GeV (filled purple squares, empty red circles and filled blue circles) and 0.2 GeV (empty black triangles). The angular velocity is set to $\Omega = 5 \times 10^{22}$ s^{-1} (filled purple squares), 2×10^{22} s^{-1} (empty red circles) and 10^{22} s^{-1} (filled blue circles and empty black triangles). The chemical potential on the rotation axis is $\mu_0 = 0.1$ GeV

Figure 4.3a shows the relative departure of the QFT energy density E_β (4.137) measured in the thermometer frame, compared to the RKT energy density $E_F = 3P_F$ (4.27). We use values of the chemical potential and angular speed relevant for heavy ion collisions, as in Sect. 4.2.2. For $k_B T_0 = 0.2$ GeV and $\Omega = 10^{22}$ s^{-1}, the relative difference is about 10^{-4} on the rotation axis. From (4.138), this value can be increased by either increasing the angular velocity Ω or decreasing the temperature T_0. We thus also consider a lower temperature relevant to the QGP, $k_B T_0 \simeq 0.13$ GeV. This enlarges the relative difference by a factor of ~ 2.4. At larger values of the angular speed, quantum corrections are close to 1% on the rotation axis. Away from the rotation axis, the relative difference $E_\beta/E_F - 1$ increases roughly as Γ^2 (4.138). This is confirmed for all regimes considered in Fig. 4.3a.

The relative difference $1 - E_F/E_\beta$ is presented in Fig. 4.3b. On the rotation axis, this ratio is negligible. As $\Gamma \to \infty$, equation (4.138) shows that the second term in the square bracket goes to 0 and thus $\lim_{\Gamma \to \infty} 1 - E_F/E_\beta \to 1$. Close to the SLS, quantum corrections therefore become the dominant contribution to the energy density E_β. The gray, dashed line in Fig. 4.3b indicates where the quantum corrections become equal to the classical contribution, $E_\beta = 2E_F$. This happens closer to the SLS when the temperature is increased or when the angular velocity is decreased.

We next consider the effect of the mass on the energy density E_β. Figure 4.4a shows a comparison between the energy densities E_β and E_F, as functions of the distance ρ from the rotation axis. When $\Omega = 5 \times 10^{22}$ s^{-1}, the SLS is located at $\rho = c/\Omega = 6$ fm. The energy density for particles of mass 0.14 GeV follows the result for the massless limit very closely, while the case with $Mc^2 = 0.548$ GeV can be distinguished from the massless limit only up to $\rho \lesssim 5.5$ fm. Figure 4.4b shows the dependence of the energy densities E_β and E_F on the Lorentz factor Γ (4.13).

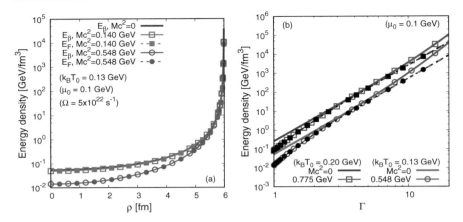

Fig. 4.4 Dependence on **a** the distance ρ, measured in fm from the rotation axis, and **b** on the Lorentz factor Γ (4.13), of the energy densities E_β and E_F obtained in QFT (empty symbols and continuous lines) and RKT (filled symbols and dashed lines) at $\mu_0 = 0.1$ GeV and $\Omega = 5 \times 10^{22}$ s^{-1}. In **a**, the temperature on the rotation axis is fixed at $k_B T_0 = 0.13$ GeV and the mass Mc^2 is set to 0 (continuous purple line, only E_β is shown), 0.140 GeV (blue squares) and 0.548 GeV (red circles). In **b**, $k_B T_0 = 0.20$ GeV (upper lines) and 0.13 GeV (lower lines). The analytic results for the massless limit are shown using continuous (QFT) and dashed (RKT) lines without symbols (purple is used for $k_B T_0 = 0.2$ GeV and blue corresponds to $k_B T_0 = 0.13$ GeV)

The RKT and QFT energy densities can be distinguished when $\Gamma \gtrsim 10$, where the higher order divergence induced by the quantum corrections becomes important. At large values of Γ, both the QFT and RKT energy densities follow their respective massless asymptotics, indicating that also in the QFT case, the corrections due to the mass terms contribute at a subleading order close to the SLS, compared with the corresponding massless limit.

Finally, we discuss the properties of quantum corrections on the rotation axis. Since the nonequilibrium terms vanish on the rotation axis, only the equilibrium quantities, E_β, P_β and Q_β need to be considered (we assume that $\varpi = 0$ here). Instead of discussing P_β, we focus on the trace of the SET. Figure 4.5 shows the properties of the quantum corrections (a) $E_\beta/E_F - 1$, (b) $(E_\beta - 3P_\beta)/(E_F - 3P_F) - 1$ and (c) $Q_\beta/Q_F - 1$, computed as relative differences between the QFT and RKT results.

Focussing on the small mass regime, it can be seen that the relative quantum corrections of the SET trace exhibit a rapid variation with respect to M. This variation can be attributed to the presence of the sign function in the SET components (4.128), which can take negative values only when $Mc^2 < \hbar\Omega/2$. In particular, the quantity $(E_\beta - 3P_\beta)/M^2c^4$ exhibits no quantum corrections with respect to the corresponding RKT quantity when $M = 0$. A rapid increase can be seen at small masses bringing the relative quantum corrections to the SET trace from zero to the values observed for the other quantities (energy and charge density). At intermediate masses, a slow increase in the relative quantum corrections of all quantities can be seen. In the large mass limit, the relative quantum corrections seem to reach a plateau value.

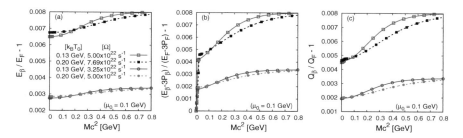

Fig. 4.5 Relative differences **a** $E_\beta/E_F - 1$, **b** $(E_\beta - 3P_\beta)/(E_F - 3P_F) - 1$, **c** $Q_\beta/Q_F - 1$, on the rotation axis ($\rho = 0$), as functions of the particle mass. The chemical potential on the rotation axis is $\mu_0 = 0.1$ GeV, and the temperature on the rotation axis is set to $k_B T_0 = 0.13$ GeV (empty symbols and continuous lines) and 0.2 GeV (filled symbols and dashed lines). We consider angular speeds Ω equal to 5×10^{22} s^{-1} (red empty squares with continuous lines and filled blue circles with dashed lines), 7.70×10^{22} s^{-1} (black filled squares with dashed lines) and 3.25×10^{22} s^{-1} (purple empty circles with continuous lines)

4.8   Rigidly Rotating Quantum Systems in Curved Space-Time

Thus far, we have focussed our attention on a quantum field in a rigidly rotating state on unbounded Minkowski space-time. We have seen that thermal states for such a setup cannot be defined if the quantum field is a scalar field [26,28]. However, it is possible to define rigidly rotating thermal states for a quantum scalar field constrained within a cylindrical reflecting boundary enclosing the axis of rotation, providing the boundary lies completely within the SLS [26,28]. In this latter situation, the rotating vacuum is identical to the nonrotating vacuum state and t.e.v.s are well-behaved. In [28], it is shown that the t.e.v.s in a co-rotating frame are very well approximated by the RKT quantities derived in Sect. 4.2, except for a region close to the boundary, where the Casimir effect becomes important.

In this chapter, we have shown that the situation on unbounded Minkowski space-time is very different for a fermion field compared to a scalar field [29], in particular we can define a rotating fermion quantum vacuum state and rigidly rotating thermal fermion states. T.e.v.s in these states are regular up to the SLS, where they diverge. A natural question is whether it is possible to consider a setup similar to that for the scalar field, namely by including a reflecting boundary. For fermions, defining reflecting boundary conditions is more involved than it is for scalars (where one can simply impose, for example, Dirichlet boundary conditions). Using either nonlocal spectral boundary conditions [45] or the local MIT-bag boundary condition [46] on a cylindrical boundary inside the SLS, the rotating fermion vacuum is identical to the nonrotating fermion vacuum [33]. Furthermore, rigidly rotating thermal states have well-defined t.e.v.s, which are computed in [33] for the case of zero chemical potential. At sufficiently high temperatures, the t.e.v.s for the bounded scenario are very well approximated by the unbounded t.e.v.s we have discussed in Sects. 4.6 and 4.7, except for a region close to the boundary. In [47] it is shown that, as well as the "bulk" mode considered in [33], the fermion field also has "edge states" localized near the boundary, which must also be taken into account. The effect of interactions for rigidly rotating fermions inside a cylindrical boundary is studied in [48,49].

In Minkowski space-time, a rigidly rotating quantum system is therefore unphysical unless an arbitrary boundary is introduced in such a way that there is no SLS. A natural question is whether rigidly rotating quantum states exist in curved space-time. One advantage of working on Minkowski space-time is that, as well as having no curvature, the space-time has maximal symmetry, which simplifies many aspects of the analysis. To explore the effect of space-time curvature on rigidly rotating quantum states, one may consider anti-de Sitter space-time (adS) [50,51]. This space-time has maximal symmetry but constant negative curvature. Furthermore, the boundary of the space-time is time-like, as is a cylindrical boundary in Minkowski space-time. In particular, appropriate conditions have to be applied to the field on the space-time boundary [52].

The properties of nonrotating thermal states on adS have been studied in the framework of RKT and QFT, for both scalars [53] and fermions [53,54], in the absence of a chemical potential. The curvature of adS space-time affects these states in a number of ways. First, the normal-ordering procedure applied in Sect. 4.5 is not valid in a general curved space-time due to the fact that v.e.v.s for the nonrotating vacuum are nonzero, for both scalars [55] and fermions [56]. Unlike our Minkowski space-time results in Sect. 4.5, the t.e.v.s for stationary states of both scalars and fermions receive quantum corrections in adS [53,54,57].

What about rigidly rotating quantum states in adS? Due to its time-like boundary, there is no SLS in adS if $\Omega \mathcal{R} < 1$, where Ω is the angular speed and $\mathcal{R}$ is the radius of curvature of the space-time. In other words, if the radius of curvature is small and the angular speed not too large, there is no SLS. Rigidly rotating quantum states on adS have been studied in much less detail than their Minkowski counterparts. For a quantum scalar field, it is known that the only possible choice of global vacuum state is the nonrotating vacuum [58], as in Minkowski space-time. One might conjecture that rigidly rotating thermal states for scalars can be defined only if there is no SLS, but this question has yet to be addressed. For a quantum fermion field, the rotating and nonrotating vacua are identical if there is no SLS, while if an SLS is present, a distinct rotating vacuum state can be defined [59]. The preliminary analysis in [59] shows that rigidly rotating thermal states have at least some features similar to those seen in Sects. 4.6 and 4.7 in Minkowski space-time, in particular the t.e.v.s diverge on the SLS (if there is one).

These results demonstrate that space-time curvature does have an effect on rigidly rotating quantum states. Asymptotically-adS space-times in particular may be relevant for studying the QGP via gauge-gravity duality (see, for example, [60–63] for reviews). In this approach, string theory on an asymptotically adS space-time is dual to a conformal quantum field theory (CFT) on the boundary of adS (which itself is conformal to Minkowski space-time). The idea is that calculations on one side of the duality may shed light on phenomena on the other side. For example, thermal states in the boundary CFT would correspond to asymptotically adS black holes in the bulk. This is because black holes emit thermal quantum radiation [3], the temperature of the radiation being known as the Hawking temperature. Asymptotically adS rotating black holes [64] can be in thermal equilibrium with radiation at the Hawking temperature provided either the black hole rotation is not too large, or the adS radius of curvature is sufficiently small [65]. These conditions ensure that there is no SLS for these black holes. A full QFT computation of the t.e.v. of the stress-energy tensor for a quantum field on a rotating asymptotically adS black hole is, however, absent from the literature.

Some of the most astrophysically important space-times with rotation are Kerr black holes [66]. These black holes are asymptotically flat, that is, far from the black hole the space-time approaches Minkowski space-time, rather than adS space-time as for the black holes discussed in the previous paragraph. Kerr black holes therefore always have an SLS, a surface on which an observer must travel at the speed of light in order to corotate with the black hole's event horizon. The quantum state describing a black hole in thermal equilibrium with radiation at the Hawking temperature is known as the Hartle–Hawking state [67]. In contrast to the situation for asymptotically adS rotating black holes, such a state cannot be defined for a quantum scalar field on an asymptotically flat Kerr black hole [5,68]. Indeed, it can be shown that any quantum state which is isotropic in a frame rigidly rotating with the event horizon of the black hole must be divergent at the SLS [69]. If the black hole is enclosed inside a reflecting mirror sufficiently close to the event horizon of the black hole, then a Hartle–Hawking state can be defined for a quantum scalar field [70]. Interestingly, this state is not exactly rigidly rotating with the angular speed of the horizon [70]. For a quantum fermion field, it is possible to define a Hartle-Hawking-like state on the Kerr black hole without the mirror present [6]. While this state is also not exactly rigidly rotating, it is nonetheless divergent on the SLS [6].

Rotating black hole space-times are much more complicated that the toy model of rigidly rotating states on Minkowski space-time that we consider in this chapter. However, the key physics remains the same in both situations. Namely, rigidly rotating states cannot be defined for a quantum scalar field if there is an SLS present. Rigidly rotating thermal states can be defined for a quantum fermion field, even when there is an SLS, but such states diverge as the SLS is approached.

4.9   Summary

In this chapter, we have considered the properties of rigidly rotating systems in QFT. Our toy models are free massive scalar and fermion fields on unbounded flat space-time. Such systems cannot be realized in nature due to the presence of the SLS, the surface outside which particles must travel faster than the speed of light in order to be rigidly rotating. Nonetheless, this approach has revealed some interesting physics which is relevant to more realistic setups, such as the QGP as formed in heavy ion collisions or quantum fields on black hole space-times.

We began the chapter by briefly reviewing the properties of rigidly rotating thermal states for scalar and fermion particles within the framework of RKT. The main feature is that, for both scalars and fermions, macroscopic quantities such as the energy and pressure diverge on the SLS but are regular inside it.

Next we constructed rigidly rotating thermal states within the canonical quantization approach to QFT on unbounded Minkowski space-time. Here there is a significant difference between scalar and fermion fields. In particular, rigidly rotating thermal states for scalars cannot be defined. The quantization of the fermion field is less constrained than that of the scalar field, and as a result we are able to define rigidly rotating thermal states for fermions. We computed the t.e.v.s of the FC, CC, AC and SET in these states. All t.e.v.s diverge on the SLS but are regular inside it. Relative to the RKT results, the quantum t.e.v.s diverge more rapidly as the SLS is approached. Quantum corrections therefore dominate close to the SLS. We stress that the advantage of the canonical quantization approach considered in this chapter is that it allows t.e.v.s to be expressed in integral form, which can then be used to obtain analytic (in the massless case) or numerical (in the massive case) results in a non-perturbative fashion, with arbitrary numerical precision, even in the regime where quantum corrections are dominant.

The toy model considered in this chapter is a good approximation to more physical rigidly rotating systems enclosed inside a reflecting boundary, except in the vicinity of the boundary. The key physics features are also shared with more complicated systems in curved space-time. We therefore conclude that our method based on canonical quantization can serve as a reliable tool to compute t.e.v.s in rigidly rotating systems of particles, in particular in setups relevant to relativistic heavy ion collisions, from the nearly classical regime to the quantum-dominated regime, with arbitrary numerical precision.

Acknowledgements The work of V. E. Ambruş is supported by a grant from the Romanian National Authority for Scientific Research and Innovation, CNCS-UEFISCDI, project number PN-III-P1-1.1-PD-2016-1423. The work of E. Winstanley is supported by the Lancaster-Manchester-Sheffield Consortium for Fundamental Physics under STFC grant ST/P000800/1 and partially supported by the H2020-MSCA-RISE-2017 Grant No. FunFiCO-777740.

References

1. Meier, D.L.: Black Hole Astrophysics: The Engine Paradigm. Springer, Berlin (2012)
2. Chandrasekhar, S.: The Mathematical Theory of Black Holes. Oxford University Press, Oxford (1985)
3. Hawking, S.W.: Particle creation by black holes. Commun. Math. Phys. **43**, 199–220 (1975)
4. Frolov, V.P., Thorne, K.S.: Renormalized stress-energy tensor near the horizon of a slowly evolving, rotating black hole. Phys. Rev. D **39**, 2125–2154 (1989)
5. Kay, B.S., Wald, R.M.: Theorems on the uniqueness and thermal properties of stationary, nonsingular, quasifree states on space-times with a bifurcate Killing horizon. Phys. Rep. **207**, 49–136 (1991)
6. Casals, M., Dolan, S.R., Nolan, B.C., Ottewill, A.C., Winstanley, E.: Quantization of fermions on Kerr space-time. Phys. Rev. D **87**, 064027 (2013)
7. Jacak, B.V., Müller, B.: The exploration of hot nuclear matter. Nature **337**, 310–314 (2012)
8. Kharzeev, D.E., Liao, J., Voloshin, S.A., Wang, G.: Chiral magnetic and vortical effects in high-energy nuclear collisions - a status report. Prog. Part. Nucl. Phys. **88**, 1–28 (2016)
9. Collaboration, S.T.A.R.: Global Λ-hyperon polarization in nuclear collisions. Nature **548**, 62–65 (2017)

10. Collaboration, S.T.A.R.: Global polarization of Λ-hyperons in Au+Au collisions at $\sqrt{s_{NN}} =$ 200GeV. Phys. Rev. C **98**, 014910 (2018)
11. Florkowski, W., Friman, B., Jaiswal, A., Speranza, E.: Relativistic fluid dynamics with spin. Phys. Rev. C **97**, 041901 (2018)
12. Becattini, F., Florkowski, W., Speranza, E.: Spin tensor and its role in non-equilibrium thermodynamics. Phys. Lett. B **789**, 419–425 (2019)
13. Buzzegoli, M., Becattini, F.: General thermodynamic equilibrium with axial chemical potential for the free Dirac field. JHEP **12**, 002 (2018)
14. Cercignani, C., Kremer, G.M.: The Relativistic Boltzmann Equation: Theory and Application. Birkhäuser Verlag, Basel (2002)
15. Becattini, F., Bucciantini, L., Grossi, E., Tinti, L.: Local thermodynamical equilibrium and the β-frame for a quantum relativistic fluid. Eur. Phys. J. C **75**, 191 (2015)
16. Becattini, F., Grossi, E.: Quantum corrections to the stress-energy tensor in thermodynamic equilibrium with acceleration. Phys. Rev. D **92**, 045037 (2015)
17. Ambruş, V.E.: Helical massive fermions under rotation. JHEP **08**, 016 (2020)
18. Ambruş, V.E., Chernodub, M.N., Helical vortical effects, helical waves, and anomalies of Dirac fermions. arXiv:1912.11034 [hep-th]
19. Jaiswal, A., Friman, B., Redlich, K.: Relativistic second-order dissipative hydrodynamics at finite chemical potential. Phys. Lett. B **751**, 548–552 (2015)
20. Wang, Q.: Global and local spin polarization in heavy ion collisions: a brief overview. Nucl. Phys. A **967**, 225–232 (2017)
21. Huang, X.-G., Koide, T.: Shear viscosity, bulk viscosity, and relaxation times of causal dissipative relativistic fluid-dynamics at finite temperature and chemical potential. Nucl. Phys. A **889**, 73–92 (2012)
22. Tanabashi, M., et al.: (Particle Data Group): The review of particle physics. Phys. Rev. D **98**, 030001 (2018)
23. Ambruş, V.E., Blaga, R.: Relativistic rotating Boltzmann gas using the tetrad formalism. Ann. West Univ. Timisoara, Ser. Phys. **58**, 89–108 (2015)
24. Florkowski, W., Ryblewski, R., Strickland, M.: Testing viscous and anisotropic hydrodynamics in an exactly solvable case. Phys. Rev. C **88**, 024903 (2013)
25. Kapusta, J.I., Landshoff, P.V.: Finite-temperature field theory. J. Phys. G: Nucl. Part. Phys. **15**, 267–285 (1989)
26. Vilenkin, A.: Quantum field theory at finite temperature in a rotating system. Phys. Rev. D **21**, 2260–2269 (1980)
27. Itzykson, C., Zuber, J.B.: Quantum Field Theory. Mcgraw-Hill, New York (1980)
28. Duffy, G., Ottewill, A.C.: The rotating quantum thermal distribution. Phys. Rev. D **67**, 044002 (2003)
29. Ambruş, V.E., Winstanley, E.: Rotating quantum states. Phys. Lett. B **734**, 296–301 (2014)
30. Letaw, J.R., Pfautsch, J.D.: The quantized scalar field in rotating coordinates. Phys. Rev. D **22**, 1345–1351 (1980)
31. Nicolaevici, N.: Null response of uniformly rotating Unruh detectors in bounded regions. Class. Quantum Grav. **18**, 5407–5411 (2001)
32. Iyer, B.R.: Dirac field theory in rotating coordinates. Phys. Rev. D **26**, 1900–1905 (1982)
33. Ambruş, V.E., Winstanley, E.: Rotating fermions inside a cylindrical boundary. Phys. Rev. D **93**, 104014 (2016)
34. Olver, F.W.J., Lozier, D.W., Boisvert, R.F., Clark, C.W.: NIST Handbook of Mathematical Functions. Cambridge University Press, New York (2010)
35. Rezzolla, L., Zanotti, O.: Relativistic Hydrodynamics. Oxford University Press, Oxford (2013)
36. Prokhorov, G.Y., Teryaev, O.V., Zakharov, V.I.: Axial current in rotating and accelerating medium. Phys. Rev. D **98**, 071901 (2018)
37. Vilenkin, A.: Macroscopic parity-violating effects: Neutrino fluxes from rotating black holes and in rotating thermal radiation. Phys. Rev. D **20**, 1807–1812 (1979)
38. Prokhorov, G.Y., Teryaev, O.V., Zakharov, V.I.: Effects of rotation and acceleration in the axial current: density operator vs Wigner function. JHEP **02**, 146 (2019)

39. Becattini, F., Chandra, V., Del Zanna, F., Grossi, E.: Relativistic distribution function for particles with spin at local thermodynamical equilibrium. Ann. Phys. **338**, 32–49 (2013)
40. Ván, P., Biró, T.S.: First order and stable relativistic dissipative hydrodynamics. Phys. Lett. B **709**, 106–110 (2012)
41. Ván, P., Biró, T.S.: Dissipation flow-frames: particle, energy, thermometer. In: Pilotelli, M., Beretta, G.P. (eds.): Proceedings of the 12th Joint European Thermodynamics Conference, Cartolibreria SNOOPY, 2013, pp. 546–551 (2013). arXiv:1305.3190 [gr-qc]
42. Bouras, I., Molnár, E., Niemi, H., Xu, Z., El, A., Fochler, O., Greiner, C., Rischke, D.H.: Investigation of shock waves in the relativistic Riemann problem: a comparison of viscous fluid dynamics to kinetic theory. Phys. Rev. C **82**, 024910 (2010)
43. Duff, M.J.: Twenty years of the Weyl anomaly. Class. Quantum Grav. **11**, 1387–1404 (1994)
44. Ambruş, V.E.: Quantum non-equilibrium effects in rigidly-rotating thermal states. Phys. Lett. B **771**, 151–156 (2017)
45. Hortacsu, M., Rothe, K.D., Schroer, B.: Zero energy eigenstates for the Dirac boundary problem. Nucl. Phys. B **171**, 530–542 (1980)
46. Chodos, A., Jaffe, R.L., Johnson, K., Thorn, C.B., Weisskopf, V.F.: A new extended model of hadrons. Phys. Rev. D **9**, 3471–3495 (1974)
47. Chernodub, M.N., Gongyo, S.: Edge states and thermodynamics of rotating relativistic fermions under magnetic field. Phys. Rev. D **96**, 096014 (2017)
48. Chernodub, M.N., Gongyo, S.: Interacting fermions in rotation: chiral symmetry restoration, moment of inertia and thermodynamics. JHEP **1701**, 136 (2017)
49. Chernodub, M.N., Gongyo, S.: Effects of rotation and boundaries on chiral symmetry breaking of relativistic fermions. Phys. Rev. D **95**, 096006 (2017)
50. Hawking, S.W., Ellis, G.F.R.: The Large-scale Structure of Space-time. Cambridge University Press, Cambridge (1973)
51. Moschella, U.: The de Sitter and anti-de Sitter sightseeing tour. Séminaire Poincaré **1**, 1–12 (2005)
52. Avis, S.J., Isham, C.J., Storey, D.: Quantum field theory in anti-de Sitter space-time. Phys. Rev. D **18**, 3565–3576 (1978)
53. Ambruş, V.E., Kent, C., Winstanley, E.: Analysis of scalar and fermion quantum field theory on anti-de Sitter spacetime. Int. J. Mod. Phys. D **27**, 1843014 (2018)
54. Ambruş, V.E., Winstanley, E.: Thermal expectation values of fermions on anti-de Sitter space-time. Class. Quant. Grav. **34**, 145010 (2017)
55. Kent, C., Winstanley, E.: Hadamard renormalized scalar field theory on anti-de Sitter spacetime. Phys. Rev. D **91**, 044044 (2015)
56. Ambruş, V.E., Winstanley, E.: Renormalised fermion vacuum expectation values on anti-de Sitter space-time. Phys. Lett. B **749**, 597–602 (2015)
57. Ambruş, V.E., Winstanley, E.: Quantum corrections in thermal states of fermions on anti-de Sitter space-time. AIP Conf. Proc. **1916**, 020005 (2017)
58. Kent, C., Winstanley, E.: The global rotating scalar field vacuum on anti-de Sitter space-time. Phys. Lett. B **740**, 188–191 (2015)
59. Ambruş, V.E., Winstanley, E.: Dirac fermions on an anti-de Sitter background. AIP Conf. Proc. **1634**, 40–49 (2015)
60. Casalderrey-Solana, J., Liu, H., Mateos, D., Rajagopal, K., Wiedemann, U.A.: Gauge/String Duality, Hot QCD and Heavy Ion Collisions. Cambridge University Press, Cambridge (2014)
61. DeWolfe, O., Gubser, S.S., Rosen, C., Teaney, D.: Heavy ions and string theory. Prog. Part. Nucl. Phys. **75**, 86–132 (2014)
62. Aharony, O., Gubser, S.S., Maldacena, J.M., Ooguri, H., Oz, Y.: Large N field theories, string theory and gravity. Phys. Rep. **323**, 183–386 (2000)
63. Ammon, M., Erdmenger, J.: Gauge/Gravity Duality: Foundations and Applications. Cambridge University Press, Cambridge (2015)
64. Carter, B.: Hamilton-Jacobi and Schrodinger separable solutions of Einstein's equations. Commun. Math. Phys. **10**, 280–310 (1968)

65. Hawking, S.W., Hunter, C.J., Taylor, M.: Rotation and the adS/CFT correspondence. Phys. Rev. D **59**, 064005 (1999)
66. Kerr, R.P.: Gravitational field of a spinning mass as an example of algebraically special metrics. Phys. Rev. Lett. **11**, 237–238 (1963)
67. Hartle, J.B., Hawking, S.W.: Path integral derivation of black hole radiance. Phys. Rev. D **13**, 2188–2203 (1976)
68. Ottewill, A.C., Winstanley, E.: The renormalized stress tensor in Kerr space-time: general results. Phys. Rev. D **62**, 084018 (2000)
69. Ottewill, A.C., Winstanley, E.: Divergence of a quantum thermal state on Kerr space-time. Phys. Lett. A **273**, 149–152 (2000)
70. Duffy, G., Ottewill, A.C.: The renormalized stress tensor in Kerr space-time: numerical results for the Hartle-Hawking vacuum. Phys. Rev. D **77**, 024007 (2008)

Particle Polarization, Spin Tensor, and the Wigner Distribution in Relativistic Systems

5

Leonardo Tinti and Wojciech Florkowski

Abstract

Particle spin polarization is known to be linked both to rotation (angular momentum) and magnetization of many particle systems. However, in the most common formulation of relativistic kinetic theory, the spin degrees of freedom appear only as degeneracy factors multiplying phase-space distributions. Thus, it is important to develop theoretical tools that allow to make predictions regarding the spin polarization of particles, which can be directly confronted with experimental data. Herein, we discuss a link between the relativistic spin tensor and particle spin polarization, and elucidate the connections between the Wigner function and average polarization. Our results may be useful for the theoretical interpretation of heavy-ion data on spin polarization of the produced hadrons.

5.1 Introduction

In relativistic heavy-ion collisions, the produced matter is formed at extreme conditions of high temperature and density [1]. The exact details of the evolution of the resulting fireball are difficult to track, however, some generally accepted concepts are typically assumed now. In particular, the evidence has been found that the strongly interacting matter behaves as an almost perfect (low viscosity) fluid (for recent reviews see, for example, Refs. [2,3]). During its expansion, when the system is diluted enough, the matter undergoes a cross-over phase transition from a strongly interacting quark-gluon plasma to a hadron gas (for systems with negligible baryon

L. Tinti (✉)
Institut für Theoretische Physik, Johann Wolfgang Goethe-Universität,
Max-von-Laue-Str. 1, 60438 Frankfurt am Main, Germany
e-mail: dr.leonardo.tinti@gmail.com

W. Florkowski
Jagiellonian University, ul. prof. St. Łojasiewicza 11, 30-348 Kraków, Poland
e-mail: wojciech.florkowski@uj.edu.pl

© Springer Nature Switzerland AG 2021
F. Becattini et al. (eds.), *Strongly Interacting Matter under Rotation*,
Lecture Notes in Physics 987,
https://doi.org/10.1007/978-3-030-71427-7_5

number density [4,5]). Shortly thereafter, the system becomes too dilute to be properly treated as a fluid, the interaction effectively ends, and produced particles freely stream to the detectors (freeze-out).

The main purpose of relativistic hydrodynamics is to solve the four-momentum conservation equation, $\partial_\mu T^{\mu\nu} = 0$, and the (baryon) charge conservation equation, $\partial_\mu J^\mu = 0$, under some realistic approximations (i.e., with suitably chosen initial conditions given by the Glauber model [6] or the theory of color glass condensate [7], and with a realistic equation of state obtained from an interpolation between the lattice QCD simulations and hadron resonance gas calculations).

In this way, one obtains the four-momentum and charge fluxes at freeze-out. These are space-time densities, which are not directly connected with the momentum distributions (and polarization) measured by different experiments. The most common way to link the stress-energy tensor and the charge flux at the freeze-out to particle spectra makes use of a classical intuition [8]; namely, one assumes that after the freeze-out the system is described well enough by the distribution functions $f(x, \mathbf{p})$ for (noninteracting) particles and the corresponding functions $\bar{f}(x, \mathbf{p})$ for antiparticles. Matching the stress-energy tensor and charge current with the corresponding formulas from the relativistic kinetic theory, one can guess the form of the distribution functions, and from that predict the final spectra.

The purpose of this contribution is to extend this formalism to include particle spin polarization (for particles with spin 1/2). In the next section, we show the limitations of the traditional kinetic theory in relation to particle polarization degrees of freedom. In Sect. 5.3, we introduce the concept of the relativistic spin tensor and discuss its relation to particle polarization for a free Fermi-field. In Sect. 5.4, we show that the appropriate generalization of the distribution function is the Wigner distribution. We summarize and conclude in Sect. 5.5. Some useful expressions and transformations are given in Appendix A. For complementary information we refer to the recent reviews [10–13].

5.2   Relativistic Kinetic Theory and Its Limitations

In the classical relativistic kinetic theory, the charge current density J^μ and the stress-energy tensor $T^{\mu\nu}$ have a rather simple connection with the phase-space distributions of particles and antiparticles [9,14],

$$
\begin{aligned}
J^\mu &= \frac{g_S}{(2\pi)^3} \int \frac{d^3 p}{E_{\mathbf{p}}} \, p^\mu \left(f(x, \mathbf{p}) - \bar{f}(x, \mathbf{p}) \right), \\
T^{\mu\nu} &= \frac{g_S}{(2\pi)^3} \int \frac{d^3 p}{E_{\mathbf{p}}} \, p^\mu p^\nu \left(f(x, \mathbf{p}) + \bar{f}(x, \mathbf{p}) \right).
\end{aligned}
\tag{5.1}
$$

Here $g_S = 2S + 1$ is the degeneracy factor with S being the spin of (massive) particles. In order to properly take into account the polarizaton degrees of freedom, one can easily notice that the framework based on Eq. (5.1) should be extended to a matrix formalism. This is so, since the standard kinetic theory essentially assumes equipartition of various spin states.

In general, the expectation values of the charge current and the stress-energy tensor, determined in a generic state of the system described by the density matrix ρ, cannot be expressed by the integrals of the form (5.1) with the integrands depending on a single momentum variable. A generic density matrix can be written as

$$\rho = \sum_i \mathsf{P}_i \, |\psi_i\rangle \, \langle\psi_i| , \tag{5.2}$$

where P_i are classical (non-interfering) probabilities normalized to one, $\sum_i \mathsf{P}_i = 1$, and $|\psi_i\rangle$ are generic quantum states. We assume that $\langle\psi_i|\psi_i\rangle = 1$ and stress that $|\psi_i\rangle$'s are not necessarily eigenstates of the total energy, linear momentum, angular momentum, or charge operators.

Starting from the definition of the charge current operator

$$\hat{J}^\mu(x) = \bar{\Psi}(x)\gamma^\mu\Psi(x), \tag{5.3}$$

expressed by the noninteracting Fermi fields Ψ

$$\Psi(x) = \sum_r \int \frac{d^3p}{(2\pi)^3\sqrt{2E_\mathbf{p}}} \left[U_r(\mathbf{p})a_r(\mathbf{p})e^{-ip\cdot x} + V_r(\mathbf{p})b_r^\dagger(\mathbf{p})e^{ip\cdot x} \right], \tag{5.4}$$

we obtain the normal-ordered expectation value

$$
\begin{aligned}
J^\mu(x) &= \langle : \hat{J}^\mu(x) : \rangle = \mathrm{tr}\left(\rho : \hat{J}^\mu(x) : \right) = \\
&= \sum_{r,s} \int \frac{d^3p\, d^3p'}{(2\pi)^6\sqrt{2E_\mathbf{p}2E_{\mathbf{p}'}}} \Big[\langle a_r^\dagger(\mathbf{p})a_s(\mathbf{p}')\rangle \bar{U}_r(\mathbf{p})\gamma^\mu U_s(\mathbf{p}')e^{i(p-p')\cdot x} \\
&\qquad - \langle b_r^\dagger(\mathbf{p})b_s(\mathbf{p}')\rangle \bar{V}_s(\mathbf{p}')\gamma^\mu V_r(\mathbf{p})e^{i(p-p')\cdot x} \\
&\qquad + \langle a_r^\dagger(\mathbf{p})b_s^\dagger(\mathbf{p}')\rangle \bar{U}_r(\mathbf{p})\gamma^\mu V_s(\mathbf{p}')e^{i(p+p')\cdot x} \\
&\qquad + \langle b_r(\mathbf{p})a_s(\mathbf{p}')\rangle \bar{V}_r(\mathbf{p})\gamma^\mu U_s(\mathbf{p}')e^{-i(p+p')\cdot x} \Big].
\end{aligned}
\tag{5.5}
$$

It is easy to check that the stress-energy tensor has an analogous structure. In general, none of the expectation values of the creation-destruction operators vanishes and the integrals over three-momenta cannot be reduced to the Dirac delta functions.

We use the Wigner representation of the Clifford algebra and we adopt the convention for the massive eigenspinors[1]

$$
U_r(\mathbf{p}) = \sqrt{E_\mathbf{p} + m} \begin{pmatrix} \phi_r \\ \frac{\sigma \cdot \mathbf{p}}{E_\mathbf{p}+m} \phi_r \end{pmatrix},
$$
$$
V_r(\mathbf{p}) = \sqrt{E_\mathbf{p} + m} \begin{pmatrix} \frac{\sigma \cdot \mathbf{p}}{E_\mathbf{p}+m} \chi_r \\ \chi_r \end{pmatrix},
$$

(5.6)

with the two component vectors ϕ_r and χ_r being the eigenstates of the matrix $\sigma_z = \text{diag}(1, -1)$,

$$
\phi_1 = \begin{pmatrix} 1 \\ 0 \end{pmatrix}, \qquad \phi_1 = \begin{pmatrix} 0 \\ 1 \end{pmatrix},
$$
$$
\chi_1 = \begin{pmatrix} 0 \\ 1 \end{pmatrix}, \qquad \chi_2 = -\begin{pmatrix} 1 \\ 0 \end{pmatrix},
$$

(5.7)

therefore, the normalization of the states $|p, r\rangle$ and the anticommutation relations between the creation-destruction operators read

$$
\{a_s(\mathbf{q}), a_r^\dagger(\mathbf{p})\} = \{b_s(\mathbf{q}), b_r^\dagger(\mathbf{p})\} = (2\pi)^3 \delta_{rs} \delta^3(\mathbf{p} - \mathbf{q}),
$$
$$
\{a_s(\mathbf{q}), b_r(\mathbf{p})\} = \{a_s(\mathbf{q}), b_r(\mathbf{p})\} = \{a_s(\mathbf{q}), b_r(\mathbf{p})\} = \{a_s(\mathbf{q}), b_r(\mathbf{p})\} = 0, \quad (5.8)
$$
$$
|p, r\rangle = \sqrt{2E_\mathbf{p}} a_r^\dagger(\mathbf{p})|0\rangle \quad \Rightarrow \quad \langle q, s|p, r\rangle = (2\pi)^3 \delta_{r,s} \delta^3(\mathbf{p} - \mathbf{q}).
$$

Differently from the current density, the total charge has a similar structure to the one used in the kinetic theory

$$
\int d^3x \, J^0(x) = \int \frac{d^3p}{(2\pi)^3} \left[\sum_r \langle a_r^\dagger(\mathbf{p}) a_r(\mathbf{p}) \rangle - \sum_r \langle b_r^\dagger(\mathbf{p}) b_r(\mathbf{p}) \rangle \right]
$$

(5.9)

which directly comes from the normalization of the bispinors (5.6)

$$
U_r^\dagger(\mathbf{p}) U_s(\mathbf{p}) = V_r^\dagger(\mathbf{p}) V_s(\mathbf{p}) = 2E_\mathbf{p} \delta_{rs},
$$
$$
U_r^\dagger(\mathbf{p}) V_s(-\mathbf{p}) = 0,
$$

(5.10)

and the integral representation of the Dirac delta functions, $\delta^{(3)}(\mathbf{p} \pm \mathbf{p}')$, used to perform the volume integrals. Following the same steps, one can compute the total four-momentum

$$
\int d^3x \, T^{0\mu}(x) = \int \frac{d^3p}{(2\pi)^3} p^\mu \left[\sum_r \langle a_r^\dagger(\mathbf{p}) a_r(\mathbf{p}) \rangle + \sum_r \langle b_r^\dagger(\mathbf{p}) b_r(\mathbf{p}) \rangle \right].
$$

(5.11)

---

[1]Note that in the Weyl representation of the Clifford algebra a different explicit formula for the massive eigenspinors is typically used, but final results remain the same.

Equations (5.9) and (5.11) should be compared to the analogous formulas obtained from the kinetic-theory definitions (5.1),

$$
\begin{aligned}
\int d^3x \, J^0 &= \int \frac{d^3p}{(2\pi)^3} \left[g_s \int d^3x f(x, \mathbf{p}) - g_s \int d^3x \bar{f}(x, \mathbf{p}) \right], \\
\int d^3x \, T^{0\mu} &= \int \frac{d^3p}{(2\pi)^3} \, p^\mu \left[g_s \int d^3x f(x, \mathbf{p}) + g_s \int d^3x \bar{f}(x, \mathbf{p}) \right].
\end{aligned}
\tag{5.12}
$$

It is important to note that for noninteracting, spin $1/2$ particles the volume integrals are time independent. The collisionless Boltzmann equation for particles reads

$$
\partial_\mu \left(p^\mu f(x, \mathbf{p}) \right) = 0,
\tag{5.13}
$$

hence, the expression $p^\mu f$ is a conserved vector current (of course, the same property holds also for the noninteracting antiparticles described by the function $\bar{f}(x, \mathbf{p})$). The integral of the divergence (5.13) over a space-time region vanishes. Therefore, as long as the spatial boundary for the volume integral lies outside of the region in which the distribution function does not vanish, the integral $E_{\mathbf{p}} \int d^3x f$ remains the same at all times. Massive particles always have a positive $E_{\mathbf{p}}$, therefore, the volume integral itself is also time independent.[2]

Moreover, the space integral equals the integral over the freeze-out hypersurface [8], with $d\Sigma_\mu$ being the generic hypersurface element (note that $d\Sigma_\mu = (d^3x, 0, 0, 0)$ for the volume integrals in the lab frame)

$$
g_s \int d\Sigma_\mu \, p^\mu f(x, \mathbf{p}) = g_s E_{\mathbf{p}} \int d^3x f(x, \mathbf{p}) = (2\pi)^3 E_{\mathbf{p}} \frac{dN}{d^3p}.
\tag{5.14}
$$

The last term here is the invariant number of particles per momentum cell, which is consistent with the formula for the total number of particles

$$
N = \int d^3x \int \frac{d^3p}{(2\pi)^3} g_s f(x, \mathbf{p}) = \int d^3x \int \frac{d^3p}{(2\pi)^3 E_{\mathbf{p}}} E_{\mathbf{p}} g_s f(x, \mathbf{p}).
\tag{5.15}
$$

We note that the factor $(2\pi)^3$ (to be replaced by $(2\pi\hbar)^3$ if the natural units are not used) is included in the momentum integration measure rather than in the definition of the phase-space distributions, and $\int d^3p/E_{\mathbf{p}}$ is the Lorentz-covariant momentum integral.

In the general case, it can be proved that the expectation value of the number operator is a nonnegative quantity and the sum over the spin states is proportional to the (anti)particle number density in momentum space, hence

$$
N = \int \frac{d^3p}{(2\pi)^3} \sum_r \langle a_r^\dagger(\mathbf{p}) a_r(\mathbf{p}) \rangle
\tag{5.16}
$$

---

[2]Massless particles have a positive energy for any nonvanishing momentum and the situation for them is quite similar.

for particles, and

$$\bar{N} = \int \frac{d^3 p}{(2\pi)^3} \sum_r \langle b_r^\dagger(\mathbf{p}) b_r(\mathbf{p}) \rangle \tag{5.17}$$

for antiparticles. For more information see Appendix in Sect. 5.5.

Consequently, even if the expectation values of the charge current and the stress-energy tensor cannot be written as momentum integrals of the phase-space distributions, it is possible to have consistent distribution functions for the particles and antiparticles in the sense that they can reproduce the correct invariant momentum (anti)particle densities

$$g_s \int d^3 x f(x, \mathbf{p}) = \sum_r \langle a_r^\dagger(\mathbf{p}) a_r(\mathbf{p}) \rangle,$$
$$g_s \int d^3 x \bar{f}(x, \mathbf{p}) = \sum_r \langle b_r^\dagger(\mathbf{p}) b_r(\mathbf{p}) \rangle. \tag{5.18}$$

For heavy-ion collisions, this implies that for any given J^μ and $T^{\mu\nu}$ at freeze-out, one can construct a pair of the distribution functions $(f, \bar{f})$ that provide the same total current, energy, and linear momentum

$$\int \frac{d^3 p}{(2\pi)^3} \left[g_s \int d^3 x f(x, \mathbf{p}) - g_s \int d^3 x \bar{f}(x, \mathbf{p}) \right] =$$
$$= \int \frac{d^3 p}{(2\pi)^3} \left[\sum_r \langle a_r^\dagger(\mathbf{p}) a_r(\mathbf{p}) \rangle - \sum_r \langle b_r^\dagger(\mathbf{p}) b_r(\mathbf{p}) \rangle \right], \tag{5.19}$$

$$\int \frac{d^3 p}{(2\pi)^3} p^\mu \left[g_s \int d^3 x f(x, \mathbf{p}) + g_s \int d^3 x \bar{f}(x, \mathbf{p}) \right] =$$
$$= \int \frac{d^3 p}{(2\pi)^3} p^\mu \left[\sum_r \langle a_r^\dagger(\mathbf{p}) a_r(\mathbf{p}) \rangle + \sum_r \langle b_r^\dagger(\mathbf{p}) b_r(\mathbf{p}) \rangle \right]. \tag{5.20}$$

We note that the distributions functions obtained from the conditions (5.19) and (5.20) provide the correct total charge and four-momentum if used in Eqs. (5.1) to define the current density and the energy-momentum tensor. However, very different distribution functions may provide (after integration) the same macroscopic quantities. In certain cases, to remove such ambiguity, one can use additional physical insights. For example, the specific forms of the distribution functions can be introduced for systems being in (local) thermodynamic equilibrium or close to such a state.

In any case, it is important to note that the total charge is sensitive to the imbalance between particles and antiparticles, and the total four- momentum is sensitive to the momentum distribution of both particles and antiparticles. Therefore, the conditions (5.19) and (5.20) provide an important constraint on the distribution functions. However, the right-hand sides in (5.19) and (5.20) are sums over the spin states. Being insensitive to polarization, they are not useful to check if a given extension of kinetic theory reproduces the average polarization in a satisfactory manner.

In the next section, we will argue that the relativistic spin tensor is sensitive to particle polarization in a very similar way as the charge current is sensitive to particle-antiparticle imbalance and the stress-energy tensor controls the average particle momentum.

5.3   The Relativistic Spin Tensor as a Polarization Sensitive Macroscopic Object

In this section, we introduce and discuss in more detail one of the main objects of our interest, namely, the relativistic spin tensor. Although it is less well-known compared to the tensors analyzed in the previous section, we are going to demonstrate that its intuitive understanding as a quantity related to particle's polarization is indeed correct.

The Noether theorem links the symmetries of the action to conserved charges and, to a lesser extent, conserved currents. If the action $\mathcal{A}$ contains only first order derivatives of the fields $\phi^a(x)$, we can write [15]

$$\mathcal{A}[\phi^a] = \int d^4x \, \mathcal{L}(\phi, \partial_\mu \phi, x), \tag{5.21}$$

where $\mathcal{L}$ is the Lagrangian density. If the action is invariant with respect to an infinitesimal transformation

$$\begin{aligned} x^\mu &\to \xi^\mu = x^\mu + \epsilon \delta x^\mu, \\ \phi^a(x) &\to \alpha^a(\xi) = \phi^a(x) + \epsilon \delta \phi^a(x) + \epsilon \delta x^\mu \partial_\mu \phi^a(x), \end{aligned} \tag{5.22}$$

one can extract a conserved current

$$Q^\mu = \left[\frac{\partial \mathcal{L}}{\partial(\partial_\mu \phi^a)} \partial_\nu \phi^a - \mathcal{L} \delta_\nu^\mu \right] \delta x^\nu - \frac{\partial \mathcal{L}}{\partial(\partial_\mu \phi^a)} \left(\delta \phi^a(x) + \delta x^\nu \partial_\nu \phi^a(x) \right),$$

$$\partial_\mu Q^\mu = 0, \tag{5.23}$$

where the summation over repeated indices is understood. Because of the vanishing divergence, the integral over a space-time region of (5.23) vanishes. Hence, if the field flux at the space boundary vanishes,[3] the space integral of Q^0 is a constant of motion. For instance, considering the action of a free, massive, spin 1/2 spinor field Ψ

$$\mathcal{A} = \int d^4x \left[\frac{i}{2} \bar{\Psi}(x) \gamma^\mu \overset{\leftrightarrow}{\partial}_\mu \Psi(x) - m \bar{\Psi}(x) \Psi(x) \right], \tag{5.24}$$

---

[3]The space-time region might be finite, with the fields going to zero at the boundary or infinite, as long as the fields decay fast enough to have a vanishing flux at infinity.

one has the internal symmetry under a global phase change of the fields with: $\delta x^\mu \equiv 0$, $\delta \Psi = i\epsilon \Psi$, and $\delta \bar{\Psi} = -i\epsilon \bar{\Psi}$. The corresponding current in this case is the charge current J^μ

$$J^\mu(x) = \bar{\Psi}(x)\gamma^\mu \Psi(x). \tag{5.25}$$

The invariance under space-time translations (with δx^μ being a constant and $\alpha(\xi) - \phi(x) = -\delta x^\mu \partial_\mu \phi$) yields the canonical stress-energy tensor $T_c^{\mu\nu}$ as the conserved current. The conserved charge in this case is the total four-momentum of the system

$$T_c^{\mu\nu}(x) = \frac{i}{2}\bar{\Psi}(x)\gamma^\mu \overleftrightarrow{\partial}^\nu \Psi(x) - g^{\mu\nu}\mathcal{L} \equiv \frac{i}{2}\bar{\Psi}(x)\gamma^\mu \overleftrightarrow{\partial}^\nu \Psi(x). \tag{5.26}$$

In the last passage, we have made use of the equations of motion of the fields.

Finally, we consider the invariance under the Lorentz group, i.e., boosts and rotations. The representation of the Lorentz group is the source of spin in quantum field theory, it is then expected that the conserved currents, in this case, are sensitive to spin polarization. In general, for an infinitesimal Lorentz transformation one has $\delta x_\mu = \omega_{\mu\nu} x^\nu$ with constant $\omega_{\mu\nu} = -\omega_{\nu\mu}$. The fields will change, according to their representation, following the rule $\delta\phi^a + \delta x^\mu \partial_\mu \phi^a = -i/2\omega_{\mu\nu}(\Sigma^{\mu\nu})_b^a \phi^b$. One obtains then the conserved angular-momentum flux density

$$M_c^{\lambda,\mu\nu} = x^\mu T_c^{\lambda\nu} - x^\nu T_c^{\lambda\mu} - i\frac{\partial\mathcal{L}}{\partial(\partial_\lambda \phi^a)}(\Sigma^{\mu\nu})_b^a \phi^b. \tag{5.27}$$

We note that the comma between the first index and the last two is used to emphasize the fact that $M_c^{\lambda,\mu\nu} = -M_c^{\lambda,\nu\mu}$ (different orders of the indices and conventions are used by different authors).

The first two terms in (5.27) depend on the canonical stress-energy tensor already obtained from the space-time translational invariance of the action and represent the orbital part of the angular momentum. The last term in (5.27) defines the spin contribution to the angular momentum and is called a canonical spin tensor. With

$$\Sigma^{\mu\nu} = \frac{i}{4}\left[\gamma^\mu, \gamma^\nu\right], \tag{5.28}$$

the canonical spin tensor reads

$$S_c^{\lambda,\mu\nu}(x) = \frac{i}{8}\bar{\Psi}(x)\left\{\gamma^\lambda, \left[\gamma^\mu, \gamma^\nu\right]\right\}\Psi(x). \tag{5.29}$$

Using the anticommutation relations $\{\gamma^\mu, \gamma^\nu\} = 2g^{\mu\nu}$, it is straightforward to check that $S_c^{\lambda,\mu\nu}$ is totally antisymmetric under the exchange of any indices.

Differently from the total charges (i.e., quantities obtained by the volume integrals), the conserved density currents given by the Noether theorem are not uniquely defined. Whatever the conserved current Q^μ is originally derived, if one builds from the fields a tensor $C^{\alpha\mu} = -C^{\mu\alpha}$, called a superpotential, the new current

$Q'^\mu = Q^\mu + \partial_\alpha C^{\alpha\mu}$ is equally conserved and provides the same conserved total charge. In the particular case of the angular momentum flux and the stress-energy tensor, there is a class of well-known transformations that leave the conserved total charges invariant (i.e., the generators of the Poincaré group). They are called pseudo-gauge transformations and have the form [19]

$$T'^{\mu\nu} = T^{\mu\nu} + \frac{1}{2}\partial_\lambda \left(\mathcal{G}^{\lambda,\mu\nu} - \mathcal{G}^{\mu,\lambda\nu} - \mathcal{G}^{\nu,\lambda\mu}\right),$$
$$S'^{\lambda,\mu\nu} = S^{\lambda,\mu\nu} - \mathcal{G}^{\lambda,\mu\nu} - \partial_\alpha \Xi^{\alpha\lambda,\mu\nu}. \tag{5.30}$$

The tensors $T^{\mu\nu}$ and $S^{\lambda,\mu\nu}$ on the right-hand side of Eq. (5.30) can be either the canonical ones or the already transformed ones. The auxiliary tensor $\mathcal{G}^{\lambda,\mu\nu}$ must be antisymmetric in the last two indices, while $\Xi^{\alpha\lambda,\mu\nu}$ should be antisymmetric in both the first two and the last two indices.

For any pair of the stress-energy and spin tensors, the following relations hold:

$$\partial_\mu T^{\mu\nu} = 0,$$
$$\partial_\lambda S^{\lambda,\mu\nu} = -\left(T^{\mu\nu} - T^{\nu\mu}\right), \tag{5.31}$$

and the total four-momentum P^μ and angular momentum $J^{\mu\nu}$ read

$$P^\mu = \int d^3x\, T^{0\mu},$$
$$J^{\mu\nu} = \int d^3x\, \left(x^\mu T^{0\nu} - x^\nu T^{0\mu} + S^{0\mu\nu}\right). \tag{5.32}$$

By construction, the last integrals are equal to those obtained with the canonical tensors, therefore, Eq. (5.31) can be equally well considered as the local four-momentum and angular momentum conservation equations. In this work, we are not going to discuss which pair is the most appropriate or convenient to represent the physical densities of the (angular) momentum of a physical system. This point is reviewed in Ref. [20].

A very special case of the transformations defined by Eq. (5.30) is the Belinfante symmetrization procedure. In this case, one starts with the canonical tensors $T_c^{\mu\nu}$ and $S_c^{\lambda,\mu\nu}$, and takes $\mathcal{G}^{\lambda,\mu\nu} = S_c^{\lambda,\mu\nu}$ and $\Xi^{\alpha\lambda,\mu\nu} = 0$. As a result, one obtains a vanishing new spin tensor $S_B^{\lambda,\mu\nu} = 0$, and the angular momentum conservation becomes just the requirement that the antisymmetric part of $T_B^{\mu\nu}$ vanishes. There is an apparent paradox here, namely, that one starts with ten independent equations for 16+24=40 degrees of freedom in (5.31)[4] and ends up with only four equations for 10 degrees of freedom; the vanishing divergence of a symmetric rank two tensors.

---

[4]This is so in the case of an arbitrary original spin tensor which is antisymmetric only in the last two indices. For the canonical spin tensor that is totally antisymmetric, the number of independent components is 16+4=20.

It is possible to resolve this paradox by writing the result of the Belinfante symmetrization in a less deceitful way, namely

$$\partial_\mu T_B^{\{\mu\nu\}} = 0,$$

$$T_B^{[\mu\nu]} = 0 \Rightarrow \partial_\lambda S_c^{\lambda\mu\nu} = -\left(T_c^{\mu\nu} - T_c^{\nu\mu}\right), \tag{5.33}$$

$$P^\mu = \int d^3x\, T_B^{\{0\mu\}}, \qquad J^{\mu\nu} = \int d^3x\, \left(x^\mu T_B^{\{0\nu\}} - x^\nu T_B^{\{0\mu\}}\right).$$

The middle line of Eq. (5.33) emphasizes an important point—although the symmetric part $T_B^{\{\mu\nu\}}$ of the Belinfante symmetrized stress-energy tensor $T_B^{\mu\nu}$ is separately conserved and both the total four-momentum and angular momentum can be expressed through $T_B^{\{\mu\nu\}}$, the requirement that the antisymmetric part of $T_B^{[\mu\nu]}$ vanishes should be treated as a complementary set of equations. Indeed, starting with the canonical tensors obtained for the Dirac field (5.26) and (5.29), and performing the Belinfante symmetrization, one obtains

$$T_B^{\mu\nu} = \frac{i}{2}\bar{\Psi}(x)\gamma^\mu \overleftrightarrow{\partial}^\nu \Psi(x) - \frac{i}{16}\partial_\lambda \left(\Psi(x)\left\{\gamma^\lambda, \left[\gamma^\mu, \gamma^\nu\right]\right\}\Psi(x)\right), \tag{5.34}$$

a formula which is not manifestly symmetric under a $\mu \leftrightarrow \nu$ exchange. In order to show that (5.34) is indeed symmetric, one has two options:

i) Solve exactly the Euler–Lagrange equations of motion for the fields (possible for a free field) and directly check the symmetry of (5.34).

ii) Make use of the angular momentum conservation for the canonical tensors, accepting them as another set of equations.

Consequently, although one needs a rank two, symmetric, and conserved tensor in order to make a comparison with kinetic theory (since the stress-energy tensor in kinetic theory is symmetric by construction, see Eq (5.1)), one can always consider the equation

$$\partial_\lambda S_c^{\lambda\mu\nu} = -\partial_\lambda S_c^{\lambda\mu\nu} = -\left(T_c^{\mu\nu} - T_c^{\nu\mu}\right), \tag{5.35}$$

which remains valid. We note that the equations for the fields are usually far from being trivial and the same property holds for the symmetrization procedure that starts from some generic $T^{\mu\nu}$ and nonvanishing $S^{\lambda,\mu\nu}$. For instance, a different Lagrangian density having the same Euler–Lagrange equation of motion for the fields, generally lead to different canonical tensors. In any case, all of these tensors lead to the same conserved charges and provide the same number of equations. For a modern and more detailed discussion over the different possible choices of $T^{\mu\nu}$ and $S^{\lambda,\mu\nu}$, and their physical consequences, we refer to Refs. [21,22].

If one excludes the particular case of quantum anomalies,[5] very similar arguments to those presented above hold also in the quantum case for the operators built from

---

[5]Neither in a free theory nor in the standard model there are anomalies in the conservation laws for the four-momentum and angular momentum. However, one has to check on a case by case basis if this is so while dealing with a generic quantum field theory.

fundamental fields. In particular, the canonical spin tensor in the quantum case still reads as in Eq. (5.29), with the only addition that one has to renormalize it (make normal ordering) to avoid infinities related to the vacuum. The macroscopic, classical, spin tensor is the expectation value of the quantum counterpart. For a generic (pure or mixed) state of the system ρ we have

$$S_c^{\lambda\mu\nu} = \text{tr}\left(\rho : \hat{S}_c^{\lambda\mu\nu} :\right) = \frac{i}{8}\text{tr}\left(\rho : \bar{\Psi}(x)\left\{\gamma^\lambda, \left[\gamma^\mu, \gamma^\nu\right]\right\}\Psi(x) :\right). \quad (5.36)$$

This form is probably the most intuitive guess for a macroscopic object embedding the particle's polarization degrees of freedom, because in the total angular momentum operator

$$\hat{J}^{\mu\nu} = \int d^3x\ \Psi^\dagger(x)\left(\frac{i}{2}x^\mu\overset{\leftrightarrow}{\partial^\nu} - \frac{i}{2}x^\nu\overset{\leftrightarrow}{\partial^\mu} + \frac{i}{8}\gamma^0\left\{\gamma^0, \left[\gamma^\mu, \gamma^\nu\right]\right\}\right)\Psi(x), \quad (5.37)$$

$\hat{S}_c^{0,\mu\nu}$ describes the last term in the round brackets, depending on the gamma matrices. It is the only term that mixes the components of the spinor fields—the part stemming from $T_c^{\mu\nu}$ depends on the gradients, hence, can be interpreted as the relativistic QFT analog of $\mathbf{x} \times \mathbf{p}$, the orbital angular momentum of classical particles.

In general, similarly to the stress-energy tensor and the current, the macroscopic spin tensor (5.36) depends on both space-time coordinates and two momentum variables. However, its volume integral can be written in terms of a single momentum variable and the expectation values of creation/destruction operators, much like we have seen in the previous section. The space integral of S_c^{0ij} can be expected to be related to the sum of the spin polarization of particles. Defining the vector of matrices Σ_i in the following way:

$$\Sigma_i = \frac{i}{4}\varepsilon_{ijk}[\gamma^j, \gamma^k] = \begin{pmatrix} \sigma_i & 0 \\ 0 & \sigma_i \end{pmatrix}, \qquad \forall i \in \{1, 2, 3\} \quad (5.38)$$

with σ_i being the 2×2 Pauli matrices, one has

$$\frac{1}{2}\varepsilon_{ijk}\int d^3x\ S_c^{0jk}$$

$$= \frac{1}{2}\sum_{r,s}\int d^3x \int \frac{d^3p\,d^3p'}{(2\pi)^6\sqrt{2E_{\mathbf{p}}2E_{\mathbf{p}'}}}\Big[\langle a_r^\dagger(\mathbf{p})a_s(\mathbf{p})\rangle U_r^\dagger(\mathbf{p})\Sigma_i U_s(\mathbf{p}')e^{i(p-p')\cdot x}$$

$$- \langle b_r^\dagger(\mathbf{p})b_s(\mathbf{p})\rangle V_s^\dagger(\mathbf{p})\Sigma_i V_r(\mathbf{p})e^{i(p-p')\cdot x}$$

$$+ \langle a_r^\dagger(\mathbf{p})b_s^\dagger(\mathbf{p})\rangle U_r^\dagger(\mathbf{p})\Sigma_i V_s(\mathbf{p}')e^{i(p+p')\cdot x}$$

$$+ \langle b_s(\mathbf{p})a_r(\mathbf{p})\rangle V_s^\dagger(\mathbf{p})\Sigma_i U_r(\mathbf{p}')e^{-i(p+p')\cdot x}\Big]$$

$$(5.39)$$

that can be rewritten as

$$\frac{1}{2}\sum_{r,s}\int\frac{d^3p}{(2\pi)^32E_{\mathbf{p}}}\left[\langle a_r^\dagger(\mathbf{p})a_s(\mathbf{p})\rangle U_r^\dagger(\mathbf{p})\Sigma_i U_s(\mathbf{p})\right.$$
$$-\langle b_r^\dagger(\mathbf{p})b_s(\mathbf{p})\rangle V_s^\dagger(\mathbf{p})\Sigma_i V_r(\mathbf{p})$$
$$+\langle a_r^\dagger(\mathbf{p})b_s^\dagger(-\mathbf{p})\rangle U_r^\dagger(\mathbf{p})\Sigma_i V_s(-\mathbf{p})e^{2iE_{\mathbf{p}}t}$$
$$\left.+\langle b_s(-\mathbf{p})a_r(\mathbf{p})\rangle V_r^\dagger(-\mathbf{p})\Sigma_i U_s(\mathbf{p})e^{-2iE_{\mathbf{p}}t}\right]. \tag{5.40}$$

Making use of the direct formulas for the massive eigenspinors[6]

$$U_r(\mathbf{p})=\sqrt{E_{\mathbf{p}}+m}\begin{pmatrix}\phi_r\\\frac{\sigma\cdot\mathbf{p}}{E_{\mathbf{p}}+m}\phi_r\end{pmatrix},$$
$$V_r(\mathbf{p})=\sqrt{E_{\mathbf{p}}+m}\begin{pmatrix}\frac{\sigma\cdot\mathbf{p}}{E_{\mathbf{p}}+m}\chi_r\\\chi_r\end{pmatrix}, \tag{5.41}$$

with the two component vectors ϕ_r and χ_r being the eigenstates of the matrix $\sigma_z = \mathrm{diag}(1,-1)$,

$$\phi_1=\begin{pmatrix}1\\0\end{pmatrix},\qquad\phi_1=\begin{pmatrix}0\\1\end{pmatrix},$$
$$\chi_1=\begin{pmatrix}0\\1\end{pmatrix},\qquad\chi_2=-\begin{pmatrix}1\\0\end{pmatrix}, \tag{5.42}$$

and the standard relations between the Pauli matrices

$$\{\sigma_i,\sigma_j\}=2\delta_{i,j},\qquad[\sigma_i,\sigma_j]=2i\,\varepsilon_{ijk}\,\sigma_k, \tag{5.43}$$

one can rewrite the matrix elements in (5.40) in the following way

$$U_r^\dagger(\mathbf{p})\Sigma_i U_s(\mathbf{p})=2m\,\phi_r\sigma_i\phi_s+\frac{2p_i}{E_{\mathbf{p}}+m}\phi_r(\mathbf{p}\cdot\sigma)\phi_s,$$
$$V_s^\dagger(\mathbf{p})\Sigma_i V_r(\mathbf{p})=2m\,\chi_s\sigma_i\chi_r+\frac{2p_i}{E_{\mathbf{p}}+m}\chi_s(\mathbf{p}\cdot\sigma)\chi_r, \tag{5.44}$$
$$U_r^\dagger(\mathbf{p})\Sigma_i V_s(-\mathbf{p})=\left(V_s^\dagger(-\mathbf{p})\Sigma_i U_r(\mathbf{p})\right)^*=-2i\sum_{j,k}\varepsilon_{ijk}\,p_j\,\phi_r\sigma_k\chi_s.$$

The first two terms do not correspond exactly to the polarization in the i'th direction of a particle or an antiparticle in an eigenstate of four-momentum p^μ, but they are very closely linked to it (we discuss this point in more detail in the next section).

---

[6]Note that in the Weyl representation of the Clifford algebra a different explicit formula for the massive eigenspinors is typically used, but the general conclusions remain the same.

In any case, the volume integral of the canonical spin tensor is strongly related to the polarization state of the fundamental excitations of the fields, before any phenomenological approximation. Because of this, it is a reasonably good candidate to study, if one wants to extend the standard treatment of hydrodynamics to include polarization degrees of freedom.

In general, the awkward mixed terms involving creation and destruction operators are present in Eq. (5.39). They introduce time dependence in the volume integral (5.39) and have no clear interpretation in terms of particle–antiparticle degrees of freedom. Regarding the time dependence, this is not completely unexpected since the canonical spin tensor is not conserved, see Eq. (5.35), hence the volume integral depends on the time at which it is done. As explained in Appendix of Sect. 5.5, the expectation values of the mixed terms $\langle a^\dagger b^\dagger \rangle$ and $\langle ab \rangle$ do not vanish only if the quantum state of the system is a superposition of states, and among them, some states differ in the number of particles/antiparticles by a single particle–antiparticle pair. How much relevant are these kind of states in a heavy-ion collision environment is yet to be understood.

If one considers non-canonical spin tensors obtained through a pseudo-gauge transformation (5.30), it is possible to remove the time dependent part. For instance, the transformation proposed in [9] has the form

$$\hat{\Xi}^{\alpha\beta,\mu\nu} = 0, \qquad \hat{\mathcal{G}}^{\lambda,\mu\nu} = -\frac{1}{8m}\bar{\Psi}(x)\left(\left[\gamma^\lambda, \gamma^\mu\right]\overset{\leftrightarrow}{\partial^\nu} - \left[\gamma^\lambda, \gamma^\nu\right]\overset{\leftrightarrow}{\partial^\mu}\right)\Psi(x),$$
(5.45)

which results in a conserved spin tensor

$$\mathcal{S}^{\lambda,\mu\nu} = S_c^{\lambda\mu\nu} - \mathcal{G}^{\lambda,\mu\nu}, \qquad \partial_\lambda \mathcal{S}^{\lambda,\mu\nu} = 0.$$
(5.46)

Hence, the flux of the spin density, described by $\mathcal{S}^{\lambda,\mu\nu}$, across the freeze-out hypersurface is equal to the volume integral of $\mathcal{S}^{0,\mu\nu}$ at later times. The latter can be computed following the same steps used for the canonical tensor. After lengthy but straightforward calculations one obtains a time-independent formula that is still strongly related to the polarization degrees of freedom

$$\frac{1}{2}\varepsilon_{ijk}\int d^3x \, \mathcal{S}^{0,jk} =$$

$$= \frac{1}{2}\sum_{r,s}\int\frac{d^3p}{(2\pi)^3}\left[\langle a_r^\dagger(\mathbf{p})a_s(\mathbf{p})\rangle\left(\frac{E_\mathbf{p}}{m}\phi_r\sigma_i\phi_s - \frac{p^i}{m(E_\mathbf{p}+m)}\phi_r(\mathbf{p}\cdot\boldsymbol{\sigma})\phi_s\right)\right.$$

$$\left. -\langle b_r^\dagger(\mathbf{p})b_s(\mathbf{p})\rangle\left(\frac{E_\mathbf{p}}{m}\chi_s\sigma_i\chi_r - \frac{p^i}{m(E_\mathbf{p}+m)}\chi_s(\mathbf{p}\cdot\boldsymbol{\sigma})\chi_r\right)\right].$$
(5.47)

It is important to note that the term in the brackets, despite reducing to the polarization of a two component spinor in the nonrelativistic limit ($m \to \infty$), does not correspond to the polarization of a relativistic particle. Therefore, the last integral must not be confused with the average (relativistic) polarization multiplied by the average number of (anti)particles. For a discussion of the spin tensor in the context of relativistic thermodynamics, we refer to [23,24], see also Becattini's review in this monograph [13].

5.4   Particle Polarization, the Wigner Distribution, and the Polarization Flux Pseudotensor

We have already shown that the spin tensor is a macroscopic object sensitive to the polarization of the excitations of a free quantum field. In this section, we show that the relativistic Wigner distribution is the appropriate extension of the distribution function, that takes into account the polarization degrees of freedom. In particular, the Wigner distribution of a generic state of the system can be linked to the average polarization of particles with fixed momentum, which is probably the most important thing from the point of view of comparisons of theory predictions with experimental data.

Our starting point is a relativistic polarization pseudovector, namely, the relativistic counterpart of the expectation value $\langle \psi | \boldsymbol{\sigma} | \psi \rangle$ used for a two component, nonrelativistic spinor. An important property to take into account is that the latter is a constant of motion for free particles (the free hamiltonian commutes with the Pauli matrices). Thus, the most straightforward way to generalize the concept of $\langle \psi | \boldsymbol{\sigma} | \psi \rangle$ is to look at the classical (non-quantum) relativistic generalization of the internal angular momentum and to apply the same reasoning to the operators in QFT.

For a classical (extended) object the angular momentum reads $\mathbf{j} = \mathbf{x} \times \mathbf{p} + \mathbf{s}$, with $\mathbf{s}$ being the intrinsic angular momentum.[7] The immediate relativistic generalization is [25,26]

$$j^{\mu\nu} = x^\mu p^\nu - x^\nu p^\mu + s^{\mu\nu}, \tag{5.48}$$

with an antisymmetric tensor $s^{\mu\nu} = -s^{\nu\mu}$. It is easy to notice that the components $(1/2) \sum_{j,k} \varepsilon_{ijk} j^{jk}$ describe the angular momentum, while the components j^{0i} are needed for relativistic covariance, to have the correct transformation rules changing the frame of reference.

The polarization pseudovector Π^μ is proportional to the dual of the angular momentum, contracted with the four-momentum[8]

$$\Pi^\mu = -\frac{1}{2m} \varepsilon^{\mu\nu\rho\sigma} j_{\nu\rho} p_\sigma = -\frac{1}{2m} \varepsilon^{\mu\nu\rho\sigma} s_{\nu\rho} p_\sigma. \tag{5.49}$$

---

[7]At the classical level, the latter corresponds to rotation with respect to an internal axis of the extended object.

[8]A very similar definition is used for the Pauli–Lubanski pseudovector. It follows the same construction procedure, but without mass in the denominator. Besides different physical dimensions, it is a very close concept which is well defined in the massless case. Since we focus on massive fields herein, we are not going to analyze it. It is useful to notice, however, that using the Pauli–Lubanski definition, one can follow the same steps in the massless case, obtaining the helicity distribution instead of the polarization one.

Here we have made use of the definition (5.48), removing the contribution of the orbital momentum since it is proportional to the four-momentum and vanishes after contraction with the Levi-Civita symbol. In any inertial reference frame comoving with the system (hence for $\mathbf{p} = 0$) the polarization pseudovector has a vanishing time component, and the space components are just the intrinsic angular momentum $\mathbf{s}$.

Since Π^μ is the contraction of an antisymmetric object with the totally antisymmetric $\varepsilon^{\mu\nu\rho\sigma}$ and the four-momentum p^μ, one needs only tree components to fully describe it, for instance, the space ones. Having these comments in mind, we obtain

$$0 = p_\mu \Pi^\mu = E_\mathbf{p} \Pi^0 - \mathbf{p} \cdot \mathbf{\Pi} \Rightarrow \Pi^0 = \frac{\mathbf{p} \cdot \mathbf{\Pi}}{E_\mathbf{p}}. \tag{5.50}$$

It is particularly useful to write the polarization pseudovector in the comoving frame, $\mathbf{\Pi}_{\text{com.}}$, in terms of the polarization in the lab frame. It will serve us later to make a direct connection between the relativistic polarization operator and the nonrelativistic one, $\langle \psi | \boldsymbol{\sigma} | \psi \rangle$.

A boost $\Lambda_\mathbf{p}$ from the lab frame to the comoving frame is characterized by the speed $\beta = \|\mathbf{p}\|/E_\mathbf{p}$ and the Lorentz gamma factor $\gamma = 1/\sqrt{1 - \beta^2} = E_\mathbf{p}/m$. The zeroth component of Π^μ must vanish after such a boost, as immediately follows from Eq. (5.50)

$$\Pi^0 \xrightarrow{\Lambda_\mathbf{p}} \Pi^0_{\text{com.}} = \gamma \left(\Pi^0 - \beta \frac{\mathbf{\Pi} \cdot \mathbf{p}}{\|\mathbf{p}\|} \right) = \gamma \left(\Pi^0 - \frac{\mathbf{\Pi} \cdot \mathbf{p}}{E_\mathbf{p}} \right) \equiv 0. \tag{5.51}$$

On the other hand the non-trivial spatial part reads

$$\begin{aligned} \mathbf{\Pi}_{\text{com.}} &= \mathbf{\Pi} - \frac{\mathbf{\Pi} \cdot \mathbf{p}}{\|\mathbf{p}\|^2} \mathbf{p} + \gamma \left(\frac{\mathbf{\Pi} \cdot \mathbf{p}}{\|\mathbf{p}\|} - \beta \Pi^0 \right) \frac{\mathbf{p}}{\|\mathbf{p}\|} \\ &= \mathbf{\Pi} - \frac{\mathbf{\Pi} \cdot \mathbf{p}}{\|\mathbf{p}\|^2} \left[1 - \frac{E_\mathbf{p}}{m} \left(1 - \frac{\|p\|^2}{E_\mathbf{p}^2} \right) \right] \mathbf{p} \\ &= \mathbf{\Pi} - \frac{\mathbf{\Pi} \cdot \mathbf{p}}{\|\mathbf{p}\|^2} \left[\frac{E_\mathbf{p} - m}{E_\mathbf{p}} \equiv \frac{\|\mathbf{p}^2\|}{E_\mathbf{p}(E_\mathbf{p} + m)} \right] \mathbf{p} \\ &= \mathbf{\Pi} - \frac{\mathbf{\Pi} \cdot \mathbf{p}}{E_\mathbf{p}(E_\mathbf{p} + m)} \mathbf{p}. \end{aligned} \tag{5.52}$$

In relativistic quantum field theory one has the operator analog of the polarization (5.49) for a massive Dirac field, namely, the operator

$$\hat{\Pi}^\mu = -\frac{1}{2m} \varepsilon^{\mu\nu\rho\sigma} : \hat{J}_{\nu\rho} :: \hat{P}_\sigma : . \tag{5.53}$$

We note that one has to use normal ordering for the two operators separately, because otherwise the expectation value of $\hat{\Pi}^\mu$ in single particle states would vanish.[9] The most general one-particle state $|\psi_1\rangle$ for a free field reads

$$|\psi_1\rangle = \sum_r \int \frac{d^3 p}{(2\pi)^3 2E_{\mathbf{p}}} \psi_1(p, r) |p, r\rangle, \tag{5.54}$$

with the normalization

$$1 = \langle\psi_1|\psi_1\rangle = \sum_r \int \frac{d^3 p}{(2\pi)^3 2E_{\mathbf{p}}} \psi_1^*(p, r) \psi_1(p, r), \tag{5.55}$$

in which we made use of the normalization of the states

$$\langle p, r|q, s\rangle = 2E_{\mathbf{p}}(2\pi)^3 \delta_{rs}\, \delta^3(\mathbf{p} - \mathbf{p}'). \tag{5.56}$$

The polarization vector then reads

$$\langle\psi_1|\hat{\Pi}^\mu|\psi_1\rangle = -\frac{1}{2m}\varepsilon^{\mu\nu\rho\sigma} \sum_{r,r'} \int \frac{d^3 p\, d^3 p'}{(2\pi)^6 2E_{\mathbf{p}} 2E_{\mathbf{p}'}} \psi_1^*(p', r')\psi_1(p, r) \times$$
$$\times \langle p', r'| : \hat{J}_{\nu\rho} :: \hat{P}_\sigma : |p, r\rangle. \tag{5.57}$$

Using the anticommutation relations

$$\{a_s(\mathbf{q}), a_r^\dagger(\mathbf{p})\} = (2\pi)^3 \delta_{rs}\delta^3(\mathbf{p} - \mathbf{q}), \tag{5.58}$$

and taking into account the definition

$$|p, r\rangle = \sqrt{2E_{\mathbf{p}}} a_r^\dagger(\mathbf{p})|0\rangle, \tag{5.59}$$

it is relatively straightforward to prove that

$$: \hat{P}_\sigma : |p, r\rangle = p_\sigma|p, r\rangle, \tag{5.60}$$

where $\hat{P}^\mu$ is the total four-momentum operator

$$: \hat{P}^\mu := \sum_s \int \frac{d^3 q}{(2\pi)^3} q^\mu \left[a_s^\dagger(\mathbf{q})a_s(\mathbf{q}) + b_s^\dagger(\mathbf{q})b_s(\mathbf{q}) \right]. \tag{5.61}$$

---

[9]If one applies the normal ordering $: \hat{J}_{\nu\rho}\hat{P}_\sigma :$ at the operator level there are two destruction operators on the left-hand side, which annihilate any single particle state. One would need at least two (anti)particle states to have a nonvanishing expectation value.

The situation is slightly more complicated for the expectation value of the angular momentum operator $\langle p', r' | : \hat{J}_{\nu\rho} : | p, r \rangle$. One can check that a non-zero contribution reads

$$
\langle \psi_1 | \hat{\Pi}^\mu | \psi_1 \rangle = -\frac{1}{2m} \varepsilon^{\mu\nu\rho\sigma} \sum_{r,r'} \int \frac{d^3 p \, d^3 p'}{(2\pi)^6 2 E_\mathbf{p} 2 E_{\mathbf{p}'}} \psi_1^*(p', r') \psi_1(p, r) p_\sigma \times
$$

$$
\times \, \langle p', r' | : \hat{J}_{\nu\rho} : | p, r \rangle =
$$

$$
= -\frac{1}{2m} \varepsilon^{\mu\nu\rho\sigma} \sum_{r,r'} \int d^3 x \int \frac{d^3 p \, d^3 p'}{(2\pi)^6 2 E_\mathbf{p} 2 E_{\mathbf{p}'}} \psi_1^*(p', r') \psi_1(p, r) p_\sigma \times
$$

$$
\times \, \frac{i}{8} U_{r'}^\dagger(\mathbf{p}') \gamma^0 \left\{ \gamma^0, \left[\gamma_\nu, \gamma_\rho \right] \right\} U_r(\mathbf{p}) e^{-i(p-p') \cdot x} = \tag{5.62}
$$

$$
= -\frac{1}{4m} \varepsilon^{\mu i j \sigma} \varepsilon_{ijk} \sum_{r,r'} \int \frac{d^3 p}{(2\pi)^3} \frac{\psi_1^*(p, r') \psi_1(p, r)}{2 E_\mathbf{p}} \frac{p_\sigma}{2 E_\mathbf{p}} U_{r'}^\dagger(\mathbf{p}) \Sigma_k U_r(\mathbf{p}).
$$

In particular, the space part of the polarization $\langle \psi_1 | \hat{\mathbf{\Pi}} | \psi_1 \rangle$ reads

$$
\langle \psi_1 | \hat{\mathbf{\Pi}} | \psi_1 \rangle = \frac{1}{4m} \sum_{r,s} \int \frac{d^3 p}{(2\pi)^3} \frac{\psi_1^*(p, r) \psi_1(p, s)}{2 E_\mathbf{p}} U_r^\dagger(\mathbf{p}) \begin{pmatrix} \sigma & 0 \\ 0 & \sigma \end{pmatrix} U_s(\mathbf{p}) =
$$

$$
= \frac{1}{2} \sum_{r,s} \int \frac{d^3 p}{(2\pi)^3} \frac{\psi_1^*(p, r) \psi_1(p, s)}{2 E_\mathbf{p}} \left[\phi_r \sigma \phi_s + \frac{\phi_r (\mathbf{p} \cdot \sigma) \phi_s}{m(E_\mathbf{p} + m)} \mathbf{p} \right]. \tag{5.63}
$$

At this point, with the help of (5.52), it is possible to highlight the link between the expected value we have just computed and the nonrelativistic polarization $\langle \psi | \sigma | \psi \rangle$. We first define the spin momentum-dependent density matrix

$$
f_{rs}(\mathbf{p}) = \frac{\psi_1^*(p, r) \psi_1(p, s)}{2 E_\mathbf{p}}, \tag{5.64}
$$

which is a two-by-two Hermitian matrix in the indices r, s for every value of the momentum $\mathbf{p}$ and describes a polarized state. It is normalized to one, i.e., its trace over the r, s indices is unitary while integrated with the measure $\int d^3 p / (2\pi)^3$ (because of the normalization of the wave function (5.55)). Taking into account a momentum eigenstate of the form $f_{rs} = (2\pi)^3 H_{rs} \delta^3(\mathbf{p} - \tilde{\mathbf{p}})$,[10] one finds the polarization

$$
\frac{1}{2} \sum_{r,s} H_{rs} \left[\phi_r \sigma \phi_s + \frac{\phi_r (\tilde{\mathbf{p}} \cdot \sigma) \phi_s}{m(E_{\tilde{\mathbf{p}}} + m)} \tilde{\mathbf{p}} \right], \tag{5.65}
$$

---

[10]This is actually forbidden, since the wave function ψ_1 is a regular distribution in momentum. However, one can have the spin density matrix factorized in a Hermitian 2×2 matrix times and arbitrarily sharp gaussian in the momentum. Such a strongly delocalized state is, for all practical purposes, equivalent to a momentum eigenstate.

with $H_{rs} = H^*_{sr}$ and $\mathrm{tr}(H) = 1$. Making use of (5.52), one finds that the polarization in the comoving frame reads

$$\frac{1}{2} \sum_{r,s} H_{rs} \phi_r \sigma \phi_s, \tag{5.66}$$

which is, indeed, the polarization of a nonrelativistic spinor. Hence, it is limited between $-1/2$ and $1/2$ in each direction.

It is important to note, however, that the polarization in the lab frame is not limited. For instance

$$H = \begin{pmatrix} 1 & 0 \\ 0 & 0 \end{pmatrix} \Rightarrow \langle \hat{\mathbf{\Pi}} \rangle = \frac{1}{2} \begin{pmatrix} \frac{\tilde{p}_z \tilde{p}_x}{m(E_{\tilde{\mathbf{p}}}+m)} \\ \frac{\tilde{p}_z \tilde{p}_y}{m(E_{\tilde{\mathbf{p}}}+m)} \\ 1 + \frac{\tilde{p}_z^2}{m(E_{\tilde{\mathbf{p}}}+m)} \end{pmatrix}, \tag{5.67}$$

which has, manifestly, arbitrarily large components as long as $\tilde{p}_z \neq 0$, maintaining the expected polarization $(0, 0, 1/2)$ in the comoving frame for a z polarized state. For a more general case, i.e., if $f_{rs} \propto \delta^3(\mathbf{p} - \tilde{\mathbf{p}})$, one must keep the relativistic corrections.

If one considers the defining relation (5.64), the spin density matrix has an immediate physical interpretation. The trace

$$\sum_r \left[\frac{f_{rr}(\mathbf{p})}{(2\pi)^3} \right] \tag{5.68}$$

is the probability density to obtain $\mathbf{p}$ in a momentum measurement, while the average polarization in the comoving frame reads

$$\frac{1}{2} \sum_{rs} \frac{\left[f_{rs}(\mathbf{p}) (\phi_r \sigma \phi_s) \right]}{\sum_t \left[f_{tt}(\mathbf{p}) \right]}. \tag{5.69}$$

We thus see that the spin density matrix is sufficient to characterize the most important experimental observables for a free spin $1/2$ particle. It is understood that all the steps can be repeated for a single antiparticle wave function to obtain the antiparticle polarization, which depends on the antiparticle spin density matrix $\bar{f}_{rs}(\mathbf{p})$. The only significant difference is an overall -1 sign, because of the corresponding sign in the spin part of the normal ordered angular momentum operator, and an exchange of the r and s indices in the χ bispinors compared to the ϕ for the particles .[11]

The appropriate generalization of the classical distribution function is expected, therefore, to produce in some limit a multiparticle generalization of the spin density

---

[11] Which can be expected, since the conventional two component spinors χ in the negative frequency solutions of the Dirac equation are taken with the opposite eigenvalue of σ_z, compared to the positive frequency solutions.

matrix that provides both the spectrum in momentum of the produced particles and their average polarization. As we have already anticipated, the desired object is the Wigner distribution. Its most convenient definition is

$$\hat{W}_{AB}(x, k) = \int \frac{d^4 v}{(2\pi)^4} e^{-ik \cdot v} \, \Psi_B^\dagger(x + v/2) \Psi_A(x - v/2), \tag{5.70}$$

that is, it is a four by four matrix obtained from thecomponents of two fields, $\Psi(y)_B^\dagger \Psi_A(z)$, Fourier transformed with respect to the relative distance $v = z - y$, and with $x = (z + y)/2$ being the middle space-time point. The inversion on the relative position of the matrix elements between the left and right-hand side of the last equation is needed in order to use the ordinary rules in matrix multiplications with respect to the A, B indices. In the remainder of this work, we will omit the matrix indices, understanding the matrix nature, and we will just write, e.g., $U_r U_r^\dagger$ without indices in a similar way, understanding the fact that it is a 4×4 matrix.

An important property of the Wigner distribution (5.70) is that it is a Hermitian matrix representing physical observables (at least in principle). Moreover, one expects that the usual causality rules apply to it. It is worth mentioning that some authors use a different sign convention or use $\bar{\Psi}_B \Psi_A$ in the definition of $W(x, k)$ [16–18], thus making the alternatively defined matrix a non-Hermitian one, so one must check which version of the Wigner distribution is actually used while comparing different works. In any case, a matrix multiplication with γ^0 and an eventual multiplication by a constant is enough to switch notation.

By the correspondence principle, the classical distribution is the expectation value of the renormalized operator

$$W(x, k) = \mathrm{tr} \left(: \hat{W}(x, k) : \right). \tag{5.71}$$

Making use of the definition (5.70), and assuming some minimal smoothness of the integrals,[12] one can rewrite the expectation value of any bilinear form in the Dirac fields using integration over the momentum k of the trace of the macroscopic Wigner distribution (5.71)

$$\mathrm{tr} \left(\rho : \bar{\Psi}(x) \gamma^{\nu_1} \cdots \gamma^{\nu_n} \frac{i}{2} \overleftrightarrow{\partial}^{\mu_1} \cdots \frac{i}{2} \overleftrightarrow{\partial}^{\mu_m} \Psi(x) : \right)$$
$$= \int d^4 k \, k^{\mu_1} \cdots k^{\mu_m} \, \mathrm{tr}_4 \left(W(x, k) \gamma^0 \gamma^{\nu_1} \cdots \gamma^{\nu_n} \right). \tag{5.72}$$

Here, the trace on the left-hand side is the usual trace over the quantum states, while tr_4 on the right-hand side denotes the trace over the matrix indices. To derive

---

[12] In order to exchange the order of the integrations and integrate by parts.

the formula (5.72), we have used the integral representation of the Dircac delta
$\delta^4(v) = \int d^4k/(2\pi)^4 \exp\{-ik \cdot v\}$ and performed the integration by parts

$$\int d^4v \, k^\mu \, e^{-ik \cdot v}[\cdots] = \int d^4v \left(i\partial^\mu_{(v)} \, e^{-ik \cdot v}\right)[\cdots] = \int d^v k \, e^{-ik \cdot v} \left(-i\partial^\mu_{(v)}\right)[\cdots].$$
(5.73)

In the next step, we can use the definition of the Dirac fields to expand (5.71) and to explicitly represent it in terms of the expectation values of the creation/destruction operators

$$W(x, k) = \sum_{rs} \int \frac{d^4v}{(2\pi)^4} e^{-ik \cdot v} \int \frac{d^3 p \, d^3 q}{(2\pi)^6 \sqrt{2E_\mathbf{p} 2E_\mathbf{q}}} \Big[$$

$$\langle a^\dagger_r(\mathbf{p}) a_s(\mathbf{q})\rangle U^\dagger_r(\mathbf{p}) U_s(\mathbf{q}) e^{i(p-q)\cdot x} e^{i\left(\frac{p+q}{2}\right)\cdot v} +$$

$$- \langle b^\dagger_r(\mathbf{p}) b_s(\mathbf{q})\rangle V^\dagger_s(\mathbf{q}) V_r(\mathbf{p}) e^{i(p-q)\cdot x} e^{-i\left(\frac{p+q}{2}\right)\cdot v} +$$ (5.74)

$$+ \langle b_s(\mathbf{q}) a_r(\mathbf{p})\rangle V^\dagger_s(\mathbf{q}) U_r(\mathbf{p}) e^{-i(p+q)\cdot x} e^{i\left(\frac{p-q}{2}\right)\cdot v} +$$

$$+ \langle a^\dagger_r(\mathbf{p}) b^\dagger_s(\mathbf{q})\rangle U^\dagger_r(\mathbf{p}) V_s(\mathbf{q}) e^{i(p+q)\cdot x} e^{i\left(\frac{p-q}{2}\right)\cdot v} \Big],$$

it is important to remind that both sides must be a matrix, therefore, the eigenspinors are not contracted but must be read, eg, $(U^\dagger_r)_B (U_s)_A$, according to (5.70). The last formula (5.74) allows us to make two important observations. The first one is that, after performing the d^4v integral, each of the four sectors is proportional to the Dirac delta function $\delta^4(k \pm (p \pm q)/2)$, with a different combination of the $+$ and $-$ signs in each sector. The momenta p^μ and q^μ are both on the mass shell but their combination, in general, is not. The pure particle/antiparticle contributions have $\delta^4(k \pm (p + q)/2)$ which can be on shell if and only if $p = q$. The mixed terms, however, include $\delta^4(k \pm (p - q)/2)$ where $(p - q)/2$ is never on shell and always space-like. This is the reason why we call k^μ a wave number vector, in order not to confuse it with the four-momentum of some particle-like degree of freedom.

The second and possibly the most important thing to notice is that $k^\mu W(x, k)$ is conserved, i.e., $k^\mu \partial_\mu W(x, k) = 0$, as one can check directly by applying $k^\mu \partial_\mu$ to the right-hand side of (5.74) and using the integration by parts in (5.73) to convert the wavenumber vector k into a derivative with respect to v. This is a consequence of the Dirac equation for the fields, which implies that the Klein-Gordon equation is satisfied as well.

Because of the conservation of $k^\mu W(x, k)$, one can use the same mathematical framework as that already used for the conserved fluxes such as $T^{\mu\nu}$ and the classical expression $p^\mu f(x, \mathbf{p})$. Here we can see the reason of the choice to define the Wigner operator as Hermitian. Being $k^\mu W(x, k)$ Hermitian too (an observable) is expected to follow causality rules. As a consequence of the Gauss theorem, the flux over the freeze-out hypersurface (or any other surface following the freeze-out) is equal to the volume integral[13]

$$\int d\Sigma_\mu \, k^\mu \, W(x,k) = \int d^3x \, k^0 \, W(x,k). \tag{5.75}$$

The analysis of the volume integral corresponds to a considerable simplification in the treatment of the Wigner distribution, like for the other quantum objects we have seen. Integrating directly (5.74) one obtains

$$\int d^3x \, k^0 \, W(x,k) = \sum_{rs} \int \frac{d^3p}{(2\pi)^3} \frac{k^0}{2E_{\mathbf{p}}} \Big[\delta^4(k-p) \, \langle a_r^\dagger(\mathbf{p}) a_s(\mathbf{p}) \rangle \, U_r^\dagger(\mathbf{p}) U_s(\mathbf{p}) +$$

$$- \delta^4(k+p) \, \langle b_r^\dagger(\mathbf{p}) b_s(\mathbf{p}) \rangle \, V_s^\dagger(\mathbf{p}) V_r(\mathbf{p}) \Big].$$

$$\tag{5.76}$$

The mixed terms vanish exactly, since the volume integral provides a $\delta^3(\mathbf{p}+\mathbf{q})$. Therefore, the Dirac function $\delta^4(k \pm (p-q)/2)$ becomes $\delta(k^0)\delta^3(\mathbf{k}\pm\mathbf{p})$ in this case. The appearance of k^0 makes these terms vanishing, as $k^0\delta(k^0)\delta^3(\mathbf{k}\pm\mathbf{p}) \equiv 0$.

The flux of the Wigner distribution (5.75) has many interesting properties. They can be identified while looking at its explicit form given by (5.76). It includes an on shell positive frequency contribution for the particles and a negative frequency (with negative momentum) contribution for the antiparticles. With k being on the mass shell, one can divide by k^0 since $\|k^0\| \geq m$ – something that cannot be done for the full distribution (5.74). Having in mind the normalization $\mathrm{tr}_4(U_r^\dagger(\mathbf{p}) \cdot U_s(\mathbf{p})) = U_r^\dagger \cdot U_s = 2E_{\mathbf{p}}\delta_{rs} = V_r^\dagger \cdot V_s = \mathrm{tr}_4(V_r^\dagger(\mathbf{p})V_s(\mathbf{p}))$, and the exact relations (see Appendix in Sect. 5.5)

$$\sum_r \frac{\langle a_r^\dagger(\mathbf{p})a_r(\mathbf{p})\rangle}{(2\pi^3)} = \frac{dN}{d^3p}, \qquad \sum_r \frac{\langle b_r^\dagger(\mathbf{p})b_r(\mathbf{p})\rangle}{(2\pi^3)} = \frac{d\bar{N}}{d^3p}, \tag{5.77}$$

it is immediate to verify that the trace of the flux (5.76) reads

$$\int d\Sigma_\mu \, k^\mu \, W(x,k) = \delta(k^0 - E_{\mathbf{k}}) \, E_{\mathbf{k}} \frac{dN}{d^3p}(\mathbf{k}) + \delta(k^0 + E_{\mathbf{k}}) \, E_{\mathbf{k}} \frac{d\bar{N}}{d^3p}(-\mathbf{k}). \tag{5.78}$$

In other words, the positive frequency contribution to the flux is directly expressed by the (invariant) spectrum of particles with momentum $\mathbf{k}$; while the negative frequency contribution is given by the invariant spectrum of antiparticles with momentum $-\mathbf{k}$.

---

[13]The hypothesis of an isolated system is important too. Being the integrand an observable, the flux over the light cone starting from the spatial boundary of an isolated system must be vanishing. Causality prevents the Wigner distribution to flow out of the light cone, as it would be a superluminal signal transfer, and the hypothesis of an isolated system prevents any signal to flow inside of the light cone. The flux over the light cone is therefore vanishing.

The structure of (5.76) is quite rich. It does not include the (anti)particle's spectra only but depends on the polarization states. By construction, the expectation values $\langle a_r^\dagger(\mathbf{p})a_s(\mathbf{p})\rangle$ and $\langle b_r^\dagger(\mathbf{p})b_s(\mathbf{p})\rangle$ have a very similar structure to the one-particle spin density matrix (5.64). They are both Hermitian matrices with respect to the indices r, s for all values of $\mathbf{p}$. Moreover, their diagonal elements $\langle a_r^\dagger(\mathbf{p})a_r(\mathbf{p})\rangle$ and $\langle b_r^\dagger(\mathbf{p})b_r(\mathbf{p})\rangle$ are always non-negative. The only difference is the normalization. Instead of being normalized to 1 they are normalized to the average number of particles $\langle N \rangle$ and antiparticles $\langle \bar{N} \rangle$

$$
\sum_r \int d^3 p\, \frac{\langle a_r^\dagger(\mathbf{p})a_r(\mathbf{p})\rangle}{(2\pi^3)} = \int d^3 p\, \frac{dN}{d^3 p} = \langle N \rangle,
$$
$$
\sum_r \int d^3 p\, \frac{\langle b_r^\dagger(\mathbf{p})b_r(\mathbf{p})\rangle}{(2\pi^3)} = \int d^3 p\, \frac{d\bar{N}}{d^3 p} = \langle \bar{N} \rangle.
$$
(5.79)

They provide therefore the desired generalization of the one-particle spin density matrix to the multiparticle case. The flux of the Wigner distribution (5.75) directly depends on them, it is therefore not surprising that one can get the average polarization density in momentum space from it. Making use of both (5.75) and (5.76) we find

$$
\frac{1}{2m}\mathrm{tr}_4\left[\left(\int d\Sigma_\mu\, k^\mu\, W(x,k)\right)\gamma^0\gamma^i\gamma_5\right] =
$$

$$
= \frac{1}{4m}\sum_{r,s}\int d^3 p \left[\delta^4(k-p)\frac{\langle a_r^\dagger(\mathbf{p})a_s(\mathbf{p})\rangle}{(2\pi^3)}U_r^\dagger(\mathbf{p})\begin{pmatrix}\sigma_i & 0\\ 0 & \sigma_i\end{pmatrix}U_s(\mathbf{p})+\right.
$$
$$
\left.\delta^4(k+p)\frac{\langle b_s^\dagger(\mathbf{p})b_s(\mathbf{p})\rangle}{(2\pi^3)}V_s^\dagger(\mathbf{p})\begin{pmatrix}\sigma_i & 0\\ 0 & \sigma_i\end{pmatrix}V_r(\mathbf{p})\right] =
$$
(5.80)

$$
= \frac{1}{2}\sum_{r,s}\left\{\delta(k^0-E_\mathbf{k})\frac{\langle a_r^\dagger(\mathbf{k})a_s(\mathbf{k})\rangle}{(2\pi^3)}\left[\phi_r\sigma_i\phi_s + \frac{\phi_r(\mathbf{k}\cdot\boldsymbol{\sigma})\phi_s}{m(E_\mathbf{k}+m)}k_i\right] + \right.
$$
$$
\left.\delta(k^0+E_\mathbf{k})\frac{\langle b_r^\dagger(-\mathbf{k})b_s(-\mathbf{k})\rangle}{(2\pi^3)}\left[\chi_s\sigma_i\chi_r + \frac{\chi_s(\mathbf{k}\cdot\boldsymbol{\sigma})\chi_r}{m(E_\mathbf{k}+m)}k_i\right]\right\},
$$

which can be immediately recognized as the average polarization of particles with momentum $\mathbf{k}$ (multiplied by the (non-invariant) spectrum $dN/d^3 p(\mathbf{k})$ for the positive frequency) minus the average polarization of antiparticles of momentum $-\mathbf{k}$ (times the spectrum $d\bar{N}/d^3 p(-\mathbf{k})$). Since the spectra can be calculated from the flux of the Wigner distribution, one can obtain the average polarizations of particles, $\langle \boldsymbol{\Pi}(\mathbf{p})\rangle$, and antiparticles, $\langle \bar{\boldsymbol{\Pi}}(\mathbf{p})\rangle$, for any momentum $\mathbf{p}$,

$$\langle \mathbf{\Pi}(\mathbf{p}) \rangle = \frac{1}{2} \sum_{r,s} \frac{\langle a_r^\dagger(\mathbf{p}) a_s(\mathbf{p}) \rangle}{\sum_t \langle a_t^\dagger(\mathbf{p}) a_t(\mathbf{p}) \rangle} \left[\phi_r \boldsymbol{\sigma} \phi_s + \frac{\phi_r(\mathbf{p} \cdot \boldsymbol{\sigma}) \phi_s}{m(E_\mathbf{p} + m)} \mathbf{p} \right],$$

$$\langle \bar{\mathbf{\Pi}}(\mathbf{p}) \rangle = -\frac{1}{2} \sum_{r,s} \frac{\langle b_s^\dagger(\mathbf{p}) b_r(\mathbf{p}) \rangle}{\sum_t \langle b_t^\dagger(\mathbf{p}) b_t(\mathbf{p}) \rangle} \left[\chi_r \boldsymbol{\sigma} \chi_s + \frac{\chi_r(\mathbf{p} \cdot \boldsymbol{\sigma}) \chi_s}{m(E_\mathbf{p} + m)} \mathbf{p} \right]. \tag{5.81}$$

Making use of (5.52) one can compute the polarization in the comoving frame as usual.

It is straightforward to check that the trace $\mathrm{tr}_4 \left[\frac{1}{2m} \left(\not{d}\Sigma_\mu \, k^\mu \, W(x,k) \right) \gamma^0 \gamma^0 \gamma_5 \right]$ corresponds to the momentum density of Π^0 (i.e., it is equal to the time component of the polarization vector – particle contribution for the positive frequency minus the antiparticle contribution for the negative frequency). The same structure of the trace is obtained with γ^0 replaced by γ^i. One can summarize all these results by making use of the definitions

$$\langle \Pi^0(\mathbf{p}) \rangle = \frac{\langle \mathbf{\Pi}(\mathbf{p}) \rangle \cdot \mathbf{p}}{E_\mathbf{p}}, \qquad \langle \bar{\Pi}^0(\mathbf{p}) \rangle = \frac{\langle \bar{\mathbf{\Pi}}(\mathbf{p}) \rangle \cdot \mathbf{p}}{E_\mathbf{p}}, \tag{5.82}$$

to complete the covariant $\langle \Pi^\mu(\mathbf{p}) \rangle$ with the correct time component.[14] Thus, the compact form of the previous results on the average polarization reads

$$\frac{1}{2m} \mathrm{tr}_4 \left[\left(\int d\Sigma_\lambda \, k^\lambda \, W(x,k) \right) \gamma^0 \gamma^\mu \gamma_5 \right] =$$

$$= \delta^4(k^0 - E_\mathbf{k}) \frac{dN}{d^3 p}(\mathbf{k}) \langle \Pi^\mu(\mathbf{k}) \rangle - \delta^4(k^0 + E_\mathbf{k}) \frac{dN}{d^3 p}(-\mathbf{k}) \langle \bar{\Pi}^\mu(-\mathbf{k}) \rangle. \tag{5.83}$$

The last equation, in conjunction with the exact result in (5.78), is enough to grant that the Wigner at the freeze-out hypersurface is sufficient to predict all the relevant experimental spectra of produced (anti)particles.

In the last section, we have seen that the spin tensor is a macroscopic observable sensitive to the microscopic polarization states. Looking at the relaitively simple trace on the left-hand side of (5.83), one may think if there is macroscopic object related to this. Taking into account the exact conversion rules (5.72), one finds that the d^4k integral of the left-hand side reads

$$\frac{1}{2m} \int d^4 k \, \mathrm{tr}_4 \left[\left(\int d\Sigma_\lambda \, k^\lambda \, W(x,k) \right) \gamma^0 \gamma^\mu \gamma_5 \right] =$$

$$= \int d\Sigma_\lambda \langle : \frac{i}{4m} \bar{\Psi} \left(\overset{\leftrightarrow}{\partial}{}^\lambda \gamma^\mu \gamma_5 \right) \Psi : \rangle. \tag{5.84}$$

---

[14]Compare with Eq. (5.50) for a quick check.

It is, therefore, the flux of (the expectation value of) the rank 2 pseudotensor

$$\frac{i}{4m} \bar{\Psi}(x) \left(\overset{\leftrightarrow}{\partial^{\lambda}} \gamma^{\mu} \gamma_5 \right) \Psi(x), \tag{5.85}$$

which we may call the polarization flux pseudotensor, given its relation to the integral of the polarization pseudovector density. It is straightforward to check that its divergence in the λ index vanishes. Hence, it is conserved, as one could expect since its flux is time independent. It does not correspond to any spin tensor, but it provides a valid alternative as a macroscopic object sensitive to the micorscopic polarization states.

5.5   Summary

In this work, we have extended the standard kinetic-theory formalism to include spin polarization for particles with spin 1/2. This has been achieved by using the spin tensor and the Wigner function. Our results can be used for the interpretation of the heavy-ion data describing spin polarization of the emitted hadrons.

Acknowledgements We would like to thank F. Becattini, B. Friman, R. Ryblewski, and E. Speranza for insightful discussions. L.T. was supported by the Deutsche Forschungsgemeinschaft (DFG, German Research Foundation) through the CRC-TR 211 "Strong-interaction matter under extreme conditions" - project number 315477589–TRR 211. W.F. was supported in part by the Polish National Science Center Grants No. 2016/23/B/ST2/00717.

Appendix: Expectation Values of Creation and Destruction Operators

In this appendix, we show details of the calculations of the expectation values of creation and destruction operators. In particular, we find an interesting and intuitive link between the average (anti)particle number and the quantum fluctuations required for the mixed terms ($\langle a^{\dagger} b^{\dagger} \rangle$ and $\langle ab \rangle$) to be nonvanishing.

The starting point is the density matrix (5.2), which reads

$$\rho = \sum_i \mathsf{P}_i |\psi_i\rangle \langle \psi_i|. \tag{5.86}$$

All P_i's are classical probabilities

$$\sum_i \mathsf{P}_i = 1. \tag{5.87}$$

The states $|\psi_i\rangle$ are proper quantum states, that is, they are normalized to one

$$\langle \psi_i | \psi_i \rangle = 1, \qquad \forall i. \tag{5.88}$$

The expectation value O of any quantum operator[15] $\hat{O}$ is a weighted average of the expectation values in the pure states, with the classical weights P_i

$$O = \mathrm{tr}\left(\rho \hat{O} \right) = \sum_i P_i \, \mathrm{tr}\left(|\psi_i\rangle \langle \psi_i | \hat{O} \right), \tag{5.89}$$

therefore, our problem reduces to the expectation value in a generic pure state. The trace must be taken over a complete set of independent states (not necessarily quantum states that are normalized to one). Since we are interested in the expectation values of the creation and destruction operators of four-momentum and polarization eigenstates, the most convenient states are N-particle and $\bar{N}$-antiparticle ones. The trace is defined as an integration over the momentum degrees of freedom and a sum over discrete polarizations, namely

$$\mathrm{tr}\left(\cdots \right) = \sum_r \int \frac{d^3 p}{(2\pi)^3 2E_{\mathbf{p}}} \langle p, r | \cdots | p, r \rangle +$$

$$+ \sum_{r,s} \int \frac{d^3 p}{(2\pi)^3 2E_{\mathbf{p}}} \frac{d^3 q}{(2\pi)^3 2E_{\mathbf{q}}} \langle p, r, q, s | \cdots | p, r, q, s \rangle + \cdots , \tag{5.90}$$

and so on, until exausting all the combinations of N particles and $\bar{N}$ antiparticles. In the last formula the standard definition is used

$$|p, r\rangle = \sqrt{2E_p} a_r^\dagger(\mathbf{p}) |0\rangle, \tag{5.91}$$

along with the analogous expressions for antiparticles and multiparticle states. The anticommutation relations have the form

$$\{a_r(\mathbf{p}), a_s^\dagger(\mathbf{p}')\} = \{b_r(\mathbf{p}), b_s^\dagger(\mathbf{p}')\} = (2\pi)^3 \delta_{rs} \, \delta^3(\mathbf{p} - \mathbf{p}'), \tag{5.92}$$

with the normalization

$$\langle p, r | q, s \rangle = 2E_{\mathbf{p}} (2\pi)^3 \delta_{rs} \, \delta^3(\mathbf{p} - \mathbf{p}'). \tag{5.93}$$

It is convenient to introduce the compact notation for multiparticle states

$$|\underline{p}, \underline{r}; \underline{\bar{q}}, \underline{\bar{s}}\rangle = |p_1, r_1, p_2, r_2, \cdots p_N, r_N; \bar{q}_1, \bar{s}_1, \bar{q}_2, \bar{s}_2, \cdots \bar{q}_{\bar{N}}, \bar{s}_{\bar{N}}\rangle,$$

$$\int [d\underline{p}]^N [d\underline{\bar{q}}]^{\bar{N}} = \int \frac{d^3 p_1}{(2\pi)^3 2E_{\mathbf{p}_1}} \cdots \frac{d^3 p_N}{(2\pi)^3 2E_{\mathbf{p}_N}} \frac{d^3 \bar{q}_1}{(2\pi)^3 2E_{\bar{\mathbf{q}}_1}} \cdots \frac{d^3 \bar{q}_{\bar{N}}}{(2\pi)^3 2E_{\bar{\mathbf{q}}_{\bar{N}}}}, \tag{5.94}$$

---

[15] In general one needs the renormalized operators. For free fields this is just the normal ordering, that is, removing the vacuum expectation value. We always assume massive free Dirac fields and normal ordering in this section.

where the bar is used to distinguish antiparticle from particle variables. In this way the trace (5.90) can be written in a more compact form as

$$\mathrm{tr}\left(\cdots\right) = \sum_{N,\bar{N}} \sum_{\underline{r},\underline{\bar{s}}} \int [d\underline{p}]^N [d\underline{\bar{q}}]^{\bar{N}} \langle \underline{p},\underline{r};\underline{\bar{q}},\underline{\bar{s}}| \cdots |\underline{p},\underline{r};\underline{\bar{q}},\underline{\bar{s}}\rangle. \tag{5.95}$$

This compact notation is useful to write the generic quantum state $|\psi\rangle$

$$|\psi\rangle = \sum_{N,\bar{N}} \sum_{\underline{r},\underline{\bar{s}}} \int [d\underline{p}]^N [d\underline{\bar{q}}]^{\bar{N}} \; \alpha_{N,\bar{N}}(\underline{p},\underline{r};\underline{\bar{q}},\underline{\bar{s}}) \, |\underline{p},\underline{r};\underline{\bar{q}},\underline{\bar{s}}\rangle, \tag{5.96}$$

where the complex functions $\alpha_{N,\bar{N}}(\underline{p},\underline{r};\underline{\bar{q}},\underline{\bar{s}})$ are partial N-particle-$\bar{N}$-antiparticle wave functions in momentum space. The normalization reads

$$1 = \langle\psi|\psi\rangle = \sum_{N,\bar{N}} \sum_{\underline{r},\underline{\bar{s}}} \int [d\underline{p}]^N [d\underline{\bar{q}}]^{\bar{N}} \; \alpha_{N,\bar{N}}^*(\underline{p},\underline{r};\underline{\bar{q}},\underline{\bar{s}})\alpha_{N,\bar{N}}(\underline{p},\underline{r};\underline{\bar{q}},\underline{\bar{s}}) =$$
$$= \sum_{N,\bar{N}} \|\alpha_{N,\bar{N}}\|^2, \tag{5.97}$$

with $\|\alpha_{N,\bar{N}}\|^2$ being a shorthand notation for the (non-negative[16]) sum of integrals

$$\|\alpha_{N,\bar{N}}\|^2 = \sum_{\underline{r},\underline{\bar{s}}} \int [d\underline{p}]^N [d\underline{\bar{q}}]^{\bar{N}} \; \alpha_{N,\bar{N}}^*(\underline{p},\underline{r};\underline{\bar{q}},\underline{\bar{s}})\alpha_{N,\bar{N}}(\underline{p},\underline{r};\underline{\bar{q}},\underline{\bar{s}}). \tag{5.98}$$

The tensor product $|\psi\rangle\langle\psi|$, that is, the projector on the quantum state $|\psi\rangle$ reads

$$|\psi\rangle\langle\psi| = \sum_{N,\bar{N}} \sum_{\underline{r},\underline{\bar{s}}} \sum_{N',\bar{N}'} \sum_{\underline{r}',\underline{\bar{s}}'} \int [d\underline{p}]^N [d\underline{\bar{q}}]^{\bar{N}} [d\underline{p}']^N [d\underline{\bar{q}}']^{\bar{N}} \times$$
$$\times \alpha_{N',\bar{N}'}^*(\underline{p}',\underline{r}';\underline{\bar{q}}',\underline{\bar{s}}')\alpha_{N,\bar{N}}(\underline{p},\underline{r};\underline{\bar{q}},\underline{\bar{s}}) \, |\underline{p},\underline{r};\underline{\bar{q}},\underline{\bar{s}}\rangle\langle\underline{p}',\underline{r}';\underline{\bar{q}}',\underline{\bar{s}}'|. \tag{5.99}$$

Making use of the normalization relations between the states, it is possible to write the trace in a pure state $|\psi\rangle$ of an operator $\hat{O}$ in the compact form

$$\mathrm{tr}\left(|\psi\rangle\langle\psi|\hat{O}\right) = \sum_{N,\bar{N}} \sum_{\underline{r},\underline{\bar{s}}} \sum_{N',\bar{N}'} \sum_{\underline{r}',\underline{\bar{s}}'} \int [d\underline{p}]^N [d\underline{\bar{q}}]^{\bar{N}} [d\underline{p}']^N [d\underline{\bar{q}}']^{\bar{N}} \times$$
$$\times \alpha_{N',\bar{N}'}^*(\underline{p}',\underline{r}';\underline{\bar{q}}',\underline{\bar{s}}')\alpha_{N,\bar{N}}(\underline{p},\underline{r};\underline{\bar{q}},\underline{\bar{s}}) \, \langle\underline{p}',\underline{r}';\underline{\bar{q}}',\underline{\bar{s}}'|\hat{O}|\underline{p},\underline{r};\underline{\bar{q}},\underline{\bar{s}}\rangle. \tag{5.100}$$

There is a couple of results that can be immediately inferred from the last formula. The first one is that the expectation values of $a^\dagger b^\dagger$ and ba, for any momentum and

---

[16]Being the sum of integrals of a real non-negative weight of the forms z^*z.

polarization combination, can be nonvanishing if and only if the quantum state of the system is in a superposition of states with different particle content. More precisely, only the quantum interference between states that differ exactly by a particle-antiparticle pair can give a nonvanishing contribution (understanding that the integral over the partial wave functions can still simplify and give a vanishing result).

The second observation is that the expectation value of $a^\dagger a$ and $b^\dagger b$ can be simplified. The only combinations that can give a contribution are the ones between states with exactly the same number of particles and the same number of antiparticles. In the following computations, we consider only the term $a^\dagger a$, understanding that the very same transformations hold for antiparticles.

As a particular case of (5.100) one can write the expectation value of $a_r^\dagger(\mathbf{p})a_s(\mathbf{p}')$

$$
\mathrm{tr}\left(|\psi\rangle\,\langle\psi|\,a_r^\dagger(\mathbf{p})a_s(\mathbf{p}')\right) = \sum_{N,\bar N}\sum_{\underline{t},\underline{t}'}\sum_{\underline{\bar u},\underline{\bar u}'}\int[d\underline{k}]^N[d\underline{k}']^N[d\underline{\bar q}]^{\bar N}[d\underline{\bar q}']^{\bar N}\times
$$
$$
\times\,\alpha^*_{N,\bar N}(\underline{k}',\underline{t}';\underline{\bar q}',\underline{\bar u}')\alpha_{N,\bar N}(\underline{k},\underline{t};\underline{\bar q},\underline{\bar u})\langle\underline{k}',\underline{t}';\underline{\bar q}',\underline{\bar u}'|a_r^\dagger(\mathbf{p})a_s(\mathbf{p}')|\underline{k},\underline{t};\underline{\bar q},\underline{\bar u}\rangle.
$$

(5.101)

It is relatively simple to obtain the final formula by making use of the standard anticommutation relations

$$
\cdots a_r^\dagger(\mathbf{p})a_s(\mathbf{p}')\sqrt{2E_{\mathbf{k}_j}}\,a_{t_j}^\dagger(\mathbf{k}_j)\cdots =
$$
$$
\cdots a_r^\dagger(\mathbf{p})\sqrt{2E_{\mathbf{k}_j}}\left(\{a_s(\mathbf{p}'),a_{t_j}^\dagger(\mathbf{k}_j)\} - a_{t_j}^\dagger(\mathbf{k}_j)a_s(\mathbf{p}')\right)\cdots =
$$
$$
\cdots\sqrt{2E_{\mathbf{k}_j}}\left(a_{t_j}^\dagger(\mathbf{k}_j)a_r^\dagger(\mathbf{p})a_s(\mathbf{p}') + a_r^\dagger(\mathbf{p})(2\pi)^3\delta_{st_j}\delta^3(\mathbf{k}_j - \mathbf{p}')\right)\cdots =
$$
$$
\cdots\left[\sqrt{2E_{\mathbf{k}_j}}a_{t_j}^\dagger(\mathbf{k}_j)\left(a_r^\dagger(\mathbf{p})a_s(\mathbf{p}')\right) + \sqrt{2E_{\mathbf{p}'}}(2\pi)^3\delta_{st_j}\delta^3(\mathbf{k}_j - \mathbf{p}')a_r^\dagger(\mathbf{p})\right]\cdots =
$$

$$
= \cdots\left[\sqrt{2E_{\mathbf{k}_j}}a_{t_j}^\dagger(\mathbf{k}_j)\left(a_r^\dagger(\mathbf{p})a_s(\mathbf{p}')\right) + \right.
$$
$$
\left. + \sqrt{\frac{2E_{\mathbf{p}'}}{2E_{\mathbf{p}}}}(2\pi)^3\delta_{st_j}\delta^3(\mathbf{k}_j - \mathbf{p}')\sqrt{2E_{\mathbf{p}}}a_r^\dagger(\mathbf{p})\right]\cdots
$$

(5.102)

In other words, even if $a_r^\dagger(\mathbf{p})a_s(\mathbf{p}')$ doesn't commute with the creation operators, it is possible to "move it to the right". However, each time we do that we have to add a new state, with a delta between the j'th degrees of freedom and the destruction operator $a_s(\mathbf{p}')$, a numerical factor $(2\pi)^3\sqrt{E_{\mathbf{p}'}/E_{\mathbf{p}}}$ and a substitution of the momentum and polarization at the j'th place with the ones related to the creation operator $a_r^\dagger(\mathbf{p})$. After moving to the right all the particle creation operators, $a_r^\dagger(\mathbf{p})a_s(\mathbf{p}')$ commutes with the creation operators of the antiparticles (if present).

In the end, after making use of the normalization of the eigenstates we find

$$
\mathrm{tr}\left(\left|\psi\right\rangle\left\langle\psi\right|a_r^\dagger(\mathbf{p})a_s(\mathbf{p}')\right) = 0+
$$

$$
+ \frac{1}{\sqrt{2E_\mathbf{p}2E_{\mathbf{p}'}}} \sum_{\bar{N},N>0} \sum_{j+1}^{N} \sum_{\underline{t}-t_j} \sum_{\underline{\bar{u}}} \int [d\underline{k}]^{(N-j)}[d\underline{\bar{q}}]^{\bar{N}} \times \qquad (5.103)
$$

$$
\times \alpha_{N,\bar{N}}^*(\underline{k}-\mathbf{k}_j,\mathbf{p},\underline{t}-t_j,r;\underline{\bar{q}},\underline{\bar{u}})\alpha_{N,\bar{N}}(\underline{k}-\mathbf{k}_j,\mathbf{p}',\underline{t}-t_j,s;\underline{\bar{q}},\underline{\bar{u}}).
$$

The notation $\sum_{\underline{t}-t_j}\int d[\underline{k}]^{(N-j)}$ means that the integral and the sum is over all the particle degrees of freedom except for the j'th. In the similar way $\alpha_{N,\bar{N}}(\underline{k}-\mathbf{k}_j,\mathbf{p},\underline{t}-t_j,r;\underline{\bar{q}},\underline{\bar{u}})$ is a shorthand notation for the (partial) wave function with the j'th degrees of freedom fixed to the momentum $\mathbf{p}$ and polarization r.

The formula (5.103) has many interesting consequences. Besides the expected vanishing expectation value for purely antiparticle states, one can immediately check that the expectation value of $a_r^\dagger(\mathbf{p})a_r(\mathbf{p})$ is nonnegative, since it is a series of integrals and sums of squares. Moreover, as one could expect, it is linked to the average number of particles. Indeed, the expression

$$
\sum_r \int \frac{d^3p}{(2\pi)^3}\mathrm{tr}\left(\left|\psi\right\rangle\left\langle\psi\right|a_r^\dagger(\mathbf{p})a_r(\mathbf{p})\right) = \sum_N N \sum_{\bar{N}} \|\alpha_{N,\bar{N}}\|^2, \qquad (5.104)
$$

exactly gives the average number of particles in the state $\left|\psi\right\rangle$ because of the normalization (5.97). More interestingly, the expectation value of $a_r^\dagger(\mathbf{p})a_s(\mathbf{p})$ (same momentum, different polarization) performs the role of a momentum-dependent spin density matrix. The momentum integral of the trace is proportional to the average number of particles, but the matrix itself is sensitive to polarization in the r,s indices and can be used to obtain the average number of particles, per momentum cell, for some polarization states.

All these arguments do not change if one reinserts the classical probabilities P_i from (5.86) and deals with mixed states. The classical fluctuations do not change the properties of the spin density matrix, like the non-negative diagonal elements and normalization of the trace (after dividing by $(2\pi)^3$ and integrating over momentum, like for the pure states) does not change the average number of particles.

References

1. Busza, W., Rajagopal, K., van der Schee, W.: Ann. Rev. Nucl. Part. Sci. **68**, 339 (2018)
2. Romatschke, P., Romatschke, U.: https://doi.org/10.1017/9781108651998, arXiv:1712.05815 [nucl-th]
3. Florkowski, W., Heller, M.P., Spalinski, M.: Rept. Prog. Phys. **81**, (2018)
4. Borsanyi, S., et al.: Wuppertal-Budapest collaboration. JHEP **1009**, 073 (2010)
5. Bazavov, A., et al.: HotQCD collaboration. Phys. Lett. B **795**, 15 (2019)

6. Miller, M.L., Reygers, K., Sanders, S.J., Steinberg, P.: Ann. Rev. Nucl. Part. Sci. **57**, 205 (2007)
7. Gelis, F., Iancu, E., Jalilian-Marian, J., Venugopalan, R.: Ann. Rev. Nucl. Part. Sci. **60**, 463 (2010)
8. Cooper, F., Frye, G.: Phys. Rev. D **10**, 186 (1974)
9. de Groot, S.R., van Leeuwen, W.A., van Weert, ChG: Relativistic Kinetic Theory. North-Holland, Amsterdam (1980)
10. Wang, Q.: Nucl. Phys. A **967**, 225 (2017)
11. Huang, X.G.: arXiv:2002.07549 [nucl-th]
12. Becattini, F., Lisa, M.A.: arXiv:2003.03640 [nucl-ex]
13. Becattini, F.: arXiv:2004.04050 [hep-th]
14. Florkowski, W.: Phenomenology of Ultra-Relativistic Heavy-Ion Collisions. World Scientific, Singapore (2010)
15. Itzykson, C., Zuber, J.B.: Quantum Field Theory. McGraw-Hill, New York (1980)
16. Florkowski, W., Kumar, A., Ryblewski, R.: Phys. Rev. C **98** (2018)
17. Weickgenannt, N., Sheng, X.L., Speranza, E., Wang, Q., Rischke, D.H.: Phys. Rev. D **100**, 56018 (2019)
18. Liu, Y.C., Mameda, K., Huang, X.G.: arXiv:2002.03753 [hep-ph]
19. Hehl, F.W.: Rep. Math. Phys. **9**, 55 (1976)
20. Florkowski, W., Ryblewski, R., Kumar, A.: Prog. Part. Nucl. Phys. **108** (2019)
21. Weickgenannt, N., Speranza, E., Sheng, X.l., Wang, Q., Rischke, D.H.: arXiv:2005.01506 [hep-ph]
22. Speranza, E., Weickgenannt, N.: arXiv:2007.00138 [nucl-th]
23. Becattini, F., Chandra, V., Del Zanna, L., Grossi, E.: Ann. Phys. **338**, 32 (2013)
24. Becattini, F., Florkowski, W., Speranza, E.: Phys. Lett. B **789**, 419 (2019)
25. Mathisson, M.: Acta Phys. Pol. **6**, 163 (1937)
26. Weyssenhoff, J., Raabe, A.: Acta Phys. Pol. **9**, 7 (1947)

Quantum Kinetic Description of Spin and Rotation

6

Yin Jiang, Xingyu Guo and Pengfei Zhuang

Abstract

Motivated by the generation of extremely strong axial vector fields, i.e., magnetic and vortical fields, and development of hadron polarization measurement in relativistic heavy ion collisions, the quantum kinetic theory becomes more and more important in the phenomenological study on the quark–gluon plasma evolution. In this chapter the quantum correction is introduced intuitively with canonical quantization and path integral approaches. The quantum kinetic theory formulated with Wigner function is discussed in detail to develop the quantum transport equations for chiral and massive fermions. As a straightforward application the polarization and anomaly effect are studied in the presence of background magnetic field and vorticity distribution. Finally the experimental results and numerical simulation frameworks are briefly reviewed.

6.1   Introduction

Transport phenomena are ubiquitous in nature. From galaxy evolution to particle motion in medium, the transport phenomena could be understood as irreversible processes of statistical nature stemming from continuous random motion of effective degrees of freedom. The study on classical transport theory could be traced

Y. Jiang
Physics Department, Beihang University, 37 Xueyuan Rd, Beijing 100191, China
e-mail: jiang_y@buaa.edu.cn

X. Guo
Guangdong Provincial Key Laboratory of Nuclear Science, Institute of Quantum Matter, South China Normal University, Guangzhou 510006, China
e-mail: guoxy@m.scnu.edu.cn

P. Zhuang (✉)
Physics Department, Tsinghua University, Beijing 100084, China
e-mail: zhuangpf@mail.tsinghua.edu.cn

© Springer Nature Switzerland AG 2021   167
F. Becattini et al. (eds.), *Strongly Interacting Matter under Rotation*,
Lecture Notes in Physics 987,
https://doi.org/10.1007/978-3-030-71427-7_6

back to eighteenth century, when even the idea of energy conservation and thus the treatment of elastic collisions are not established. Thereafter it was developed by many physicists, including Rudolf Clausius, James Clerk Maxwell, and Ludwig Boltzmann. Focusing on the single-particle distribution, the classical kinetic theory has been widely and successfully applied in physics, chemistry, and engineering by using the well-known Boltzmann equation which usually takes the form of

$$\partial_t f + \frac{\mathbf{p}}{E} \cdot \nabla f + \mathbf{F} \cdot \nabla_p f = C_{coll}, \tag{6.1}$$

where ∇ and ∇_p are three-dimensional space and momentum derivatives, $\mathbf{F}$ is the force field acting on the particle, and C_{coll} is the term describing the effect of collisions of two or more particles in medium, usually called collision terms or collision cores. The classical single-particle distribution $f(x, \mathbf{p})$ is defined in phase space, it is the most commonly used one of the key quantities for describing out-of-equilibrium.

Motivated by recent experimental results on hadron polarization in relativistic heavy ion collisions, quantum corrections to classical transport theory are urged to be involved in phenomenological studies. For quantum systems, it is not enough to consider only the classical number distribution f, instead one has to take into account quantum corrections self-consistently with Wigner function. As an unusual quantization procedure, the Wigner function formalism can be semi-classically expanded in $\hbar$ and used to study quantum corrections systematically. Such a quantum transport theory for partons and hadrons in hot medium is firstly investigated in a covariant version, see the review paper [1] and the recent development [2], and then extended to the equal-time version to solve quantum corrections as an initial value problem [3–7]. The first-order correction, which can be extracted alternatively from the path integral approach [8], could be observed on top of a background magnetic field by generating the imbalance of chiral fermions, which is known as the chiral magnetic effect (CME). This is a new response relation between magnetic field and induced electric current and has been identified as a deep relation to the chiral anomaly and a nontrivial topology of degenerated eigenstates in momentum space. In condensed matter physics the similar semi-classical studies have been done from the canonical quantization approach and realized in graphene and ferro- or antiferro-magnetic ordering crystal with time-reversal symmetry breaking [9]. Besides the CME, subsequent researches show that the vorticity would induce particle current as well which is known as the chiral vortical effect (CVE). All the quantization approaches, including canonical quantization, path integration, and Wigner function formalism, indicate that these novel chiral effects could not emerge without a background axial vector field. The semi-classical kinetic theory, including CME and CVE corrections for massless fermions, is named as chiral kinetic theory (CKT) and is adopted to understand the anisotropy of final hadron distributions in heavy ion collisions. In this chapter we will review the quantum kinetic theory including CKT briefly from different theoretical approaches.

Although all the approaches are able to give the same $\mathcal{O}(\hbar)$ order correction in chiral limit, the Wigner function formalism is more suitable for systematic studies on the mass dependence and higher order corrections. Different from the chiral modes in novel condensed materials, even light quarks are not perfect chiral fermions. As a consequence, it is necessary to take the mass dependence into account self-consistently in order to understand experimental data quantitatively. We will show the mass correction to the usual CKT. It is found that the spin current would be entangled with the vector and axial vector currents and serve as an important complement in the theory [10].

For near-equilibrium systems, such as the quark–gluon plasma (QGP) created in heavy ion collisions, simulations with transport approaches are time-consuming and hence impractical because of the complexity of collision terms. Instead, the classical hydrodynamics is adopted as a more realistic and precise-enough framework to complete the simulation. By averaging over the momentum space hydrodynamics is known to be a proper coarse-grained result of the usual kinetic theory. Therefore it is a natural question what one would obtain if the same procedure is applied to a quantum kinetic theory. We will briefly review the development of spin hydrodynamics which is expected to be a potential framework to study the hadron polarization consistently.

Despite the chiral symmetry related effects exhibit several novel phenomena, it is worthwhile to mention that they are all $\mathcal{O}(\hbar)$ corrections to the classical transport theory. It means that the background of such a system should be controlled by the ordinary Boltzmann equations. In the last section of this chapter we will briefly review several numerical works on the simulation of the equation, including a multiphase transport (AMPT) model which focus on the parton cascade and a Boltzmann approach to multiparton scatterings (BAMPS) which mainly emphasizes gluon collision and transportation. For more information on effective models readers could refer to the chapter on transport models.

6.2 Semi-classical Approaches

Chiral kinetic equations were firstly discussed by condensed matter physicists when studying the transport properties of quasi-particles with zero mass [11,12]. Near the Fermi surface the semi-classical equation is derived. Starting from the Hamiltonian for Weyl fermions and introducing the Berry connection $i\mathbf{A}_p = u_p^\dagger \nabla_p u_p$, the Berry curvature is the curl of the Berry connection $\mathbf{b} = \nabla_p \times \mathbf{A}_p$, and the action for the fermion system is

$$S = \int dt \left[\mathbf{p} \cdot \dot{\mathbf{x}} + \mathbf{A}_x \cdot \dot{\mathbf{x}} - \mathbf{A}_p \cdot \dot{\mathbf{p}} - \epsilon_p - A_x^0 \right]. \tag{6.2}$$

Corresponding to the electromagnetic vector potential $\mathbf{A}_x$, $\mathbf{A}_p$ can be viewed as a vector potential in momentum space. From the action, the Poisson brackets for phase space coordinates x_i and p_i can be derived as

$$\{p_i, p_j\} = -\frac{\epsilon_{ijk} B_k}{1 + \mathbf{B} \cdot \mathbf{b}},$$

$$\{x_i, x_j\} = \frac{\epsilon_{ijk} b_k}{1 + \mathbf{B} \cdot \mathbf{b}},$$

$$\{p_i, x_j\} = \frac{\delta_{ij} + b_i B_j}{1 + \mathbf{B} \cdot \mathbf{b}}, \tag{6.3}$$

where $\mathbf{B} = \nabla \times \mathbf{A}$ is the background magnetic field. These relations are very different from the usual ones. With these relations, the equation of motion can be derived, and solving it gives the result of $\dot{\mathbf{x}}$ and $\dot{\mathbf{p}}$. Substituting the solution into the Boltzmann equation

$$\partial_t f + \dot{\mathbf{x}} \cdot \nabla f + \dot{\mathbf{p}} \cdot \nabla_p f = 0 \tag{6.4}$$

gives the chiral kinetic equation.

Following References [8, 12] we continue the procedure in a simplest case. Starting from the Hamiltonian for chiral fermions,

$$H = \boldsymbol{\sigma} \cdot \mathbf{p}, \tag{6.5}$$

in order to write down the semi-classical equation of motion for a chiral fermion, which should include the quantum correction to the order of $\hbar$, it is intuitive to take the path integral quantization. The transition amplitude between the initial and final spin states i and f is

$$\langle f | e^{i H (t_f - t_i)} | i \rangle = \int \mathcal{D}_x \mathcal{D}_p \mathcal{P} e^{i \int_{t_i}^{t_f} dt (\mathbf{p} \cdot \dot{\mathbf{x}} - \boldsymbol{\sigma} \cdot \mathbf{p})} \Big|_{fi}. \tag{6.6}$$

In order to eliminate the non-diagonal elements, a unitary transformation V_p is introduced as

$$V_p^\dagger \boldsymbol{\sigma} \cdot \mathbf{p} V_p = |\mathbf{p}| \sigma_3. \tag{6.7}$$

In this way the amplitude could be reduced as

$$\langle f | e^{i H (t_f - t_i)} | i \rangle = V_{p_f} \int \mathcal{D}_x \mathcal{D}_p \mathcal{P} e^{i \int_{t_i}^{t_f} dt (\mathbf{p} \cdot \dot{\mathbf{x}} - |\mathbf{p}| \sigma_3 - \mathbf{A}_p \cdot \dot{\mathbf{p}})} \Big|_{fi} V_{p_i}^\dagger. \tag{6.8}$$

Focusing on particles with positive helicity the classical action becomes

$$I = \int_{t_i}^{t_f} dt \left(\mathbf{p} \cdot \dot{\mathbf{x}} - |\mathbf{p}| - \mathbf{A}_p \cdot \dot{\mathbf{p}} \right). \tag{6.9}$$

As a result the semi-classical equation of motion for the chiral fermions with positive helicity in electromagnetic fields $\mathbf{E}$ and $\mathbf{B}$ are

$$\dot{\mathbf{x}} = \hat{\mathbf{p}} + \hbar\dot{\mathbf{p}} \times \mathbf{b},$$
$$\dot{\mathbf{p}} = \mathbf{E} + \dot{\mathbf{x}} \times \mathbf{B} \tag{6.10}$$

with $\hat{\mathbf{p}} = \mathbf{p}/|\mathbf{p}|$ and $\mathbf{b} = \hat{\mathbf{p}}/(2\mathbf{p}^2)$. Obviously the $\hbar$ term is the leading order of the quantum corrections and would vanish when the background fields disappear. To derive the classical transport equation, we consider the Liouville equation for the invariant measure of the phase space integration

$$\partial_t \rho + \nabla \cdot (\rho\dot{\mathbf{x}}) + \nabla_p \cdot (\rho\dot{\mathbf{p}}) = 2\pi \mathbf{E} \cdot \mathbf{B} f \delta^3(\mathbf{p}) \tag{6.11}$$

with $\rho = (1 + \mathbf{b} \cdot \mathbf{B}) f$ and the ordinary distribution function f. In the fully thermalized state the equation gives the well-known CME current [13, 14] induced by the background magnetic field

$$\mathbf{J}_{CME} = \frac{\mu}{2\pi^2}\mathbf{B}, \tag{6.12}$$

where μ is the chemical potential of the fermions.

It is also possible to reach CKT using other field theory approaches. For example, in the on-shell effective field theory (OSEFT) [15], assuming that the physics we are interested in is dominated by contribution from the on-shell particles, the Lagrangian can be written as the sum of different on-shell fields,

$$\mathcal{L} = \bar{\chi}_v(x) \left(iv \cdot D + i\slashed{D}_\perp \frac{1}{2E + i\tilde{v} \cdot D} i\slashed{D}_\perp \right) \frac{\slashed{\tilde{v}}}{2} \chi_v(x)$$
$$+ \bar{\xi}_{\tilde{v}}(x) \left(i\tilde{v} \cdot D + i\slashed{D}_\perp \frac{1}{-2E + iv \cdot D} i\slashed{D}_\perp \right) \frac{\slashed{v}}{2} \xi_{\tilde{v}}(x), \tag{6.13}$$

where χ_v and $\xi_{\tilde{v}}$ are on-shell fields, $v = (1, \mathbf{v})$ and $\tilde{v} = (1, -\mathbf{v})$ are the four velocities, and $\slashed{D}$ is defined as $\slashed{D} = (g^{\mu\nu} - \frac{1}{2}v^\mu\tilde{v}^\nu + v^\nu\tilde{v}^\mu)\gamma_\mu D_\nu$. The Wigner function is expressed as the Fourier transformation of the two-point Green's function $G_{E,v}(x, y) = \langle\bar{\chi}_v(y)\gamma^\mu\tilde{v}_\mu\chi_v(x)\rangle/2$. Considering the fact that to the first order in $\hbar$ the Wigner function for chiral fermions is on the shell, the OSEFT is able to reproduce the chiral kinetic theory.

Another example is the world-line formalism [16, 17], which is a one-loop effective action for a Dirac fermion coupling to vector and axial vector gauge fields. The fermion part of the action is

$$S = \int d^4x\,\bar{\psi}\theta\psi,$$
$$\theta = \left(i\partial_\mu + A_\mu + \gamma_5 B_\mu\right)\gamma^\mu \tag{6.14}$$

with the covariant vector potential A_μ and magnetic field $B_\mu = \epsilon_{\mu\nu\sigma\rho}n^\nu F^{\sigma\rho}/2$ which depends on the reference frame n^μ and the electromagnetic tensor $F^{\mu\nu} = \partial^\mu A^\nu - \partial^\nu A^\mu$. The integral can be performed to give the fermion effective action

$$-W[A, B] = \log \det(\theta)$$
$$= \frac{1}{2}Tr\log(\theta^\dagger\theta) + i\arg\det[\theta]. \tag{6.15}$$

The main idea of the world-line formalism is that both the real and imaginary parts of W can be represented by path integral of some bosonic and fermionic fields. In Minkowskian space, the effective Lagrangian becomes

$$\mathcal{L} = \frac{\dot{x}^2}{2\epsilon} + \frac{i}{2}\psi^\mu\dot{\psi}_\mu + \frac{i}{2}\psi_5\dot{\psi}_5 + \frac{i}{2}\psi_6\dot{\psi}_6 + \dot{x}_\mu A^\mu(x) - \frac{i\epsilon}{2}\psi^\mu F_{\mu\nu}\psi^\nu, \tag{6.16}$$

where x_μ and $\psi_a(a = 1, ..., 6)$ are the effective bosonic and fermionic fields, respectively. Taking the pseudo-classical limit, the equations of motion can be derived and again CKT can be acquired.

For a systematic approach some basic concepts should be changed. Firstly, due to the uncertainty principle, the quantum version of the distribution function $f(x, \mathbf{p})$ cannot be simply interpreted as the probability to find a particle at the coordinate $x\delta x$ and momentum $p\delta p$, as this becomes meaningless when both δx and δp approaches zero. Moreover, the idea of an on-shell particle is not always accurate enough in quantum physics. In a general case, the quasi-particle approximation may fail, and one should consider off-shell effects. The situation could become more complicated if one wants to further discuss the spin degree of freedom, especially for massless fermions. The definition of spin and orbital angular momenta is not trivial and firmly connected to the Lorentz transformation which requires the theory to be Lorentz covariant.

Taking all these into consideration, one reaches the conclusion that it would be better to construct a quantum kinetic theory directly from quantum field theory. Wigner function [18] is one of the ways to meet this requirement. We will use it in the following to discuss the quantum kinetic theory:

6.3 Wigner Function Formalism

For a Dirac fermion field ψ, the covariant Wigner operator is defined as

$$\hat{W}(x, p) = \int \frac{d^4y}{(2\pi)^4} e^{-\frac{i}{\hbar}p\cdot y}\psi(x + \frac{y}{2})\bar{\psi}(x - \frac{y}{2}), \tag{6.17}$$

and the covariant Wigner function is the expectation of this operator,

$$W(x, p) = \langle\hat{W}(x, p)\rangle. \tag{6.18}$$

If fermions are coupled to an electromagnetic field A_μ, the above definition is not gauge invariant. One has to modify it as [5,7,19]

$$
\hat{W}(x, p) = \int \frac{d^4 y}{(2\pi)^4} e^{-\frac{i}{\hbar} p \cdot y} \psi(x + \frac{y}{2}) U(x, y) \bar{\psi}(x - \frac{y}{2}),
$$

$$
U(x, y) = e^{-\frac{iq}{\hbar} y^\mu \int_0^1 ds A_\mu (x - \frac{y}{2} + sy)}, \tag{6.19}
$$

where q is the charge of the fermion and U is the gauge link to ensure gauge invariance.

Different from the classical Boltzmann distribution which is a scalar function, the Wigner function is a 4×4 matrix, it is not Hermitian but satisfies the relation

$$
\gamma_0 W^\dagger \gamma_0 = W \tag{6.20}
$$

and can be decomposed in terms of the 16 independent generators of the Clifford algebra $\{I, \gamma_\mu, i\gamma_5, \gamma_\mu \gamma_5, \sigma_{\mu\nu}\}$, which is often called spin decomposition,

$$
W(x, p) = \frac{1}{4}(F + i\gamma_5 P + \gamma_\mu V^\mu + \gamma_\mu \gamma_5 A^\mu + \frac{1}{2}\sigma_{\mu\nu} S^{\mu\nu}), \tag{6.21}
$$

where the scalar, pseudoscalar, vector, pseudovector, and tensor components $F, P, V^\mu, A^\mu, S^{\mu\nu}$ are all real functions of x and p.

The Wigner function can also be defined in its equal-time version [3,7],

$$
W(x, \mathbf{p}) = \langle \hat{W}(x, \mathbf{p}) \rangle,
$$

$$
\hat{W}(x, \mathbf{p}) = \int \frac{d^3 y}{(2\pi)^3} e^{-\frac{i}{\hbar} \mathbf{p} \cdot \mathbf{y}} \psi(t, \mathbf{x} + \frac{\mathbf{y}}{2}) U(x, y) \psi^\dagger(t, \mathbf{x} - \frac{\mathbf{y}}{2}). \tag{6.22}
$$

Similar to the covariant one, the equal-time Wigner function can also be decomposed in spin space

$$
W(x, \mathbf{p}) = \frac{1}{4}(f_0 + \gamma_5 f_1 - i\gamma_0\gamma_5 f_2 + \gamma_0 f_3
$$

$$
+ \gamma_5\gamma_0 \cdot \mathbf{g}_0 + \gamma_0 \cdot \mathbf{g}_1 - i \cdot \mathbf{g}_2 - \gamma_5 \cdot \mathbf{g}_3). \tag{6.23}
$$

The covariant Wigner function cannot be solved as an initial value problem, since its initial value is related to the fields at all times, see the time integration in its definition. The equal-time Wigner function is, however, well-defined as an initial value problem, and all the spin components f_i and $\mathbf{g}_i$, $(i = 0, 1, 2, 3)$ can be directly interpreted as physics distributions observed in the final state [3]. The disadvantage of the equal-time Wigner function is clearly the lack of the Lorentz covariance for kinetic equations.

The two Wigner functions and then their spin components are connected by the energy average [5,7],

$$W(x, \mathbf{p}) = \int dp_0 W(x, p)\gamma_0,$$

$$f_0(x, \mathbf{p}) = \int dp_0 V_0(x, p),$$

$$f_1(x, \mathbf{p}) = -\int dp_0 A_0(x, p),$$

$$f_2(x, \mathbf{p}) = \int dp_0 P(x, p),$$

$$f_3(x, \mathbf{p}) = \int dp_0 F(x, p),$$

$$g_{0i}(x, \mathbf{p}) = -\int dp_0 A_i(x, p),$$

$$g_{1i}(x, \mathbf{p}) = \int dp_0 V_i(x, p),$$

$$g_{2i}(x, \mathbf{p}) = -\int dp_0 S_{0i}(x, p),$$

$$g_{3i}(x, \mathbf{p}) = -\frac{1}{2} \sum_{jk} \epsilon_{ijk} \int dp_0 S_{jk}(x, p). \tag{6.24}$$

It is easy to see that, when the particles are on the energy shell, namely, $W(x, \mathbf{p}) = W(x, p)\delta(p_0 \mp E_p)$, the covariant and equal-time Wigner functions are equivalent to each other. However, in general case for off-shell particles, the two are fundamentally different [4], one should consider all the energy moments $W_n(x, \mathbf{p}) = \int dp_0 p_0^n W(x, p)\gamma_0$, $n = 0, 1, ...\infty$, and only full collection of them is equivalent to the covariant Wigner function.

From the above relation between the two Wigner functions, one can start from covariant kinetic equations for $W(x, p)$, and then integrate them over p_0 to get the corresponding equal-time kinetic equations for $W(x, \mathbf{p})$. To be simple and compact, we will focus on the covariant form but also present the equal-time one when it is necessary.

If one knows the equations of motion the fermion and gauge field operators ψ and A_μ, by combining them with the definition of the fermion Wigner function, one can get the kinetic equation for the Wigner function. Considering the interaction between fermion and gauge fields, the set of kinetic equations for fermions is not closed, but coupled to the gauge field. When the gauge field is an external field or even in the case without gauge interaction, the self-interaction among the fermions will couple the Wigner function to the higher order correlation functions as well. For example, if we take account of four-fermion interactions usually used in Nambu–Jona–Lasinio (NJL) model, the equation of motion for ψ would involve a term of $\bar{\psi}\psi\psi$ and the kinetic equation for the Wigner function would have terms like $\psi\bar{\psi}\psi\psi$. In this case, one has to include equations for the new terms or make some more approximations, such as cut-off of certain terms, to close the kinetic equation [4,20].

Let us now consider a simpler case with only fermions coupling to a classical electromagnetic field which can be realized in heavy ion collisions. The Lagrangian density of the system is

$$\mathcal{L} = \bar{\psi}[i\gamma^{\mu}(\partial_{\mu} + iqA_{\mu}) - m]\psi. \tag{6.25}$$

The equations of motion for the field operators are just the Dirac equations with a background electromagnetic field,

$$i\gamma^{\mu}(\partial_{\mu} + iqA_{\mu})\psi = m\psi,$$
$$i(\partial_{\mu} - iqA_{\mu})\bar{\psi}\gamma^{\mu} = -m\bar{\psi} \tag{6.26}$$

from which one can get the kinetic equation for the covariant Wigner function [19],

$$\gamma^{\mu}\left(\Pi_{\mu} + \frac{1}{2}i\hbar D_{\mu}\right)W(x, p) = mW(x, p) \tag{6.27}$$

with the electromagnetic operators in phase space,

$$\Pi_{\mu}(x, p) = p_{\mu} - iq\hbar\int_{-\frac{1}{2}}^{\frac{1}{2}} ds\, s\, F_{\mu\nu}(x - i\hbar s\partial_p)\partial_p^{\nu},$$
$$D_{\mu}(x, p) = \partial_{\mu} - q\int_{-\frac{1}{2}}^{\frac{1}{2}} ds\, F_{\mu\nu}(x - i\hbar s\partial_p)\partial_p^{\nu}. \tag{6.28}$$

The operators $D_{\mu}(x, p)$ and $\Pi_{\mu}(x, p)$ in phase space are gauge covariant extension of the partial derivative ∂_{μ} and the momentum p_{μ}. They both are self-adjoint, $D_{\mu}^{\dagger} = D_{\mu}$ and $\Pi_{\mu}^{\dagger} = \Pi_{\mu}$. Since the electromagnetic field is an external field, and we have neglected the interaction between fermions, there is no collision term on the right-hand side of the kinetic equation.

Note that the kinetic equation for the Wigner function contains real and imaginary parts, one is the transport equation, and the other is the constraint equation which is the quantum extension of the on-shell condition in classical limit [4,19]. After the spin decomposition, the two equations are decomposed into 32 real equations for the 16 spin components [21–23],

$$\Pi^{\mu}V_{\mu} = mF,$$
$$\hbar D^{\mu}A_{\mu} = 2mP,$$
$$\Pi_{\mu}F - \frac{1}{2}\hbar D^{\nu}S_{\nu\mu} = mV_{\mu},$$
$$-\hbar D_{\mu}P + \epsilon_{\mu\nu\sigma\rho}\Pi^{\nu}S^{\sigma\rho} = 2mA_{\mu},$$
$$\frac{1}{2}\hbar(D_{\mu}V_{\nu} - D_{\nu}V_{\mu}) + \epsilon_{\mu\nu\sigma\rho}\Pi^{\sigma}A^{\rho} = mS_{\mu\nu},$$

$$\hbar D^\mu V_\mu = 0,$$

$$\Pi^\mu A_\mu = 0,$$

$$\frac{1}{2}\hbar D_\mu F + \Pi^\nu S_{\nu\mu} = 0,$$

$$\Pi_\mu P + \frac{\hbar}{4}\epsilon_{\mu\nu\sigma\rho}D^\nu S^{\sigma\rho} = 0,$$

$$\Pi_\mu V_\nu - \Pi_\nu V_\mu - \frac{\hbar}{2}\epsilon_{\mu\nu\sigma\rho}D^\sigma A^\rho = 0. \tag{6.29}$$

These equations are difficult to be solved. A common method is to semi-classically expand the operators and components as series of $\hbar$ and solve them order by order. The expansion for the electromagnetic operators can be expressed as

$$D_\mu = \partial_\mu - qF_{\mu\nu}\partial_p^\nu + \frac{q}{24}\hbar^2\Delta^2 F_{\mu\nu}\partial_p^\nu + O(\hbar^4),$$

$$\Pi_\mu = p_\mu - \frac{1}{12}\hbar^2\Delta F_{\mu\nu}\partial_p^\nu + O(\hbar^4),$$

$$\Delta = \partial_p \cdot \partial_x. \tag{6.30}$$

For massless particles, the Dirac fermions are replaced by Weyl fermions, and the correlation between the particle momentum and spin reduces the independent number of the spin components from 16 to 4 and thus makes the physics picture very different [2,24–28]. When the mass terms disappear, kinetic equations for the spin components become

$$\Pi^\mu V_\mu = 0,$$

$$\hbar D^\mu A_\mu = 0,$$

$$\Pi_\mu F - \frac{1}{2}\hbar D^\nu S_{\nu\mu} = 0,$$

$$-\hbar D_\mu P + \epsilon_{\mu\nu\sigma\rho}\Pi^\nu S^{\sigma\rho} = 0,$$

$$\frac{1}{2}\hbar(D_\mu V_\nu - D_\nu V_\mu) + \epsilon_{\mu\nu\sigma\rho}\Pi^\sigma A^\rho = 0,$$

$$\hbar D^\mu V_\mu = 0,$$

$$\Pi^\mu A_\mu = 0,$$

$$\frac{1}{2}\hbar D_\mu F + \Pi^\nu S_{\nu\mu} = 0,$$

$$\Pi_\mu P + \frac{\hbar}{4}\epsilon_{\mu\nu\sigma\rho}D^\nu S^{\sigma\rho} = 0,$$

$$\Pi_\mu V_\nu - \Pi_\nu V_\mu - \frac{\hbar}{2}\epsilon_{\mu\nu\sigma\rho}D^\sigma A^\rho = 0. \tag{6.31}$$

One discovers that the vector and axial vector components V_μ and A_μ decouple from other components. It is then convenient to introduce the chiral components for Weyl fermions,

$$V_\chi^\mu = \frac{1}{2}(V^\mu + \chi A^\mu), \tag{6.32}$$

where $\chi = \pm 1$ correspond to the right- and left-handed fermions. The kinetic equations for V_χ^μ are

$$\Pi_\mu V_\chi^\mu = 0,$$
$$D_\mu V_\chi^\mu = 0,$$
$$\chi(\Pi^\mu V_\chi^\nu - \Pi^\nu V_\chi^\mu) + \frac{\hbar}{2}\epsilon^{\mu\nu\rho\sigma} D_\rho V_{\chi\sigma} = 0. \tag{6.33}$$

Since V_+^μ and V_-^μ are decoupled to each other, we can solve them separately. Up to the first order in $\hbar$, the general solution is

$$V_\chi^\mu = \delta(p^2)[p^\mu f_\chi - \hbar\frac{\chi q}{p^2}\tilde{F}^{\mu\nu}p_\nu f_\chi - \hbar\frac{\chi}{p\cdot n}\epsilon^{\mu\nu\lambda\rho}n_\nu p_\lambda D_\rho^{(0)} f_\chi] \tag{6.34}$$

with the dual field tensor $\tilde{F}^{\mu\nu} = \frac{1}{2}\epsilon^{\mu\nu\sigma\rho}F_{\sigma\rho}$. Unlike the massive case, now an additional time-like vector n_μ is needed to formulate the solution. This is related to the fact that a massless particle does not have a rest frame. One natural concern is whether the solution depends on the choice of n_μ. It is proven [24] that with different n_μ, the third term only differs by a quantity proportional to p_μ, which can then be absorbed into the first term. Thus the value of the distribution function f_χ depends on the choice of n_μ. As the frame dependence of the equilibrium distribution function is known, a straightforward choice of n_μ is the average four-velocity of a small space grid or the fluid velocity in hydrodynamics. From now on we will take this choice. The change in the reference frame means an effective boost to n_μ, which leads to a change in the distribution f_χ. This is known as the side-jump effect, and the third term of the above solution is exactly the same as the side-jump term. More discussions can be found in [29, 30].

Putting the solution V_χ^μ back into the kinetic equations, one derives the transport equation for f_χ,

$$\delta\left(p^2 - \hbar\frac{\chi q}{p\cdot n}p_\mu\tilde{F}^{\mu\nu}n_\nu\right)$$
$$\times\left\{p^\mu D_\mu^{(0)} + \hbar\frac{\chi}{2(p\cdot n)^2}\left[(\partial_\mu n_\sigma)p^\sigma - q F_{\mu\alpha}n^\alpha\right]\epsilon^{\mu\nu\lambda\rho}n_\nu p_\lambda D_\rho^{(0)}\right.$$
$$\left. +\hbar\frac{\chi}{2p\cdot n}\epsilon^{\mu\nu\lambda\rho}(\partial_\mu n_\nu)p_\lambda D_\rho^{(0)} + \hbar\frac{\chi q}{2p\cdot n}p_\lambda\partial_\sigma\tilde{F}^{\lambda\nu}n_\nu\partial_p^\sigma\right\} = 0. \tag{6.35}$$

It shows clearly that up to the first order in $\hbar$, the mass shell is shifted. The energy shift can be interpreted as the interaction between the magnetic moment of the fermion and the magnetic field. Note that, while we do not require the vorticity $\omega^{\mu\nu} = \partial^\mu n^\nu -$

$\partial^\nu n^\mu$ to be zero, there is no energy shift corresponding to the interaction between the spin and vorticity. This shows that the magnetic field and vorticity have many similarities, but they sometimes behave differently.

If we first assume the vorticity to be zero, the transport equation is reduced to

$$\delta \left(p^2 - \hbar \frac{\chi q}{p^\nu n_\nu} B^\mu p_\mu \right)$$
$$\times \left\{ p^\mu D_\mu^{(0)} - \hbar \frac{\chi q}{2 p^\nu n_\nu} \left[\epsilon^{\mu\nu\lambda\rho} E_\mu n_\nu p_\lambda D_\rho^{(0)} + p_\lambda \partial_\rho B^\lambda \partial_p^\rho \right] \right\} f_\chi = 0 \quad (6.36)$$

with the covariant electric and magnetic field $E_\mu = F_{\mu\nu} n^\nu$, $B_\mu = \epsilon_{\mu\nu\sigma\rho} n_\nu F^{\sigma\rho}$. The δ-function here shows that to the first order in $\hbar$ the chiral fermions are still on the mass shell, but the shell is shifted from zero to a nonzero value. It is worth noting that fermions with different chirality have different mass shift. In this case, the vector and axial vector components, being the mixing of left-hand and right-handed components, cannot be described by a single mass shell even in massless case. With the help of the δ-function, the energy integration of the above covariant equation can be done easily, and we obtain the equal-time transport equation

$$\left\{ \partial_t + \frac{1}{\sqrt{G}} \left(\tilde{\mathbf{v}} + \hbar \epsilon q (\tilde{\mathbf{v}} \cdot \mathbf{b}) \mathbf{B} + \hbar \epsilon q \tilde{\mathbf{E}} \times \mathbf{b} \right) \cdot \nabla \right.$$
$$\left. + \frac{\epsilon q}{\sqrt{G}} \left(\tilde{\mathbf{E}} + \tilde{\mathbf{v}} \times \mathbf{B} + \hbar \epsilon q (\tilde{\mathbf{E}} \cdot \mathbf{B}) \mathbf{b} \right) \cdot \nabla_p \right\} f_\chi^\epsilon (x, \mathbf{p}) = 0 \qquad (6.37)$$

which is usually called the chiral kinetic equation [11], where $\epsilon = \pm 1$ correspond to the positive and negative energy fermions. $\mathbf{b} = \chi \mathbf{p}/(2|\mathbf{p}|^3)$ is the Berry curvature, and

$$\sqrt{G} = 1 + \hbar \epsilon q \mathbf{b} \cdot \mathbf{B},$$
$$E_p = |\mathbf{p}|(1 - \hbar \epsilon q \mathbf{B} \cdot \mathbf{b}),$$
$$\tilde{\mathbf{E}} = \mathbf{E} - \frac{1}{\epsilon q} \nabla E_p,$$
$$\tilde{\mathbf{v}} = \nabla_p E_p = \frac{\mathbf{p}}{|\mathbf{p}|}(1 + 2\hbar \epsilon q \mathbf{B} \cdot \mathbf{b}) - \hbar \epsilon q b \mathbf{B} \qquad (6.38)$$

are, respectively, the phase space measurement, fermion energy, effective electric field, and group velocity.

For massive fermions, in classical limit ($\hbar = 0$) the general kinetic equations (6.29) are reduced to

$$p^\mu V_\mu^{(0)} = m F^{(0)},$$
$$0 = m P^{(0)},$$
$$p_\mu F^{(0)} = m V_\mu^{(0)},$$

$$\epsilon_{\mu\nu\sigma\rho}p^{\nu}S^{(0)\sigma\rho} = 2mA_{\mu}^{(0)},$$

$$\epsilon_{\mu\nu\sigma\rho}p^{\sigma}A^{(0)\rho} = mS_{\mu\nu}^{(0)},$$

$$p^{\mu}A_{\mu}^{(0)} = 0,$$

$$p^{\nu}S_{\nu\mu}^{(0)} = 0,$$

$$p_{\mu}P^{(0)} = 0,$$

$$p_{\mu}V_{\nu}^{(0)} - p_{\nu}V_{\mu}^{(0)} = 0 \qquad (6.39)$$

for the components $F^{(0)}$, $P^{(0)}$, $V_{\mu}^{(0)}$, $A_{\mu}^{(0)}$ and $S_{\mu\nu}^{(0)}$. They can be analytically solved with the solution

$$P^{(0)} = 0,$$

$$F^{(0)} = mf\delta(p^2 - m^2),$$

$$V_{\mu}^{(0)} = p_{\mu}f\delta(p^2 - m^2),$$

$$A_{\mu}^{(0)} = (p^2 g_{\mu\nu} - p_{\mu}p_{\nu})K^{\nu}\delta(p^2 - m^2),$$

$$S_{\mu\nu}^{(0)} = \frac{1}{m}\epsilon_{\mu\nu\sigma\rho}p^{\sigma}A^{(0)\rho} = m\epsilon_{\mu\nu\sigma\rho}p^{\sigma}K^{\rho}\delta(p^2 - m^2), \qquad (6.40)$$

where $f(x, p)$ and $K_{\mu}(x, p)$ are arbitrary scalar and four-vector functions in phase space.

It is clear that in the classical limit the relations among the spin components are not affected by the background field and perfectly on the mass shell. Among the 16 spin components, the number of independent components is 4. This can be seen more clearly from the solution in equal-time formalism,

$$f_1^{(0)\pm} = \pm\frac{\mathbf{p}}{E_p}\cdot\mathbf{g}_0^{(0)\pm},$$

$$f_2^{(0)\pm} = 0,$$

$$f_3^{(0)\pm} = \pm\frac{m}{E_p}f_0^{(0)\pm},$$

$$\mathbf{g}_1^{(0)\pm} = \pm\frac{\mathbf{p}}{E_p}f_0^{(0)\pm},$$

$$\mathbf{g}_2^{(0)\pm} = \frac{\mathbf{p}\times\mathbf{g}_0^{(0)\pm}}{m},$$

$$\mathbf{g}_3^{(0)\pm} = \mp\frac{E_p^2\mathbf{g}_0^{(0)\pm} - (\mathbf{p}\cdot\mathbf{g}_0^{(0)\pm})\mathbf{p}}{mE_p} \qquad (6.41)$$

with the fermion energy $E_p = \sqrt{m^2 + p^2}$ and the sign $\pm$ corresponding to the positive and negative fermions determined by the δ function. Here the particle number density f_0 (V_0 in covariant formalism) and the spin density $\mathbf{g}_0$ ($\mathbf{A}$ in covariant formalism) are taken as independent components. Obviously it is also possible to choose other set of independent components [21].

By putting the classical solution into the linear order in $\hbar$ of the general kinetic equations (6.29), one derives the transport equations for the independent variables,

$$D_\mu^{(0)} V^{(0)\mu} = 0,$$
$$p^\nu D_\nu^{(0)} A_\mu^{(0)} - q F_{\mu\nu} A^{\nu(0)} = 0 \tag{6.42}$$

in covariant formalism or

$$\left(D_t^{(0)} \pm \frac{\mathbf{p}}{E_p} \cdot \mathbf{D}^{(0)} \right) f_0^{(0)\pm} = 0, \tag{6.43}$$
$$\left(D_t^{(0)} \pm \frac{\mathbf{p}}{E_p} \cdot \mathbf{D}^{(0)} \right) \mathbf{g}_0^{(0)\pm} = \frac{1}{E_p^2} \left(\mathbf{p} \times (\mathbf{E} \times \mathbf{g}_0^{(0)\pm}) \mp E_p \mathbf{B} \times \mathbf{g}_0^{(0)\pm} \right)$$

in equal-time formalism, where the classical operators $D_\mu^{(0)}$, $D_t^{(0)}$ and $\mathbf{D}^{(0)}$ are defined as

$$D_\mu^{(0)}(x, p) = \partial_\mu - q F_{\mu\nu}(x)\partial_p^\nu,$$
$$D_t^{(0)}(x, \mathbf{p}) = \partial_t + q\mathbf{E}(x) \cdot \nabla_p,$$
$$\mathbf{D}^{(0)}(x, \mathbf{p}) = \nabla + q\mathbf{B}(x) \times \nabla_p. \tag{6.44}$$

The first equation is the Boltzmann equation for the particle number density f_0 in external electromagnetic field which is hidden in the operators D_t and $\mathbf{D}$. The second one is the generalized Bargmann–Michel–Telegdi equation [31] in phase space, the effective collision terms on the right-hand side are from the spin interaction with the electromagnetic field.

The linear order in $\hbar$ of the generalized equations (6.29) gives not only the transport equations for the classical components but also the constraint equations for the first-order quantum corrections to the Wigner function $W^{(1)}$ or the components $F^{(1)}$, $P^{(1)}$, $V_\mu^{(1)}$, $A_\mu^{(1)}$ and $S_{\mu\nu}^{(1)}$,

$$p^\mu V_\mu^{(1)} = m F^{(1)},$$
$$\frac{1}{2} D_\mu^{(0)} A^{(0)\mu} = m P^{(1)},$$
$$p^\mu F^{(1)} - \frac{1}{2} D_\nu^{(0)} S^{(0)\nu\mu} = m V^{(1)\mu},$$
$$-D_\mu^{(0)} P^{(0)} + \epsilon_{\mu\nu\sigma\rho} p^\nu S^{(1)\sigma\rho} = 2m A_\mu^{(1)},$$
$$\frac{1}{2} \left(D_\mu^{(0)} V_\nu^{(0)} - D_\nu^{(0)} V_\mu^{(0)} \right) + \epsilon_{\mu\nu\sigma\rho} p^\sigma A^{(1)\rho} = m S_{\mu\nu}^{(1)},$$
$$D_\mu^{(0)} V^{(0)\mu} = 0,$$
$$p^\mu A_\mu^{(1)} = 0,$$
$$\frac{1}{2} D_\mu^{(0)} F^{(0)} + p^\nu S_{\nu\mu}^{(1)} = 0$$

$$p^\mu P^{(1)} + \frac{1}{4}\epsilon^{\mu\nu\sigma\rho} D^{(0)}_\nu S^{(0)}_{\sigma\rho} = 0,$$

$$p^\mu V^{(1)\nu} - p^\nu V^{(1)\mu} - \frac{1}{2}\epsilon^{\mu\nu\sigma\rho} D^{(0)}_\sigma A^{(0)}_\rho = 0. \qquad (6.45)$$

With the known classical components, the first-order quantum corrections can be expressed in terms of them

$$P^{(1)} = \frac{1}{2m} D^{(0)}_\mu A^{(0)\mu},$$

$$F^{(1)} = m\tilde{f}\delta(p^2 - m^2) - \frac{1}{2m(p^2 - m^2)}\epsilon^{\mu\nu\sigma\rho} p_\mu D^{(0)}_\nu p_\sigma A^{(0)}_\rho,$$

$$V^{(1)}_\mu = p_\mu \tilde{f}\delta(p^2 - m^2) + \frac{1}{2}\epsilon_{\mu\nu\sigma\rho} D^{(0)\nu} p^\sigma K^\rho \delta(p^2 - m^2)$$
$$\quad - \frac{qp_\mu}{2}\epsilon_{\lambda\nu\sigma\rho} F^{\lambda\nu} p^\sigma K^\rho \delta'(p^2 - m^2),$$

$$A^{(1)}_\mu = (p^2 g_{\mu\nu} - p_\mu p_\nu)\tilde{K}^\nu \delta(p^2 - m^2) - \frac{1}{2(p^2 - m^2)}\epsilon_{\mu\nu\sigma\rho} p^\nu D^{(0)\sigma} V^{(0)\rho},$$

$$S^{(1)}_{\mu\nu} = \frac{1}{2m}\left(D^{(0)}_\mu V^{(0)}_\nu - D^{(0)}_\nu V^{(0)}_\mu\right) + \frac{1}{m}\epsilon_{\mu\nu\sigma\rho} p^\sigma A^{(1)\rho} \qquad (6.46)$$

with the new scalar and four-vector functions $\tilde{f}(x, p)$ and $\tilde{K}_\mu(x, p)$.

Considering the terms with $\delta'(p^2 - m^2)$, while all the components are still on their own mass shells, the quantum corrections to the classical mass shell, namely, the shell shifts, are different for different components. Therefore, there is no longer a common mass shell for all the components to the first order in $\hbar$, and the Wigner function cannot be factorized like $W(x, p)\delta(p^2 - m^2 + \hbar E_p \Delta E)$.

Nevertheless, up to the first order in $\hbar$, all the terms in the Wigner function contain $\delta(p^2)$ or $\delta'(p^2)$. In this case, the relationship between covariant and equal-times formalisms is still quite straightforward, the simple integration over p_0 leads to the equal-time relations

$$f^{(1)\pm}_1 = \pm\frac{\mathbf{p}\cdot\mathbf{g}^{(1)\pm}_0}{E_p} \pm \frac{\mathbf{p}\cdot\mathbf{B}}{2E_p^3}f^{(0)\pm}_0,$$

$$f^{(1)\pm}_2 = -\frac{\mathbf{D}^{(0)}\cdot\mathbf{g}^{(0)\pm}_0}{2m} + \frac{\mathbf{p}\cdot(\mathbf{p}\cdot\mathbf{D}^{(0)})\mathbf{g}^{(0)\pm}_0}{2mE_p^2} - \frac{(\mathbf{B}\times\mathbf{p})\cdot\mathbf{g}^{(0)\pm}_0}{mE_p^2} \mp \frac{\mathbf{E}\cdot\mathbf{g}^{(0)\pm}_0}{2mE_p},$$

$$f^{(1)\pm}_3 = \pm\frac{mf^{(1)}_0}{E_p} \mp \frac{(\mathbf{p}\times\mathbf{D}^{(0)})\cdot\mathbf{g}^{(0)\pm}_0}{2mE_p} + \frac{\mathbf{p}\cdot(\mathbf{E}\times\mathbf{g}^{(0)\pm}_0)}{2mE_p^2} \mp \frac{\mathbf{B}\cdot\mathbf{g}^{(0)\pm}_0}{2mE_p}$$
$$\quad \mp \frac{(\mathbf{B}\cdot\mathbf{p})(\mathbf{p}\cdot\mathbf{g}^{(0)\pm}_0)}{2mE_p^3},$$

$$\mathbf{g}^{(1)\pm}_1 = \pm\frac{\mathbf{p}}{E_p}f^{(1)}_0 \pm \frac{1}{2E_p}\mathbf{D}^{(0)}\times\mathbf{g}^{(0)}_0 + \frac{\mathbf{E}}{2E_p^2}\times\mathbf{g}^{(0)\pm}_0 \pm \frac{\mathbf{B}(\mathbf{p}\cdot\mathbf{g}^{(0)\pm}_0)}{2E_p^3},$$

$$\mathbf{g}_2^{(1)\pm} = \frac{\mathbf{p} \times \mathbf{g}_0^{(1)\pm}}{m} \pm \left(\frac{\mathbf{p}(\mathbf{p} \cdot \mathbf{E})}{2m E_p^3} - \frac{\mathbf{E}}{2m E_p} \right) f_0^{(0)\pm} + \frac{\mathbf{p}}{2m E_p^2} \mathbf{p} \cdot \mathbf{D}^{(0)} f_0^{(0)\pm}$$

$$- \frac{1}{2m} \mathbf{D}^{(0)} f_0^{(0)\pm},$$

$$\mathbf{g}_3^{(1)\pm} = \mp \left(\frac{E_p}{m} \mathbf{g}_0^{(1)\pm} - \frac{\mathbf{p} \cdot \mathbf{g}_0^{(1)\pm}}{m E_p} \mathbf{p} \right) + \left(\frac{\mathbf{E} \times \mathbf{p}}{2m E_p^2} \mp \frac{m\mathbf{B}}{2 E_p^3} \right) f_0^{(0)\pm}$$

$$\mp \frac{1}{2m E_p} \mathbf{p} \times \mathbf{D}^{(0)} f_0^{(0)\pm}. \tag{6.47}$$

It is clear that the equal-time equations are not Lorentz invariant or covariant, these particular equations above hold in the local rest frame.

The transport equations for the independent components $f_0^{(1)}$ and $\mathbf{g}_0^{(1)}$ can be derived from the second order in $\hbar$ of the general kinetic equations (6.29). Combining with the classical transport equations, one obtains finally the transport equations for the independent distributions to the first order in $\hbar$, $f_0^\pm = f_0^{(0)\pm} + \hbar f_0^{(1)\pm}$ and $\mathbf{g}_0^\pm = \mathbf{g}_0^{(0)\pm} + \hbar \mathbf{g}_0^{(1)\pm}$,

$$\left(D_t^{(0)} \pm \frac{\mathbf{p}}{E_p} \cdot \mathbf{D}^{(0)} \right) f_0^\pm = \frac{\hbar \mathbf{E}}{2 E_p^2} \cdot \left(\mathbf{D}^{(0)} \times \mathbf{g}_0^{(0)\pm} \right) \mp \frac{\hbar}{2 E_p^3} \mathbf{B} \cdot \left(\mathbf{p} \cdot \mathbf{D}^{(0)} \right) \mathbf{g}_0^{(0)\pm}$$

$$+ \frac{\hbar (\mathbf{B} \times \mathbf{p})}{E_p^4} \cdot \left(\mathbf{E} \times \mathbf{g}_0^{(0)} \right), \tag{6.48}$$

$$\left(D_t^{(0)} \pm \frac{\mathbf{p}}{E_p} \cdot \mathbf{D}^{(0)} \right) \mathbf{g}_0^\pm = \frac{1}{E_p^2} \left(\mathbf{p} \times \mathbf{E} \mp E_p \mathbf{B} \right) \times \mathbf{g}_0^\pm$$

$$\mp \hbar \left(\frac{\mathbf{B}}{2 E_p^3} \pm \frac{\mathbf{E} \times \mathbf{p}}{2 E_p^4} \right) \mathbf{p} \cdot \mathbf{D}^{(0)} f_0^{(0)\pm}$$

$$\mp \hbar \left(\frac{(\mathbf{p} \cdot \mathbf{E})(\mathbf{E} \times \mathbf{p})}{E_p^5} \pm \frac{\mathbf{p} \times (\mathbf{B} \times \mathbf{E})}{2 E_p^4} \right) f_0^{(0)\pm}.$$

Unlike the classical transport equations where the two independent components $f_0^{(0)}$ and $\mathbf{g}_0^{(0)}$ are decoupled from each other, the components f_0 and $\mathbf{g}_0$ to the first-order quantum corrections are coupled to each other due to the electromagnetic interaction.

6.4   Spin Polarization in Transport Theory

The physics of the 16 spin components in covariant or equal-time versions can be obtained from the conservation laws of the system. With the charge conservation and total angular momentum conservation, f_0 and $\mathbf{g}_0$ are, respectively, interpreted as the particle number density and spin density.

The transport equation for the three-dimensional spin density $\mathbf{g}_0^{(0)}$ is equivalent to the covariant Bargmann–Michel–Telegdi [31] equation for a spinning particle in an external field,

$$m\frac{ds^\mu}{d\tau} = qF^{\mu\nu}(\tau)s_\nu(\tau), \tag{6.49}$$

where $s_\mu = A_\mu/(A^\nu A_\nu)^{1/2}$ is the covariant spin phase space density.

We can also start from the canonical definition of the spin tensor [32]

$$S^{\lambda,\mu\nu} = \frac{\hbar}{4}Tr\left[\{\sigma^{\mu\nu},\gamma^\lambda\}W(x,k)\right] = \frac{\hbar}{2}\epsilon^{\sigma\lambda\mu\nu}A_\sigma(x,k), \tag{6.50}$$

with which the observed polarization density in a given reference frame characterized by the vector n_μ can be expressed as

$$L^\mu = -\frac{1}{2}\epsilon^{\mu\nu\alpha\beta}\int\frac{d^4k}{(2\pi)^4}\Pi_\nu S_{\lambda,\alpha\beta}n^\lambda = \frac{\hbar}{2}\int\frac{d^4k}{(2\pi)^4}\left(\Pi^\nu n_\nu\right)A^\mu. \tag{6.51}$$

This explains why $\mathbf{g}_0$ is called the spin density.

We can try to discuss the conditions for spin polarization in a few characteristic cases. We firstly consider a trivial configuration where there is no electromagnetic field. In this case the two transport equations (6.48) for the number density and spin density to the first order in $\hbar$ become identical to each other, which means that there can be arbitrary polarization without a preferred direction. Also, due to the absence of collision terms, the averaged polarization will not relax to zero, opposed to the equilibrium case.

The second example is with a constant magnetic field. This is one of the situations where the global equilibrium is known to exist. In this case the quantum transport equation for the spin density (6.48) becomes

$$\left(D_t^{(0)} \pm \frac{\mathbf{p}}{E_p}\cdot\mathbf{D}^{(0)}\right)\mathbf{g}_0^\pm = \pm\frac{1}{E_p}\mathbf{B}\times\mathbf{g}_0^\pm \mp \hbar\frac{\mathbf{B}}{2E_p^3}\mathbf{p}\cdot\mathbf{D}^{(0)}f_0^{(0)\pm}. \tag{6.52}$$

The first term on the right-hand side corresponds to the spin precession, and the second term is related to the polarization in non-equilibrium state. Note that it is still possible for $\mathbf{g}_0$ to be zero. From the Dirac equation one can check that a free fermion will not be polarized by a constant magnetic field, i.e., the direction of the particle spin will not turn toward the direction of the magnetic field. A multi-particle system can be polarized, because the magnetic field changes the energy distribution among different spin states. Therefore, in equilibrium case the possibility for one spin direction can be larger than other directions. As there is no collision term in our current model, the process from non-equilibrium toward equilibrium can not be described, while we can still study the equilibrium case itself. One can find an obvious solution,

$$\mathbf{g}_0 = a\frac{\mathbf{B}}{B}f_0^{(0)}. \tag{6.53}$$

This shows clearly that the averaged spin of particles is polarized along the direction of the magnetic field, but the degree of polarization, which is proportional to the constant a, is not determined, as it depends on the detail of the interaction.

One could have already noticed that the vorticity or the derivative of velocity does not explicitly appear in the kinetic equations. However, this does not mean that massive fermions are not polarized by the vorticity field. We can look at the axial current A_μ as an example. If we set the electromagnetic field to be zero, the only restriction on A_μ is

$$\Pi^\mu A_\mu = 0. \tag{6.54}$$

Thus we can include a term $\epsilon^{\mu\nu\alpha\beta}\Pi_\alpha n_\beta \partial_\nu f_V$ in A^μ. Assuming f_V to be the Fermi–Dirac distribution, this term will introduce the vorticity. In fact, to take the limit of $m \to 0$ to be consistent with an original fermion system, a term with this form is necessary [23].

From the discussions above, one can see that the collision terms are actually very important, as when using transport theory we are usually more interested in non-equilibrium systems, and the collision terms are necessary in this case. What is more, if we want to study a transport theory with spin, a spin-dependent collision term is required. Although up to now it is still not very clear how to self-consistently derive the collision terms in Wigner function formalism from field theory, there are many discussions about this topic [33–35].

6.5   Anomaly Induced Transport Theory

From the chiral vector current V_μ^χ for Weyl fermions in (6.34), one can reconstruct the vector and axial vector components,

$$V^\mu = \left(p^\mu f_V - \hbar \frac{q}{p^2} \tilde{F}^{\mu\nu} p_\nu f_A - \hbar \frac{1}{p \cdot n} \epsilon^{\mu\nu\lambda\rho} n_\nu p_\lambda D_\rho f_A \right) \delta(p^2),$$

$$A^\mu = \left(p^\mu f_A - \hbar \frac{q}{p^2} \tilde{F}^{\mu\nu} p_\nu f_V - \hbar \frac{1}{p \cdot n} \epsilon^{\mu\nu\lambda\rho} n_\nu p_\lambda D_\rho f_V \right) \delta(p^2) \tag{6.55}$$

with $f_V = f_+ + f_-$ and $f_A = f_+ - f_-$ and the particle current and axial current

$$J^\mu(x) = \int d^4 p V^\mu(x, p),$$

$$J_5^\mu(x) = \int d^4 p A^\mu(x, p). \tag{6.56}$$

It is easy to see that the currents J^μ and J_5^μ are coupled to each other. To compare with the usual expression for anomaly transport, one has to go into an equilibrium state. We choose the normal Fermi–Dirac distribution $f_\chi = [e^{sgn(p \cdot n)(p \cdot n - \mu_\chi)/T} + 1]^{-1}$.

Note that with this choice f_V and f_A do not take the Fermi–Dirac distribution. A straightforward calculation leads to the vector and axial vector currents

$$J^\mu = \frac{2\pi\mu}{3}\left(\pi^2 T^2 + \frac{3\mu^2 + \mu_5^2}{4}\right) n^\mu + \hbar\pi\mu_5 B^\mu + 3\hbar\mu\mu_5\omega^\mu,$$

$$J_5^\mu = \frac{2\pi\mu_5}{3}\left(\pi^2 T^2 + \frac{\mu^2 + 3\mu_5^2}{4}\right) n^\mu + \hbar\pi\mu B^\mu$$

$$+ \hbar\pi T^2 \left(2\pi^2 T^2 + \frac{3\mu^2 + 3\mu_5^2}{2}\right)\omega^\mu \qquad (6.57)$$

with $\omega^\mu = \epsilon^{\mu\nu\sigma\rho} n_\nu \partial_\sigma n_\rho / 2$, $\mu = \mu_+ + \mu_-$ and $\mu_5 = \mu_+ - \mu_-$. Apart from the normal current proportional to the frame vector n^μ, there are contributions proportional to the magnetic field B^μ and vorticity ω^μ, corresponding to CME and CVE. This result, up to a difference of an overall coefficient, is consistent with the results obtained by other methods. The transport theory is used in a lot of studies to derive various transport coefficients related to magnetic field and vorticity in hydrodynamics [25,30,36,37].

It is possible to consider quantum correction to the distribution function f_χ at the order of $O(\hbar)$. For example, considering the interaction between spin and vorticity, one can add the correction $f_\chi^{(1)} = \hbar\epsilon^{\mu\nu\sigma\rho} p_\mu n_\nu \partial_\sigma n_\rho / (4Tp \cdot n)$ to f_χ. This will change the coefficient for CVE [30].

In heavy ion collisions, the global equilibrium description is not very sufficient, since the magnetic field, vorticity, and particle density all have spatial and temporal dependence. Also, with the existence of axial anomaly, the axial charge is not conserved, and μ_5 is not a well-defined quantity. Strictly speaking, we must discuss anomalous transport in non-equilibrium states. On the other hand, unlike hydrodynamic models, transport theories are supposed to work in out-of-equilibrium cases.

Although there may be not global and consistent transport coefficients, it has been shown that starting from a non-equilibrium initial condition, albeit with nonzero axial charge, the chiral kinetic theory can produce separation of particle density over time [26]. This allows the CKT to be used in a much wider range of studies.

6.6   Degenerate to Hydrodynamics

Usually one can integrate out the momentum dependence of the kinetic equations to obtain the hydrodynamic equations. For QED the vector current to the first order in $\hbar$ in coordinate space can be expressed as [21]

$$J_\mu = \int d^4 p V_\mu(x, p) \qquad (6.58)$$

$$= \int d^4 p \delta(p^2 - m^2)$$

$$\times \left[p_\mu \left(V_0^{(0)} + \hbar V_0^{(1)} \right) + \frac{\hbar}{2} \partial^\nu \Sigma_{\mu\nu}^{(0)} A_0^{(0)} + \frac{\hbar}{4} F^{\alpha\beta} \partial_{p\mu} \left(\Sigma_{\alpha\beta}^{(0)} A_0^{(0)} \right) \right],$$

where $\Sigma_{\mu\nu}^{(0)}$ is the fermion selfenergy at classical level. The corresponding energy-momentum tensor is

$$\begin{aligned} T^{\mu\nu} &= \left\langle : \frac{\partial \mathcal{L}}{\partial(\partial_\mu \psi)} \partial^\nu \psi + \partial^\nu \bar\psi \frac{\partial \mathcal{L}}{\partial(\partial_\mu \bar\psi)} + \frac{\partial \mathcal{L}}{\partial(\partial_\mu A_\alpha)} \partial^\nu A_\alpha - g^{\mu\nu} \mathcal{L} : \right\rangle \\ &= T_{mat}^{\mu\nu} + T_{int}^{\mu\nu} + T_{em}^{\mu\nu}, \end{aligned} \tag{6.59}$$

where we have separated the total energy-momentum tensor into three parts: the gauge-invariant matter part $T_{mat}^{\mu\nu}$, the contribution from the interaction between gauge potential and matter current, $T_{int}^{\mu\nu}$, and the electromagnetic part $T_{em}^{\mu\nu}$,

$$\begin{aligned} T_{mat}^{\mu\nu} &= \frac{\hbar}{2} \left\langle : \bar\psi \gamma^\mu (i \overrightarrow{D}^\nu - i \overleftarrow{D}^{\dagger\nu}) \psi : \right\rangle = \int d^4 p \, p^\nu V^\mu, \\ T_{int}^{\mu\nu} &= A^\nu \left\langle : \bar\psi \gamma^\mu \psi : \right\rangle = A^\nu \int d^4 p \, V^\mu, \\ T_{em}^{\mu\nu} &= \frac{1}{4} g^{\mu\nu} F^{\alpha\beta} F_{\alpha\beta} - F^{\mu\alpha} \partial^\nu A_\alpha. \end{aligned} \tag{6.60}$$

Note that none of these is symmetric under the change $\mu \leftrightarrow \nu$. While the total energy-momentum tensor is conserved

$$\partial_\nu T^{\mu\nu}(x) = 0 \tag{6.61}$$

which can be checked using the Dirac and Maxwell equations, the matter part is not conserved,

$$\partial_\mu T_{mat}^{\mu\nu}(x) = F^{\nu\alpha}(x) J_\alpha(x). \tag{6.62}$$

Besides currents and energy-momentum tensor, one should take the angular momentum tensor into account when the spin and vorticity are considered [38–40]

$$\begin{aligned} J^{\lambda,\mu\nu} &= x^\mu T^{\lambda\nu} - x^\nu T^{\lambda\mu} + \frac{\hbar}{4} \left\langle : \bar\psi \{\gamma^\lambda, \sigma^{\mu\nu}\} \psi : \right\rangle \\ &\quad - (F^{\lambda\mu} A^\nu - F^{\lambda\nu} A^\mu). \end{aligned} \tag{6.63}$$

The first two terms, $x^\mu T^{\lambda\nu} - x^\nu T^{\lambda\mu}$, can be interpreted as the orbital angular momentum tensor, and the remaining terms constitute the spin angular momentum tensor which can be further separated into a matter and a field part. The matter part can be defined as [41]

$$S_{mat}^{\lambda,\mu\nu}(x) = \frac{1}{4} \left\langle : \bar\psi \{\gamma^\lambda, \sigma^{\mu\nu}\} \psi : \right\rangle = -\frac{1}{2} \epsilon^{\lambda\mu\nu\rho} \int d^4 p \, A_\rho(x, p). \tag{6.64}$$

With the help of the covariant kinetic equations, we find, to any order in $\hbar$,

$$\hbar \partial_\lambda S_{mat}^{\lambda,\mu\nu}(x) = T_{mat}^{\nu\mu}(x) - T_{mat}^{\mu\nu}(x). \tag{6.65}$$

In the case with a background vorticity, the thermal distribution for massless fermions satisfies the chiral kinetic equation

$$\left[p^\mu \partial_\mu \pm \hbar (\partial_\mu \frac{\epsilon^{\mu\nu\rho\sigma} p_\rho n_\sigma}{2n \cdot p}) \partial_\nu \right] f_\pm = 0 \tag{6.66}$$

to the first order in $\hbar$. The solution could be analytically written as

$$f_\pm(p) = \frac{1}{e^{(p \cdot n - \mu_\pm)/T \pm \hbar \epsilon^{\mu\nu\rho\sigma} \omega_{\mu\nu} n_\rho p_\sigma/(4n \cdot p)} + 1}, \tag{6.67}$$

and the corresponding currents and energy-momentum tensor are derived as

$$
\begin{aligned}
J_\pm^\mu &= \frac{\mu_\pm}{6}\left(T^2 + \frac{\mu_\pm^2}{\pi^2} \right) n^\mu \pm \frac{\hbar}{2}\left(\frac{T^2}{6} + \frac{\mu_\pm^2}{2\pi^2} \right) \omega^\mu, \\
T^{\mu\nu} &= \left(\frac{7\pi^2 T^4}{45} + \frac{2T^2(\mu_V^2 + \mu_A^2)}{3} + \frac{\mu_V^4 + 6\mu_V^2\mu_A^2 + \mu_A^4}{3\pi^2} \right)\left(n^\mu n^\nu - \frac{g^{\mu\nu}}{4} \right) \\
&\quad + \frac{\hbar\mu_A}{12}\left(T^2 + \frac{3\mu_V^2 + \mu_A^2}{\pi^2} \right)\left(8\omega^\mu u^\nu + T\epsilon^{\mu\nu\sigma\lambda}\omega_{\sigma\lambda} \right).
\end{aligned} \tag{6.68}
$$

The viscous terms could be obtained by considering the out-of-equilibrium contribution as

$$\delta f^\pm = f_{eq}^\pm \left(1 - f_{eq}^\pm \right)\left(\lambda_\Pi^\pm \Pi + \lambda_v^\pm v_\pm^\mu p_\mu + \lambda_\pi^\pm \pi^{\mu\nu} p_\mu p_\nu \mp \hbar\lambda_\Omega^\pm \frac{\Omega^\pm \cdot p}{n \cdot p} \right) \tag{6.69}$$

with the dissipative terms defined as

$$
\begin{aligned}
\Pi &= -\frac{1}{3}\int d^4 p \Delta^{\mu\nu} p_\mu p_\nu (\delta f_+ + \delta f_-), \\
v_\pm^\mu &= \int d^4 p \Delta_\alpha^\mu p^\alpha \delta f_\pm, \\
\pi^{\mu\nu} &= \int d^4 p \Delta_{\alpha\beta}^{\mu\nu} p^\alpha p^\beta (\delta f_+ + \delta f_-), \\
\Omega_\pm^\mu &= \pm \int d^4 p \Delta_\alpha^\mu p^\alpha (u \cdot p) \delta f_\pm.
\end{aligned} \tag{6.70}
$$

The studies on the viscous spin fluid are still processing.

6.7   Experiments and Numerical Simulations

Although light quarks are good candidates of observing the chiral anomaly effects, signals in heavy ion collision are not clear enough because of the large fluctuations in background and short lifetime of the initial magnetic field. The experimental and theoretical studies on identifying the CME in heavy ion collisions are still in process [42]. Fortunately robust signals are found in so-called Weyl metal systems. The first observation of chiral magnetic effect was reported in [43] through the measurement of magneto-transport in zirconium pentatelluride, $ZrTe_5$ (see Fig. 6.1). The angle-resolved photoemission spectroscopy experiments show that the electronic structure is consistent with a three-dimensional Dirac semi-metal. A large negative magnetoresistance is observed when the magnetic field is parallel to the current. The measured quadratic field dependence of the magnetoconductance is a clear indication of the chiral magnetic effect. The observed phenomena stem from the effective transmutation of Dirac semi-metal into a Weyl semi-metal induced by the parallel electric and magnetic fields which behave as the topologically nontrivial gauge field background. After this work a series of phenomena have been discovered in similar condensed matter systems [44–46].

The spin current induced by the vorticity polarization has also been observed in a well-designed $Ga_{62}In_{25}Sn_{13}$(GaInSn) system, which is another chemically stable liquid metal, the spin-orbit coupling may be weaker than that in Hg because all atoms in GaInSn are lighter than Hg [47].

In heavy ion collisions, for the vorticity-polarization effect, the first measurement of an alignment between the angular momentum of a non-central collision and the spin of emitted particles was done by STAR [48]. It reveals that the fluid produced in non-central heavy ion collisions is by far the most vortical system ever observed. It is found that Λ and $\bar{\Lambda}$ hyperons show a positive polarization of the order of a few percents, consistent with some hydrodynamic predictions (see Fig. 6.2). The previous measurement that reported a null result at higher collision energies seems

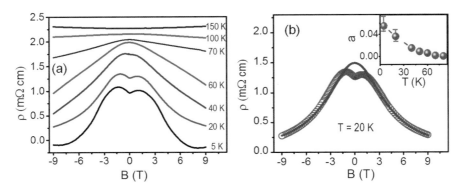

Fig. 6.1 Magnetoresistance in field parallel to current (**B** ∥ **a**) in $ZrTe_5$. **a** MR at various temperatures. For clarity, the resistivity curves were shifted by 1.5 mΩcm (150 K), 0.9 mΩcm (100 K), 0.2 mΩcm (70 K) and −0.2 mΩcm (5 K). **b** MR at 20K fitted with the CME curve; inset: temperature dependence of the fitting parameter $a(T)$ in units of S/(cm T²). The figure is taken from Ref. [43]

Fig. 6.2 The averaged polarization $\overline{P}_H$ (H = Λ or $\overline{Λ}$) in 20–50% central Au+Au collisions is plotted as a function of collision energy. The results of the present study ($\sqrt{s_{NN}} < 40$ GeV) are shown together with those reported earlier [49] for 62.4 and 200 GeV collisions, for which only statistical errors are plotted. Boxes indicate systematic uncertainties. The figure is taken from Ref. [48]

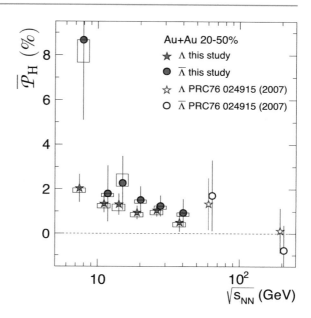

to be consistent with the trend of the new observations, though with larger statistical uncertainties. These data provide the first experimental access to the vortical structure of the perfect fluid created in heavy ion collision.

There are numerous approaches to solve the kinetic equations numerically. The most straightforward way is solving the integrodifferential equation by replacing the derivative and integral with proper numerical algorithms on the grids. However the realistic collision terms are always too complicated and thus make the task almost impossible. In particle scenario people have developed several successful simulation frameworks to solve different systems. These approaches are also roughly divided into the so-called Boltzmann–Uehling–Uhlenbeck (BUU) approaches and the molecular dynamics (MD) approaches [50]. In [50] authors have done very comprehensive studies on different simulation packages, i.e., 15 such codes, to understand the origins of discrepancies between different widely used transport codes. For more details of different algorithms readers could refer to this reference.

If the collision term is simple enough, it is possible to solve kinetic equations straightforwardly by discretizing the system and complete the derivative and integral on grids. This has been done in the ϕ^4 theory, pure gluon system with the small-angle approximation and the near-equilibrium systems with relaxation-time approximation (RTA). As a typical example in [51] authors have performed the first phenomenological study of the CME-induced charge separation during the pre-thermal stage in heavy ion collisions by solving the chiral kinetic equation in the RTA. With this intuitive solution on the net charge dipole moment its dependence on various ingredients in the modeling has been studied.

A successful MD model is AMPT (A Multi-Phase Transport). It is a Monte Carlo transport model for nuclear collisions at relativistic energies [52]. Although the anomaly effect is missed in its original version, it is a successful phenomenological model which includes both initial partonic and final hadronic interactions and the transition between the two phases of matter. As it has integrated almost all the stages of a heavy ion collision, it is a very popular phenomenological model to test and understand different effects in experiments. And it has also been extended to describe some of chiral effects recently [53]. More details of the AMPT model are in the chapter on Transport models.

In the MD scenario the test particle method is a widely used and systematic algorithm to solve the kinetic equations. It is able to be proven mathematically that the method could surely approach the strict solution with more and more computing resource. BAMPS (Boltzmann Approach of MultiParton Scatterings) [54] is one of such simulation packages which adopt this method. It is a microscopical and relativistic transport model which solves the Boltzmann equation for partons produced in ultrarelativistic heavy ion collisions. Its early version only considered gluons and their interactions including elastic process $gg \leftrightarrow gg$ and bremsstrahlung process $gg \leftrightarrow ggg$. Although the total scattering cross section calculated in the frame of perturbative quantum chromodynamics is only a few mb, it is enough to drive a gluon system toward full thermal equilibrium [54] and to generate sufficiently large elliptic flow v_2 [55]. The goal of the development of the model is to understand the mechanism for the fast equilibration of the gluon system created in the early stage of heavy ion collisions. Recently the model has been generalized to a charged particle system to compute its magnetic field at the early stage of the QGP and the field-modified transport coefficients. In the work with BAMPS [56], the authors used the Kubo formulas to calculate the anisotropic transport coefficients (shear viscosity and electric conductivity) for an ultrarelativistic Boltzmann gas in the presence of a magnetic field. The results are compared with those recently obtained by using the Grad's approximation. It is found that the good agreement between both results confirms the general use of the Kubo formulas for calculating anisotropic transport coefficients of QGP in a magnetic field. As a general algorithm of solving Boltzmann equation, BAMPS has the advantage on studying the vorticity distribution of the fireball especially at the early stage of a gluon system and solving the electromagnetic field self-consistently during the whole QGP evolution. For the chiral effects in [57] authors have solved the chiral kinetic equation with the test particle method. Using an anomalous transport model for massless quarks and antiquarks, the authors focused on the effect of a magnetic field on the elliptic flow of quarks and antiquarks in relativistic heavy ion collisions. It has been found that an appreciable electric quadrupole moment in the transverse plane of a heavy ion collision could be obtained with initial conditions from a blast wave model and assuming a strong and long-lived enough magnetic field in non-central heavy ion collisions. Obviously the electric quadrupole moment subsequently would lead to a splitting between the elliptic flows of quarks and antiquarks which could be inherited by the final hadron distributions.

6.8 Summary

The quantum kinetic theory in the Wigner function formalism is a powerful framework to study the quantum corrections to the classical transport theory systematically. In this chapter we mainly reviewed the full quantum kinetic equations for fermionic fields in chiral and massive cases. In chiral limit besides the well-known semi-classical corrections we have shown a natural derivation of the CME, CVE, and side-jump effect as well. In real world with massive fermions the mixture of particle number and spin density could not be disentangled. The mixing terms are proportional to the mass of fermion as expected. With the quantum kinetic equations the particle polarization, anomaly effects, and spin hydrodynamics have been studied as straightforward applications. Finally the related experiments and numerical simulation methods are briefly reviewed. For more details of phenomenological topics readers could refer to the relevant chapters.

Acknowledgements We thank Anping Huang, Jinfeng Liao, Ziwei Lin, Shuzhe Shi and Ziyue Wang for collaborations and discussions. The work is supported in part by the National Natural Science Foundation of China under Grant Nos. 11875002, 11890712, 11905066, 12075129 and Science and Technology Program of Guangzhou under Grant No. 2019050001.

References

1. Elze, H.-T., Heinz, U.: Phys. Rep. **183**, 81 (1989). ISSN 0370-1573
2. Chen, J.-W., Pu, S., Wang, Q., Wang, X.-N.: Phys. Rev. Lett. **110** (2013)
3. Bialynicki-Birula, I., Górnicki, P., Rafelski, J.: Phys. Rev. D **44**, 1825 (1991). ISSN 0556-2821
4. Zhuang, P., Heinz, U.: Ann. Phys. **245**, 311 (1996). ISSN 0003-4916
5. Zhuang, P., Heinz, U.: Phys. Rev. D **53**, 2096 (1996). ISSN 0556-2821, 1089-4918
6. Zhuang, P., Heinz, U.: Phys. Rev. D **57**, 6525 (1998). ISSN 0556-2821, 1089-4918
7. Ochs, S., Heinz, U.: Ann. Phys. **266**, 351 (1998). ISSN 0003-4916
8. Stephanov, M.A., Yin, Y.: Phys. Rev. Lett. **109** (2012)
9. Xiao, D., Chang, M.C., Niu, Q.: Rev. Mod. Phys. **82**, 1959 (2010)
10. Wang, Z., Guo, X., Zhuang, P.: [arXiv:2009.10930 [hep-th]]
11. Son, D.T., Yamamoto, N.: Phys. Rev. Lett. **109** (2012)
12. Son, D.T., Yamamoto, N.: Phys. Rev. D **87**, 085016 (2013). ISSN 1550-7998, 1550-2368
13. Fukushima, K., Kharzeev, D.E., Warringa, H.J.: Phys. Rev. D **78** (2008)
14. Kharzeev, D.E.: Ann. Phys. **325**, 205 (2010). ISSN 00034916, arXiv: 0911.3715
15. Carignano, S., Manuel, C., Torres-Rincon, J.M.: Phys. Rev. D **98** (2018). ISSN 2470-0010, 2470-0029, arXiv:1806.01684
16. Mueller, N., Venugopalan, R.: arXiv:1712.04057 [hep-ph, physics:hep-th, physics:nucl-th] (2017).
17. Mueller, N., Venugopalan, R.: Phys. Rev. D **96** (2017). ISSN 2470-0010, 2470-0029, arXiv: 1702.01233
18. Wigner, E.: Phys. Rev. **40**, 749 (1932). ISSN 0031-899X
19. Vasak, D., Gyulassy, M., Elze, H.-T.: Ann. Phys. **173**, 462 (1987). ISSN 0003-4916
20. Guo, X., Zhuang, P.: Phys. Rev. D **98**, 016007 (2018). ISSN 2470-0010, 2470-0029
21. Weickgenannt, N., Sheng, X.-l., Speranza, E., Wang, Q., Rischke, D.H.: arXiv:1902.06513 [hep-ph] (2019).
22. Gao, J.-H., Liang, Z.-T.: Phys. Rev. D **100**, 056021 (2019). ISSN 2470-0010, 2470-0029, arXiv: 1902.06510

23. Hattori, K., Hidaka, Y., Yang, D.-L.: arXiv:1903.01653 [hep-ph, physics:hep-th, physics:nucl-th] (2019)
24. Huang, A., Shi, S., Jiang, Y., Liao, J., Zhuang, P.: Phys. Rev. D **98** (2018)
25. Gao, J.-H., Pang, J.-Y., Wang, Q.: arXiv:1810.02028 [nucl-th] (2018)
26. Huang, A., Jiang, Y., Shi, S., Liao, J., Zhuang, P.: Phys. Lett. B **777**, 177 (2018). ISSN 0370-2693, arXiv: 1703.08856
27. Gao, J.-H., Liang, Z.-T., Wang, Q.: arXiv:1910.11060 [hep-ph, physics:hep-th, physics:nucl-th] (2019)
28. Gao, J.-H., Pang, J.-Y., Wang, Q.: Phys. Rev. D **100**, 016008 (2019). ISSN 2470-0010, 2470-0029
29. Hidaka, Y., Pu, S., Yang, D.-L.: Phys. Rev. D **95**, 091901 (2017). ISSN 2470-0010, 2470-0029
30. Yang, D.-L.: Phys. Rev. D **98** (2018). ISSN 2470-0010, 2470-0029, arXiv: 1807.02395
31. Bargmann, V., Michel, L., Telegdi, V.L.: Phys. Rev. Lett. **2**, 435 (1959). ISSN 0031-9007
32. De Groot, S.R., Van Leeuwen, W.A., Van Weert, C.G.: Relativistic Kinetic Theory. Principles and Applications. North-Holland (1980)
33. Hidaka, Y., Pu, S., Yang, D.-L.: arXiv:1807.05018 [hep-ph, physics:hep-th, physics:nucl-th] (2018)
34. Li, S., Yee, H.-U.: Phys. Rev. D **100**, 056022 (2019). ISSN 2470-0010, 2470-0029
35. Carignano, S., Manuel, C., Torres-Rincon, J.M.: Phys. Rev. D **98**(7), 076005 (2018). https://doi.org/10.1103/PhysRevD.98.076005, [arXiv:1806.01684 [hep-ph]]
36. Florkowski, W., Kumar, A., Ryblewski, R.: Phys. Rev. C **98**, 044906 (2018). ISSN 2469-9985, 2469-9993
37. Hidaka, Y., Pu, S., Yang, D.-L.: Phys. Rev. D **97** (2018). ISSN 2470-0010, 2470-0029, arXiv:1710.00278
38. Florkowski, W., Friman, B., Jaiswal, A., Speranza, E.: Phys. Rev. C **97** (2018)
39. Hattori, K., Hongo, M., Huang, X.-G., Matsuo, M., Taya, H.: arXiv:1901.06615 [cond-mat, physics:hep-ph, physics:hep-th, physics:nucl-th] (2019)
40. Kumar, A.: In: Proceedings of XIII Quark Confinement and the Hadron Spectrum, PoS(Confinement2018), p. 281. Maynooth University, Ireland, Sissa Medialab (2019)
41. Becattini, F., Chandra, V., Del Zanna, L., Grossi, E.: Ann. Phys. **338**, 32 (2013). ISSN 00034916
42. Kharzeev, D.E., Liao, J., Voloshin, S.A., Wang, G.: Progr. Particle Nuclear Phys. **88**, 1 (2016). ISSN 0146-6410
43. Fedorov, A.V., Zhang, C., Kharzeev, D.E., Gu, G.D., Pletikosi, I., Schneeloch, J.A., Li, Q., Zhong, R.D., Valla, T., Huang, Y.: Nat. Phys. **12**, 550 (2016). ISSN 1745-2481
44. Huang, X., Zhao, L., Long, Y., Wang, P., Chen, D., Yang, Z., Liang, H., Xue, M., Weng, H., Fang, Z., et al.: Phys. Rev. X **5**, 031023 (2015). ISSN 2160-3308
45. Xiong, J., Kushwaha, S.K., Liang, T., Krizan, J.W., Hirschberger, M., Wang, W., Cava, R.J., Ong, N.P.: Science **350**, 413 (2015). ISSN 0036-8075, 1095-9203
46. Arnold, F., Shekhar, C., Wu, S.-C., Sun, Y., Dos Reis, R.D., Kumar, N., Naumann, M., Ajeesh, M.O., Schmidt, M., Grushin, A.G., et al.: Nat. Commun. **7**, 11615 (2016). ISSN 2041-1723
47. Takahashi, R., Matsuo, M., Ono, M., Harii, K., Chudo, H., Okayasu, S., Ieda, J., Takahashi, S., Maekawa, S., Saitoh, E.: Nat. Phys. **12**, 52 (2016). ISSN 1745-2473, 1745-2481
48. The STAR Collaboration: Nature **548**, 62 (2017). ISSN 0028-0836, 1476-4687
49. Abelev, B.I., Aggarwal, M.M., Ahammed, Z., Anderson, B.D., Arkhipkin, D., Averichev, G.S., Bai, Y., Balewski, J., Barannikova, O., Barnby, L.S., et al.: Phys. Rev. C **76** (2007)
50. Zhang, Y.X., Wang, Y.J., Colonna, M., Danielewicz, P., Ono, A., Tsang, B., Wolter, H., Xu, J., Chen, L.W., Cozma, D., et al.: Phys. Rev. C **97**(3) (2018). https://doi.org/10.1103/PhysRevC.97.034625
51. Huang, A.-P., Jiang, Y., Shi, S.-Z., Liao, J.-F., Zhuang, P.-F.: Phys. Lett. B **777**, 177 (2018)
52. Lin, Z.-W., Ko, C.M., Li, B.-A., Zhang, B., Pal, S.: Phys. Rev. C **72** (2005)
53. Huang, L., Nie, M.W., Ma, G.L.: Phys. Rev. C **101**(2) (2020). https://doi.org/10.1103/PhysRevC.101.024916
54. Xu, Z., Greiner, C.: Phys. Rev. C **71** (2005)

55. Uphoff, J., Senzel, F., Fochler, O., Wesp, C., Xu, Z., Greiner, C.: Phys. Rev. Lett. **114**, 112301 (2015). ISSN 0031-9007, 1079-7114
56. Chen, Z., Greiner, C., Huang, A., Xu, Z.: arXiv:1910.13721 [hep-ph] (2019)
57. Sun, Y., Ko, C.M., Li, F.: Phys. Rev. C **94**(4) (2016)

Global Polarization Effect and Spin-Orbit Coupling in Strong Interaction

7

Jian-Hua Gao, Zuo-Tang Liang, Qun Wang and Xin-Nian Wang

Abstract

In non-central high energy heavy ion collisions, the colliding system posses a huge orbital angular momentum in the direction opposite to the normal of the reaction plane. Due to the spin-orbit coupling in strong interaction, such huge orbital angular momentum leads to the polarization of quarks and antiquarks in the same direction. This effect, known as the global polarization effect, has been recently observed by STAR Collaboration at RHIC that confirms the theoretical prediction made more than ten years ago. The discovery has attracted much attention to the study of spin effects in heavy ion collision. It opens a new window to study properties of QGP and a new direction in high energy heavy ion physics—Spin Physics in Heavy Ion Collisions. In this chapter, we review the original ideas

J.-H. Gao
Shandong Provincial Key Laboratory of Optical Astronomy and Solar-Terrestrial Environment, Institute of Space Sciences, Shandong University, Weihai, Shandong 264209, China
e-mail: gaojh@sdu.edu.cn

Z.-T. Liang (✉)
Key Laboratory of Particle Physics and Particle Irradiation (MOE), Institute of Frontier and Interdisciplinary Science, Shandong University, Qingdao, Shandong 266237, China
e-mail: liang@sdu.edu.cn

Q. Wang
Department of Modern Physics, University of Science and Technology of China, Hefei, Anhui 230026, China
e-mail: qunwang@ustc.edu.cn

X.-N. Wang
Key Laboratory of Quark and Lepton Physics (MOE) and Institute of Particle Physics, Central China Normal University, Wuhan 430079, China
e-mail: xnwang@lbl.gov

Nuclear Science Division, MS 70R0319, Lawrence Berkeley National Laboratory, Berkeley, CA 94720, USA

© Springer Nature Switzerland AG 2021
F. Becattini et al. (eds.), *Strongly Interacting Matter under Rotation*,
Lecture Notes in Physics 987,
https://doi.org/10.1007/978-3-030-71427-7_7

and calculations that lead to the predictions. We emphasize the role played by spin-orbit coupling in high energy spin physics and discuss the new opportunities and challenges in this connection.

7.1 Introduction

Recently, the global polarization effect (GPE) of Λ and $\bar{\Lambda}$ hyperons in heavy ion collisions (HIC) has been observed [1] by the STAR Collaboration at the Relativistic Heavy Ion Collider (RHIC) in Brookhaven National Laboratory (BNL). The discovery confirms the theoretical prediction [2] made more than ten years ago and has attracted much attention on the study of spin effects in HIC. This opens a new window to study properties of QGP and a new direction in high energy heavy ion physics—Spin Physics in HIC. New experiments along this line are being carried out and/or planned. It is, therefore, timely to summarize the original ideas and theoretical calculations [2–4] that lead to the predictions and discuss new opportunities and challenges.

Spin, as a fundamental degree of freedom of elementary particles, plays a very important role in modern physics and often brings us surprises. There are many well-known examples in the field of particle and nuclear physics. The anomalous magnetic moments of nucleons are usually regarded as one of the first clear signatures for the existence of the inner structure of nucleon. The explanation of these anomalous magnetic moments in the 1960s was one of the great successes of the quark model that lead us to believe that it provides us the correct picture for hadron structure.

High energy spin physics experiments started since 1970s. Soon after the beginning, a series of striking spin effects have been observed that were in strong contradiction to the theoretical expectations at that time and have been pushing the studies to move forward. The most famous ones might be classified as following:

(i) Proton's "spin crisis": Measurements of spin-dependent structure functions in deeply inelastic lepton-nucleon scatterings, started by E80 and E143 Collaborations at SLAC [5,6] and later on by the European Muon Collaboration (EMC) at CERN [7, 8], seem to suggest that the contribution of the sum of spins of quarks and antiquarks to proton spin is consistent with zero. This has triggered the so-called spin crisis of the proton and the intensive study on the spin structure of nucleon [9].

(ii) Single spin left-right asymmetry (SSA): It has been observed [10–13] that in inclusive hadron-hadron collisions with singly transversely polarized beams or targets, the produced hadron has a large azimuthal angle dependence characterized by the left-right asymmetry. The observed asymmetry can be as large as 40% but the theoretical expectation at the quark level using pQCD at the leading order was close to zero.

(iii) Transverse hyperon polarization: It has been observed [14–18] that hyperons produced in unpolarized hadron-hadron and hadron-nucleus collisions are transversely polarized with respect to the production plane. The observed polarization can reach a magnitude as high as 40% but the leading order pQCD expectation was again close to zero.

(iv) Spin asymmetries in elastic pp-scattering: It has been observed [18–21] that the azimuthal dependence, called the spin analyzing power, in scattering with single-transversely polarized proton and doubly polarized asymmetries are very significant, much larger than theoretical expectations available at that time.

Such striking spin effects came out often as such a shock to the field of strong interaction physics that lead to the famous comment by Bjorken [22] in a QCD workshop that "Polarization phenomena are often graveyards of fashionable theories. ...". In the last decades, the study on such spin effects leads to one of the most active fields in strong interaction or QCD physics.

At the same time, high energy HIC physics has become the other active field in strong interaction physics in particular after the quark-gluon plasma (QGP) has been discovered at RHIC [23,24]. The study on properties of QGP in HIC is the core of high energy HIC physics currently.

We recall that RHIC is not only the first relativistic heavy ion collider in the world, but also the first polarized high energy proton-proton collider. It is, therefore, natural to ask whether we can do spin physics in HIC.

Spin physics in HIC was however used to be regarded as difficult or impossible because the polarization of the nucleon in a heavy nucleus is very small even if the nucleus is completely polarized. The breakthrough came out in 2005, when it was realized that [2] there is, however, a great advantage to study spin and/or angular momentum effects in HIC, i.e., the reaction plane in a HIC can be determined experimentally by measuring flows and/or spectator nucleons and there exists a huge orbital angular momentum for the participating system in a non-central HIC with respect to the reaction plane! It provides a unique place in high energy reactions to study the mutual exchange of orbital angular momentum and the spin polarization. The discovery of GPE leads to an active field of Spin Physics in HIC [25].

In this chapter, we review the original ideas and calculations [2–4] that lead to the prediction of GPE in HIC. We present also a rough comparison to data available and an outlook for future studies. The rest of the chapter is arranged as follows: In Sect. 7.2, we present the orbital angular momentum of the colliding system in non-central HIC and the resulting gradient in momentum or rapidity distribution. In Sect. 7.3, we recall the origin of spin-orbit coupling and the famous example in electromagnetic and strong interaction systems. In Sect. 7.4, we present calculations and results for the global polarization in HIC, and finally, a short summary and outlook is presented in Sect. 7.5.

7.2 Orbital Angular Momenta of QGP in HIC

7.2.1 The Reaction Plane in HIC

We consider two colliding nuclei with the projectile of beam momentum per nucleon $\mathbf{p}_{in}$. For a non-central collision, there is a transverse separation between the centers of the two colliding nuclei. The impact parameter $\mathbf{b}$ is defined as the transverse vector pointing from the target to the projectile. The reaction plane of a HIC is usually

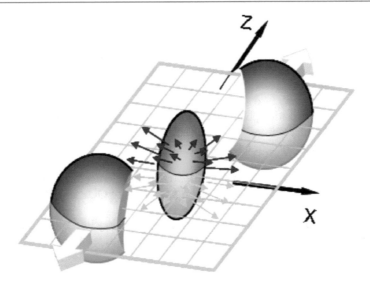

Fig. 7.1 Illustration diagram for the reaction plane in a non-central heavy ion collision. In contrast to high energy pp or e^+e^- collisions, the reaction plane in a high energy heavy collision can be determined experimentally

defined by $\mathbf{b}$ and $\mathbf{p}_{in}$ and is illustrated in Fig. 7.1. The overlap parts, hereafter referred to as the colliding system, interact with each other and form the system denoted by the red core in the middle, while the other parts, denoted by the blue parts in the figure, are just spectators and move apart in the original directions.

The geometry and the coordinate system are further specified in Fig. 7.2. The beam direction of the colliding nuclei is taken as the z axis, as illustrated in the upper-left panel in the figure. The transverse separation is called the impact parameter $\mathbf{b}$ defined as the transverse distance of the projectile from the target nucleus and is taken as in the x-direction. The normal of the reaction plane is given by

$$\mathbf{n} \equiv \mathbf{p}_{in} \times \mathbf{b}/|\mathbf{p}_{in} \times \mathbf{b}|, \tag{7.1}$$

and is taken as the y-direction, where $\mathbf{p}_{in}$ is the momentum per nucleon in the incident nucleus A.

Usually in a high energy reaction such as a hadron-hadron, or lepton-hadron or e^+e^- annihilation, the size of the reaction region is typically less than $1\,\mathrm{fm}$. The reaction plane in such collisions can be defined theoretically but can not be determined experimentally. However, in a HIC, the reaction region is usually much larger and colliding parts give rise to a quark matter system with very high temperature and high density and expand violently while the spectators just leave the region in the original directions. Since the colliding system is not isotropic, the pressures in different directions are also different in different directions, thus leading to a system that expands non-isotropically. In the transverse directions, they behave like an ellipse as illustrated in the lower-right panel in Fig. 7.2. Such a non-isotropy is described

Fig. 7.2 Illustration of the geometry and coordinate system for the non-central HIC with impact parameter **b**. The global angular momentum of the produced matter is along the minus y direction, opposite to the reaction plane. This figure is taken from [2]

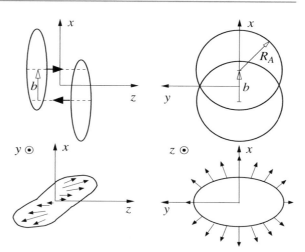

by the elliptic flow v_2 and the directed flow v_1 that can be measured experimentally (see, e.g., [26,27]). Clearly, by measuring v_2, one can determine the reaction plane and further determine the direction of the plane by measuring the directed flow v_1.

In experiments, the reaction plane in a HIC can not only be determined by measuring v_2 and v_1 but also determined by measuring the sidewards deflection of the forward- and backward-going fragments and particles in the beam–beam counter detectors [1]. This is quite unique in different high energy reactions.

7.2.2   The Global Orbital Angular Momentum

Just as illustrated in Figs. 7.1 and 7.2, in a non-central HIC, there is a transverse separation between the overlapping parts of the two colliding nuclei in the same direction as the impact parameter **b**. Hence the whole system that takes part in the reaction, i.e., the colliding system carries a finite orbital angular momentum L_y along the direction orthogonal to the reaction plane. We call L_y the global orbital angular momentum. The magnitude of this global orbital angular momentum L_y can be calculated by

$$L_y = -p_{in} \int x \, dx \left(\frac{d N_{\text{part}}^P}{dx} - \frac{d N_{\text{part}}^T}{dx} \right), \tag{7.2}$$

where $d N_{\text{part}}^{P,T} / dx$ is the transverse distributions (integrated over y and z) of participant nucleons in each nucleus A along the x-direction, the superscript P or T denotes projectile or target, respectively. These transverse distributions are given by

$$\frac{dN_{\text{part}}^{P,T}}{dx} = \int dy dz \, \rho_A^{P,T}(x, y, z, b), \tag{7.3}$$

where $\rho_A^{P,T}(x, y, z, b)$ is the number density of participant nucleons in nucleus A in the coordinate system defined in Fig. 7.2.

The number density $\rho_A^{P,T}(x, y, z, b)$ of participant nucleons in nucleus A can easily be calculated if we take a hard-sphere distribution of nucleons in the nucleus A. In this model, the overlapping area has a clear boundary and the participant nucleon density is given by the overlapping area of two hard spheres, as illustrated in the upper-right panel of Fig. 7.2, i.e.

$$\rho_{A,HS}^{P,T}(x, y, z, b) = f_{A,HS}^{P,T}(x, y, z, b) \, \theta \left(R_A - \sqrt{(x \pm b/2)^2 + y^2 + z^2} \right), \tag{7.4}$$

where $f_{A,HS}^{P,T}(x, y, z, b)$ is the hard-sphere nuclear distribution in A that is given by

$$f_{A,HS}^{P,T}(x, y, z, b) = \frac{3A}{4\pi R_A^3} \theta \left(R_A - \sqrt{(x \mp b/2)^2 + y^2 + z^2} \right), \tag{7.5}$$

where $R_A = 1.12 A^{1/3}$ fm is the nuclear radius and A the atomic number.

If we take the Woods-Saxon nuclear distribution, i.e.

$$f_{A,WS}^{P,T}(x, y, z, b) = C_0 \left(1 + \exp \frac{\sqrt{(x \mp b/2)^2 + y^2 + z^2} - R_A}{a} \right)^{-1}, \tag{7.6}$$

there is no clear boundary of the overlapping region and the participant nucleon number density is calculated using the Glauber model and is given by

$$\rho_{A,WS}^{P,T}(x, y, z, b) = f_{WS}^{P,T}(x, y, z, b) \left\{ 1 - \exp\left[-\sigma_{NN} \int dz f_{WS}^{T,P}(x, y, z, b) \right] \right\}, \tag{7.7}$$

where σ_{NN} is the total cross section of nucleon-nucleon scatterings, C_0 is the normalization constant

$$C_0 = A/4\pi \int r^2 dr \left(1 + e^{(r - R_A)/a} \right)^{-1}, \tag{7.8}$$

and a is the width parameter set to $a = 0.54$ fm.

The calculations have been carried out in [2,4]. The obtained results are shown in Fig. 7.3. From the results shown in Fig. 7.3, we see that though there are significant differences between two nuclear geometry models the global orbital angular momentum L_y of the overlapped parts of two colliding nuclei is huge and is of the order of 10^4 at most impact parameters.

Fig. 7.3 Global orbital angular momentum of the colliding system in the non-central HIC as a function of the impact parameter obtained from the Woods-Saxon and hard-sphere distributions, respectively. This figure is taken from [4]

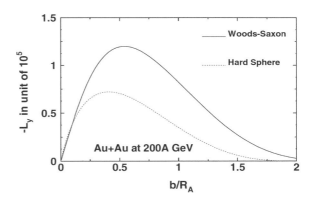

7.2.3 The Transverse Gradient of the Momentum Distribution and the Local Orbital Angular Momentum

How the global orbital angular momentum discussed above is transferred to the final state particles depends on the equation of state (EOS) of the dense matter. At low energies, the final state is expected to be the normal nuclear matter with an EOS of rigid nuclei. In such cases, a rotating compound nucleus can be formed when the colliding energy is comparable or smaller than the nuclear binding energy. The finite value of the global orbital angular momentum of the non-central collision at such low energies provides a useful tool for the study of the properties of super-deformed nuclei under such rotation [28].

At high colliding energies such as those at RHIC, the dense matter is expected to be partonic with an EOS of QGP. Given such a soft EOS, the global orbital angular momentum would probably not lead to the global rotation of the dense matter system. Instead, the global angular momentum could be distributed across the overlapped region of nuclear scattering and is manifested in the shear of the longitudinal flow leading to a finite value of local vorticity density. Under such longitudinal fluid shear, a pair of scattering partons will on average carry a finite value of relative orbital angular momentum that will be referred to as the local orbital angular momentum in the opposite direction to the reaction plane as defined in Eq. (7.1).

By momentum conservation, the average initial collective longitudinal momentum at any given transverse position can be calculated as the total momentum difference between participating projectile and target nucleons. Since the total multiplicity in HIC is proportional to the number of participant nucleons [29], we can make the same assumption for the produced partons with a proportionality constant fixed at a given center of mass energy $\sqrt{s}$. How the global angular momentum is distributed to the longitudinal flow shear and the magnitude of the local relative orbital angular momentum depends on the parton production mechanism and their longitudinal momentum distributions. We consider two different scenarios: the Landau fireball and the Bjorken scaling model.

7.2.3.1 Results from the Landau Fireball Model

In the Landau fireball model, we assume that the produced partons thermalize quickly and have a common longitudinal flow velocity at a given transverse position in the overlapped region. The average collective longitudinal momentum per parton can be written as

$$p_z(x, b, \sqrt{s}) = p_0 R_N(x, b, \sqrt{s}), \qquad (7.9)$$

where $p_0 = \sqrt{s}/2c(s)$ is an energy dependent constant, $\sqrt{s}$ is the center of mass energy of a colliding nucleon pair, $c(s)$ is the average number of partons produced per participating nucleon; and $R_N(x, b, \sqrt{s})$ is the ratio defined as

$$R_N(x, b, \sqrt{s}) = \left(\frac{dN_{\text{part}}^P}{dx} - \frac{dN_{\text{part}}^T}{dx} \right) \Big/ \left(\frac{dN_{\text{part}}^P}{dx} + \frac{dN_{\text{part}}^T}{dx} \right) \qquad (7.10)$$

It is clear that in the symmetric AA collision (where the beam and target nuclei are the same), the ratio $R_N(x, b, \sqrt{s})$ thus the distribution $p_z(x, b, \sqrt{s})$ is an odd function in both x and b, and therefore, vanishes at $x = 0$ or $b = 0$. In Fig. 7.4, $p_z(x, b, \sqrt{s})$ is plotted as a function of x at different impact parameters b. We see

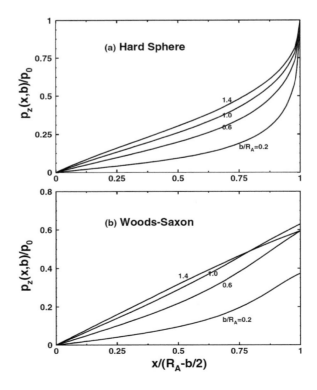

Fig. 7.4 The average longitudinal momentum distribution $p_z(x, b, \sqrt{s})$ in unit of $p_0 = \sqrt{s}/[2c(s)]$ as a function of $x/(R_A - b/2)$ for different values of b/R_A with the hard-sphere (upper panel) and Woods-Saxon (lower panel) nuclear distributions. This figure is taken from [4]

clearly that $p_z(x, b, \sqrt{s})$ is a monotonically increasing function of x until the edge of the overlapped region $|x \pm b/2| = R_A$ beyond which it drops to zero (gradually for Woods-Saxon geometry).

From $p_z(x, b, \sqrt{s})$, one can compute the transverse gradient of the average longitudinal collective momentum per parton dp_z/dx which is an even function of x and vanishes at $b = 0$. One can then estimate the longitudinal momentum difference Δp_z between two neighboring partons in QGP. On average, the relative orbital angular momentum for two colliding partons separated by Δx in the transverse direction is

$$l_y \equiv -(\Delta x)^2 \frac{dp_z}{dx}. \tag{7.11}$$

With the hard-sphere nuclear distribution, l_y is proportional to

$$\frac{dp_0}{dx} \equiv \frac{p_0}{R_A} = \frac{\sqrt{s}}{2c(s)R_A}. \tag{7.12}$$

This provides a measure of order of magnitude of dp_z/dx. In $Au + Au$ collisions at $\sqrt{s} = 200$ GeV, the number of charged hadrons per participating nucleon is about 15 [29]. Assuming the number of partons per (meson dominated) hadron is about 2, we have $c(s) \simeq 45$ (including neutral hadrons). Given $R_A = 6.5$ fm, $dp_0/dx \simeq 0.34$ GeV/fm and we obtain a value of $l_0 \equiv -(\Delta x)^2 dp_0/dx \simeq -1.7$ for $\Delta x = 1$ fm.

In Fig. 7.5, we show the average local orbital angular momentum l_y given by Eq. (7.11) for two neighboring partons separated by $\Delta x = 1$ fm as a function of x for different impact parameter b for both Woods-Saxon and hard-sphere nuclear distributions. We see that l_y is in general of the order of 1 and is comparable or larger than the spin of a quark. It is expected that $c(s)$ should depend logarithmically on the colliding energy $\sqrt{s}$, therefore l_y should increases with growing $\sqrt{s}$.

7.2.3.2 Results from the Bjorken Scaling Model

In a three dimensional expanding system, there could be a strong correlation between longitudinal flow velocity and spatial coordinate of the fluid cell. The most simplified picture is the Bjorken scaling scenario [30] in which the longitudinal flow velocity is identical to the spatial velocity $\eta = \log[(t + z)/(t - z)]$. With such correlation, the local interaction and thermalization require that a parton only interacts with other partons in the same region of longitudinal momentum or rapidity Y. The width of such region in rapidity is determined by the half-width of the thermal distribution $f_{th}(Y, p_T) = \exp[-p_T \cosh(Y - \eta)/T]$ [31], which is approximately $\Delta_Y \approx 1.5$ (with $\langle p_T \rangle \approx 2T$ and T is the local temperature). The relevant measure of the local relative orbital angular momentum between two interacting partons is, therefore, the difference in parton rapidity distributions at a transverse distance of the order of the average interaction range.

The variation of the rapidity distributions with respect to the transverse coordinate can be described by the normalized rapidity distribution $f_p(Y, x)$ at given x

$$f_p(Y, x, b, \sqrt{s}) = \frac{d^2N}{dxdY} \Big/ \frac{dN}{dx}, \tag{7.13}$$

Fig. 7.5 The average orbital angular momentum $l_y \equiv -(\Delta x)^2 dp_z/dx$ of two neighboring partons separated by $\Delta x = 1$ fm as a function of the scaled transverse coordinate $x/(R_A - b/2)$ for different values of the impact parameter b/R_A with the hard-sphere (upper panel) and Woods-Saxon (lower panel) nuclear distributions. This figure is taken from [4]

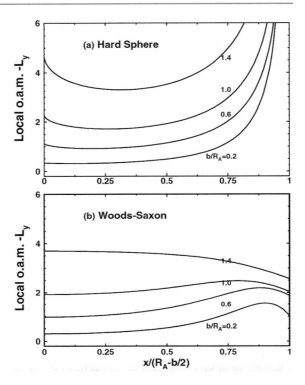

where $d^2N/dxdY$ denotes the number density of particles produced with respect to x and Y and $dN/dx \equiv \int dY d^2N/dxdY$ is the distribution of particles with respect to x. At a given x, the overall average value of the rapidity is given by

$$\langle Y(x, b, \sqrt{s}\,)\rangle = \int Y dY f_p(Y, x, b, \sqrt{s}\,). \qquad (7.14)$$

$\langle Y(x, b, \sqrt{s}\,)\rangle$ just corresponds to $p_z(x, b, \sqrt{s})$ given by Eq. (7.9) discussed in the Landau fireball model. It measures the overall behavior of the rapidity distribution of partons at given transverse coordinate x. To further quantify such longitudinal fluid shear, one can calculate the average rapidity within an interval Δ_Y at a given rapidity Y, i.e.

$$\langle Y_l(Y, x, b, \sqrt{s}\,)\rangle \approx Y + \frac{\Delta_Y^2}{12}\frac{1}{f_p}\frac{\partial f_p}{\partial Y} = Y + \frac{\Delta_Y^2}{12}\frac{\partial \ln f_p}{\partial Y}. \qquad (7.15)$$

Here, we use the subscript l to denote that this is the average of Y in a localized interval $[Y - \Delta_Y/2, Y + \Delta_Y/2]$ to differentiate it from the overall average $\langle Y(x, b, \sqrt{s})\rangle$ given by Eq. (7.14). The average rapidity shear or the difference in average rapidity for two partons separated by a unit of transverse distance Δx is then given by

$$\frac{\partial}{\partial x}\langle Y_l(Y, x, b, \sqrt{s}\,)\rangle \approx \frac{\Delta_Y^2}{12}\frac{\partial^2 \ln f_p}{\partial Y \partial x}. \tag{7.16}$$

The averaged longitudinal momentum is

$$\langle p_z \rangle \approx p_T \sinh\langle Y_l \rangle \approx p_T \left(\sinh Y + \cosh Y \frac{\Delta_Y^2}{12}\frac{\partial \ln f_p}{\partial Y}\right). \tag{7.17}$$

The corresponding local relative longitudinal momentum shear is given by

$$\frac{\partial\langle p_z \rangle}{\partial x} \approx p_T \cosh Y \frac{\partial\langle Y_l \rangle}{\partial x} \approx p_T \cosh Y \frac{\Delta_Y^2}{12}\frac{\partial^2 \ln f_p}{\partial Y \partial x}. \tag{7.18}$$

The corresponding local orbital angular momentum l_y for two partons separated by a transverse separation Δx at a given rapidity Y is $\langle l_y(Y)\rangle = -\Delta x \Delta\langle p_z\rangle = -(\Delta x)^2 \partial\langle p_z\rangle/\partial x$. We transform it into the co-moving frame or the center of mass frame of the two partons and obtain

$$\langle l_y^*(Y, x, b, \sqrt{s}\,)\rangle = -\Delta x \langle p_z^*\rangle \approx -(\Delta x)^2 p_T \frac{\Delta_Y^2}{24}\frac{\partial^2 \ln f_p}{\partial Y \partial x}. \tag{7.19}$$

We see that they are all determined by a key quantity

$$\xi_p(Y, x, b, \sqrt{s}\,) \equiv \frac{\partial^2 \ln f_p(Y, x, b, \sqrt{s})}{\partial Y \partial x}, \tag{7.20}$$

that is determined by $d^2N/dxdY$. In terms of $\xi_p(Y, x, b, \sqrt{s}\,)$, we have

$$\frac{\partial\langle Y_l \rangle}{\partial x} \approx \frac{\Delta_Y^2}{12}\xi_p, \tag{7.21}$$

$$\frac{\partial\langle p_z \rangle}{\partial x} \approx \frac{\Delta_Y^2}{12}\xi_p\, p_T \cosh Y, \tag{7.22}$$

$$\langle l_y^*(Y, x, b, \sqrt{s}\,)\rangle \approx -\frac{\Delta_Y^2}{24}\xi_p\,(\Delta x)^2 p_T. \tag{7.23}$$

The Y-dependence averaged over the transverse separation x is determined by the average value of $\xi_p(Y, x, b, \sqrt{s})$ defined by

$$\langle \xi_p \rangle = \int dx\,\xi_p(Y, x, b, \sqrt{s}\,)\frac{d^2N}{dxdY}\Big/\frac{dN}{dY}, \tag{7.24}$$

where $dN/dY = \int dx(d^2N/dxdY)$ is the rapidity distribution of partons produced in a AA collision at the given impact parameter b. In the binary approximation

$$\frac{dN}{dY} = N_{\text{part}}\frac{dN_{pp}}{dY}. \tag{7.25}$$

Fig. 7.6 The average
rapidity $\langle Y \rangle$ of the final state
particles as a function of the
transverse coordinate x from
HIJING Mont Carlo
simulations [32,33] of
non-central $Au + Au$
collisions at $\sqrt{s} = 200$ GeV.
This figure is taken from [4]

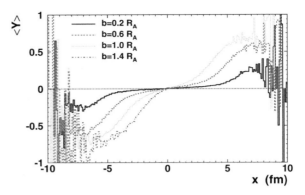

To proceed with numerical calculations, one needs a dynamical model to estimate the local rapidity distribution $d^2 N/dx dY$ of produced partons. For this purpose, two models, the HIJING Monte Carlo model [32,33] and the model proposed by Brodsky, Gunion, and Kuhn (denoted as BGK model) [34], have been used [4,35]. We present the results [4,35] obtained in the following, respectively.

(i) *Results Obtained Using HIJING*

In [4], the HIJING Monte Carlo model [32,33] was used to calculate the hadron rapidity distributions at different transverse coordinate x and assume that parton distributions of the dense matter are proportional to the final hadron spectra. We show the results obtained in this way in [4] in the following:

Shown in Fig. 7.6 is the average rapidity of particles in the final state as a function of the transverse coordinate x for different values of the impact parameter b. We see that, besides the edge effects, the distributions have exactly the same qualitative features as given by the wounded nucleon model in Fig. 7.4.

In Fig. 7.7, we see the results of normalized rapidity distributions $f_p(Y, x, b, \sqrt{s})$ at different values of the transverse coordinate x. We see that at finite values of x, $f_p(Y, x, b, \sqrt{s})$ evidently peak at larger values of rapidity $|Y|$. The shift in the shape of the rapidity distributions will provide the local longitudinal fluid shear or finite relative orbital angular momentum for two interacting partons in the local co-moving frame at any given rapidity Y. The fluid shear in the local co-moving frame at given rapidity Y is finite and peaks at a large value of rapidity $|Y| \approx 2$. It is also generally smaller than the averaged fluid shear in the center of mass frame of two colliding nuclei in the Landau fireball model.

Shown in Fig. 7.8 is the average rapidity shear $\partial \langle Y_l \rangle / \partial x$ as a function of the rapidity Y at different values of the transverse coordinate x for $\Delta_Y = 1$. As we can see, the average rapidity shear has a positive and finite value in the central rapidity region. As given by Eq. (7.18), the corresponding local relative longitudinal momentum shear $\partial \langle p_z \rangle / \partial x$ is determined by this rapidity shear multiplied by $p_T \cosh Y$. With $\langle p_T \rangle \approx 2T \sim 0.8$ GeV, we have $\partial \langle p_z \rangle / \partial x \sim 0.003$ GeV/fm in the central rapidity region of a non-central $Au + Au$ collision at the RHIC energy given by the HIJING simulations, which is smaller than that from a Landau fireball model estimate.

(ii) *Results Obtained Using the BGK Model*

Fig. 7.7 The normalized rapidity distribution $f_p(Y, x, b, \sqrt{s})$ (in unit of 1/fm) of particles at different transverse position x from HIJING simulations of non-central $Au + Au$ collisions at $\sqrt{s} = 200$ GeV. This figure is taken from [4]

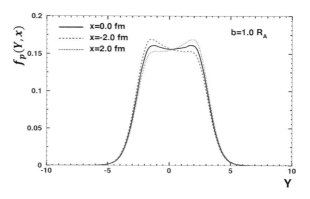

Fig. 7.8 (Color online) The average rapidity shear $\partial\langle Y_l\rangle/\partial x$ within a window $\Delta_Y = 1$ as a function of the rapidity Y at different transverse position x from HIJING calculation of non-central $Au + Au$ collisions at $\sqrt{s} = 200$ GeV. This figure is taken from [4]

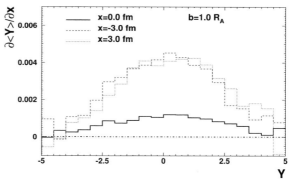

In a recent paper [35], a simple model [34] instead of HIJING [32,33] was used to repeat these calculations. Here, in this simple BGK model [34], the rapidity distribution of produced hadrons is given by that in pp-collision, dN_{pp}/dY, multiplied by the following Y linearly dependent factor, i.e.

$$\frac{d^3N}{dx\,dy\,dY} = \frac{dN_{pp}}{dY}\left[T_A^P(x, y, b)\frac{Y_L + Y}{2Y_L} + T_A^T(x, y, b)\frac{Y_L - Y}{2Y_L}\right], \qquad (7.26)$$

where $T_A^{P/T}$ is the thickness function for the projectile or target nucleus given by

$$T_A^{P,T}(x, y, b) = \int dz\, \rho_A^{P,T}(x, y, z, b), \qquad (7.27)$$

$Y_L \approx \ln(\sqrt{s}/2m_N)$ is the maximum of the rapidity of the produced hadron; dN_{pp}/dY of hadrons produced in a pp-collision is taken as a modified Gaussian

$$\frac{dN_{pp}}{dY} = a_1 \exp(-Y^2/a_2)/\sqrt{1 + a_3 \cosh^4 Y}, \qquad (7.28)$$

where a_1, a_2, and a_3 are parameters depending on the collision energy. They are determined by fitting the results obtained from PYTHIA8.2 [36] for pp collisions. A few examples obtained in [35] is given in Table 7.1.

Table 7.1 The parameters a_1, a_2 and a_3 for the rapidity distribution dN_{pp}/dY given by Eq. (7.28) determined from PYTHIA8.2 [36]. These numbers are taken from [35]

$\sqrt{s}$ (GeV)	a_1	a_2	a_3
200	4.584	26.112	9.70×10^{-8}
130	4.096	25.896	5.61×10^{-7}
62.4	3.862	18.911	9.75×10^{-6}
39	3.420	18.779	6.61×10^{-5}
27	3.421	13.555	2.50×10^{-4}
11.5	2.784	10.488	5.90×10^{-3}

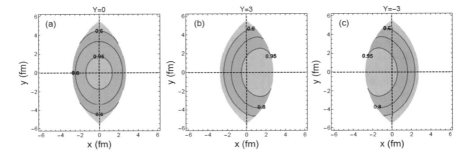

Fig. 7.9 Contour plots for distributions of hadrons obtained in BGK model [34] with a hard-sphere nuclear distribution in the transverse plane for non-central $Au + Au$ collisions at $\sqrt{s} = 200$ GeV at $b = 1.2R_A$ and different rapidities. The number on the contour line denotes the value on the line normalized by that at the origin. This figure is taken from [35]

One great advantage to take this simple model [34] is that we have analytical expressions for all the quantities need so the calculations are quite simplified so that the physical significance can be easily demonstrated. In Ref. [35], different results obtained using a hard-sphere or Woods-Saxon nuclear distribution are given. In the following, we show those obtained using a hard-sphere distribution as an example. Those obtained using Woods-Saxon are similar.

Shown in Fig. 7.9 are the contour plots for distributions of hadrons in the transverse plane with different rapidities. This provides us a very intuitive picture of how particles are distributed in the transverse plane at different rapidities. We see that at $Y = 0$, the distributions are symmetric with respect to x while at $Y = -3$ the center shifts to positive x and at $Y = -3$ shifts to negative x. But they are all symmetric or even function of y.

We integrate over the transverse coordinates and obtain

$$\frac{d^2N}{dxdY} = \frac{dN_{pp}}{dY}\left(\frac{dN_{part}^P}{dx}\frac{Y_L + Y}{2Y_L} + \frac{dN_{part}^T}{dx}\frac{Y_L - Y}{2Y_L}\right), \qquad (7.29)$$

$$\frac{dN}{dx} = \frac{1}{2}\langle N_{pp}\rangle\left(\frac{dN_{part}^P}{dx} + \frac{dN_{part}^T}{dx}\right), \qquad (7.30)$$

where $\langle N_{pp} \rangle = \int dY \, (dN_{pp}/dY)$ is the average total number of particles produced in the pp collision. The normalized rapidity distribution at given x is given by

$$f_p(Y, x, b, \sqrt{s}\,) = \frac{dN_{pp}}{\langle N_{pp} \rangle dY} \left[1 + \frac{Y}{Y_L} R_N(x, b, \sqrt{s}\,) \right], \qquad (7.31)$$

where the ratio $R_N(x, b, \sqrt{s})$ is defined by Eq. (7.10).

The overall average value of Y at a given x is given by

$$\langle Y(x, b, \sqrt{s}\,) \rangle = \frac{\langle Y^2 \rangle}{Y_L} R_N(x, b, \sqrt{s}\,), \qquad (7.32)$$

where $\langle Y^2 \rangle = \int Y^2 dY (dN_{pp}/dY)/\langle N_{pp} \rangle$ is the average value of Y^2 in pp collision.

Comparing Eq. (7.30) with Eq. (7.9), we see that $\langle Y(x, b, \sqrt{s}\,) \rangle$ in this model has exactly the same behavior as $p_z(x, b, \sqrt{s}\,)$ in the Landau fireball model.

Figure 7.10 shows the average values of Y as functions of x plotted in the same format as that in Fig. 7.6. We see that, besides those in the edge regions where the calculations need to be modified, the results exhibit the same qualitative features as those in Fig. 7.6, though the quantitative results show slight differences. Figure 7.11 shows the corresponding normalized distributions $f_p(Y, x, b, \sqrt{s}\,)$. The right panel is to compare with Fig. 7.7 where HIJING monte Carlo model was used. We see in particular a clear shift of the peak to positive Y for $x > 0$ and to negative Y for $x < 0$.

To show the rapidity dependence of the local orbital angular momentum or momentum shear, Ref. [35] also calculated $\langle \xi_p \rangle$ defined in Eq. (7.20) as a function of Y at different energies. The obtained results are shown in Fig. 7.12. From this figure, we see that the rapidity dependence of $\langle \xi_p \rangle$ is quite weak except at the limiting region when Y reaches its maximum. This represents the characteristics of the

Fig. 7.10 The average rapidity $\langle Y \rangle$ of the final state particles as a function of the transverse coordinate x from BGK [34] with a hard-sphere nuclear distribution in non-central $Au + Au$ collisions at $\sqrt{s} = 200$ GeV. This figure is taken from [35]

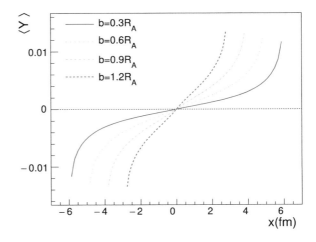

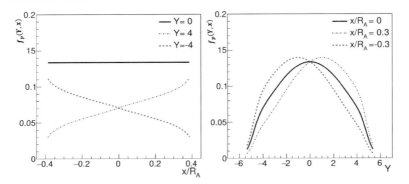

Fig. 7.11 The normalized distribution $f_p(Y, x)$ of hadrons in BGK model [34] with a hard-sphere nuclear distribution in the transverse plane for non-central $Au + Au$ collisions at $\sqrt{s} = 200$ GeV and $b = 1.2R_A$ as a function of x at different rapidity Y (left panel), and as a function of Y at different x (right panel). This figure is taken from [35]

Fig. 7.12 The averaged $\langle \xi_p \rangle = \langle \partial^2 \ln f_p / \partial Y \partial x \rangle$ as a function of rapidity Y of final state particles in BGK model [34] with a hard-sphere nuclear distribution for non-central $Au + Au$ collisions at different energies and impact parameter $b = 1.2R_A$. This figure is taken from [35]

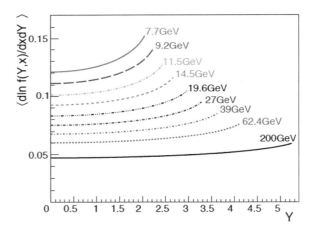

rapidity dependence of the microscopic local momentum shear and may also reflect the rapidity dependence of the corresponding macroscopic observable effects.

7.3 Spin-Orbit Coupling in a Relativistic Quantum System

The spin-orbit coupling is a well-known effect in a quantum system. Here, we present a short discussion of the origin and a brief review of related phenomena.

7.3.1 Dirac Equation and Spin-Orbit Coupling

The spin-orbit coupling is an intrinsic property for a relativistic fermionic quantum system. This is derived explicitly from the Dirac equation. A number of characteristics of the Dirac equation show that it describes particles of spin-1/2, and the spin and orbital angular momentum couple to each other intrinsically even for free particles. Here, we recall a few of such characteristics in the following:

First of all, it is well known that, even for a free Dirac particle, the Hamiltonian $\hat{H}$ does not commute with the orbital angular momentum $\hat{\mathbf{L}}$ and the spin $\boldsymbol{\Sigma}$ separately, but commutes with the total angular momentum $\hat{\mathbf{J}} = \hat{\mathbf{L}} + \boldsymbol{\Sigma}/2$, i.e., $[\hat{H}, \hat{\mathbf{L}}] = -i\boldsymbol{\alpha} \times \hat{\mathbf{p}}$, $[\hat{H}, \boldsymbol{\Sigma}] = 2i\boldsymbol{\alpha} \times \hat{\mathbf{p}}$, but $[\hat{H}, \hat{\mathbf{J}}] = 0$. This clearly shows that the spin and orbital angular momentum couple to each other and transform from one to another in a relativistic fermionic quantum system, though the strength of the spin-orbit coupling can be different for an electromagnetic or a strongly interacting system.

Second, the magnetic momentum of a Dirac particle with electric charge e is obtained simply by replacing the classical expression $\mathbf{M} = e\mathbf{r} \times \mathbf{v}/2$ with operators, i.e., $\hat{\mathbf{M}} = e\mathbf{r} \times \boldsymbol{\alpha}/2$. In an eigenstate $|\psi\rangle$ of $\hat{H}$, if we take the nonrelativistic approximation $E \approx m$, we obtain immediately that [37]

$$\langle \mathbf{M} \rangle \approx \frac{e}{2m} \langle \varphi | (\hat{\mathbf{L}} + \boldsymbol{\sigma}) | \varphi \rangle, \tag{7.33}$$

where φ is the upper component of ψ. This is just the well-known result for point-like spin-1/2 particles where the Landre factors are $g_L = 1$ and $g_s = 2$.

If we consider a Dirac particle moving in a central potential, the stationary state is the eigenstate of $\hat{H}$, $\hat{\mathbf{J}}^2$, $\hat{J}_z$ and the parity $\hat{\mathcal{P}}$ with eigenvalues $(\varepsilon, j, m, \mathcal{P})$, i.e.,

$$\psi_{\varepsilon j m \mathcal{P}}(r, \theta, \phi, s) = \begin{pmatrix} f_{\varepsilon l}(r) \Omega^l_{jm}(\theta, \phi) \\ (-1)^{\frac{1}{2}(l-l'+1)} g_{\varepsilon l'}(r) \Omega^{l'}_{jm}(\theta, \phi) \end{pmatrix}, \tag{7.34}$$

where $\Omega^l_{jm}(\theta, \phi)$ is the 2×1 spheric harmonic wave function in the nonrelativistic case, $f_{\varepsilon l}(r)$ and $g_{\varepsilon l'}(r)$ are the radial parts, $j = l \pm 1/2 = l' \mp 1/2$ and $\mathcal{P} = (-1)^l$. In the ground state $\varepsilon = \varepsilon_0$, $j = 1/2$, $\mathcal{P} = +$, the magnetic moment is given by [37]

$$\langle \varepsilon_0, 1/2, m, + | \hat{\mathbf{M}} | \varepsilon_0, 1/2, m, + \rangle = \mu_q \langle \xi(m) | \boldsymbol{\sigma} | \xi(m) \rangle, \tag{7.35}$$

where $\mu_q = -2e \int r^3 dr f_{00}(r) g_{01}(r)/3$ is a constant determined by ground state radial wave functions, $\xi(m)$ is the eigenstate of σ_z and is a Pauli spinor. Equation 7.35 has exactly the same form as that for a quark at rest. This explains why the static quark model works well in describing the magnetic moment of baryon although we know that the quark mass is small and the relativistic treatment has to be used.

Third, we consider a Dirac particle moving in a magnetic field with potential $A = (\phi, \mathbf{A})$. By replacing $\hat{p}$ with $\hat{p} - eA$ in the Dirac equation and taking the nonrelativistic approximation, we obtain immediately

$$\hat{H}_{\mathrm{nr}} = \frac{1}{2m} (\hat{\mathbf{p}} - e\mathbf{A})^2 - e\phi - \frac{1}{4m^2} \frac{d\phi}{r dr} \hat{\mathbf{L}} \cdot \boldsymbol{\sigma}, \tag{7.36}$$

where the spin-orbit coupling is obtained automatically.

7.3.2   Spin-Orbit Coupling in Systems Under Electromagnetic Interactions

Intuitively, the spin-orbit coupling in systems under electromagnetic interactions has a very clear physical picture and also leads to many well-known effects. The most famous textbook example might be the fine structure of atomic light spectra. Here, we consider the electron moving in the electromagnetic field induced by the hydrogen atom, we take the extra $1/2$ factor due to Thomas precession into account and obtain immediately

$$V_{ls}(\mathbf{r}) = -\frac{1}{2}\boldsymbol{\mu} \cdot \mathbf{B} = \frac{e}{4m}\boldsymbol{\sigma} \cdot \mathbf{v} \times \mathbf{E} = \frac{e}{4m^2}\frac{d\phi}{r\,dr}\boldsymbol{\sigma} \cdot \mathbf{L}. \qquad (7.37)$$

This is exactly the same as that in Eq. (7.36) derived from Dirac equation.

The spin-orbit coupling plays also a very important in modern spintronics in condensed matter physics where spin transport in the electromagnetically interacting system is studied. There are also examples in electromagnetically interacting systems where spin polarization (magnetization) and orbital angular momentum (rotation) are transferred from one to the other. Earlier examples may even be traced back to Einstein and de Haas [38] and Barnet [39]. It was known as the Einstein–de Haas effect, where the rotation is caused by magnetization and the Barnett effect that is the gyromagnetic effect where magnetization is caused by rotation.

7.3.3   Spin-Orbit Coupling in Systems Under Strong Interactions

In systems under strong interactions, the spin-orbit coupling also leads to many distinguished effects. One of such famous examples is the nuclear shell model developed by Mayer and Jensen [40–42] where the spin-orbit coupling plays a crucial role to produce the magic numbers of atomic nuclei.

There is no such clear intuitive picture for the spin-orbit interaction in systems under strong interactions as that for electromagnetic interactions so the strength cannot be derived explicitly. Usually, in the covariant relativistic formalism, the spin-orbit coupling does appears explicitly. However, the role that it plays can be seen whenever one separates spin and orbital angular momentum from each other. Besides the famous example in the nuclear shell model, another explicit example is the heavy quarkonium spectra where spin-orbit coupling has to be taken into account [43].

Even more interesting is that, in the frontier of high energy spin physics, it seems that spin-orbit coupling plays a key role in understanding all the four classes of striking spin effects mentioned in Sect. 7.1 observed in experiments since 1970s. The simplest argument that orbital angular momentum contributes significantly to proton spin is that discussed in the first point in Sect. 7.3.1 where it has been shown that the orbital angular momentum for a Dirac particle is not a good quantum number. Hence, even if a quark is in the ground states in a central potential as given by Eq. (7.34), the average value of the orbital angular momentum is not zero. If we, e.g., consider a quark in the ground state in a spheric potential well with infinite depth such as in the MIT bag model, the orbital angular momentum contributes $\sim 35\%$ to the total angular momentum.

Both phenomenological model [37,44] and pQCD calculations [45] indicate that orbital angular momentum of quarks in a polarized nucleon and the initial or final state interactions are responsible for SSA observed [10–13] in inclusive hadron-hadron collisions. It has also been shown that transverse hyperon polarization observed [14–18] in unpolarized hadron-hadron collisions are closely related to SSA thus has the same physical origins [46]. The spin analyzing power observed [18–21] in elastic pp scattering is due to color magnetic interaction during the scattering [47] thus originates also from the orbital angular momentum of the constituents in the polarized proton. The study of the role played by the orbital angular momentum is one of the core issues currently in high energy spin physics. See recent reviews such as [9,48–51].

7.4 Theoretical Predictions on the Global Polarization Effect of QGP in HIC

It has been shown [2] that due to spin-orbit interactions in a strongly interacting system such as QGP, the orbital angular momentum can be transferred to the polarization of the constituents in the system such as the quarks and antiquarks.

7.4.1 Global Quark Polarization in QGP in HIC

In Sect. 7.2, we have seen that in a non-central AA collision, there is a huge global orbital angular momentum for the colliding system. Such a global angular momentum leads to the longitudinal fluid shear in the produced system of partons. A pair of interacting partons will have a finite value of relative orbital angular momentum along the direction opposite to the normal of the reaction plane. We have also seen in Sect. 7.3 that spin-orbit coupling is an intrinsic property of a relativistic system. It is thus natural to ask whether the orbital angular momentum or momentum shear leads to the polarization of partons in the system.

There is no field theoretical calculation that can be applied directly to answer this question because usually the calculations are in the momentum space where the momentum shear with respect to x coordinate cannot be taken into account. To achieve this, Ref. [2] took the approach by considering parton scattering with impact parameter in the preferred direction and reach the positive conclusion. We summarize the studies of Refs. [2,4] in this section.

7.4.1.1 Quark Scattering at Fixed Impact Parameter
To be explicit, we consider the scattering $q_1(p_1) + q_2(p_2) \rightarrow q_1(p_3) + q_2(p_4)$ of two quarks with different flavors. The scattering matrix element in momentum space

is given by

$$S_{fi} = \langle f|\hat{S}|i\rangle = \mathcal{M}_{fi}(q)(2\pi)^4\delta^4(p_1 + p_2 - p_3 - p_4), \qquad (7.38)$$

where $p_i = (E_i, \mathbf{p}_i)$ is the four-momentum of the quark, $q = p_1 - p_3 = p_4 - p_2$ is the four-momentum transfer and $\mathcal{M}_{fi}(q)$ is the scattering amplitude in momentum space. The incident momenta are taken as in z or $-z$ direction and the transverse momentum is denoted as $\mathbf{p}_T = \mathbf{p}_{3T} = -\mathbf{p}_{4T}$. The differential cross section in the momentum space is given by

$$d\sigma = \frac{c_{qq}}{F} \frac{|S_{fi}(q)|^2}{TV} \frac{d^3 p_3}{(2\pi)^3 2E_3} \frac{d^3 p_4}{(2\pi)^3 2E_4}, \qquad (7.39)$$

where T and V are interaction time and volume of the space, $c_{qq} = 2/9$ is the color factor, and $F = 4\sqrt{(p_1 \cdot p_2)^2 - m_1^2 m_2^2}$ is the flux factor. Here, just for clarity of equations, we omit the spin indices and will pick them up later in the following:

It can easily be verified that

$$S_{fi} = \int d^2 x_T \int \frac{d^2 q_\perp}{(2\pi)^2} \mathcal{M}_{fi}(q) e^{-i(\mathbf{q}_T + \mathbf{p}_T)\cdot\mathbf{x}_T} (2\pi)^4 \delta^4(p_1 + p_2 - p_3 - p_4), \qquad (7.40)$$

where we use $\mathbf{x}_T$ to denote the impact parameter of the two scattering quarks to distinguish it from the impact parameter $\mathbf{b} = b\,\mathbf{e}_x$ of the two nuclei. By inserting Eq. (7.40) into (7.39), we obtain

$$d\sigma = \frac{c_{qq}}{F} \int d^2 x_T \int \frac{d^2 q_\perp}{(2\pi)^2} \frac{d^2 k_\perp}{(2\pi)^2} e^{-i(\mathbf{q}_T - \mathbf{k}_T)\cdot\mathbf{x}_T} \frac{\mathcal{M}_{fi}(q)}{\Lambda(q)} \frac{\mathcal{M}_{fi}^*(k)}{\Lambda(k)}, \qquad (7.41)$$

where $\mathcal{M}_{fi}(q)$ and $\mathcal{M}_{fi}(k)$ are scattering amplitudes in momentum space with four-momentum transfer $q = (q_0, \mathbf{q}_T, q_z)$ and $k = (k_0, \mathbf{k}_T, k_z)$, respectively; $\Lambda(q)$ is a kinematic factor obtained in carrying out the integration and is given by

$$\Lambda^{-2}(q) = \int \delta^4(p_1 + p_2 - p_3 - p_4)\delta^2(\mathbf{q}_T + \mathbf{p}_T) \frac{d^3 p_3}{2E_3} \frac{d^3 p_4}{2E_4} = \frac{1}{(E_1 + E_2)p_{3z}}, \qquad (7.42)$$

where p_{3z} is the positive solution of $\sqrt{q_T^2 + p_{3z}^2 + m_3^2} + \sqrt{q_T^2 + p_{3z}^2 + m_4^2} = E_1 + E_2$. Here, in obtaining Eq. (7.41), we have taken the symmetric form with the exchange of q and k to guarantee the integrand of $d^2 x_T$ to be positive definite.

We pick up the spin indices and suppose that we are interested in the polarization of quark q_1 after the scattering. We, therefore, average over the spins of initial quarks and sum over the spin of quark q_2 in the final state. In this case, we have

$$\frac{d^2\sigma_{\lambda_3}}{d^2 x_T} = \frac{c_{qq}}{16F} \sum_{\lambda_1,\lambda_2,\lambda_4} \int \frac{d^2 q_T}{(2\pi)^2} \frac{d^2 k_T}{(2\pi)^2} e^{i(\mathbf{k}_T - \mathbf{q}_T)\cdot\mathbf{x}_T} \frac{\mathcal{M}(q)}{\Lambda(q)} \frac{\mathcal{M}^*(k)}{\Lambda(k)}. \qquad (7.43)$$

We define

$$\frac{d^2 \Delta \sigma}{d^2 x_T} = \frac{d^2 \sigma_+}{d^2 x_T} - \frac{d^2 \sigma_-}{d^2 x_T}, \tag{7.44}$$

$$\frac{d^2 \sigma}{d^2 x_T} = \frac{d^2 \sigma_+}{d^2 x_T} + \frac{d^2 \sigma_-}{d^2 x_T}, \tag{7.45}$$

where $\lambda_3 = +$ or $-$ denotes that the spin of q_1 after the scattering is in the positive or negative direction of the normal $\mathbf{n}$ of the reaction plane; $d^2 \sigma / d^2 x_T$ is just the unpolarized cross section at the fixed impact parameter.

Suppose that the impact parameter $\mathbf{x}_T$ has a given distribution $f_{qq}(\mathbf{x}_T, b, Y, \sqrt{s})$, we can calculate the polarization in the following way:

$$\langle \Delta \sigma \rangle = \int d^2 x_T f_{qq}(\mathbf{x}_T, b, Y, \sqrt{s}) \frac{d^2 \Delta \sigma}{d^2 x_T}, \tag{7.46}$$

$$\langle \sigma \rangle = \int d^2 x_T f_{qq}(\mathbf{x}_T, b, Y, \sqrt{s}) \frac{d^2 \sigma}{d^2 x_T}, \tag{7.47}$$

and the polarization of the quark q_1 after the scattering is given by

$$P_q = \langle \Delta \sigma \rangle / \langle \sigma \rangle. \tag{7.48}$$

As discussed in Sect. 7.2, the average relative orbital angular momentum l of two scattering quarks is in the opposite direction of the normal of the reaction plane in non-central AA collisions. Since a given direction of l corresponds to a given direction of $\mathbf{x}_T$, there should be a preferred direction of $\mathbf{x}_T$ at a given direction of the nucleus-nucleus impact parameter $\mathbf{b}$. The distribution $f_{qq}(\mathbf{x}_T, b, Y, \sqrt{s})$ of $\mathbf{x}_T$ at given $\mathbf{b}$ depends on the collective longitudinal momentum distribution shown in Sect. 7.2. Clearly, it depends on the dynamics of QGP and that of AA collisions.

To see the qualitative features of the physical consequences explicitly, Refs. [2,4] took a simplified $f_{qq}(\mathbf{x}_T, b, Y, \sqrt{s})$ as an example, i.e., a uniform distribution of $\mathbf{x}_T$ in the upper half xy-plane with $x > 0$, i.e.

$$f_{qq}(\mathbf{x}_T, b, Y, \sqrt{s}) \propto \theta(x), \tag{7.49}$$

so that

$$\langle \Delta \sigma \rangle \approx \int_0^\infty dx \int_{-\infty}^\infty dy \frac{d^2 \Delta \sigma}{d^2 x_T}, \tag{7.50}$$

$$\langle \sigma \rangle \approx \int_0^\infty dx \int_{-\infty}^\infty dy \frac{d^2 \sigma}{d^2 x_T}. \tag{7.51}$$

7.4.1.2 Quark Scattering by a Static Potential

To see the characteristics of the physical consequences clearly, in [2], we considered first a quark scattering by a static potential. Here, it is envisaged that a quark incident in z-direction and is scattered by an effective static potential induced by other constituents of QGP. In this case, we obtain

$$\mathcal{M}_{fi}(q) = \bar{u}_\lambda(p+q)\,\slashed{A}(q)\,u(p), \tag{7.52}$$

where $A(q) = (A_0(q), \mathbf{0})$ and $A_0(q) = g/(q^2 + \mu_D^2)$ is the screened static potential with Debye screen mass μ_D [52]. It follows that:

$$\mathcal{M}_{fi}(q)\mathcal{M}_{fi}^*(k) = A_0(q)A_0(k)\bar{u}_\lambda(p+q)(\slashed{\tilde{p}}+m_q)u_\lambda(p+k), \tag{7.53}$$

where $\tilde{p} \equiv (E, -\mathbf{p})$. We choose $\mathbf{n}$ as the quantization axis of spin and denote the eigenvalue by $\lambda = \pm 1$. For small angle scattering, $q_T, k_T \sim \mu_D \ll E$, we obtain,

$$\mathcal{M}_{fi}(q)\mathcal{M}_{fi}^*(k) \approx 4E^2 A_0(q)A_0(k)\left[1 - i\lambda\frac{(\mathbf{q}_T - \mathbf{k}_T)\cdot(\mathbf{n}\times\mathbf{p})}{2E(E+m_q)}\right], \tag{7.54}$$

and the cross sections are given by

$$\frac{d^2\sigma}{d^2x_T} = \frac{g^4 c_T}{4}\int\frac{d^2q_T}{(2\pi)^2}\frac{d^2k_T}{(2\pi)^2}\frac{e^{i(\mathbf{k}_T-\mathbf{q}_T)\cdot\mathbf{x}_T}}{(q_T^2+\mu_D^2)(k_T^2+\mu_D^2)}, \tag{7.55}$$

$$\frac{d^2\Delta\sigma}{d^2x_T} = i\frac{g^4 c_T}{8\mathbf{p}^2}\int\frac{d^2q_T}{(2\pi)^2}\frac{d^2k_T}{(2\pi)^2}\frac{(\mathbf{n}\times\mathbf{p})\cdot(\mathbf{k}_T-\mathbf{q}_T)\,e^{i(\mathbf{k}_T-\mathbf{q}_T)\cdot\mathbf{x}_T}}{(q_T^2+\mu_D^2)(k_T^2+\mu_D^2)}. \tag{7.56}$$

where c_T is the color factor. It is interesting to note that, under such approximation, these two parts of the cross section are related to each other

$$\frac{d^2\Delta\sigma}{d^2x_T} = \frac{1}{2\mathbf{p}^2}(\mathbf{n}\times\mathbf{p})\cdot\nabla\frac{d^2\sigma}{d^2x_T}. \tag{7.57}$$

Completing the integrations over d^2q_T and d^2k_T by using the integration formulae

$$\int\frac{d^2q_T}{(2\pi)^2}\frac{e^{i\mathbf{q}_T\cdot\mathbf{x}_T}}{q_T^2+\mu_D^2} = \int\frac{q_T dq_T}{2\pi}\frac{J_0(q_T x_T)}{q_T^2+\mu_D^2} = \frac{1}{2\pi}K_0(\mu_D x_T), \tag{7.58}$$

we obtain from Eqs. (7.55) and (7.56) that [2],

$$\frac{d^2\sigma}{d^2x_T} = \alpha_s^2 c_T K_0^2(\mu_D x_T), \tag{7.59}$$

$$\frac{d^2\Delta\sigma}{d^2x_T} = \alpha_s^2 c_T\left[(\mathbf{p}\times\mathbf{n})\cdot\hat{\mathbf{x}}_T/\mathbf{p}^2\right]\mu_D K_0(\mu_D x_T)K_1(\mu_D x_T). \tag{7.60}$$

where J_0 and K_0 are the Bessel and modified Bessel functions respectively and $x_T = |\mathbf{x}_T|$. The unpolarized cross section just corresponds to $d^2\sigma/d^2q_T = 4\pi\alpha_s^2 c_T/(q_T^2 + \mu_D^2)^2$ in the momentum space.

It is evident from Eq. (7.60) that parton scattering polarizes quarks along the direction opposite to the normal of the parton reaction plane determined by the impact parameter $\mathbf{x}_T$, i.e., along the direction of the relative orbital angular momentum. This is essentially the manifest of spin-orbit coupling in QCD. Ordinarily, the polarized cross section along a fixed direction $\mathbf{n}$ vanishes when averaged over all possible direction of the parton impact parameter $\mathbf{x}_T$. However, in non-central HIC the local relative orbital angular momentum $\langle l_y \rangle$ provides a preferred average reaction plane for parton collisions. This leads to a quark polarization opposite to the normal of the reaction plane of HIC. This conclusion should not depend on our perturbative treatment of parton scattering as far as the effective interaction is mediated by the vector coupling in QCD.

Averaging over the relative angle between parton $\mathbf{x}_T$ and nuclear impact parameter $\mathbf{b}$ from $-\pi/2$ to $\pi/2$ and over x_T, one can obtain the global quark polarization

$$P_q = -\pi\mu_D|\mathbf{p}|/2E(E + m_q) \tag{7.61}$$

via a single scattering for given E.

If one takes the nonrelativistic limit, $E \sim m_q \gg |\mathbf{p}|, \mu_D$, one obtains

$$P_q \approx -\pi\mu_D|\mathbf{p}|/4m_q^2. \tag{7.62}$$

One of the advantages in this limit is that one can check effects due to spin-orbit coupling explicitly. Here, the spin-orbit coupling is given by Eq. (7.36). The corresponding energy is roughly given by $\langle E_{ls} \rangle \sim \langle \mathbf{l} \cdot \mathbf{s} \, dV/rdr/m^2 \rangle$. Given the interaction range is $r \sim 1/\mu_D$, $\langle dV/rdr \rangle \sim -\langle V \rangle \mu_D^2$; $\langle \mathbf{l} \cdot \mathbf{s} \rangle \sim \langle l \rangle/2 \sim |\mathbf{p}|/2\mu_D$. The quark polarization is $P_q \sim \langle E_{ls} \rangle/\langle V \rangle$. We obtain $P_q \sim -\mu_D|\mathbf{p}|/m^2$ that is just the result given by Eq. (7.62).

If one takes the ultra-relativistic limit $m_q = 0$ and $|\mathbf{p}| \gg \mu_D$, one expects from Eq. (7.61) that $P_q \sim -\pi\mu_D/2E$. However, given $dp_0/dx = 0.34$ GeV/fm for semiperipheral ($b = R_A$) collisions at RHIC, and an average range of interaction $\Delta x^{-1} \sim \mu_D \sim 0.5$ GeV, $\Delta p_z \sim 0.1$ GeV is smaller than the typical transverse momentum transfer μ_D. In this case, one has to go beyond small angle approximation.

We also note that the cross sections can be written in a general form as

$$\frac{d^2\sigma}{d^2x_T} = F(x_T, E),$$

(7.63)

$$\frac{d^2\Delta\sigma}{d^2x_T} = \mathbf{n} \cdot (\mathbf{x}_T \times \mathbf{p}) \, \Delta F(x_T, E),$$

(7.64)

where $F(x_T, E)$ and $\Delta F(x_T, E)$ are scalar functions of both $x_T \equiv |\mathbf{x}_T|$ and the c.m. energy E of the two quarks. We would like to emphasize that Eqs. (7.63) and (7.64) are in fact the most general forms of the two parts of the cross sections under parity conservation in the scattering process. The unpolarized part of the cross section should be independent of any transverse direction thus can only take the form as given by Eq. (7.63), i.e. it depends only on the magnitude of x_T but not on the direction. For the spin-dependent part, the only scalar that we can construct from the available vectors is $\mathbf{n} \cdot (\mathbf{p} \times \mathbf{x}_T)$. Hence, $d^2\Delta\sigma/d^2x_T$ can only take the form given by Eq. (7.64).

We also note that $\mathbf{x}_T \times \mathbf{p}$ is nothing but the relative orbital angular momentum of the two-quark system, $\boldsymbol{l} = \mathbf{x}_T \times \mathbf{p}$. Therefore, the polarized cross section takes its maximum when $\mathbf{n}$ is parallel or antiparallel to the relative orbital angular momentum, depending on whether ΔF is positive or negative. This corresponds to quark polarization in the direction $\boldsymbol{l}$ or $-\boldsymbol{l}$.

7.4.1.3 Quark-Quark Scattering in a Thermal Medium

The quark-quark scattering amplitude in a thermal medium can be calculated by using the Hard Thermal Loop (HTL) resummed gluon propagator [53,55]

$$\Delta^{\mu\nu}(q) = \frac{P_T^{\mu\nu}}{-q^2 + \Pi_T(\xi)} + \frac{P_L^{\mu\nu}}{-q^2 + \Pi_L(\xi)} + (\alpha - 1)\frac{q^\mu q^\nu}{q^4},$$

(7.65)

where q denotes the gluon four-momentum and α is the gauge fixing parameter, $x = \omega/\sqrt{-\tilde{q}^2}$ and $\omega = q \cdot u, \tilde{q} = q - \omega u$, u is the fluid velocity of the local medium. The longitudinal and transverse projectors $P_{T,L}^{\mu\nu}$ are defined by

$$P_L^{\mu\nu} = \frac{1}{q^2\tilde{q}^2}(\omega q^\mu - q^2 u^\mu)(\omega q^\nu - q^2 u^\nu),$$

(7.66)

$$P_T^{\mu\nu} = \tilde{g}^{\mu\nu} - \frac{\tilde{q}^\mu \tilde{q}^\nu}{\tilde{q}^2},$$

(7.67)

where $\tilde{g}_{\mu\nu} = g_{\mu\nu} - u_\mu u_\nu$. Π_L and Π_T are the transverse and longitudinal self-energies and are given by [53]

$$\Pi_L(\xi) = \mu_D^2 \left[1 - \frac{\xi}{2} \ln\left(\frac{1+\xi}{1-\xi}\right) + i\frac{\pi}{2}\xi \right](1 - \xi^2),$$

(7.68)

$$\Pi_T(\xi) = \mu_D^2 \left[\frac{\xi^2}{2} + \frac{\xi}{4}(1 - \xi^2) \ln\left(\frac{1+\xi}{1-\xi}\right) - i\frac{\pi}{4}\xi(1 - \xi^2) \right],$$

(7.69)

where the Debye screening mass is $\mu_D^2 = g^2(N_c + N_f/2)T^2/3$.

With the above HTL gluon propagator, the quark-quark scattering amplitude $\mathcal{M}_{fi}(q)$ in the momentum space can be expressed as

$$\mathcal{M}_{fi}(q) = \bar{u}_{\lambda_3}(p_3)\gamma_\mu u_{\lambda_1}(p_1)\Delta^{\mu\nu}(q)\bar{u}_{\lambda_4}(p_4)\gamma_\nu u_{\lambda_2}(p_2). \qquad (7.70)$$

The product $\mathcal{M}_{fi}(q)\mathcal{M}^*_{fi}(k)$ can be converted to the following trace form:

$$\sum_{\lambda_1,\lambda_2} \mathcal{M}_{fi}(q)\mathcal{M}^*_{fi}(k) = \Delta^{\mu\nu}(q)\Delta^{\alpha\beta*}(k)\mathrm{Tr}[u_{\lambda_3}(p_1-k)\bar{u}_{\lambda_3}(p_1+q)\gamma_\mu(\not{p}_1+m_1)\gamma_\alpha]$$

$$\times \mathrm{Tr}[u_{\lambda_4}(p_2-k)\bar{u}_{\lambda_4}(p_2-q)\gamma_\nu(\not{p}_2+m_2)\gamma_\beta]. \qquad (7.71)$$

In calculations of transport coefficients such as jet energy loss parameter [54] and thermalization time [55] that generally involve cross sections weighted with transverse momentum transfer, the imaginary part of the HTL propagator in the magnetic sector is enough to regularize the infrared behavior of the transport cross sections. However, in the calculation of quark polarization, the total parton scattering cross section is involved. The contribution from the magnetic part of the interaction has, therefore, infrared divergence that can only be regularized through the introduction of non-perturbative magnetic screening mass $\mu_m \approx 0.255\sqrt{N_c/2}g^2T$ [56].

Since we have neglected the thermal momentum perpendicular to the longitudinal flow, the energy transfer $\omega = 0$ in the center of mass frame of the two colliding partons. This corresponds to setting $x = 0$ in the HTL resummed gluon propagator in Eq. (7.65). In this case, the center of mass frame of scattering quarks coincides with the local co-moving frame of QGP and the fluid velocity is $u = (1, 0, 0, 0)$. The corresponding HTL effective gluon propagator in Feynman gauge that contributes to the scattering amplitudes reduces to

$$\Delta^{\mu\nu}(q) = \frac{g^{\mu\nu} - u^\mu u^\nu}{q^2 + \mu_m^2} + \frac{u^\mu u^\nu}{q^2 + \mu_D^2}. \qquad (7.72)$$

The spin-dependent part determines the polarization of the final state quark q_1 via the scattering. The calculation is much involved. A detailed study is given in [4]. We summarize part of the key results in the following:

(i) *Small Angle Approximation*

We only consider light quarks and neglect their masses. Carrying out the traces in Eq. (7.71), we can obtain the expression of the cross section with HTL gluon propagators. The results are much more complicated than those as obtained in Sect. 7.4.1.3 using a static potential model [2]. However, if we consider small transverse momentum transfer and use the small angle approximation, the results are still very simple.

In this case, with $q_z \sim 0$ and $q_T \equiv |\mathbf{q}_T| \ll p$, we obtain

$$\frac{d^2\sigma}{d^2x_T} = \frac{g^4 c_{qq}}{8} \int \frac{d^2q_T}{(2\pi)^2} \frac{d^2k_T}{(2\pi)^2} e^{i(\mathbf{k}_T - \mathbf{q}_T) \cdot \mathbf{x}_T}$$

$$\times \left(\frac{1}{q_T^2 + \mu_m^2} + \frac{1}{q_T^2 + \mu_D^2} \right) \left(\frac{1}{k_T^2 + \mu_m^2} + \frac{1}{k_T^2 + \mu_D^2} \right), \qquad (7.73)$$

$$\frac{d^2\Delta\sigma}{d^2x_T} = -i\frac{g^4 c_{qq}}{16p^2} \int \frac{d^2q_T}{(2\pi)^2} \frac{d^2k_T}{(2\pi)^2} e^{i(\mathbf{k}_T - \mathbf{q}_T) \cdot \mathbf{x}_T} \left[(\mathbf{k}_T - \mathbf{q}_T) \cdot (\mathbf{p} \times \mathbf{n}) \right]$$

$$\times \left(\frac{1}{q_T^2 + \mu_m^2} + \frac{1}{q_T^2 + \mu_D^2} \right) \left(\frac{1}{k_T^2 + \mu_m^2} + \frac{1}{k_T^2 + \mu_D^2} \right). \qquad (7.74)$$

We note that there exist the same relationship between the polarized and unpolarized cross section as that given by Eq. (7.57) obtained in the case of static potential model under the same small angle approximation. Completing the integration over d^2q_T and d^2k_T by using the formulae given by Eqs. (7.58), we obtain

$$\frac{d^2\sigma}{d^2x_T} = \frac{c_{qq}}{2} \alpha_s^2 \left[K_0(\mu_m x_T) + K_0(\mu_D x_T) \right]^2, \qquad (7.75)$$

$$\frac{d^2\Delta\sigma}{d^2x_T} = \frac{c_{qq}\alpha_s^2}{2p^2} \left[(\mathbf{p} \times \mathbf{n}) \cdot \hat{\mathbf{x}}_T \right] \left[K_0(\mu_m x_T) + K_0(\mu_D x_T) \right]$$

$$\times \left[\mu_m K_1(\mu_m x_T) + \mu_D K_1(\mu_D x_T) \right], \qquad (7.76)$$

where $\hat{\mathbf{x}}_T = \mathbf{x}_T / x_T$ is the unit vector of $\mathbf{x}_T$. We compare the above results with those given by Eqs. (7.59) and (7.60) obtained in the screened static potential model where one also made the small angle approximation. We see that the only difference between the two results is the additional contributions from magnetic gluons, whose contributions are absent in the static potential model.

(ii) *Beyond Small Angle Approximation*

Now we present the complete results for the cross section in impact parameter space using HTL gluon propagators without small angle approximation. The unpolarized and polarized cross section can be expressed as

$$\frac{d\sigma}{d^2x_T} = \frac{g^4 c_{qq}}{16\hat{s}} \int \frac{d^2q_T}{(2\pi)^2} \frac{d^2k_T}{(2\pi)^2} e^{i(\mathbf{k}_T - \mathbf{q}_T) \cdot \mathbf{x}_T} \frac{f(q, k)}{\Lambda(q)\Lambda(k)}, \qquad (7.77)$$

$$\frac{d\Delta\sigma}{d^2x_T} = i\frac{g^4 c_{qq}}{8\hat{s}^2} \int \frac{d^2q_T}{(2\pi)^2} \frac{d^2k_T}{(2\pi)^2} e^{i(\mathbf{k}_T - \mathbf{q}_T) \cdot \mathbf{x}_T} \frac{\Delta f(q, k)}{\Lambda(q)\Lambda(k)}, \qquad (7.78)$$

where $\hat{s}$ is the c.m. energy squared of the quark-quark system, $f(q, k)$ and $\Delta f(q, k)$ are given by

$$f(q, k) = \sum_{a,b} \frac{A_{ab}(q, k)}{(q^2 + \mu_a^2)(k^2 + \mu_b^2)}, \tag{7.79}$$

$$\Delta f(q, k) = (\mathbf{p} \times \mathbf{n}) \cdot \sum_{ab} \frac{\Delta \mathbf{A}_{ab}(q, k)}{(q^2 + \mu_a^2)(k^2 + \mu_b^2)}, \tag{7.80}$$

where the subscript a or b denotes m or D representing the magnetic or electric part and the sum runs over all possibilities of (a, b). A_{ab} are Lorentz scalar functions of (q, k) given by

$$A_{mm}(k, q) = \hat{s}[\hat{s} - (q + k)^2] + (q \cdot k)^2, \tag{7.81}$$

$$A_{DD}(q, k) = (\hat{s} - q^2 - k^2)[\hat{s} - (q + k)^2] + (q \cdot k)^2, \tag{7.82}$$

$$A_{mD}(q, k) = A_{Dm}(k, q) = \hat{s}[\hat{s} - k^2 - (q + k)^2] + (k^2 - k \cdot q)^2 + \frac{k^2 q^2}{\hat{s}}(q + k)^2, \tag{7.83}$$

$\Delta \mathbf{A}_{ab}(q, k)$ is a vector in the momentum space and can be written as

$$\Delta \mathbf{A}_{ab}(q, k) = \Delta g_{ab}^{(q)}(q, k)\, \mathbf{q}_T - \Delta g_{ab}^{(k)}(q, k)\, \mathbf{k}_T, \tag{7.84}$$

where $\Delta g_{ab}^{(q)}(q, k)$ and $\Delta g_{ab}^{(k)}(q, k)$ are Lorentz scalar functions given by

$$\Delta g_{mm}^{(q)}(q, k) = \Delta g_{mm}^{(k)}(k, q) = \hat{s}(\hat{s} - q \cdot k) - (\hat{s} + q^2 + k^2 - q \cdot k)k^2, \tag{7.85}$$

$$\Delta g_{DD}^{(q)}(q, k) = \Delta g_{DD}^{(k)}(k, q) = (\hat{s} - q^2 - k^2 - q \cdot k)(\hat{s} - k^2), \tag{7.86}$$

$$\Delta g_{mD}^{(q)}(q, k) = \Delta g_{Dm}^{(k)}(k, q) = \hat{s}(\hat{s} - 2k^2 - q \cdot k) - (k^2 - q \cdot k - \frac{q^2 k^2}{\hat{s}})k^2, \tag{7.87}$$

$$\Delta g_{mD}^{(k)}(q, k) = \Delta g_{Dm}^{(q)}(k, q) = \hat{s}(\hat{s} + q^2 - k^2 - q \cdot k) + (q^2 - q \cdot k - \frac{q^2 k^2}{\hat{s}})q^2, \tag{7.88}$$

We note that $A_{ab}(q, k) = A_{ab}(k, q)$, $\Delta \mathbf{A}_{ab}(q, k) = -\Delta \mathbf{A}_{ab}(k, q)$ so that $f(q, k) = f(k, q)$ and $\Delta f(q, k) = -\Delta f(k, q)$, i.e., they are symmetric or anti-symmetric w.r.t. the two variables, respectively. Hence, the integration result in Eq. (7.77) is real, while that in Eq. (7.78) is pure imaginary so that the cross section is real.

We also note that $f(q, k)$ and $\Delta g_{\alpha\beta}^{(q/k)}(q, k)$ are all functions of Lorentz invariants $\hat{s}, q^2, k^2$ and $q \cdot k$. Furthermore, $A_{ab}(k, q) = \sum_{n=0-2} g_{ab}^{(n)}(\hat{s}, q^2, k^2)(\mathbf{q}_T \cdot \mathbf{k}_T)^n$, and $\Delta g_{ab}^{(q/k)}(k, q) = \sum_{n=0,1} \Delta g_{ab}^{(q/k,n)}(\hat{s}, q^2, k^2)(\mathbf{q}_T \cdot \mathbf{k}_T)^n$. The angular parts of the integrations in Eqs. (7.77) and (7.78) can be carried out. For this purpose, we note that, e.g., for any scalar function f_s of $(\hat{s}, q^2, k^2)$, we have

$$
\int \frac{d^2 q_T}{(2\pi)^2} \frac{d^2 k_T}{(2\pi)^2} e^{i(\mathbf{k}_T - \mathbf{q}_T) \cdot \mathbf{x}_T} f_s(\hat{s}, q^2, k^2)
$$

$$
= \int \frac{dq_T^2}{4\pi} \frac{dk_T^2}{4\pi} J_0(q_T x_T) J_0(k_T x_T) f_s(\hat{s}, q^2, k^2) \equiv F^{(0)}(x_T, \hat{s}), \qquad (7.89)
$$

$$
\int \frac{d^2 q_T}{(2\pi)^2} \frac{d^2 k_T}{(2\pi)^2} e^{i(\mathbf{k}_T - \mathbf{q}_T) \cdot \mathbf{x}_T} (\mathbf{q}_T \cdot \mathbf{k}_T) f_s(\hat{s}, q^2, k^2)
$$

$$
= \int \frac{dq_T^2}{4\pi} \frac{dk_T^2}{4\pi} q_T k_T J_0'(q_T x_T) J_0'(k_T x_T) f_s(\hat{s}, q^2, k^2) \equiv F^{(1)}(x_T, \hat{s}), \; (7.90)
$$

$$
\int \frac{d^2 q_T}{(2\pi)^2} \frac{d^2 k_T}{(2\pi)^2} e^{i(\mathbf{k}_T - \mathbf{q}_T) \cdot \mathbf{x}_T} g_s(\hat{s}, q^2, k^2) \mathbf{q}_T = -i \hat{\mathbf{x}}_T G^{(0)}(x_T, \hat{s}),
$$

$$
G^{(0)}(x_T, \hat{s}) = \int \frac{dq_T^2}{4\pi} \frac{dk_T^2}{4\pi} J_0'(q_T x_T) J_0(k_T x_T) g_s(\hat{s}, q^2, k^2), \qquad (7.91)
$$

$$
\int \frac{d^2 q_T}{(2\pi)^2} \frac{d^2 k_T}{(2\pi)^2} e^{i(\mathbf{k}_T - \mathbf{q}_T) \cdot \mathbf{x}_T} g_s(\hat{s}, q^2, k^2) (\mathbf{q}_T \cdot \mathbf{k}_T) \mathbf{q}_T = -i \hat{\mathbf{x}}_T G^{(1)}(x_T, \hat{s}),
$$

$$
G^{(1)}(x_T, \hat{s}) = \int \frac{dq_T^2}{4\pi} \frac{dk_T^2}{4\pi} q_T k_T J_0''(q_T x_T) J_0'(k_T x_T) g_s(\hat{s}, q^2, k^2). \qquad (7.92)
$$

Hence, we see clearly that

$$
\frac{d\sigma}{d^2 x_T} = \frac{g^4 c_{qq}}{16\hat{s}} \sum_{a,b} F_{ab}(x_T, \hat{s}), \qquad (7.93)
$$

$$
\frac{d\Delta\sigma}{d^2 x_T} = \frac{g^4 c_{qq}}{8\hat{s}^2} (\mathbf{p} \times \mathbf{n}) \cdot \hat{\mathbf{x}}_T \sum_{a,b} \Delta F_{ab}(x_T, \hat{s}). \qquad (7.94)
$$

The scalar functions $F_{ab}(x_T, \hat{s})$ and $\Delta F_{ab}(x_T, \hat{s})$ are rather involved. However, if we take the simple form of $f_{qq}(x_T, Y, b, \sqrt{s})$ given by Eq. (7.49) and calculate σ and $\Delta\sigma$ using Eqs. (7.50) and (7.51), we may first carry out the integration over x_T. In this case we obtain

$$
\langle \sigma \rangle = \frac{g^4 c_{qq}}{32\hat{s}} \int_{q_T \le p} \frac{d^2 q_T}{(2\pi)^2} \frac{f(q, q)}{\Lambda^2(q)}, \qquad (7.95)
$$

$$
\langle \Delta\sigma \rangle = -\frac{g^4 c_{qq}}{8\hat{s}^2} \int_{-E}^{E} \frac{dq_y}{2\pi} \int_{-\sqrt{E^2 - q_y^2}}^{\sqrt{E^2 - q_y^2}} \frac{dq_x}{2\pi} \int_{-\sqrt{E^2 - q_y^2}}^{\sqrt{E^2 - q_y^2}} \frac{dk_x}{2\pi} \frac{\Delta f(q_x, q_y; k_x, q_y)}{(k_x - q_x)\Lambda(q)\Lambda(k)}. \quad (7.96)
$$

These equations can be further simplified to the form suitable for carrying out numerical calculations. Details are given in Ref. [4] where different cases are also studied. Here, we present only the result of the quark polarization P_q as function of c,m, energy of the quark-quark system $\sqrt{\hat{s}}/T$ in Fig. 7.13.

From Fig. 7.13, we see that the quark polarization changes drastically with $\sqrt{\hat{s}}/T$. It increases to some maximum values and then decreases with the growing energy, approaching the result of small angle approximation in the high-energy limit. This

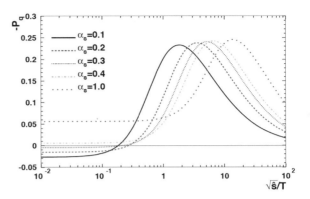

Fig. 7.13 Quark polarization $-P_q$ as a function of $\sqrt{\hat{s}}/T$ for different α_s's obtained in quark-quark scattering with a hard thermal loop propagator. This figure is taken from [4]

structure is caused by the interpolation between the high-energy and low-energy behavior dominated by the magnetic part of the interaction in the weak coupling limit $\alpha_s < 1$. Therefore, the position of the maxima in $\sqrt{\hat{s}}$ should approximately scale with the magnetic mass μ_m.

7.4.1.4 Conclusions and Discussions on Global Quark Polarization

Although approximations and/or models have to be used in the calculations presented above, the physical picture and consequence are very clear. It is confident that after the scattering of two constituents in QGP, the orbital angular momentum will be transferred partly to the polarization of quarks and antiquarks in the system due to spin-orbit coupling in QCD. Such a polarization is very different from those that we meet usually in high energy physics such as the longitudinal or the transverse polarization. The longitudinal polarization refers to the helicity or the polarization in the direction of the momentum, whereas the transverse polarization refers to directions perpendicular to the momentum, either in the production plane or along the normal of the production plane. These directions are all defined by the momentum of the individual particle and are in general different for different particles in the same collision event. In contrast, the polarization discussed here refers to the normal of the reaction plane. It is a fixed direction for one collision event and is independent of any particular hadron in the final state. Hence, in Ref. [2], this polarization was given a new name—the global polarization, and the QGP was referred to as the globally polarized QGP in non-central HIC. We illustrate this in Fig. 7.14.

The following three points should be addressed in this connection:

(i) The results presented above are mainly a summary of those obtained in the original papers [2,4], where the global orbital angular momentum for the colliding system in HIC was first pointed out and the GPE was first predicted. These results are for a single quark-quark scattering. In a realistic HIC where QGP is created, such quark-quark scatterings may take place for a few times before they hadronize into hadrons. The calculations presented above or in [2,4] provide the theoretical basis for GPE. They do not provide the final results of global quark polarizations.

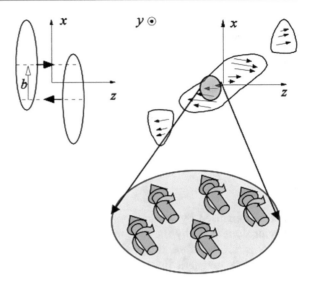

Fig. 7.14 Illustration of the global quark polarization effect in non-central heavy ion collisions

(ii) The numerical results on quark polarization presented above are based on the approximation by taking the simple form of $f_{qq}(\mathbf{x}_T, Y, b, \sqrt{s})$ given by Eq. (7.49). They provide practical guidance for the magnitude of the quark polarization but cannot give us the relationship between the polarization and the local orbital angular momentum. Further studies along this line are necessary. In practice, to describe the evolution of the global quark polarization, one can invoke a dynamical model of QGP evolution or effectively a dynamical model for $f_{qq}(\mathbf{x}_T, Y, b, \sqrt{s})$.

(iii) If we consider QGP as a fluid, the momentum shear distribution discussed in Sec. 7.2 implies a nonvanishing vorticity $\boldsymbol{\omega} = (1/2)\nabla \times \mathbf{v}$. The spin-orbit coupling can be replaced by spin-vortical coupling. This provides a good opportunity to study spin-vortical effects in strongly interacting system and has attracted much attention [59–67]. See chapter on this topic in this series.

7.4.2   A Kinetic Approach for Quark Polarization Rate

The global polarization in heavy ion collisions arises from scattering processes of partons or hadrons with spin-orbit couplings. In a 2-to-2 particle scattering at a fixed impact parameter, one can calculate the polarized cross section arising from the spin-orbit coupling. In a thermal medium, however, momenta of incident particles are randomly distributed and particles participating in the scattering are located at different space-time points. In order to obtain observables, we have to take an ensemble average over random momenta of incident particles and treat scatterings at different space-time points properly. To this end, a microscopic model was proposed for the polarization from the first principle through the spin-orbit coupling in particle scatterings in a thermal medium with a shear flow [68]. It is based on scatterings of particles as wave packets, an effective method to deal with particle

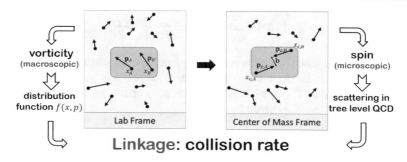

Fig. 7.15 A collision or scattering in the Lab frame (left) and center of mass frame (right)

scatterings at specified impact parameters. The polarization is then the consequence of particle collisions in a nonequilibrium state of spins. The spin-vorticity coupling naturally emerges from the spin-orbit one encoded in polarized scattering amplitudes of collisional integrals when one assumes local equilibrium in momentum but not in spin.

As an illustrative example, we have calculated the quark polarization rate per unit volume from all 2-to-2 parton (quark or gluon) scatterings in a locally thermalized quark-gluon plasma. It can be shown that the polarization rate for antiquarks is the same as that for quarks because they are connected by the charge conjugate transformation. This is consistent with the fact that the rotation does not distinguish particles and antiparticles. The spin-orbit coupling is hidden in the polarized scattering amplitude at specified impact parameters. We can show that the polarization rate per unit volume is proportional to the vorticity as the result of particle scatterings. Thus, we build up a nonequilibrium model for the global polarization.

7.4.2.1 Collision Rate for Spin-0 Particles in a Multi-particle System

We aim to derive the spin polarization rate in a thermal medium with a shear flow from particle scatterings through spin-orbit couplings (Fig. 7.15). Before we do it in the next section, let us first look at the collision rate of spin-zero particles. It is easy to generalize it to the spin polarization rate for spin-1/2 particles.

In the center of mass frame (CMS) of the incident particle A and B, the collision rate (the number of collisions per unit time) per unit volume is given by

$$R_{AB \to 12} = n_A n_B |v_A - v_B| \sigma \frac{d^3 p_A}{(2\pi)^3} \frac{d^3 p_B}{(2\pi)^3} f_A(x_A, p_A) f_B(x_B, p_B) |v_A - v_B| \Delta \sigma,$$

$$(7.97)$$

where $v_A = |\mathbf{p}_A|/E_A$ and $v_B = -|\mathbf{p}_B|/E_B$ are the velocity of A and B respectively with $\mathbf{p}_A = -\mathbf{p}_B$, f_A and f_B are the phase space distributions for A and B, respectively, and $\Delta \sigma$ denotes the infinitesimal element of the cross section which is given by

$$\Delta\sigma = \frac{1}{C_{AB}} d^4 x_A d^4 x_B \delta(\Delta t)\delta(\Delta x_L) \frac{d^3 p_1}{(2\pi)^3 2E_1} \frac{d^3 p_2}{(2\pi)^3 2E_2} \frac{1}{(2E_A)(2E_B)} K,$$
(7.98)

where we assumed that the scattering takes place at the same time and the same longitudinal position in the CMS (these conditions are represented by two delta functions), the constant C_{AB} makes $\Delta\sigma$ have the correct dimension whose definition will be given later, and K is given by

$$K = (2E_A)(2E_B)|_{\text{out}}\langle p_1 p_2 | \phi_A(x_A, p_A)\phi_B(x_B, p_B)\rangle_{\text{in}}|^2,$$
(7.99)

with $(i = A, B)$

$$|\phi_i(x_i, p_i)\rangle_{\text{in}} = \int \frac{d^3 k_i}{(2\pi)^3} \frac{1}{\sqrt{2E_{i,k}}} \phi_i(\mathbf{k}_i - \mathbf{p}_i)e^{-i\mathbf{k}_i \cdot \mathbf{x}_i}|\mathbf{k}_i\rangle_{\text{in}},$$
(7.100)

being the wave packets for incident particles. If incoming particles are described by two plane waves, there is no initial angular momentum. This is why we should use wave packets for incoming particles. Normally one can choose a Gaussian form for the wave packet amplitude

$$\phi_i(\mathbf{k}_i - \mathbf{p}_i) = \frac{(8\pi)^{3/4}}{\alpha_i^{3/2}} \exp\left[-\frac{(\mathbf{k}_i - \mathbf{p}_i)^2}{\alpha_i^2}\right],$$
(7.101)

where α_i denote the width of the wave packet. For simplicity, we use plane waves to represent outgoing particles.

Now we consider the scattering process in Fig. 7.16. The incoming particles are located at x_A and x_B. We can use new variables $X = (x_A + x_B)/$ and $y = x_A - x_B$ to replace x_A and x_B. We then define $C_{AB} \equiv \int d^4 X = t_X \Omega_{\text{int}}$, where t_X and Ω_{int} are the local time and space volume for the interaction. The local collision rate from Eq. (7.97) can be written as

$$\frac{d^4 N_{AB\to 12}}{dX^4} = \frac{1}{(2\pi)^4} \int \frac{d^3 p_A}{(2\pi)^3 2E_A} \frac{d^3 p_B}{(2\pi)^3 2E_B} \frac{d^3 p_1}{(2\pi)^3 2E_1} \frac{d^3 p_2}{(2\pi)^3 2E_2}$$

$$\times |v_A - v_B| G_1 G_2 \int d^3 k_A d^3 k_B d^3 k'_A d^3 k'_B$$

$$\times \phi_A(\mathbf{k}_A - \mathbf{p}_A)\phi_B(\mathbf{k}_B - \mathbf{p}_B)\phi_A^*(\mathbf{k}'_A - \mathbf{p}_A)\phi_B^*(\mathbf{k}'_B - \mathbf{p}_B)$$

$$\times \delta^{(4)}(k'_A + k'_B - p_1 - p_2)\delta^{(4)}(k_A + k_B - p_1 - p_2)$$

$$\times \mathcal{M}(\{k_A, k_B\} \to \{p_1, p_2\})\mathcal{M}^*(\{k'_A, k'_B\} \to \{p_1, p_2\})$$

$$\times \int d^2 \mathbf{b} f_A\left(X + \frac{y_T}{2}, p_A\right) f_B\left(X - \frac{y_T}{2}, p_B\right) \exp\left[i(\mathbf{k}'_A - \mathbf{k}_A) \cdot \mathbf{b}\right], \quad (7.102)$$

where $N_{AB\to 12}$ is the number of collisions and G_i $(i = 1, 2)$ denote the distribution factors which depends on the particle types in the final state. We have $G_i = 1$ for the Boltzmann particles and $G_i = 1 \pm f_i(p_i)$ for bosons (upper sign) and fermions (lower sign).

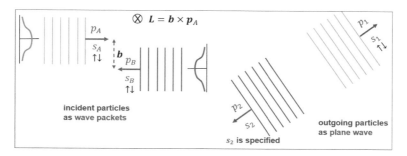

Fig. 7.16 Scattering of two particles in the center of mass frame

7.4.2.2 Polarization Rate for Spin-1/2 Particles from Collisions

Based on the collision rate for spin-zero particles in the above section, we now consider spin-1/2 particles. We assume that particle distributions are independent of spin states, so the spin dependence comes only from scatterings of particles carrying the spin degree of freedom. In this section, we will distinguish quantities in the CMS from those in the lab frame, we will put an index c for a CMS quantity.

If the system has reached local equilibrium in momentum, we can make an expansion of $f_A f_B$ in $y_{c,T} = (0, \mathbf{b}_c)$, and thus

$$
f_A \left(X_c + \frac{y_{c,T}}{2}, p_{c,A}\right) f_B \left(X_c - \frac{y_{c,T}}{2}, p_{c,B}\right)
$$
$$
= f_A (X, p_A) f_B (X, p_B) + \frac{1}{2} y^\mu_{c,T} [\Lambda^{-1}]^\nu_\mu \frac{\partial(\beta u_\rho)}{\partial X^\nu}
$$
$$
\times \left[p^\rho_A f_B (X, p_B) \frac{df_A (X, p_A)}{d(\beta u \cdot p_A)} - p^\rho_B f_A (X, p_A) \frac{df_B (X, p_B)}{d(\beta u \cdot p_B)} \right],
$$
$$(7.103)$$

where we have used the defination of the Lorentz transformation matrix $\partial X^\nu / \partial X^\mu_c = [\Lambda^{-1}]^\nu_\mu = \Lambda^{\ \nu}_\mu$, and the scalar invariance $f_A (X, p_A) = f_A (X_c, p_{c,A})$ and $f_B (X, p_B) = f_B (X_c, p_{c,B})$. From Eq. (7.103), we see that the local vorticity $\partial(\beta u_\rho)/\partial X^\nu$ shows up. We look closely at the term $y^\mu_{c,T} [\partial(\beta u_{c,\rho})/\partial X^\mu_c] p^\rho_{c,A}$,

$$
y^\mu_{c,T} p^\rho_{c,A} \frac{\partial(\beta u_\rho)}{\partial X^\mu_c} = \frac{1}{4} y^{[\mu}_{c,T} p^{\rho]}_{c,A} \left[\frac{\partial(\beta u_{c,\rho})}{\partial X^\mu_c} - \frac{\partial(\beta u_{c,\mu})}{\partial X^\rho_c} \right]
$$
$$
+ \frac{1}{4} y^{\{\mu}_{c,T} p^{\rho\}}_{c,A} \left[\frac{\partial(\beta u_{c,\rho})}{\partial X^\mu_c} + \frac{\partial(\beta u_{c,\mu})}{\partial X^\rho_c} \right]
$$
$$
= -\frac{1}{2} y^{[\mu}_{c,T} p^{\rho]}_{c,A} \varpi^{(c)}_{\mu\rho} + \frac{1}{4} y^{\{\mu}_{c,T} p^{\rho\}}_{c,A} \left[\frac{\partial(\beta u_{c,\rho})}{\partial X^\mu_c} + \frac{\partial(\beta u_{c,\mu})}{\partial X^\rho_c} \right]
$$
$$
= -\frac{1}{2} L^{\mu\rho}_{(c)} \varpi^{(c)}_{\mu\rho} + \frac{1}{4} y^{\{\mu}_{c,T} p^{\rho\}}_{c,A} \left[\frac{\partial(\beta u_{c,\rho})}{\partial X^\mu_c} + \frac{\partial(\beta u_{c,\mu})}{\partial X^\rho_c} \right],
$$
$$(7.104)$$

where $[\mu\rho]$ and $\{\mu\rho\}$ denote the anti-symmetrization and symmetrization of two indices respectively, $L_{(c)}^{\mu\rho} \equiv y_{c,T}^{[\mu} p_{c,A}^{\rho]}$ is the OAM tensor, and $\varpi_{\mu\rho}^{(c)} \equiv -(1/2) [\partial_\mu^{X_c}(\beta u_{c,\rho}) - \partial_\rho^{X_c}(\beta u_{c,\mu})]$ is the thermal vorticity. We see that the coupling term of the OAM and vorticity appear in Eq. (7.103). The second term in last line of Eq. (7.104) is related to the Killing condition required by the thermal equilibrium of the spin.

Now we consider the scattering process $A + B \to 1 + 2$ where incoming and outgoing particles are in the spin state labeled by s_A, s_B, s_1 and s_2 ($s_i = \pm 1/2$, $i = A, B, 1, 2$), respectively. For simplicity, we sum over s_A, s_B, s_1, and leave s_2 open. Defining the direction of the reaction plane in the CMS as $\mathbf{n}_c = \hat{\mathbf{b}}_c \times \hat{\mathbf{p}}_{c,A}$, we have, from Eq. (7.102), the polarization rate of particle 2 per unit time and unit volume is

$$
\frac{d^4\mathbf{P}_{AB\to 12}(X)}{dX^4} = -\frac{1}{(2\pi)^4}\int \frac{d^3p_A}{(2\pi)^3 2E_A}\frac{d^3p_B}{(2\pi)^3 2E_B}\frac{d^3p_{c,1}}{(2\pi)^3 2E_{c,1}}\frac{d^3p_{c,2}}{(2\pi)^3 2E_{c,2}}
$$

$$
\times |v_{c,A} - v_{c,B}| \int d^3k_{c,A}d^3k_{c,B}d^3k'_{c,A}d^3k'_{c,B}
$$

$$
\times \phi_A(\mathbf{k}_{c,A} - \mathbf{p}_{c,A})\phi_B(\mathbf{k}_{c,B} - \mathbf{p}_{c,B})\phi_A^*(\mathbf{k}'_{c,A} - \mathbf{p}_{c,A})\phi_B^*(\mathbf{k}'_{c,B} - \mathbf{p}_{c,B})
$$

$$
\times \delta^{(4)}(k'_{c,A} + k'_{c,B} - p_{c,1} - p_{c,2})\delta^{(4)}(k_{c,A} + k_{c,B} - p_{c,1} - p_{c,2})
$$

$$
\times \frac{1}{2}\int d^2\mathbf{b}_c \exp\left[i(\mathbf{k}'_{c,A} - \mathbf{k}_{c,A})\cdot\mathbf{b}_c\right]\mathbf{b}_{c,j}[\Lambda^{-1}]_j^v \frac{\partial(\beta u_\rho)}{\partial X^v}
$$

$$
\times [p_A^\rho - p_B^\rho] f_A(X, p_A) f_B(X, p_B) \Delta I_M^{AB\to 12}\mathbf{n}_c, \tag{7.105}
$$

where $\mathbf{P}_{AB\to 12}$ denotes the polarization vector. In the derivation of Eq. (7.105), we have used Boltzmann distributions for $f_A(X, p_A) f_B(X, p_B)$ with $G_1 G_2 = 1$. The quantity $\Delta I_M^{AB\to 12}$ is defined as

$$
\Delta I_M^{AB\to 12} = \sum_{s_A, s_B, s_1, s_2}\sum_{color} 2s_2 \mathcal{M}\left(\{s_A, k_{c,A}; s_B, k_{c,B}\} \to \{s_1, p_{c,1}; s_2, p_{c,2}\}\right)
$$

$$
\times \mathcal{M}^*\left(\{s_A, k'_{c,A}; s_B, k'_{c,B}\} \to \{s_1, p_{c,1}; s_2, p_{c,2}\}\right). \tag{7.106}
$$

Since we consider the polarization of quarks, there are seven processes involved as shown in Fig. 7.17. Evaluate all these diagrams will give more than 5000 terms. However, all these terms are spin-orbit coupling ones [2,4] that have four types of structures: $(\mathbf{n} \times \mathbf{p}_1) \cdot \hat{\mathbf{k}}_A$, $(\mathbf{n} \times \mathbf{p}_1) \cdot \hat{\mathbf{k}}'_A$, $(\mathbf{n} \times \hat{\mathbf{k}}_A) \cdot \hat{\mathbf{k}}'_A$ and $(\mathbf{p}_1 \times \hat{\mathbf{k}}_A) \cdot \hat{\mathbf{k}}'_A$.

7.4.2.3 Numerical Results for Quark/antiquark Polarization Rate

Finally, the polarization rate of quarks per unit time and unit volume in Eq. (7.105) can be put into a compact form

$$
\frac{d^4\mathbf{P}_q(X)}{dX^4} = \frac{\pi}{(2\pi)^4}\frac{\partial(\beta u_\rho)}{\partial X^v}\sum_{A,B,1}\int \frac{d^3p_A}{(2\pi)^3 2E_A}\frac{d^3p_B}{(2\pi)^3 2E_B}|v_{c,A} - v_{c,B}|
$$

$$
\times [\Lambda^{-1}]_j^v \mathbf{e}_{c,i}\epsilon_{ikh}\hat{\mathbf{p}}_{c,A}^h f_A(X, p_A) f_B(X, p_B)\left(p_A^\rho - p_B^\rho\right)\Theta_{jk}(\mathbf{p}_{c,A})
$$

$$
\equiv \frac{\partial(\beta u_\rho)}{\partial X^v}\mathbf{W}^{\rho v}, \tag{7.107}
$$

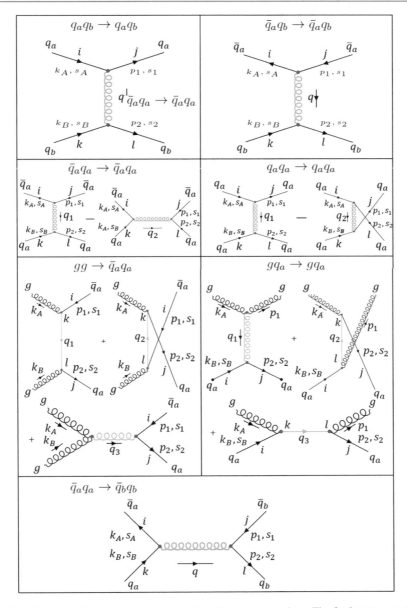

Fig. 7.17 Tree level Feynman diagrams of all 2-to-2 parton scatterings. The final states contain at least one quark. Here a and b denote the quark flavor, $s_i = \pm 1/2$ ($i = A, B, 1, 2$) denote the spin states, k_i ($i = A, B, 1, 2$) denote the momenta, q, q_1, q_2, q_3 denote the momenta in propagators. The processes for antiquark are similar

where the tensor $\mathbf{W}^{\rho\nu}$, defined in the last line, contains 64 components, and each of its component is 16 dimensional integration.

This is a major challenge in the numerical calculation. To handle this high dimension integration, we split the integration into two parts: a 10-dimension (10D) integration over $(\mathbf{p}_{c,1}, \mathbf{p}_{c,2}, \mathbf{k}^T_{c,A}, \mathbf{k}'^T_{c,A})$ and a 6-dimension (6D) integration over $(\mathbf{p}_A, \mathbf{p}_B)$. We first carry out the 10D integration by ZMCintegral-3.0, a Monte Carlo integration package that we have newly developed and runs on multi-GPUs [69]. Then we save this 10D result $\Theta_{jk}(\mathbf{p}_{c,A})$ as a function of $\mathbf{p}_{c,A}$ (and $\mathbf{p}_{c,B} = -\mathbf{p}_{c,A}$). Finally, we perform the 6D integration using the pre-calculated 10D integral. The main parameters are set to following values: the quark mass $m_q = 0.2$ GeV for all flavors $(u, d, s, \bar{u}, \bar{d}, \bar{s})$, the gluon mass $m_g = 0$ for the external gluon, the internal gluon mass (Debye screening mass) $m_g = m_D = 0.2$ GeV in gluon propagators in the t and u channel to regulate the possible divergence, the width $\alpha = 0.28$ GeV of the Gaussian wave packet, and the temperature $T = 0.3$ GeV.

The numerical results are shown in Fig. 7.18, from which we see an explicit form of $\mathbf{W}^{\rho\nu}$ as

$$
\mathbf{W}^{\rho\nu} = \begin{pmatrix} 0 & 0 & 0 & 0 \\ 0 & 0 & W\mathbf{e}_z & -W\mathbf{e}_y \\ 0 & -W\mathbf{e}_z & 0 & W\mathbf{e}_x \\ 0 & W\mathbf{e}_y & -W\mathbf{e}_x & 0 \end{pmatrix}, \tag{7.108}
$$

or in a compact form

$$
\mathbf{W}^{\rho\nu} = W\epsilon^{0\rho\nu j}\mathbf{e}_j. \tag{7.109}
$$

Therefore, Eq. (7.107) becomes

$$
\frac{d^4\mathbf{P}_q(X)}{dX^4} = \epsilon^{0j\rho\nu}\frac{\partial(\beta u_\rho)}{\partial X^\nu}W\mathbf{e}_j = 2\epsilon_{jkl}\omega_{kl}W\mathbf{e}_j = 2W\nabla_X \times (\beta\mathbf{u}), \tag{7.110}
$$

where $\varpi_{\rho\nu} = -(1/2)[\partial^X_\rho(\beta u_\nu) - \partial^X_\nu(\beta u_\rho)]$.

7.4.2.4 Summary and Discussions of This Approach

We have constructed a microscopic model for the global polarization from particle scatterings in a many-body system. The core of the idea is the scattering of particles as wave packets so that the orbital angular momentum is present in the initial state of the scattering which can be converted to the spin polarization of final state particles. As an illustrative example, we have calculated the quark/antiquark polarization in a QGP. The quarks and gluons are assumed to obey the Boltzmann distribution which simplifies the heavy numerical calculation. There is no essential difficulty to treat quarks and gluons as fermions and bosons, respectively.

To simplify the calculation, we also assume that the quark distributions are the same for all flavors and spin states. As a consequence, the inverse process is absent that one polarized quark is scattered by a parton to two final state partons as wave packets. So the relaxation of the spin polarization cannot be described without inverse processes and spin-dependent distributions. We will extend our model by including

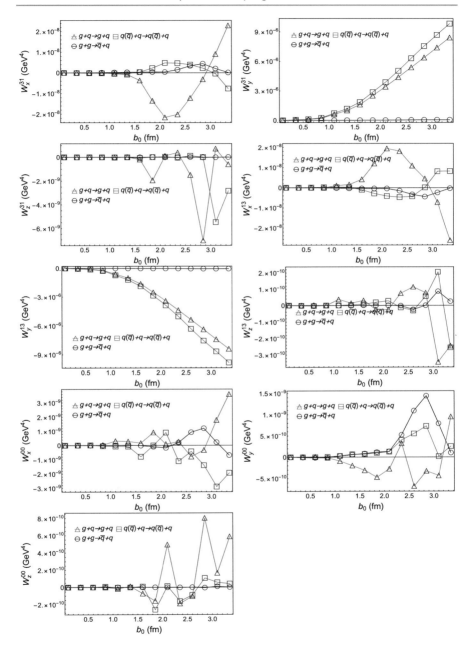

Fig. 7.18 Numerical results for components of $\mathbf{W}^{\rho\nu}$. Here b_0 is the cut-off of the impact parameter in the CMS of the scattering

the inverse process in the future. In Ref. [70], local and nonlocal collision terms in the Boltzmann equation for massive spin-1/2 particles in the Wigner function approach [71] have been derived for spin-dependent distributions. The equilibration of spin degrees of freedom can be fully described by such a spin Boltzmann equation. Nonlocal collision terms are found to be responsible for the conversion of orbital into spin angular momentum. It can be shown that collision terms vanish in global equilibrium and that the spin potential is equal to the thermal vorticity. Such a Boltzmann equation can be applied to parton collisions in quark matter.

7.4.3 Global Hadron Polarization in HIC

The global polarization of quarks and antiquarks in QGP produced in non-central HIC has different direct consequences. The most obvious and measurable effects is the global polarization of hadrons produced after the hadronization of QGP. In [2], the global polarization of produced hyperons has been given. The spin alignment of vector mesons has been calculated in [3].

It is clear that the global hadron polarization depends not only on the global quark polarization, but also on the hadronization mechanism. In the following, we discuss the results obtained in quark combination and fragmentation, respectively.

7.4.3.1 Global Hyperon Polarization

For all hyperons belong to the $J^P = (1/2)^+$ baryon octet except Σ^0, the polarization can be measured via the angular distribution of decay products in the corresponding weak decay. Such decay process is often called "spin self analyzing parity violating weak decay". Because of this, hyperon polarizations are widely studied in the field of high energy spin physics.

(i) *Hyperon Polarization in the Quark Combination*

Different aspects of experimental data suggest that hadronization of QGP proceeds via combination of quarks and/or antiquarks. This mechanism is phrased as "quark re-combination", or "quark coalescence" or simply as "quark combination". We simply refer it as "the quark combination mechanism" and use it to calculate the hyperon polarization in the following:

In the quark combination mechanism, it is envisaged that quarks and antiquarks evolve into constituent quarks and antiquarks and combine with each other to form hadrons. We choose the minus direction of the normal of the reaction plane $-\mathbf{n}$ as the quantization axis. The spin density matrix of quark or antiquark is given by

$$\hat{\rho}_q = \frac{1}{2} \begin{pmatrix} 1 + P_q & 0 \\ 0 & 1 - P_q \end{pmatrix}. \tag{7.111}$$

We do not consider the correlation between the polarizations of different quarks and/or antiquarks, hence the spin density matrix for a $q_1 q_2 q_3$ is given by

$$\hat{\rho}_{q_1 q_2 q_3} = \hat{\rho}_{q_1} \otimes \hat{\rho}_{q_2} \otimes \hat{\rho}_{q_3}. \tag{7.112}$$

Table 7.2 Polarization of hyperons directly produced in the quark combination or fragmentation mechanism. The results for fragmentation are for the leading hadrons only, where n_s and f_s in fragmentation are the strange quark abundances relative to up or down quarks in QGP and quark fragmentation, respectively. These results are taken from [2]

hyperon	Λ	Σ^+	Σ^0	Σ^-	Ξ^0	Ξ^-
combination	P_s	$\frac{4P_u-P_s}{3}$	$\frac{2(P_u+P_d)-P_s}{3}$	$\frac{4P_d-P_s}{3}$	$\frac{4P_s-P_u}{3}$	$\frac{4P_s-P_d}{3}$
fragmentation	$\frac{n_s P_s}{n_s+2f_s}$	$\frac{4f_s P_u-n_s P_s}{3(2f_s+n_s)}$	$\frac{2f_s(P_u+P_d)-n_s P_s}{3(2f_s+n_s)}$	$\frac{4f_s P_d-n_s P_s}{3(2f_s+n_s)}$	$\frac{4n_s P_s-f_s P_u}{3(2n_s+f_s)}$	$\frac{4n_s P_s-f_s P_d}{3(2n_s+f_s)}$

Suppose a hyperon H is produced via the combination of $q_1 q_2 q_3$, we obtain

$$\rho_H(m', m) = \frac{\sum_{m_i,m_i'} \rho_{q_1q_2q_3}(m_i', m_i)\langle j_H, m'|m_1', m_2', m_3'\rangle\langle m_1, m_2, m_3|j_H, m\rangle}{\sum_{m,m_i,m_i'} \rho_{q_1q_2q_3}(m_i', m_i)\langle j_H, m|m_1', m_2', m_3'\rangle\langle m_1, m_2, m_3|j_H, m\rangle},$$
(7.113)

where $|j_H, m\rangle$ is the spin wave function of H in the constituent quark model, and $\langle j_H, m|m_1, m_2, m_3\rangle$ is the Clebsh-Gordon coefficient. The polarization of H is

$$P_H = \rho_H(1/2, 1/2) - \rho_H(-1/2, -1/2).$$
(7.114)

Since $\hat{\rho}_q$ is diagonal so is $\hat{\rho}_{q_1q_2q_3}$, i.e., $\rho_{q_1q_2q_3}(m_i', m_i) = \Pi_i(1 + \tilde{P}_{q_i})\delta_{m_i,m_i'}/8$, where $\tilde{P}_{q_i} \equiv \text{sign}(m_i) P_{q_i}$, Eq. (7.113) reduces to

$$\rho_H(m', m) = \frac{\sum_{m_i} \Pi_j(1 + \tilde{P}_{q_j})\langle j_H, m'|m_1, m_2, m_3\rangle\langle m_1, m_2, m_3|j_H, m\rangle}{\sum_{m,m_i} \Pi_j(1 + \tilde{P}_{q_j})|\langle j_H, m|m_1, m_2, m_3\rangle|^2}.$$
(7.115)

The remaining calculations are straight forward and we list the results in table 7.2. It is also obvious that if $P_u = P_d = P_s \equiv P_q$, we obtain $P_H = P_q$ for all hyperons.

(ii) *Hyperon Polarization in the Quark Fragmentation*

In the high p_T region, hadron production is dominated by the quark fragmentation mechanism, described by quark fragmentation functions defined via the quark-quark correlator such as

$$D_1(z) = \sum_{S_h} \int \frac{d\xi^-}{2\pi} e^{-i\xi^- p_h^+/z} \, \text{Tr}\, \gamma^+ \langle 0|\mathcal{L}(0, +\infty)\psi(0)|p_h, S_h, X\rangle$$

$$\times \langle p_h, S_h, X|\bar{\psi}(\xi)\mathcal{L}(\xi, +\infty)|0\rangle,$$
(7.116)

which is the number density of hadron h produced in the fragmentation process $q \to h + X$; $z = p_h^+/p^+$ is the momentum fraction of quark q carried by hadron h, where p and p_h denote the momenta of q and h, respectively. Here the light cone coordinate is used and the superscript + denotes the + component. $\mathcal{L}$ is the gauge

Fig. 7.19 Longitudinal polarization of Λ in $e^+e^- \rightarrow \Lambda + X$ as described by using a parameterization of $G_{1L}(z)$. The data points are from experiments at LEP [72,73]. This figure is taken from [78]

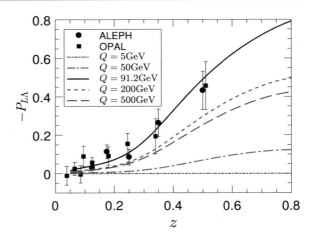

link that originates from the multiple gluon scattering and guarantees the gauge invariance. The polarization transfer is described by

$$G_1(z) = \int \frac{d\xi^-}{2\pi} e^{-i\xi^- p_h^+/z} \, \text{Tr} \, \gamma_5 \gamma^+ \langle 0|\psi(0)|p_h, +, X\rangle\langle p_h, +, X|\bar{\psi}(\xi)|0\rangle, \qquad (7.117)$$

$$H_{1T}(z) = \int \frac{d\xi^-}{2\pi} e^{-i\xi^- p_h^+/z} \, \text{Tr} \, \gamma_T \gamma^+ \langle 0|\psi(0)|p_h, +_T, X\rangle\langle p_h, +_T, X|\bar{\psi}(\xi)|0\rangle, \qquad (7.118)$$

for the longitudinal and transverse polarization, respectively; the $+$ or $+_T$ in $|p_h, S_h, X\rangle$ represents that the spin of h is in the $S_{hz} = +1/2$ or $S_{hT} = +1/2$ state and gauge links are omitted for clarity of equations. The presence of γ_5 or $\gamma_T = \boldsymbol{\gamma} \cdot \mathbf{n}_T$ introduces the dependence on the spin of the fragmenting quark q.

Fragmentation functions are best studied in e^+e^- annihilations. They cannot be calculated using pQCD, so currently, we have to rely on parameterizations or models. There are still not much data available yet. For longitudinal polarization, we have data from LEP at CERN for Λ polarization [72,73]. A recent parameterization of G_1 can be found in [78]. For the transversely polarized case, little data and no parameterization of H_{1T} is available.

To get a feeling of the z-dependence of the spin transfer in quark fragmentations, we show the fit obtained in [78] to the LEP data in Fig. 7.19. We see that, although the accuracy, still, needs to be improved, it is definite that there is a strong z-dependence of G_1 and the spin transfer G_1/D_1 is usually significantly smaller than unity. This implies that the hyperon polarization obtained in the fragmentation mechanism should be much smaller than that obtained in the combination case.

In [2], a model estimation was made for the polarization of the leading hyperon produced in the fragmentation of a polarized quark. It was assumed that two unpolarized quarks are created in the fragmentation and they combine with the polarized q to form the leading hyperon. In this case, we obtain the results as given in Table 7.2. We see if $n_s = f_s$ the result from fragmentation is just $1/3$ of the corresponding result from the combination, i.e., much smaller than the latter even for the leading hyperon.

7.4.3.2 Global Spin Alignment of Vector Mesons

Vector meson spin alignment can also be measured via angular distribution of decay products in the strong two body decay $V \rightarrow 1 + 2$ into two spinless mesons. Hence, it is also frequently studied in high energy spin physics.

(i) *Vector Meson Alignment in the Quark Combination*

Similar to $q_1 q_2 q_3$, we do not consider the correlation between polarizations of quarks and antiquarks, and obtain the spin density matrix for a $q_1 \bar{q}_2$-system as

$$\hat{\rho}_{q_1 \bar{q}_2} = \hat{\rho}_{q_1} \otimes \hat{\rho}_{\bar{q}_2}. \tag{7.119}$$

The spin density matrix for a vector meson V produced via the combination of $q_1 \bar{q}_2$ is given by

$$\rho_{m'm}^V = \frac{\sum_{m_i, m_i'} \rho_{q_1 \bar{q}_2}(m_i', m_i) \langle j_V, m' | m_1', m_2' \rangle \langle m_1, m_2 | j_V, m \rangle}{\sum_{m, m_i, m_i'} \rho_{q_1 \bar{q}_2}(m_i', m_i) \langle j_V, m | m_1', m_2' \rangle \langle m_1, m_2 | j_V, m \rangle}, \tag{7.120}$$

where $|j_V, m\rangle$ is the spin wave function of V in the constituent quark model. For diagonal $\hat{\rho}_q$ and $\hat{\rho}_{\bar{q}}$, we have

$$\rho_{m'm}^V = \frac{\sum_{m_i} (1 + \tilde{P}_{q_1})(1 + \tilde{P}_{\bar{q}_2}) \langle j_V, m' | m_1, m_2 \rangle \langle m_1, m_2 | j_V, m \rangle}{\sum_{m, m_i} (1 + \tilde{P}_{q_1})(1 + \tilde{P}_{\bar{q}_2}) | \langle j_V, m | m_1, m_2 \rangle |^2}, \tag{7.121}$$

The spin alignment is described by ρ_{00}^V and is obtained as [3],

$$\rho_{00}^V = \frac{1 - P_{q_1} P_{\bar{q}_2}}{3 + P_{q_1} P_{\bar{q}_2}}. \tag{7.122}$$

From Eq. (7.122), we see clearly that the global vector meson spin alignment ρ_{00}^V obtained in quark combination should be less than $1/3$. We also see that in contrast to the hyperon polarization P_H, ρ_{00}^V is a quadratic effect of P_q.

(ii) *Vector Meson Spin Alignment in the Quark Fragmentation*

To define the fragmentation functions for spin-1 hadrons in $q \rightarrow V + X$, one usually decomposes the 3×3 spin density matrix ρ in terms of the 3×3 representation of the spin operator Σ^i and $\Sigma^{ij} = \frac{1}{2}(\Sigma^i \Sigma^j + \Sigma^j \Sigma^i) - \frac{2}{3} \mathbf{1} \delta^{ij}$, i.e.

$$\rho = \frac{1}{3}(\mathbf{1} + \frac{3}{2} S^i \Sigma^i + 3 T^{ij} \Sigma^{ij}), \tag{7.123}$$

where the spin polarization tensor $T^{ij} = \text{Tr}(\rho \Sigma^{ij})$ and is parameterized as

$$\mathbf{T} = \frac{1}{2} \begin{pmatrix} -\frac{2}{3} S_{LL} + S_{TT}^{xx} & S_{TT}^{xy} & S_{LT}^x \\ S_{TT}^{xy} & -\frac{2}{3} S_{LL} - S_{TT}^{xx} & S_{LT}^y \\ S_{LT}^x & S_{LT}^y & \frac{4}{3} S_{LL} \end{pmatrix}. \tag{7.124}$$

The spin alignment ρ_{00} is directly related to S_{LL} by $\rho_{00} = (1 - 2S_{LL})/3$ and $S_{LL} = 3\langle\Sigma_z^2\rangle/2 - 1$ is a Lorentz scalar. The complete set of fragmentation functions for spin-1 hadrons can be found in [78]. The S_{LL}-dependence is given by

$$D_{1LL}(z) = \sum_{\lambda}(-1)^{\lambda+1} \int \frac{d\xi^-}{4\pi} e^{-i\xi^- p^+} \text{Tr } \gamma^+ \langle 0|\psi(0)|p_h\lambda X\rangle\langle p_h\lambda X|\bar{\psi}(\xi)|0\rangle \quad (7.125)$$

where $\lambda = \pm 1, 0$ represents the spin of the vector meson. It is very interesting to see that $D_{1LL}(z)$, in fact, does not depends on the spin of the fragmenting quark q.

There are data available on the vector meson spin alignment from experiments at LEP [74–76]. A parameterization of $D_{1LL}(z)$ is given in [79,80] and the fit to the data is shown in Fig. 7.20.

From Fig. 7.20, we see clearly that, in contrast to the quark combination mechanism, ρ_{00} obtained in fragmentation is larger than $1/3$. This indicates that the spin of $\bar{q}$ produced in the fragmentation $q \to h + X$ has larger probability to be in the opposite direction as q. For the leading meson, a parameterization of $P_{\bar{q}} = -\beta P_q$ (where $\beta \sim 0.5$) for the antiquark $\bar{q}$ produced in the fragmentation process and combine with the fragmenting quark to form the vector meson was obtained [77] to fit the data [74,75]. Ref. [3] also made an estimation for such leading vector mesons in fragmentation based on this empirical relation and obtained that

$$\rho_{00}^V = (1 + \beta P_q^2)/(3 - \beta P_q^2). \quad (7.126)$$

We see that the spin alignment ρ_{00}^V obtained this way is indeed larger than $1/3$.

7.4.3.3 Decay Contributions

It is clear that final state hadrons in a high energy reaction usually contain the contributions from decays of heavier resonances, in particular, those from strong and electromagnetic decays. To compare with the data, we need to take such decay contributions into account.

Fig. 7.20 Spin alignment of K^* in $e^+e^- \to K^* + X$ as described by using a parameterization of $D_{1LL}(z)$. The data points are from experiments at LEP [74,75]. This figure is taken from [79]

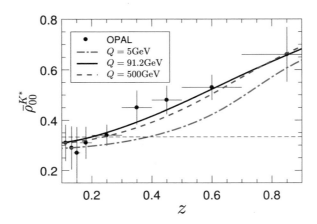

The decay contributions have influences both on the momentum distribution and on the polarization of final hadrons. Such influences have been discussed repeatedly in literature calculating hyperon polarizations in high energy reactions (see, e.g., [81–84] and recently in HIC [85,86]). For hadrons consisting of light flavors of quarks, we usually consider only the production of $J^P = (1/2)^+$ octet and $J^P = (3/2)^+$ decuplet baryons, and $J^P = 0^-$ pseudo-scalar and $J^P = 1^-$ vector mesons. In this case, there is no decay contribution to vector mesons. We only need to consider those to hyperons and most of them are just two body decay $H_j \rightarrow H_i + M$, where H_j and H_i are two hyperons and M is a pseudo-scalar meson. We limit our discussions to this process in the following:

To be explicit, we consider the fragmentation mechanism and study decay contributions to fragmentation functions. For quark combination, we need only to replace the fragmentation function by the corresponding distribution function and z by the corresponding variable. We start with the unpolarized case and the contribution from $H_j \rightarrow H_i + M$ to the unpolarized fragmentation function of H_i is given by

$$D_1^{ij}(z_i, \mathbf{p}_{Ti}) = \text{Br}(H_i, H_j) \int dz_j d^2 p_{Tj} K_{ji}(z_i, \mathbf{p}_{Ti}; z_j, \mathbf{p}_{Tj}) D_1^j(z_j, \mathbf{p}_{Tj}),$$

(7.127)

where $\text{Br}(H_i, H_j)$ is the decay branch ratio. $K_{ji}(z_i, \mathbf{p}_{Ti}; z_j, \mathbf{p}_{Tj})$ is a kernel function representing the probability for a H_j with $(z_j, \mathbf{p}_{Tj})$ to decay into a H_i with $(z_i, \mathbf{p}_{Ti})$. It is just the normalized distribution of H_i from $H_j \rightarrow H_i + M$ and should be determined by the dynamics of the decay process. However, in the unpolarized case, for two body decay, it is determined completely by the energy momentum conservation. From energy conservation, we obtain that, in the rest frame of H_j,

$$E_i^* = (M_j^2 + M_i^2 - M_m^2)/2M_j \equiv E_0^*,$$

(7.128)

$$|\mathbf{p}^*| = \lambda^{1/2}(M_j^2, M_i^2, M_m^2)/2M_j \equiv p_0^*,$$

(7.129)

where the λ-function is $\lambda(x, y, z) = x^2 + y^2 + z^2 - 2xy - 2yz - 2zx$. We see that the magnitude of $\mathbf{p}^*$ is completely fixed. Furthermore, because there is no specified direction in the initial state, the decay product should be distributed isotropically. Hence, in the Lorentz invariant form, the distribution of H_i from $H_j \rightarrow H_i + M$ is given by

$$E_i \frac{d^3 N}{d^3 p_i} = \frac{M_j^2}{\pi \lambda^{1/2}(M_j^2, M_i^2, M_m^2)} \delta\left((p_j - p_i)^2 - M_m^2\right).$$

(7.130)

By replacing variables $\mathbf{p}$ with z and $\mathbf{p}_T$, we obtain the kernel function K_{ji} as

$$K_{ji}(z_i, \mathbf{p}_{Ti}; z_j, \mathbf{p}_{Tj}) = \frac{d^3 N}{dz_i d^2 p_{Ti}}$$

$$= \frac{z_j M_j^2}{\pi \lambda^{1/2}(M_j^2, M_i^2, M_m^2)} \delta\left((\frac{\mathbf{p}_{Tj}}{z_j} - \frac{\mathbf{p}_{Ti}}{z_i})^2 + (\frac{M_j}{z_j} - \frac{M_i}{z_i})^2 - \frac{\Delta M^2 - M_m^2}{z_i z_j}\right), \quad (7.131)$$

where $\Delta M \equiv M_j - M_i$ is the mass difference between the two hyperons.

In practice, we often use the following approximation: We note that the Lorentz transformation of the four-momentum of H_i from the rest frame of H_j to the laboratory frame is given by

$$E_i = (E_j E_i^* + \mathbf{p}_j \cdot \mathbf{p}_i^*)/M_j, \tag{7.132}$$

$$\mathbf{p}_i = \mathbf{p}_i^* + \frac{\mathbf{p}_j \cdot \mathbf{p}_i^* + (E_j - M_j)E_i^*}{M_j(E_j - M_j)} \mathbf{p}_j, \tag{7.133}$$

We take the average over the distribution of $\mathbf{p}_i^*$ at given $\mathbf{p}_j$, and obtain

$$\langle p_i \rangle = p_j \, \xi_{ij}, \qquad \xi_{ij} = (M_j^2 + M_i^2 - M_m^2)/2M_j^2. \tag{7.134}$$

In the case that $\Delta M \ll M_j \sim M_i$ and $p_0^* \ll |\mathbf{p}_i|$, one can simply neglect the distribution and take, $p_i \approx \langle p_i \rangle = p_j \xi_{ij}$ so that $z_i \approx z_j \xi_{ij}$, $\mathbf{p}_{Ti} \approx \mathbf{p}_{Tj} \xi_{ij}$ and

$$K_{ij}(z_i, \mathbf{p}_{Ti}; z_j, \mathbf{p}_{Tj}) \approx \delta(z_j - z_i/\xi_{ij}) \, \delta^2(\mathbf{p}_{Tj} - \mathbf{p}_{Ti}/\xi_{ij}), \tag{7.135}$$

$$D_1^{ij}(z_i, \mathbf{p}_{Ti}) \approx \mathrm{Br}(H_i, H_j) D_1^j(z_i/\xi_{ij}, \mathbf{p}_{Ti}/\xi_{ij}). \tag{7.136}$$

In the polarized case, we need also to consider the polarization transfer t_D^{ij}. In general, in the rest frame of H_j, t_D^{ij} may depend on the momentum $\mathbf{p}_i^*$ of H_i. By transforming it to the Lab frame, we should obtain a result depending on $(z_i, \mathbf{p}_{Ti})$ and $(z_j, \mathbf{p}_{Tj})$ and it is different for the longitudinal and transverse polarization. This is much involved. In practice, we often take the approximation by neglecting the momentum dependence and calculate t_D^{ij} in the rest frame of H_j. In this case, it is the same for the longitudinal and transverse polarization. E.g., for the longitudinal polarized case, we have

$$G_{1L}^{ij}(z_i, \mathbf{p}_{Ti}) = \mathrm{Br}(H_i, H_j) \, t_D^{ij} \int dz_j d^2 p_{Tj} K_{ij}(z_i, \mathbf{p}_{Ti}; z_j, \mathbf{p}_{Tj}) G_{1L}^j(z_j, \mathbf{p}_{Tj}). \tag{7.137}$$

Under the approximation given by Eq. (7.135), we have

$$G_{1L}^{ij}(z_i, \mathbf{p}_{Ti}) \approx \mathrm{Br}(H_i, H_j) \, t_D^{ij} \, G_{1L}^j(z_i/\xi_{ij}, \mathbf{p}_{Ti}/\xi_{ij}), \tag{7.138}$$

For parity conserving decays, the polarization transfer factor t_D^{ij} can easily be calculated from angular momentum conservation. The results are given in Table 7.3. For the weak decay $\Xi \to \Lambda\pi$, $t_D = (1 + \gamma)/2$ where γ is a decay parameter that can be found in Review of Particle Properties (see, e.g., [87]).

If we taken only $J^P = (1/2)^+$ hyperons into account and use spin counting for relative production weights, we obtain

$$P_\Lambda^{final} = P_\Lambda^{direct}[2 + 3\lambda(1 + \gamma)]/6(1 + \lambda), \tag{7.139}$$

where λ is the strangeness suppression factor for s-quarks. This leads to a reduction factor between 0.33 and 0.44 for $\lambda = 0$ and 1, respectively. In this sense, it is more sensitive to study polarization of $\Sigma^\pm$ or Ξ where decay influences are negligible.

Table 7.3 The decay spin transfer factor in parity conserving two body decay $H_j \rightarrow H_i + M$. The first column specifies the spin and parity J^P of hadrons

$H_j \rightarrow H_i + M$	Relative orbital angular momentum	$t_D^{ij} = P_{H_i}/P_{H_j}$
$1/2^+ \rightarrow 1/2^+ + 0^-$	$l = 1$ (P-wave decay)	$-1/3$
$1/2^- \rightarrow 1/2^+ + 0^-$	$l = 0$ (S-wave decay)	1
$3/2^+ \rightarrow 1/2^+ + 0^-$	$l = 1$ (P-wave decay)	1
$3/2^- \rightarrow 1/2^+ + 0^-$	$l = 2$ (D-wave decay)	$-3/5$

7.4.4 Comparison with Experiments

The novel predictions [2,3] on GPE attracted immediate attention, both experimentally and theoretically. A new preprint [88] only three days after the first prediction [2] attempted to extend the idea to other reactions. Experimentalists in the STAR Collaboration had started measurements shortly after the publication of theoretical predictions [2,3], both on the global Λ hyperon polarization and on spin alignments of K^* and ϕ [89–95]. Studies on both aspects have advantages and disadvantages. Hyperon polarization is a linear effect where the polarization for directly produced Λ is equal to that of quarks. The spin alignment of vector meson is a quadratic effect proportional to the square of the quark polarization. Hence, the magnitude of the latter should be much smaller than that of the former. However, to measure the polarization of hyperon, one has to determine the direction of the normal of the reaction plane, which is not needed for measurements of vector meson spin alignments. Also, the contamination effects due to decay contributions to vector mesons are negligible but not for Λ hyperons.

Although there were some promising indications, the results obtained in the early measurements [94,95] by the STAR Collaboration were consistent with zero within large errors. STAR measurements continued during the beam energy scan (BES) experiments and positive results were obtained in lower energy region with improved accuracies [1]. The obtained value averaged over energy is $1.08 \pm 0.15 \pm 0.11$ per cent and $1.38 \pm 0.30 \pm 0.13$ per cent for Λ and $\bar{\Lambda}$, respectively. With much higher statistics, the STAR Collaboration has repeated measurements [96] in Au-Au collisions at 200AGeV and obtained positive result of $P_\Lambda \sim -0.003$ with much higher accuracies.

To compare with experiments at this stage, we start with the following rough estimations: (i) From both Figs. 7.8 and 7.12 obtained using HIJING and BGK, respectively, we obtain at $Y \sim 0$, $\Delta p \sim 0.002$GeV for $\Delta x \sim 1$fm. If we take $T \sim 140$ MeV, $\Delta p/T \sim 0.015$. From Fig. 7.13, we see that the quark polarization P_q is unfortunately in the small and rapidly changing region. Nevertheless, the order of magnitude is in the same range of STAR data [96]. (ii) If we take $\omega \sim \partial u_z/\partial x$, $u_z \sim \langle p_z \rangle / p_T$, we obtain, $\omega \sim \Delta_Y^2 \cosh Y \, \xi_p/12$ from Eq. (7.22). By using the results for $\langle \xi_p \rangle$ shown in Fig. 7.8 or Fig. 7.12 and $T \sim 140$ MeV, we obtain $P_q \sim -0.003$ at $\sqrt{s} = 200$GeV that is consistent with STAR experimental results [96]. (iii) If we take the result at nonrelativistic limit given by Eq. (7.62), and note that $\delta u \sim |\mathbf{p}|/m_q, \delta x \sim$

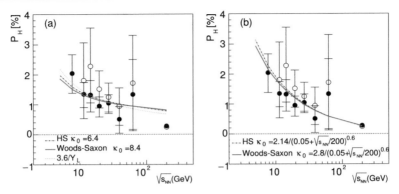

Fig. 7.21 Energy dependence of the global polarization of Λ obtained by taking κ as a constant or an energy dependent form. The data points are taken from [1,96]. This figure is taken from [35]

$1/\mu_D$, so that $\omega \sim \delta u/\delta x \sim \mu_D |\mathbf{p}|/m_q$, and quark polarization is $P_q \sim \pi \omega/4m_q$. If we take an effective quark mass $m_q \sim 200$ MeV at the hadronization, this is clearly also of the same order of magnitude as ω/T.

Such rough estimations are rather encouraging. We continue with more realistic estimations. We note that quark polarization is given by Eqs. (7.46–7.48) and $d\sigma$ and $d\Delta\sigma$ take the general form given by Eqs. (7.63) and (7.64). Before we construct a dynamical model for $f_{qq}(\mathbf{x}_T, b, Y, \sqrt{s})$, we present the following qualitative discussion.

It is clear that at $b = 0$, $f_{qq}(\mathbf{x}_T, 0, Y, \sqrt{s})$ should be independent of the direction of $\mathbf{x}_T$. The $\hat{\mathbf{x}}_T$-dependent term should given by $\hat{\mathbf{x}}_T \cdot \mathbf{b}$. We take the linearly dependent term into account and have

$$f_{qq}(\mathbf{x}_T, b, Y, \sqrt{s}) = f_{qq}(x_T, 0, Y, \sqrt{s}) + f_{qq}(x_T, b, Y, \sqrt{s}) \, \hat{\mathbf{x}}_T \cdot \mathbf{b}, \quad (7.140)$$

We insert Eq. (7.140) into (7.63) and (7.64) and obtain immediately that $P_q \propto \langle l_y^* \rangle$, i.e.

$$P_q = \alpha(b, Y, \sqrt{s}) \langle l_y^*(b, Y, \sqrt{s}) \rangle. \quad (7.141)$$

We insert the result of $\langle l_y^* \rangle$ given by Eq. (7.23) into (7.141), average over the impact parameter b and obtain

$$P_q = -\kappa(Y, \sqrt{s}) \langle p_T \rangle \langle \xi_p \rangle. \quad (7.142)$$

where $\kappa = \alpha(\Delta x)^2 \Delta_Y^2/24$. The proportional coefficient α in Eq. (7.141) hence also κ in Eq. (7.142) are very involved. They are determined by the dynamics in QGP formation and evolution. Averaged over b, κ can still be dependent of Y and $\sqrt{s}$. In [35], the simplest choice, i.e., κ is taken as a constant independent of $\sqrt{s}$ at $Y = 0$, was first considered and obtained the energy dependence of P_q shown in Fig. 7.21a. Taking an energy dependent κ, Ref. [35] made a better fit to the data [1,96] available as shown in Fig. 7.21b.

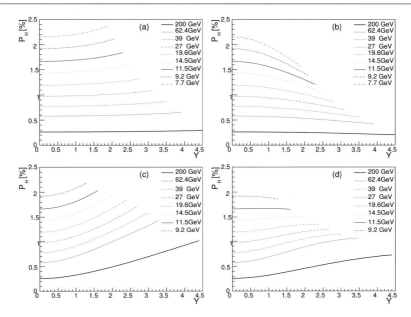

Fig. 7.22 Rapidity dependence of the global polarization of Λ obtained in four different cases **a** neither κ nor $\langle p_T \rangle$ depends on Y; **b** κ depends on Y but $\langle p_T \rangle$ does not; **c** κ does not but $\langle p_T \rangle$ depends on Y; **d** both κ and $\langle p_T \rangle$ depend on Y. This figure is taken from [35]

Figure 7.22 shows the rapidity dependence of the polarization at different energies obtained in [35] in the different cases. The Y-dependence of κ was obtained by assuming that the dependence is mediated by the chemical potential. The Y-dependence of $\langle p_T \rangle$ was taken empirically [35]. See [35] for details.

7.5   Summary and Outlook

To summarize, high energy HIC is usually non-central, thus the colliding system and the produced partonic system QGP carries a huge global orbital angular momentum as large as $10^5 \hbar$ in Au-Au collisions at RHIC energies. Due to the spin-orbit coupling in QCD, such huge orbital angular momentum can be transferred to quarks and antiquarks thus leads to a globally polarized QGP. The global polarization of quarks and antiquarks manifest itself as the global polarization of hadrons such as hyperons and vector mesons produced in HIC.

The early theoretical prediction [2] and discovery by the STAR Collaboration [1] open a new window to study properties of QGP and a new direction in high energy heavy ion physics. Similar measurements have been carried in other experiments such as those by ALICE Collaboration at the Large Hadron Collider (LHC) in Pb-Pb collisions [97]. Other efforts have also been made on measurements of vector meson spin alignments [98,99]. The STAR Collaboration has just finished major detector upgrades and started the beam energy scan at phase II (BES II). The successful detector upgrade with improved inner time projection chamber (iTPC) and event plane detector (EPD) will be crucial to the measurements of global hadron polarizations. The STAR BES II will provide an excellent opportunity to study GPE in HIC and we expect new results with higher accuracies in next years.

The experimental efforts in turns further inspire theoretical studies. The rapid progresses and continuous studies along this line lead to a very active research direction—the Spin Physics in HIC in the field of high energy nuclear physics. Among the most active aspects, we have, in particular, the following:

(i) *GPE Phenomenology*

This includes different model approaches [35, 59, 60, 100–111] to numerical calculations of GPE in HIC and its dependences on different kinematic variables. The model approaches are basically divided into two categories, i.e., microscopic approaches based on the spin-orbit (or spin-vorticity) coupling and hydrodynamic approaches based on equilibrium assumptions. The various dependences of GPE are studied on kinematic variables describing (a) the initial state such as energy, centrality (impact parameter), different incident nuclei even pA collisions, etc.; (b) the produced hadron such as transverse momentum, rapidity, azimuthal angle, different types of hyperons and/or vector mesons; (c) other related measurable effects such as longitudinal polarization, the interplay with other effects, and so on. Short summaries can, e.g., be found in plenary talks given at recent Quark Matter conferences [112, 113].

(ii) *Spin-Vortical Effects in Strong Interacting System*

If we can treat QGP as a vortical ideal fluid consisting of quarks and antiquarks, the global polarization of hadrons is directly related to the vorticity of the system [85]. The fluid vorticity may be estimated from the data [1] on GPE of Λ hyperon using the relation given in the hydrodynamic model, and it leads to a vorticity $\omega \approx (9 \pm 1) \times 10^{21} s^{-1}$. This far surpasses the vorticity of all other known fluids. It was, therefore, concluded that QGP created in HIC is the most vortical fluid in nature observed yet. GPE in HIC, therefore, provides a very special place to study spin-vortical effects in strong interaction and attracts many studies [59–67]. See chapter on this topic for discussions in this aspect.

(iii) *Spin-Magnetic Effects in HIC*

Because of the huge orbital angular momentum, there exists also a very strong magnetic field for the colliding system in HIC. In Au-Au collision at RHIC, it can reach at least instantaneously the order $10^{14} - 10^{16}$ Tesla. Such a strong magnetic field can manifest itself in different aspects and lead to different measurable effects. The most frequently discussed currently are the following three aspects:

(a) The fine structure of GPE of different hadrons. The spin-orbit coupling in QCD predicts e.g. the same polarization of quarks and antiquarks thus also the same for hyperons and anti-hyperons. The strong magnetic field can lead to differences between the polarization of quarks and that of antiquarks thus lead to difference in the polarization of hyperons and anti-hyperons. Indeed, the STAR data in Ref. [1] suggests such a fine structure pattern, and if errorbars are ignored, would indicate $B \sim 10^{14}$ T. However, much smaller uncertainties—available with the new BES

II data—will be needed to resolve the issue. Also the magnetic field may lead to different behavior of vector meson spin alignment [117].

(b) Chiral magnetic effect. In Ref. [114,115], a novel electromagnetic spin effect—the chiral magnetic effect was proposed. It was argued that such effects have deep connection to P and CP violation. Clearly, if they exist, strong magnetic field in HIC provides good opportunity to detect such effects [116]. This has attracted much attention both experimentally and theoretically. See a number of reviews such as [118–123], plenary talks at QM2019 by Xu-guang Huang and Jin-feng Liao [124,125] and related chapter in this series.

(c) Spin-electromagnetic effects in ultra-peripheral collisions (UPC) in HIC. From a field theoretical point of view, the electromagnetic coupling for a HIC is enhanced by a factor Z (number of protons in the nucleus). Hence, many electromagnetic effects become visible in UPC with nuclei of large A. This provides a good place to study the spin-electromagnetic effects and develop the theoretical methodology in particular those developed in studying nucleon structure in the small-x region. See, e.g., the plenary talk at QM2019 by Zhangbu Xu [126] for a brief summary.

(iv) *Spin Transport Theory in Relativistic Quantum System*

Theoretically, a very challenging task is to derive GPE, describe the spin transport, calculate the polarization and other related spin effects directly from QCD. This is rather involved since, to describe orbital angular momentum or vorticity of the system, not only momentum, but also space coordinate are needed. It seems that quantum kinetic theory based on the Wigner function formalism [127–130] is very promising, and thus has attracted much attention recently. Many progresses have been made. Besides others, the local polarization effect has been first derived [131] and a disentanglement theorem [133] in the massless case has been proposed. It has now extended to massive case [135–138] and has been shown that different spin effects can indeed be derived. See chapter on this topic for more discussions in this aspect.

Acknowledgements We thank in particular many collaborators for excellent collaborations on this subject. This work was supported in part by the National Natural Science Foundation of China (Nos. 11890713, 11675092, 11535012, 11935007, 11861131009 and 11890714), and by the Director, Office of Energy Research, Office of High Energy and Nuclear Physics, Division of Nuclear Physics, of the U.S. Department of Energy under grant No. DE-AC02-05CH11231, the National Science Foundation (NSF) under grant No. ACI-1550228 within the framework of the JETSCAPE Collaboration.

References

1. Adamczyk, L., et al.: STAR Collaboration. Nature **548**, 62 (2017)
2. Liang, Z.T., Wang, X.N.: Phys. Rev. Lett. **94**, 102301 (2005) Erratum: [Phys. Rev. Lett. **96**, 039901 (2006)]
3. Liang, Z.T., Wang, X.N.: Phys. Lett. B **629**, 20 (2005)
4. Gao, J.H., Chen, S.W., Deng, W.T., Liang, Z.T., Wang, Q., Wang, X.N.: Phys. Rev. C **77** (2008)
5. Baum, G., et al.: SLAC E80. Phys. Rev. Lett. **45**, 2000 (1980)

6. Baum, G., et al.: SLAC E130. Phys. Rev. Lett. **51**, 1135 (1983)
7. Ashman, J., et al.: European Muon Collaboration. Phys. Lett. B **206**, 364 (1988)
8. Ashman, J., et al.: European Muon Collaboration. Nucl. Phys. B **328**, 1 (1989)
9. For a recent review see e.g., C. A. Aidala, S. D. Bass, D. Hasch and G. K. Mallot, Rev. Mod. Phys. **85**, 655 (2013)
10. Klem, R.D., et al.: Phys. Rev. Lett. **36**, 929 (1976)
11. Dragoset, W.H., et al.: Phys. Rev. D **18**, 3939 (1978)
12. Adams, D.L., et al.: FNAL-E704 Collaboration. Phys. Lett. B **264**, 462 (1991)
13. For a short review, see e.g. Z. t. Liang and C. Boros, Int. J. Mod. Phys. A **15**, 927 (2000)
14. Lesnik, A., et al.: Phys. Rev. Lett. **35**, 770 (1975)
15. Bunce, G., et al.: Phys. Rev. Lett. **36**, 1113 (1976)
16. Bensinger, J., et al.: Phys. Rev. Lett. **50**, 313 (1983)
17. Gourlay, S.A., et al.: Phys. Rev. Lett. **56**, 2244 (1986)
18. For a recent review, see e.g., A. D. Krisch, Eur. Phys. J. A **31**, 417 (2007)
19. O'Fallon, J.R., et al.: Phys. Rev. Lett. **39**, 733 (1977)
20. Crabb, D.G., et al.: Phys. Rev. Lett. **41**, 1257 (1978)
21. Cameron, P.R., et al.: Phys. Rev. D **32**, 3070 (1985)
22. Bjorken, J.D., Sci, N.A.T.O.: Ser. B **197**, 1 (1987)
23. Gyulassy, M., McLerran, L.: Nucl. Phys. A **750**, 30 (2005)
24. Adams, J., et al.: STAR Collaboration. Nucl. Phys. A **757**, 102 (2005)
25. Liang, Z.T., Lisa, M.A., Wang, X.N.: Nucl. Phys. News **30**(2), 10–16 (2020)
26. Ackermann, K.H., et al.: STAR Collaboration. Phys. Rev. Lett. **86**, 402 (2001)
27. Adams, J., et al.: Phys. Rev. Lett. **92** (2004)
28. Cederwall, B., et al.: Phys. Rev. Lett. **72**, 3150 (1994)
29. Back, B.B. et al.: [PHOBOS Collaboration], arXiv:nucl-ex/0301017; Nouicer, R. et al.: [PHOBOS Collaboration], J. Phys. G **30**, S1133 (2004)
30. Bjorken, J.D.: Phys. Rev. D **27**, 140 (1983)
31. Levai, P., Muller, B., Wang, X.N.: Phys. Rev. C **51**, 3326 (1995)
32. Wang, X.N., Gyulassy, M.: Phys. Rev. D **44**, 3501 (1991)
33. Wang, X.N.: Phys. Rep. **280**, 287 (1997)
34. Brodsky, S.J., Gunion, J.F., Kuhn, J.H.: Phys. Rev. Lett. **39**, 1120 (1977)
35. Liang, Z.T., Song, J., Upsal, I., Wang, Q., Xu, Z.B.: Chin. Phys. C **45**(1), 014102 (2021) https://doi.org/10.1088/1674-1137/abc065
36. Sjöstrand, T., et al.: Comput. Phys. Commun. **191**, 159 (2015)
37. Liang, Z.T., Meng, T.C.: Phys. Z. A **344**, 171 (1992)
38. Einstein, A., de Haas, W.J.: DFG Verhandlungen **17**, 152 (1915)
39. Barnett, S.J.: Rev. Mod. Phys. **7**, 129 (1935)
40. Mayer, M.G., Jensen, J.H.D.: Elementary Theory of Nuclear Shell Structure. Wiley, New York and Chapman Hall, London (1955)
41. Mayer, M.G.: Phys. Rev. **75**, 1969 (1949)
42. Haxel, O., Jensen, J.H.D., Suess, H.E.: Phys. Rev. **75**(11), 1766 (1949)
43. Brambilla, N., Pineda, A., Soto, J., Vairo, A.: Rev. Mod. Phys. **77**, 1423 (2005)
44. Boros, C., Liang, Z.T., Meng, T.C.: Phys. Rev. Lett. **70**, 1751 (1993); Boros, C., Liang, Z.T.: Phys. Rev. D **53**, 2279-2283 (1996)
45. Brodsky, S.J., Hwang, D.S., Schmidt, I.: Phys. Lett. B **530**, 99 (2002)
46. Liang, Z.T., Boros, C.: Phys. Rev. Lett. **79**, 3608 (1997)
47. Liang, Z.T., Meng, T.C.: Phys. Rev. D **42**, 2380 (1990)
48. Bass, S.D.: Rev. Mod. Phys. **77**, 1257 (2005)
49. D'Alesio, U., Murgia, F.: Prog. Part. Nucl. Phys. **61**, 394 (2008)
50. Liang, Z.T.: Plenary talk at the 20th International Conference on Spin Physics, Oct. 2014, Beijing; published in Int. J. Mod. Phys. Conf. Ser. **40**, 1660008 (2016)
51. Chen, K.B., Wei, S.Y., Liang, Z.T.: Front. Phys. (Beijing) **10**, 6 (2015)
52. Gyulassy, M., Wang, X.N.: Nucl. Phys. B **420**, 583 (1994)
53. Weldon, H.A.: Phys. Rev. D **26**, 1394 (1982)

54. Wang, X.-N.: Phys. Lett. B **485**, 157 (2000)
55. Heiselberg, H., Wang, X.-N.: Nucl Phys. **B462**, 389 (1996)
56. Biró, T.S., Müller, B.: Nucl. Phys. A **561**, 477 (1993)
57. Muller, B., Rajagopal, K.: Eur. Phys. J. C **43**, 15 (2005)
58. Li, S.Y., Wang, X.N.: Phys. Lett. B **527**, 85 (2002)
59. Betz, B., Gyulassy, M., Torrieri, G.: Phys. Rev. C **76** (2007)
60. Becattini, F., Piccinini, F., Rizzo, J.: Phys. Rev. C **77** (2008)
61. Deng, W.T., Huang, X.G.: Phys. Rev. C **93**, 6 (2016)
62. Fang, R.H., Pang, L.G., Wang, Q., Wang, X.N.: Phys. Rev. C **94**, 2 (2016)
63. Pang, L.G., Petersen, H., Wang, Q., Wang, X.N.: Phys. Rev. Lett. **117** (2016)
64. Li, H., Petersen, H., Pang, L.G., Wang, Q., Xia, X.L., Wang, X.N.: Nucl. Phys. A **967**, 772 (2017)
65. Xia, X.L., Li, H., Tang, Z.B., Wang, Q.: Phys. Rev. C **98** (2018)
66. Florkowski, W., Kumar, A., Ryblewski, R.: Phys. Rev. C **98**, 4 (2018)
67. Wei, D.X., Deng, W.T., Huang, X.G.: Phys. Rev. C **99**, 1 (2019)
68. Zhang, J.J., Fang, R.H., Wang, Q., Wang, X.N.: Phys. Rev. C **100**, 6 (2019)
69. Wu, H.Z., Zhang, J.J., Pang, L.G., Wang, Q.: Comput. Phys. Commun. (2019)
70. Weickgenannt, N., Speranza, E., Sheng, X., Wang, Q., Rischke, D.H.: arXiv:2005.01506 [hep-ph]
71. De Groot, S.R., Van Leeuwen, W.A., Van Weert, C.G.: Amsterdam, p. 417p. North-holland, Netherlands (1980)
72. Buskulic, D., et al.: ALEPH Collaboration. Phys. Lett. B **374**, 319 (1996)
73. Ackerstaff, K., et al.: OPAL Collaboration. Eur. Phys. J. C **2**, 49 (1998)
74. Ackerstaff, K., et al.: OPAL Collaboration. Phys. Lett. B **412**, 210 (1997)
75. Abreu, P., et al.: DELPHI Collaboration. Phys. Lett. B **406**, 271 (1997)
76. Ackerstaff, K., et al.: OPAL Collaboration. Z. Phys. C **74**, 437 (1997)
77. Xu, Q.H., Liu, C.X., Liang, Z.T.: Phys. Rev. D **63** (2001)
78. Chen, K.B., Yang, W.H., Wei, S.Y., Liang, Z.T.: Phys. Rev. D **94**, 3 (2016)
79. Chen, K.B., Yang, W.H., Zhou, Y.J., Liang, Z.T.: Phys. Rev. D **95**, 3 (2017)
80. Chen, K.B., Liang, Z.T., Song, Y.K., Wei, S.Y.: Phys. Rev. D **102**, 3 (2020)
81. Gatto, R.: Phys. Rev. **109**(2), 610 (1958)
82. Gustafson, G., Hakkinen, J.: Phys. Lett. B **303**, 350 (1993)
83. Boros, C., Liang, Z.T.: Phys. Rev. D **57**, 4491 (1998)
84. Liu, C.-X., Liang, Z.-T.: Phys. Rev. D **62** (2000)
85. Becattini, F., Karpenko, I., Lisa, M., Upsal, I., Voloshin, S.: Phys. Rev. C **95** (2017)
86. Xia, X.L., Li, H., Huang, X.G., Huang, H.Z.: Phys. Rev. C **100**, 1 (2019)
87. Tanabashi, M., et al.: Particle Data Group. Phys. Rev. D **98**, 3 (2018)
88. Voloshin, S.A.: nucl-th/0410089
89. Selyuzhenkov, I.V.: STAR Collaboration. Rom. Rep. Phys. **58**, 049 (2006). [nucl-ex/0510069]
90. Selyuzhenkov, I.V.: STAR Collaboration. J. Phys. G **32**, S557 (2006)
91. Selyuzhenkov, I.V.: STAR Collaboration. AIP Conf. Proc. **870**(1), 712 (2006)
92. Selyuzhenkov, I.V.: STAR Collaboration. J. Phys. G **34**, S1099 (2007)
93. Chen, J.H.: STAR Collaboration. J. Phys. G **34**, S331 (2007)
94. Abelev, B.I., et al.: STAR Collaboration. Phys. Rev. C **76** (2007)
95. Abelev, B.I., et al.: STAR Collaboration. Phys. Rev. C **77** (2008)
96. Adam, J., et al.: STAR Collaboration. Phys. Rev. C **98** (2018)
97. Acharya, S. et al.: [ALICE Collaboration], arXiv:1909.01281 [nucl-ex]
98. Zhou, C.: Nucl. Phys. A **982**, 559 (2019)
99. Acharya, S. et al.: [ALICE Collaboration], arXiv:1910.14408 [nucl-ex]
100. Ipp, A., Di Piazza, A., Evers, J., Keitel, C.H.: Phys. Lett. B **666**, 315 (2008)
101. Barros, C.D.C.Jr, Hama, Y.: Phys. Lett. B **699**, 74 (2011)
102. Xie, Y., Glastad, R.C., Csernai, L.P.: Phys. Rev. C **92**, 6 (2015)
103. Jiang, Y., Lin, Z.W., Liao, J.: Phys. Rev. C **94**(4), 044910 (2016) Erratum: [Phys. Rev. C **95**(4), 049904 (2017)]

104. Montenegro, D., Tinti, L., Torrieri, G.: Phys. Rev. D **96**(5), 056012 (2017) Addendum: [Phys. Rev. D **96**(7), 079901 (2017)]
105. Xie, Y., Wang, D., Csernai, L.P.: Phys. Rev. C **95**, 3 (2017)
106. Li, H., Pang, L.G., Wang, Q., Xia, X.L.: Phys. Rev. C **96**, 5 (2017)
107. Sun, Y., Ko, C.M.: Phys. Rev. C **96**, 2 (2017)
108. Yang, Y.G., Fang, R.H., Wang, Q., Wang, X.N.: Phys. Rev. C **97**, 3 (2018)
109. Pang, L.G., Petersen, H., Wang, X.N.: Phys. Rev. C **97**, 6 (2018)
110. Hirono, Y., Kharzeev, D.E., Sadofyev, A.V.: Phys. Rev. Lett. **121**, 14 (2018)
111. Sun, Y., Ko, C.M.: Phys. Rev. C **99**, 1 (2019)
112. Liang, Z.T.: plenary talk given at the19th International Conference on Ultra-Relativistic nucleus-nucleus collisions (Quark Matter 2006), Shanghai, China, November 14-20, 2006, published in J. Phys. G **34**, S323 (2007)
113. Wang, Q.: Plenary talk at 26th International Conference on Ultra-relativistic Nucleus-Nucleus Collisions (Quark Matter 2017), Chicago, Illinois, USA, February 5-11, 2017, published in Nucl. Phys. A **967**, 225 (2017)
114. Kharzeev, D.E., McLerran, L.D., Warringa, H.J.: Nucl. Phys. A **803**, 227 (2008). https://doi.org/10.1016/j.nuclphysa.2008.02.298, arXiv:0711.0950 [hep-ph]
115. Fukushima, K., Kharzeev, D.E., Warringa, H.J.: Phys. Rev. D **78** (2008). https://doi.org/10.1103/PhysRevD.78.074033, arXiv:0808.3382 [hep-ph]
116. Kharzeev, D.E., Son, D.T.: Phys. Rev. Lett. **106** (2011)
117. Sheng, X.L., Oliva, L., Wang, Q.: arXiv:1910.13684 [nucl-th]
118. Huang, X.G.: Rept. Prog. Phys. **79**, 7 (2016)
119. Kharzeev, D.E., Liao, J., Voloshin, S.A., Wang, G.: Prog. Part. Nucl. Phys. **88**, 1 (2016)
120. Florkowski, W., Ryblewski, R., Kumar, A.: Prog. Part. Nucl. Phys. **108** (2019)
121. Zhao, J., Wang, F.: Prog. Part. Nucl. Phys. **107**, 200 (2019)
122. Liu, Y.C., Huang, X.G.: Nucl. Sci. Tech. **31**(6), 56 (2020)
123. Wang, F.Q., Zhao, J.: Nucl. Sci. Tech. **29**(12), 179 (2018)
124. Huang, X.-G.: Plenary talk at the 28th International Conference on Ultra-Relativistic Nucleus-Nucleus Collisions (Quark Matter 2019), Wuhan, China, November 4 to 9, 2019
125. Liao, J.-F.: Plenary talk at the 28th International Conference on Ultra-Relativistic Nucleus-Nucleus Collisions (Quark Matter 2019), Wuhan, China, November 4 to 9, 2019
126. Xu, Z.: Plenary talk at the 28th International Conference on Ultra-Relativistic Nucleus-Nucleus Collisions (Quark Matter 2019), Wuhan, China, November 4 to 9, 2019
127. Heinz, U.W.: Phys. Rev. Lett. **51**, 351 (1983)
128. Elze, H.T., Gyulassy, M., Vasak, D.: Nucl. Phys. B **276**, 706 (1986)
129. Vasak, D., Gyulassy, M., Elze, H.T.: Annals Phys.(N.Y.) **173**, 462 (1987)
130. Zhuang, P., Heinz, U.W.: Ann. Phys. **245**, 311 (1996)
131. Gao, J.H., Liang, Z.T., Pu, S., Wang, Q., Wang, X.N.: Phys. Rev. Lett. **109** (2012)
132. Chen, J.W., Pu, S., Wang, Q., Wang, X.N.: Phys. Rev. Lett. **110**, 26 (2013)
133. Gao, J.H., Liang, Z.T., Wang, Q., Wang, X.N.: Phys. Rev. D **98**, 3 (2018)
134. Gao, J.h., Pang, J.Y., Wang, Q.: Phys. Rev. D **100**(1), 016008 (2019)
135. Gao, J.H., Liang, Z.T.: Phys. Rev. D **100**, 5 (2019)
136. Weickgenannt, N., Sheng, X.L., Speranza, E., Wang, Q., Rischke, D.H.: Phys. Rev. D **100**, 5 (2019)
137. Li, S., Yee, H.U.: Phys. Rev. D **100**, 5 (2019)
138. Gao, J.H., Ma, G.L., Pu, S., Wang, Q.: Nucl. Sci. Tech. **31**(9), 90 (2020)

Vorticity and Polarization in Heavy-Ion Collisions: Hydrodynamic Models

8

Iurii Karpenko

Abstract

Fluid dynamic approach is a workhorse for modelling collective dynamics in relativistic heavy-ion collisions. The approach has been successful in describing various features of the momentum distributions of hadrons produced in the heavy-ion collisions, such as p_T spectra and flow coefficients v_n. As such, the description of the phenomenon of polarization of Λ hyperons in heavy-ion collisions has to be incorporated into the hydrodynamic approach. We start this chapter by introducing different definitions of vorticity in relativistic fluid dynamics. Then we present a derivation of the polarization of spin 1/2 fermions in the relativistic fluid. The latter is directly applied to compute the spin polarization of the Λ hyperons, which are produced from the hot and dense medium, described with fluid dynamics. It is followed by a review of the existing calculations of global or local polarization of Λ hyperons in different hydrodynamic models of relativistic heavy-ion collisions. We particularly focus on the explanations of the collision energy dependence of the global Λ polarization from the different hydrodynamic models, the polarization component in the beam direction as well as on the origins of the global and local Λ polarization.

8.1 Introduction: Vorticities in a Fluid

Heavy-ion collisions at ultra-relativistic energies create a strongly interacting system characterized by extremely high temperature and energy density. For a large fraction of its lifetime, the system shows strong collective effects and can be described by relativistic hydrodynamics. In particular, the large elliptic flow observed in such collisions indicates that the created quasi-macroscopic system is strongly coupled

I. Karpenko (✉)
Czech Technical University in Prague, Břehová 7, 11519 Prague 1, Czech Republic
e-mail: yu.karpenko@gmail.com

© Springer Nature Switzerland AG 2021 247
F. Becattini et al. (eds.), *Strongly Interacting Matter under Rotation*,
Lecture Notes in Physics 987,
https://doi.org/10.1007/978-3-030-71427-7_8

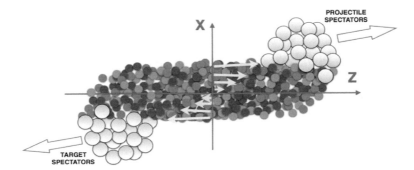

Fig. 8.1 Schematic view of the collision. Arrows indicate the flow velocity field. The $+\hat{y}$ direction is out of the page; both the orbital angular momentum and the magnetic field point into the page

and has an extremely low viscosity to entropy ratio. From the very success of the hydrodynamic description, one can also conclude that the system might possess an extremely high vorticity, likely the highest ever made under the laboratory conditions.

A simple estimate of the non-relativistic vorticity, defined as

$$\boldsymbol{\omega} = \frac{1}{2} \nabla \times \mathbf{v}, \tag{8.1}$$

[1]can be made based on a very schematic picture of the collision depicted in Fig. 8.1. As the projectile and target spectators move in the opposite direction with the velocity close to the speed of light, the z component of the collective velocity in the system close to the projectile spectators and that close to the target spectators are expected to be different. Assuming that this difference is a fraction of the speed of light, e.g. 0.1 (in units of the speed of light), and that the transverse size of the system is about 5 fm, one concludes that the vorticity in the system is of the order $0.02 \, \text{fm}^{-1} \approx 10^{22} \, \text{s}^{-1}$.

Unlike in classical hydrodynamics, where vorticity is the curl of the velocity field $\mathbf{v}$, several vorticities can be defined in relativistic hydrodynamics which can be useful in different applications (for more on that, we refer the reader to [1]). We list and discuss the different vorticity definitions as follows:

The Kinematical Vorticity

The kinematical vorticity is defined as

$$\omega_{\mu\nu} = \frac{1}{2}(d_\nu u_\mu - d_\mu u_\nu) = \frac{1}{2}(\partial_\nu u_\mu - \partial_\mu u_\nu), \tag{8.2}$$

where d_μ is a covariant derivative (different from the ordinary derivative ∂_μ) and u is the four-velocity field. This tensor includes both the acceleration A and the relativistic

---

[1]Sometimes the vorticity is defined without the factor $1/2$; we use the definition that gives the vorticity of the fluid rotating as a whole with a constant angular velocity Ω, to be $\omega = \Omega$.

extension of the angular velocity pseudo-vector ω_μ in the usual decomposition of an antisymmetric tensor field into a polar and pseudo-vector fields:

$$\omega_{\mu\nu} = \epsilon_{\mu\nu\rho\sigma}\omega^\rho u^\sigma + \frac{1}{2}(A_\mu u_\nu - A_\nu u_\mu)$$
$$A_\mu = 2\omega_{\mu\nu}u^\nu = u^\nu d_\nu u_\mu \equiv Du_\mu$$
$$\omega_\mu = -\frac{1}{2}\epsilon_{\mu\rho\sigma\tau}\,\omega^{\rho\sigma},u^\tau \tag{8.3}$$

where $\epsilon_{\mu\nu\rho\sigma}$ is the Levi-Civita symbol. Using the transverse (to u) projector:

$$\Delta^{\mu\nu} \equiv g^{\mu\nu} - u^\mu u^\nu,$$

and the usual definition of the orthogonal derivative

$$\nabla_\mu \equiv \Delta_\mu^\alpha d_\alpha = d_\mu - u_\mu D,$$

where $D = u^\alpha d_\alpha$ is a so-called co-moving derivative, it is convenient to define also a transverse kinematical vorticity as

$$\omega_{\mu\nu}^\Delta = \Delta_{\mu\rho}\Delta_{\nu\sigma}\omega^{\rho\sigma} = \frac{1}{2}(\nabla_\nu u_\mu - \nabla_\mu u_\nu). \tag{8.4}$$

Using the above definition in the decomposition (8.3), it can be shown that

$$\omega_{\mu\nu}^\Delta = \epsilon_{\mu\nu\rho\sigma}\omega^\rho u^\sigma \tag{8.5}$$

that is ω^Δ is the tensor formed with the angular velocity vector only. As we will show in the next subsection, only ω^Δ shares the "conservation" property of the classical vorticity for an ideal barotropic fluid.

The T-Vorticity

This is defined as

$$\Omega_{\mu\nu} = \frac{1}{2}\left[\partial_\nu(Tu_\mu) - \partial_\mu(Tu_\nu)\right], \tag{8.6}$$

and it is particularly useful for a relativistic uncharged fluid, such as the QCD plasma formed in nuclear collisions at very high energy. This is because from the basic thermodynamic relations when the temperature is the only independent thermodynamic variable, the ideal relativistic equation of motion $(\varepsilon + p)A_\mu = \nabla_\mu p$ can be recast in the simple form:

$$u^\mu \Omega_{\mu\nu} = \frac{1}{2}(TA_\nu - \nabla_\nu T) = 0 \tag{8.7}$$

The above (8.7) is also known as Carter–Lichnerowicz equation [2] for an ideal uncharged fluid and it entails conservation properties which do not hold for the kinematical vorticity. This can be better seen in the the language of differential forms, rewriting the definition of the T-vorticity as the exterior derivative of a the vector field (1-form) Tu, that is $\Omega = \mathbf{d}(Tu)$. Indeed, Eq. (8.7) implies—through the Cartan identity—that the Lie derivative of Ω along the vector field u vanishes, that is

$$\mathcal{L}_u \, \Omega = u \cdot \mathbf{d}\Omega + \mathbf{d}(u \cdot \Omega) = 0 \qquad (8.8)$$

because Ω is itself the external derivative of the vector field Tu and $\mathbf{dd} = 0$. Equation (8.8) states that the T-vorticity is conserved along the flow and, thus, if it vanishes at an initial time, it will remain so at all times. This can be made more apparent by expanding the Lie derivative definition in components:

$$(\mathcal{L}_u \, \Omega)^{\mu\nu} = D\Omega^{\mu\nu} - \partial_\sigma u^\mu \Omega^{\sigma\nu} - \partial_\sigma u^\nu \Omega^{\sigma\mu} = 0. \qquad (8.9)$$

The above equation is in fact a differential equation for Ω precisely showing that if $\Omega = 0$ at the initial time then $\Omega \equiv 0$. Thereby, the T-vorticity has the same property as the classical vorticity for an ideal barotropic fluid, such as the Kelvin circulation theorem, so the integral of Ω over a surface enclosed by a circuit co-moving with the fluid will be a constant.

One can write the relation between the T-vorticity and the kinematical vorticity by expanding the definition (8.6):

$$\Omega_{\mu\nu} = \frac{1}{2} \left[(\partial_\nu T) \, u_\mu - (\partial_\mu T) \, u_\nu \right] + T\omega_{\mu\nu}$$

implying that the double-transverse projection of Ω:

$$\Delta_{\mu\rho} \Delta_{\nu\sigma} \Omega^{\rho\sigma} \equiv \Omega^\Delta_{\mu\nu} = T\omega^\Delta_{\mu\nu}.$$

Hence, the tensor ω^Δ shares the same conservation properties of Ω^Δ, namely it vanishes at all times if it is vanishing at the initial time. Conversely, the mixed projection of the kinematical vorticity:

$$u^\rho \omega_{\rho\sigma} \Delta^{\sigma\nu} = \frac{1}{2} A_\sigma$$

does not. It then follows that for an ideal uncharged fluid with $\omega^\Delta = 0$ at the initial time, the kinematical vorticity is simply

$$\omega_{\mu\nu} = \frac{1}{2}(A_\mu u_\nu - A_\nu u_\mu). \qquad (8.10)$$

The Thermal Vorticity
The thermal vorticity is defined as [3]

$$\varpi_{\mu\nu} = \frac{1}{2}(\partial_\nu \beta_\mu - \partial_\mu \beta_\nu), \qquad (8.11)$$

where β is the temperature four-vector. This vector is defined as $(1/T)u$ once a four-velocity u, that is a hydrodynamical frame, is introduced, but it can also be taken as a primordial quantity to define a velocity through $u \equiv \beta/\sqrt{\beta^2}$ [4]. The thermal vorticity features two important properties: it is adimensional in natural units (in cartesian coordinates) and it is the actual constant vorticity at the global equilibrium with rotation [5] for a relativistic system, where β is a Killing vector field whose expression in Minkowski space–time is $\beta_\mu = b_\mu + \varpi_{\mu\nu}x^\nu$ being b and ϖ constant. In this case, the magnitude of thermal vorticity is—with the natural constants restored—simply $\hbar\omega/k_B T$ where ω is a constant angular velocity. In general, (replacing ω with the classical vorticity defined as the curl of a proper velocity field) it can be readily realized that the adimensional thermal vorticity is a tiny number for most hydrodynamical systems, though it can be significant for the plasma formed in relativistic nuclear collisions.

8.2 Polarization of Particles in the Fluid

Particles produced in relativistic heavy-ion collisions are expected to be polarized in peripheral collisions because of angular momentum conservation. At finite impact parameter, the QGP has a finite angular momentum perpendicular to the reaction plane and some fraction thereof may be converted into a spin of final-state hadrons. Therefore, measured particles may show a finite mean *global* polarization along the angular momentum direction. In a fluid at local thermodynamic equilibrium, the polarization can be calculated by using the principle of quantum statistical mechanics, that is assuming that the spin degrees of freedom are at local thermodynamical equilibrium at the hadronization stage, much the same way as the momentum degrees of freedom.

The crucial role in the calculation of the polarization for the fluid produced in relativistic heavy-ion collisions is played by the density operator. For a system at Local Thermodynamic Equilibrium (LTE), this reads [4]

$$\widehat{\rho}_{\mathrm{LE}} = (1/Z) \exp\left[-\int_\Sigma d\Sigma_\mu \left(\widehat{T}^{\mu\nu}\beta_\nu - \zeta\widehat{j}^\mu\right)\right], \qquad (8.12)$$

where $\beta = (1/T)u$ is the four-temperature vector, $\widehat{T}$ the stress-energy tensor, $\widehat{j}$ a conserved current—like the baryon number—and $\zeta = \mu/T$. The mean value of a local operator $\widehat{O}(x)$ (such as, for instance the stress-energy tensor $\widehat{T}$, or the current $\widehat{j}$) at LTE:

$$O(x) = \mathrm{tr}(\widehat{\rho}_{\mathrm{LE}}\,\widehat{O}(x)) \qquad (8.13)$$

and if the fields β,ζ vary significantly over a distance which is much larger than the typical microscopic length (indeed the *hydrodynamic limit*), then they can be Taylor expanded in the density operator starting from the point x where the mean value

$O(x)$ is to be calculated. The leading terms in the exponent of (8.12) then become [4]

$$
\widehat{\rho}_{\text{LE}} \simeq \frac{1}{Z_{\text{LE}}} \exp\left[-\beta_\nu(x)\widehat{P}^\nu + \xi(x)\widehat{Q} - \frac{1}{4}(\partial_\nu\beta_\lambda(x) - \partial_\lambda\beta_\nu(x))\widehat{J}_x^{\lambda\nu} \right.
$$
$$
\left. + \frac{1}{2}(\partial_\nu\beta_\lambda(x) + \partial_\lambda\beta_\nu(x))\widehat{L}_x^{\lambda\nu} + \nabla_\lambda\xi(x)\,\widehat{d}_x^\lambda \right], \qquad (8.14)
$$

where the last two terms with the shear tensor and the gradient of ζ are dissipative and vanish at equilibrium. The ∇_λ operator stands for

$$
\nabla_\lambda = \partial_\lambda - u_\lambda u \cdot \partial
$$

as usual in relativistic hydrodynamics. The term which is responsible for a non-vanishing polarization is the one involving the angular momentum-boosts operators $\widehat{J}_x$.

The polarization of particles in a fluid at LTE can in principle be obtained by calculating matrices like

$$
W_{\sigma,\sigma'} = \text{tr}(\widehat{\rho}_{\text{LE}}a^\dagger(p)_\sigma a(p')_{\sigma'}),
$$

where $a(p)_\sigma$ are the destruction operators of final-state particles of four-momentum p and σ is the spin state index. Nevertheless, the exact calculation of W is a difficult one even with the expansion of $\widehat{\rho}_{\text{LE}}$ and the mean polarization was obtained in Ref. [6] by means of a different method, involving the spin tensor and an *ansatz* about the form of the covariant Wigner function at LTE (see also [7]). As a result, the mean spin vector of $1/2$ particles with four-momentum p turns out to be

$$
S^\mu(x, p) = -\frac{1}{8m}(1 - f(x, p))\epsilon^{\mu\rho\sigma\tau} p_\tau \varpi_{\rho\sigma}, \qquad (8.15)
$$

where $f(x, p) = (1 + \exp[\beta(x) \cdot p - \mu(x)Q/T(x)] + 1)^{-1}$ is the Fermi–Dirac distribution and $\varpi(x)$ is the *thermal vorticity*, that is

$$
\varpi_{\mu\nu} = -\frac{1}{2}\left(\partial_\mu\beta_\nu - \partial_\nu\beta_\mu\right) \qquad (8.16)
$$

In the hydrodynamic picture of heavy-ion collisions, particles with a given momentum are produced across the entire particlization hypersurface (see the next Subsection for details). Therefore to calculate the relativistic mean spin vector of a given particle species with given momentum, one has to integrate the above expression over the particlization hypersurface Σ [6]:

$$
S^\mu(p) = \frac{\int d\Sigma_\lambda p^\lambda f(x, p)S^\mu(x, p)}{\int d\Sigma_\lambda p^\lambda f(x, p)}. \qquad (8.17)
$$

With the expression for S^μ, Eq. (8.17) can be expanded as follows:

$$S^\mu(p) = -\frac{1}{8m}\epsilon^{\mu\rho\sigma\tau} p_\tau \frac{\int d\Sigma_\lambda p^\lambda f(x,p)(1 - f(x,p))\varpi_{\rho\sigma}}{\int d\Sigma_\lambda p^\lambda f(x,p)}. \tag{8.18}$$

The mean (i.e. momentum average) spin vector of all particles of given species can be expressed as

$$S^\mu = \frac{1}{N} \int \frac{d^3\mathrm{p}}{p^0} \int d\Sigma_\lambda p^\lambda f(x,p) S^\mu(x,p), \tag{8.19}$$

where $N = \int \frac{d^3\mathrm{p}}{p^0} \int d\Sigma_\lambda p^\lambda f(x,p)$ is the average number of particles produced at the particlization surface.

In the experiment, the Λ polarization is measured in its rest frame, therefore one can derive the expression for the mean polarization vector in the rest frame from Eq. (8.19) taking into account Lorentz invariance of most of the terms in it:

$$S^{*\mu} = \frac{1}{N} \int \frac{d^3\mathrm{p}}{p^0} \int d\Sigma_\lambda p^\lambda f(x,p) S^{*\mu}(x,p), \tag{8.20}$$

where asterisk denotes a quantity in the rest frame of particle.

Equations 8.18 and 8.20 have been used in all numerical calculations of polarization, either based on the hydrodynamic model discussed in the following subsections or transport approaches (next Section), and a good agreement with the data is observed. A crucial feature of Eq. 8.18, and more in general of this effect, is that it predicts an almost equal polarization of particles and anti-particles (if quantum statistics effect are not important) for it is a statistical thermodynamic effect driven by local equilibrium and not by an external C-odd field like the electromagnetic field. This distinctive feature is confirmed—modulo small deviations—by the experimental measurements described in Chap. 10.

Non-relativistic Limit of Eq. 8.15

It is instructive to check that Eq. (8.15) yields, in the non-relativistic and global equilibrium limit, the formulae obtained in the first part of this Section. First of all, at low momentum, in Eq. (8.15), one can keep only the term corresponding to $\tau = 0$ and $p_0 \simeq m$, so that $S^0 \simeq 0$ and

$$S^\mu(x,p) \simeq -\epsilon^{\mu\rho\sigma 0} \frac{1 - f(x,p)}{8} \varpi_{\rho\sigma}. \tag{8.21}$$

Then, the condition of global equilibrium makes the thermal vorticity field constant and equal to the ratio of a constant angular velocity ω and a constant temperature T [5] that is

$$-\frac{1}{2}\epsilon^{ijk0}\varpi_{jk} = \frac{1}{T_0}\omega^i. \tag{8.22}$$

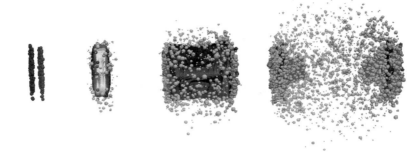

Fig. 8.2 Different stages of relativistic heavy-ion collision. From left to right: (1) two Lorentz-contracted nuclei right before the collision, (2) formation of dense, hydrodynamically expanding matter at around 1–2 fm/c after the collision, (3) hydrodynamic expansion of the dense core, surrounded by hadronic corona (the particles on the plot represent individual hadrons), (4) final-state hadronic interactions and decoupling of the fireball. Images taken from an animation by MADAI (http://madai.phy.duke.edu/indexaae2.html?page_id=503)

Finally, in the Boltzmann statistics limit $1 - n_F \simeq 1$ and one finally gets the spin 3-vector as

$$S(x, p) \simeq \frac{1}{4} \frac{\omega}{T}. \tag{8.23}$$

8.3   Hydrodynamic Modelling of Heavy-Ion Collisions

Let us start the section by outlining the established paradigm of hydrodynamic modelling of heavy-ion collisions. Hydrodynamic approximation is not used to describe all stages of a heavy-ion collision; instead, a multi-component approach is generally adopted in the field. The approach also reflects different dominant physics processes, which happen at different stages of the heavy-ion collision.

The first stage of heavy-ion collision comprises the primary nucleon–nucleon scatterings, which—at top RHIC or LHC energy—take less than a fraction of fm/c, due to a strong Lorentz contraction of the incoming nuclei. At this stage, a dense parton (at lower energies—hadron) system is formed. Within the first fermi/c, the system is assumed to reach enough degree of local equilibration, so that the subsequent evolution is described by relativistic hydrodynamics of ideal or viscous fluid.

The modelling of the next, hydrodynamic stage of collision became more sophisticated over the last decades. Successful interpretation of the early results from heavy-ion collisions at the RHIC collider within the hydrodynamic picture and the associated discovery of the nearly perfect fluid at RHIC led to a boom in hydrodynamic modelling. The simulations evolved from 1+1D to 2+1D ideal fluid to 2+1D and 3+1D viscous fluid approximation.

As the fireball expands, its density decreases, and so the mean free path of the constituents of the medium becomes larger. When the mean free path becomes comparable to the size of the fireball, the hydrodynamic picture does not apply anymore.

At this point, a so-called particlization (see [8] for more details) takes place: the fluid medium decouples into particles (hadrons). In the state-of-the-art models, the process of particlization typically takes place at a hypersurface of constant temperature or constant local rest frame energy density. Such three-dimensional hypersurface in four-dimensional space–time is reconstructed from a full hydrodynamic solution evolved till large enough time. To convert the fluid dynamic degrees of freedom to hadrons, a Cooper–Frye prescription, first introduced in [9], is often used. Practically speaking, in some of the hydrodynamic models discussed below, the spin polarization is calculated at the hypersurface of particlization, using Eqs. 8.17, 8.19 or 8.20—which are nothing more but modified Cooper–Frye formulas. The latter makes the computation relatively straightforward. Other hydrodynamic models take a simpler way to compute the spin polarization on a hypersurface of constant proper time $\tau = const$, even if it does not coincide with the hypersurface of particlization in the model.

However, the cross-sections of hadron scatterings are still not small right after the particlization. Therefore, both inelastic scatterings—which change the composition of hadrons in the event—and elastic reactions which only change hadron's momenta take place. An effective moment when the inelastic reaction ceases is known as a *chemical freeze-out*, whereas the moment where also elastic scatterings cease is known as *kinetic freeze-out*. As those processes happen gradually, the post-hydrodynamic phase is often modelled using a hadronic cascade, sometimes called a hadronic afterburner.

8.4 Hydrodynamic Calculations at $\sqrt{s_{\mathrm{NN}}} = 7 \dots 62$ GeV

Hydrodynamic modelling of heavy-ion collisions at very high energies, such as $\sqrt{s_{\mathrm{NN}}} = 200$ GeV at RHIC or $\sqrt{s_{\mathrm{NN}}} = 2.76, 5.02$ TeV at the LHC, can be numerically simplified by taking into account a strong Lorentz contraction of the colliding nuclei. This practically means that as long as observables in the central rapidity slice $y \approx 0$ are concerned, such as transverse momentum distributions of produced hadrons, their flow coefficients $v_n(p_T)$ and the initial state for the hydrodynamic expansion at $t = t_0$ can be approximated by a thin disk with thickness $z_0 \approx t_0$, with initial longitudinal flow $v_z = z/t_0$. The hydrodynamic solution will then have a symmetry with respect to Lorentz boosts in the longitudinal direction. Then the dynamics in the longitudinal direction can be integrated out analytically, leaving only the transverse expansion to the numerics. Likewise, widely used initial state models, such as CGC (Colour Glass Condensate), IP-Glasma and Monte Carlo Glauber, only evaluate the initial energy/entropy density profiles in the direction, transverse to the beam axis.

Modelling of relativistic heavy-ion collisions at collision energies $\sqrt{s_{\mathrm{NN}}}$ from a few to a hundred GeV is more challenging as compared to the high-energy regime. Many of the hydrodynamical and hybrid models used to model collisions at top RHIC and LHC energies are not directly applicable to collisions at the lower energies. The simplifying approximations of boost invariance and zero net baryon density are not valid, and different kinds of non-equilibrium effects play a larger role.

At such collision energies, the colliding nuclei do not resemble thin disks because of a weaker Lorentz contraction; also, the partonic models of the initial state (CGC, IP-Glasma) gradually lose their applicability in this regime. The longitudinal boost invariance is not a good approximation anymore, therefore one needs to simulate a three-dimensional hydrodynamic expansion.

Historically, the first full-fledged hydrodynamic model applied to study the polarization of hadrons—Λ hyperons—in heavy-ion collisions at $\sqrt{s_{\text{NN}}} = 7\ldots62$ GeV is UrQMD+vHLLE model [10]. The second hydrodynamic calculation of Λ polarization for this collision energy range was performed in PICR model [11]. More recently, Λ polarization was calculated in three-Fluid Dynamics (3FD) model [12]. We proceed by describing the details of the abovementioned hydrodynamic models.

Initial states in the hydrodynamic calculations. In the UrQMD+vHLLE calculation, the UrQMD string/hadronic cascade is used to describe the primary collisions of the nucleons and to create the initial state of the hydrodynamical evolution. The two nuclei are initialized according to Woods–Saxon distributions and the initial binary interactions proceed via string or resonance excitations, the former process being dominant in ultra-relativistic collisions (including the BES collision energies). All the strings are fragmented into hadrons before the transition to fluid phase (fluidization) takes place, although not all hadrons are yet fully formed at that time, i.e. they do not yet have their free-particle scattering cross-sections, and thus do not yet interact at all. The hadrons before conversion to fluid should not be considered physical hadrons, but rather marker particles to describe the flow of energy, momentum and conserved charges during the pre-equilibrium evolution of the system. The use of UrQMD to initialize the system allows us to describe some of the pre-equilibrium dynamics and dynamically generates event-by-event fluctuating initial states for hydrodynamical evolution.

The interactions in the pre-equilibrium UrQMD evolution are allowed until a hypersurface of constant Bjorken proper time $\tau_0 = \sqrt{t^2 - z^2}$ is reached, since the hydrodynamical code is constructed using the Milne coordinates (τ, x, y, η), where $\tau = \sqrt{t^2 - z^2}$ [13]. The UrQMD evolution, however, proceeds in Cartesian coordinates (t, x, y, z), and thus evolving the particle distributions to constant τ means evolving the system until large enough time t_l in such a way that the collisional processes and decays are only allowed in the domain $\sqrt{t^2 - z^2} < \tau_0$. The resulting particles on $t = t_l$ surface are then propagated backwards in time to the $\tau = \tau_0$ surface along straight trajectories to obtain an initial state for the hydrodynamic evolution.

The lower limit for the starting time of the hydrodynamic evolution depends on the collision energy according to

$$\tau_0 = 2R/\sqrt{(\sqrt{s_{\text{NN}}}/2m_N)^2 - 1}, \qquad (8.24)$$

which corresponds to the average time, when two nuclei have passed through each other, i.e. all primary nucleon–nucleon collisions have happened. This is the earliest possible moment in time, where approximate local equilibrium can be assumed.

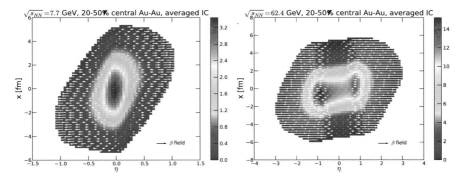

Fig. 8.3 Initial energy density profiles for the hydrodynamic stage with arrows depicting initial beta field superimposed. The hydrodynamic evolutions start from averaged initial state corresponding to 20–50% central Au-Au collisions at $\sqrt{s_{NN}} = 7.7$ (top row) and 62.4 GeV (bottom row)

To perform event-by-event hydrodynamics using fluctuating initial conditions, every individual UrQMD event is converted to an initial state profile. As mentioned, the hadron transport does not lead to an initial state in full local equilibrium, and the thermalization of the the system at $\tau = \tau_0$ has to be artificially enforced. The energy and momentum of each UrQMD particle at τ_0 is distributed to the hydrodynamic cells ijk assuming Gaussian density profiles

$$\Delta P_{ijk}^{\alpha} = P^{\alpha} \cdot C \cdot \exp\left(-\frac{\Delta x_i^2 + \Delta y_j^2}{R_{\perp}^2} - \frac{\Delta \eta_k^2}{R_{\eta}^2} \gamma_{\eta}^2 \tau_0^2\right) \tag{8.25}$$

$$\Delta N_{ijk}^0 = N^0 \cdot C \cdot \exp\left(-\frac{\Delta x_i^2 + \Delta y_j^2}{R_{\perp}^2} - \frac{\Delta \eta_k^2}{R_{\eta}^2} \gamma_{\eta}^2 \tau_0^2\right), \tag{8.26}$$

where Δx_i, Δy_j and $\Delta \eta_k$ are the differences between particle's position and the coordinates of the hydrodynamic cell $\{i, j, k\}$, and $\gamma_{\eta} = \cosh(y_p - \eta)$ is the longitudinal Lorentz factor of the particle as seen in a frame moving with the rapidity η. The normalization constant C is calculated from the condition that the discrete sum of the values of the Gaussian in all neighbouring cells equals one. The resulting ΔP^{α} and ΔN^0 are transformed into Milne coordinates and added to the energy, momentum and baryon number in each cell. This procedure ensures that in the initial transition from transport to hydrodynamics, the energy, momentum and baryon number are conserved.

In Fig. 8.17, the initial energy density profiles are visualized for two selected collision energies: $\sqrt{s_{NN}} = 7.7$ and 62.4 GeV. To produce this figure, two single hydrodynamic calculations with averaged initial conditions from 100 initial UrQMD simulations each were run. At $\sqrt{s_{NN}} = 62.4$ GeV, because of the baryon transparency effect, the x, z components of beta vector at mid-rapidity are small and do not have a regular pattern, therefore the distribution of ϖ_{xz} in the hydrodynamic cells close to particlization energy density includes both positive and negative parts, as it is seen

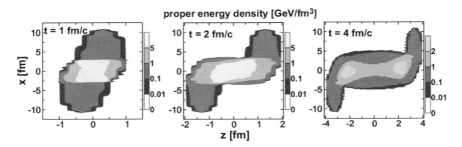

Fig. 8.4 Snapshots of the energy density profile in the $x - z$ plane in a 3-fluid dynamic simulation of a semi-central Au-Au collision at $\sqrt{s_{NN}} = 19.6$ GeV. Plots are taken from [15]

on the corresponding plot in the right column. At $\sqrt{s_{NN}} = 7.7$ GeV, baryon stopping results in a shear flow structure, which leads to the same (positive) sign of the ϖ_{xz}.

In the PICR model, the physics picture of the initial state is a Yang–Mills field, stretched between Lorentz-contracted streaks after impact [14]. Such initial state also produces torqued initial state of the fireball with finite angular momentum.

The 3-fluid dynamic model is somewhat different from the UrQMD+vHLLE and PICR models. In the 3FD, the evolution starts with two nuclei right before the moment of impact, which are represented by two blobs of cold baryon-rich fluid [16]. The process of nucleus–nucleus collision is then modelled as an inter-penetration of the baryon-rich fluids, which leads to friction between the fluids. The fluids lose energy and momentum via the friction terms, which leads to a creation of the third fluid, which is baryon-free. Similarly to the UrQMD+vHLLE or PICR models, the friction between the baryon-rich fluids leads to a total energy density profile which is tilted in the $x - z$ plane, as shown in Fig. 8.4. The friction also produces the velocity shear, which corresponds to a finite angular momentum of the participant system.

Hydrodynamic stage in UrQMD+vHLLE hybrid is simulated with a (3+1)-dimensional viscous hydrodynamical code vHLLE, which is described in full detail in Ref. [13]. The code solves the local energy-momentum conservation equations:

$$d_v T^{\mu\nu} = 0, \tag{8.27}$$

$$d_v N_{B,Q}^v = 0, , \tag{8.28}$$

where N_B^v and N_Q^v are the net baryon and electric charge currents, respectively, and we remind that d_v denotes a covariant derivative. The calculation[2] is done in Milne coordinates (τ, x, y, η), where $\tau = \sqrt{t^2 - z^2}$ and $\eta = 1/2 \ln[(t + z)/(t - z)]$.

In the Israel–Stewart framework of causal dissipative hydrodynamics [17], the dissipative currents are independent variables. For the calculations of Λ polarization, $\zeta/s = 0$ is set in the UrQMD+vHLLE calculation. The code works in the Landau

---

[2]Typical grid spacing used in the calculations: $\Delta x = \Delta y = 0.2$ fm, $\Delta \eta = 0.05 - 0.15$ and timestep $\Delta \tau = 0.05 - 0.1$ fm/c depending on the collision energy. A finer grid with $\Delta x = \Delta y = 0.125$ fm was taken to simulate peripheral collisions.

frame, where the energy diffusion flow is zero, and the baryon and charge diffusion currents are neglected for simplicity, which is equivalent to zero heat conductivity. For the shear-stress evolution we choose the relaxation time $\tau_\pi = 5\eta/(Ts)$ and the coefficient $\delta_{\pi\pi} = 4/3\tau_\pi$, and approximate all the other higher-order coefficients by zero. The following evolution equations are solved for the shear-stress tensor $\pi^{\mu\nu}$:

$$\langle u^\gamma d_\gamma \pi^{\mu\nu}\rangle = -\frac{\pi^{\mu\nu} - \pi_{NS}^{\mu\nu}}{\tau_\pi} - \frac{4}{3}\pi^{\mu\nu}\partial_{;\gamma}u^\gamma, \tag{8.29}$$

where the brackets denote the traceless and orthogonal to u^μ part of the tensor and $\pi_{NS}^{\mu\nu}$ is the Navier–Stokes value of the shear-stress tensor.

Another necessary ingredient for the hydrodynamic stage is the Equation of State (EoS) of the medium. In UrQMD+vHLLE, the chiral model EoS [18] features correct asymptotic degrees of freedom, i.e. quarks and gluons in the high temperature and hadrons in the low-temperature limits, crossover-type transition between confined and deconfined matter for all values of μ_B and qualitatively agrees with lattice QCD data at $\mu_B = 0$.

Different from that, both PICR and 3FH models feature ideal fluid approximation. The hydrodynamic evolution is simulated with the Relativistic Particle-in-Cell (PICR) method. Both in the initial state and subsequent CFD simulation, a classic 'Bag Model' EoS was applied: $P = c_0^2 e^2 - \frac{4}{3}B$, with constant $c_0^2 = \frac{1}{3}$ and a fixed Bag constant B. The energy density takes the form: $e = \alpha T^4 + \beta T^2 + \gamma + B$, where α, β and γ are constants arising from the degeneracy factors for (anti-)quarks and gluons.

Final conditions for the hydrodynamic stage. In the UrQMD+vHLLE hybrid, the fluid-to-particle transition, or particlization, is performed using the conventional Cooper–Frye prescription [9]. The Cooper–Frye prescription is applied at a hypersurface of constant local rest frame energy density. The hadrons, generated at the particlization, are then re-scattered with the UrQMD cascade. This particlization hypersurface is reconstructed during the hydrodynamic evolution based on the criterion of a fixed energy density $\epsilon = \epsilon_{se}$ and using the Cornelius routine [8]. The default value for the particlization energy density is $\epsilon_{sw} = 0.5$ GeV/fm^3, which in the chiral model EoS corresponds to $T \approx 175$ MeV at $\mu_B = 0$. At this energy density, the crossover transition is firmly on the hadronic side, but the density is still a little higher than the chemical freeze-out energy density suggested by the thermal models (for the topic of thermal models, we refer the reader to [19]).

In the PICR model, the particlization is set to happen at a fixed time t in the laboratory frame.

As given by the Cooper–Frye prescription, the hadron distribution on each point of the hypersurface is

$$p^0\frac{d^3N_i(x)}{d^3p} = d\Sigma_\mu p^\mu f(p \cdot u(x), T(x), \mu_i(x)). \tag{8.30}$$

The phase-space distribution function f is usually assumed to be the one corresponding to a noninteracting hadron resonance gas in or close to the local thermal equilibrium.

In a standard calculation in UrQMD+vHLLE, the Cooper–Frye formula (8.30) is used as a probability density to sample ensembles of hadrons with the Monte Carlo method. Further on, the sampled hadrons are passed to the UrQMD cascade to simulate inelastic and elastic interactions in the dilute post-hydrodynamic stage. However, for the calculation of the polarization, the Monte Carlo hadron sampling is replaced with a direct calculation based on Eq. (8.17), applied on particlization surfaces from the event-by-event hydrodynamics. For that, one can realize that the formula for the mean polarization of spin 1/2 hadrons (8.18) looks similar to the Cooper–Frye formula, except for the factor $(1 - f(x, p))\varpi_{\rho\sigma}$ under the integral.

The 3-fluid dynamic model has again somewhat different final conditions for its hydrodynamic stage, as compared to UrQMD+vHLLE or PICR. The distributions of hadrons at the particlization are computed not with the Cooper–Frye but with Milekhin formula [20], and the criterion for the particlization is a fixed combined energy density of all 3 fluids in a given space–time point:

$$\varepsilon_{\text{tot}} = \left(T_{\text{proj}}^{00} + T_{\text{targ}}^{00} + T_{\text{B-free}}^{00} \right)_{\text{rest frame}} < \varepsilon_{\text{frz}} , \qquad (8.31)$$

The 3-fluid dynamic model does not feature the final-state hadronic cascade, therefore the particlization in 3FD is the same as the freeze-out.

It is important to note that all the abovementioned models had been tuned to reproduce the basic hadronic observables prior to the calculations of polarization. In particular, a reasonable reproduction of the experimental data—(pseudo)rapidity distributions, transverse momentum spectra and elliptic flow coefficients—has been achieved in UrQMD+vHLLE with the parameter values depending monotonically on the collision energy as it is shown in Table 8.1. This was obtained when the particlization energy density was fixed to $\epsilon_{\text{sw}} = 0.5\,\text{GeV/fm}^3$ for the whole collision

Table 8.1 Collision energy dependence of the UrQMD+vHLLE parameters chosen to reproduce the experimental data in the RHIC BES range: $\sqrt{s_{\text{NN}}} = 7.7 - 200\,\text{GeV}$

$\sqrt{s_{\text{NN}}}$ [GeV]	τ_0 [fm/c]	$R_\perp$ [fm]	R_η [fm]	η/s
7.7	3.2	1.4	0.5	0.2
8.8 (SPS)	2.83	1.4	0.5	0.2
11.5	2.1	1.4	0.5	0.2
17.3 (SPS)	1.42	1.4	0.5	0.15
19.6	1.22	1.4	0.5	0.15
27	1.0	1.2	0.5	0.12
39	0.9	1.0	0.7	0.08
62.4	0.7	1.0	0.7	0.08
200	0.4	1.0	1.0	0.08

Fig. 8.5 Coordinate system used for the components of Λ polarization vector. The reaction plane is xz, and y coordinate is opposite to the vector of the global angular momentum of the system, which points upwards and perpendicular to the reaction plane. The plot is taken from [3]

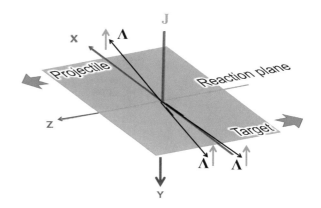

energy range. In 3FD, a typical choice of the freeze-out energy density to reproduce a broad set of experimental data is $\varepsilon_{frz} \simeq 0.2$ GeV/fm^3 [16].

Patterns of Λ Polarization in the UrQMD+vHLLE and PICR Models

Before starting to refer to the components of the spin polarization vector of Λ hyperons, it is worth to sketch the coordinate system used. The coordinate system is shown in Fig. 8.5: the x axis is parallel to the vector of the impact parameter of the heavy-ion collision; the z axis is parallel to the beam direction, thus xz is the so-called reaction plane. The y axis points perpendicular to the reaction plane and is directed opposite to the vector of the total angular momentum of the fireball.

Already an early calculation [1] of the Λ polarization vector as a function of the p_T of the Λ at mid-rapidity performed with 3+1 dimensional hydrodynamic code ECHO-QGP [21] has shown quite an assorted pattern, see Fig. 8.6. The ECHO-QGP calculation has been made for Au-Au collisions at fixed impact parameter $b = 11.6$ fm (corresponding to peripheral collisions) at the top RHIC energy $\sqrt{s_{NN}} = 200$ GeV; hydrodynamic calculations for top RHIC and LHC energies will be discussed in a next subsection. At large transverse momenta and at $|p_x| = |p_y|$, the polarization vector component along the beam axis, P^z (marked as Π_0^z on Fig. 8.6, also marked as $P_{||}$ on some of the plots below) has the largest amplitude. The component along the impact parameter, P^x (marked as Π_0^x on Fig. 8.6, also marked as P_b on some of the plots below) has a quadrupole pattern similar to P^z but with a smaller amplitude. However, because of symmetry of the system, the $\mathbf{p}_T$ integrated P^x and P^z integrate out to zero, and the only nonzero component remaining is P^y (Π_0^y on Fig. 8.6), which is opposite to the direction of the total angular momentum $\mathbf{J}$ of the fireball.

Very similar patterns for the components of the polarization vector were observed later in the UrQMD+vHLLE calculations for the Beam Energy Scan energies. On Fig. 8.7, the transverse momentum dependence of the components of Λ polarization vector is shown for 40–50% central Au-Au collisions (impact parameter range $b = 9.3 - 10.4$ fm) at collision energy $\sqrt{s_{NN}} = 19.6$ GeV, which is located in the middle of the Beam Energy Scan range. 1000 event-by-event hydrodynamic simulations were executed, then the event-averaged denominator and numerator of Eq. 8.17 were computed as a function of p_x and p_y, in order to produce Fig. 8.7.

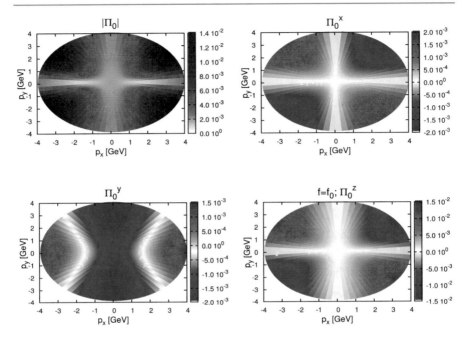

Fig. 8.6 Components and modulus of the Λ polarization vector as a function of p_T of the Λ, in ECHO-QGP simulation of Au-Au collisions with fixed impact parameter $b = 11.6$ fm at $\sqrt{s_{NN}} = 200$ GeV. The picture is taken from [1]

The polarization patterns in the $p_x p_y$ plane reflect the corresponding patterns of the components of thermal vorticity over the particlization hypersurface. In particular, it was found in [22] that the leading contribution to P^x stems from the term $\varpi_{tz} p_y$ in Eq. 8.15. In turn, ϖ_{tz} shown in left panel of Fig. 8.8 is a result of the interplay of $\partial_t \beta_z$ (acceleration of longitudinal flow and temporal gradients of temperature—conduction) and $\partial_z \beta_t$ (convection and conduction), according to Eq. (8.16). The P^y component has a leading contribution from the term $\varpi_{xz} p_0$ (which is also the only non-vanishing contribution at $p_T = 0$), and ϖ_{xz} has a rather uniform profile over the mid-rapidity slice of the freeze-out hypersurface, and the leading contribution to it comes from $\partial_x u_z$ (shear flow in z direction).

The PICR calculation provides a transverse momentum pattern of the y component of polarization (P^y), which is different from the UrQMD+vHLLE calculation, see the left panel of Figure 8.9. At the $p_y = 0$ line, the polarization changes sign between large $|p_x|$ and zero p_x. As for now, it is not clear why the transverse momentum patterns are different in the two models.

Centrality and Collision Energy Dependence of the Polarization

Figure 8.10 shows the collision energy dependence of the global polarization of Λ in UrQMD+vHLLE and PICR models. To follow recent STAR measurements, in UrQMD+vHLLE calculation, the 20–50% centrality bin was constructed by a correspondingly chosen range of impact parameters for the initial state UrQMD

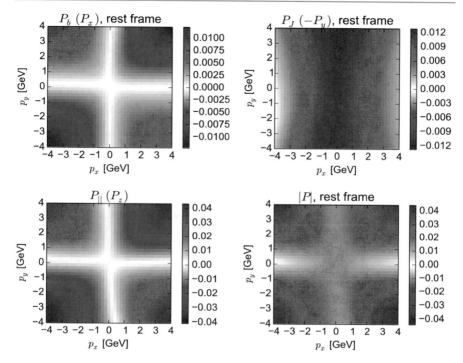

Fig. 8.7 Components of spin polarization vector of primary Λ baryons produced at mid-rapidity in UrQMD+vHLLE calculation for 40–50% central Au-Au collisions at $\sqrt{s_{NN}} = 19.6$ GeV. The polarization is calculated in the rest frame of Λ

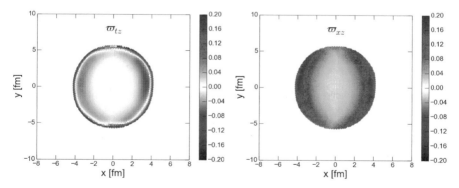

Fig. 8.8 Components of thermal vorticity ϖ_{tz} (left) and ϖ_{xz} (right) on the mid-rapidity slice of particlization hypersurface, projected on the xy plane. The UrQMD+vHLLE calculation with an averaged initial state corresponds to 40–50% central Au-Au collisions at $\sqrt{s_{NN}} = 19.6$ GeV

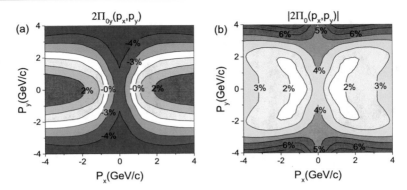

Fig. 8.9 The y component (left) and the modulus (right) of the Λ polarization at $p_z = 0$ in PICR model, for the Au+Au reaction at $\sqrt{s_{NN}} = 11.5$ GeV. The figure is in the frame of the Λ. The impact parameter $b = 0.7 b_m = 0.7 \times 2R$, where R is the radius of Au and $b_m = 2R$ is the maximum value of b. The freeze-out time is $6.25 = (2.5 + 4.75)$ fm/c, including 2.5 fm/c for initial state and 4.75 fm/c for hydro-evolution

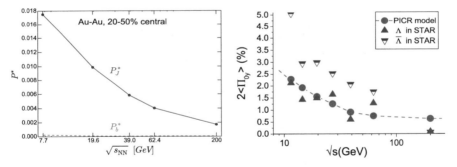

Fig. 8.10 Collision energy dependence of the $P_J \equiv P^y$ and $P_b \equiv P^x$ components of polarization vector of Λ, calculated in its rest frame, in UrQMD+vHLLE (left panel) and PICR (right panel) models for 20–50% central Au-Au collisions

calculation. We observe that the polarization component along **J**, the P^y decreases by about one order of magnitude as collision energy increases from $\sqrt{s_{NN}} = 7.7$ GeV to full RHIC energy, where it turns out to be consistent with an early calculation of the global hyperon polarization at the top RHIC energy in [1]. In the PICR calculation, the impact parameter $b_0 = 0.7$, which corresponds to centrality $c = 49\%$, was chosen to simulate the 20–50% centrality bin. For comparison, the data of Λ and $\bar{\Lambda}$ polarization from STAR (RHIC) were inserted into the right panel of Fig. 8.10 with blue triangle symbols.

In the UrQMD+vHLLE calculation, the fall of the out-of-plane component P^y is not directly related to a change in the out-of-plane component of the total angular momentum of the fireball. In fact, the total angular momentum increases as the collision energy increases, which can be seen on the top panel of Fig. 8.11. However, the total angular momentum is not an intensive quantity like polarization, so, to have a better benchmark, we took the ratio between the total angular momentum and the

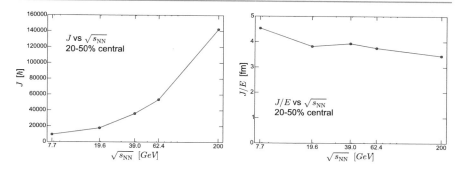

Fig. 8.11 Total angular momentum of the fireball (left) and total angular momentum scaled by the total energy of the fireball (right) as a function of collision energy, in UrQMD+vHLLE calculation for 20–50% central Au-Au collisions

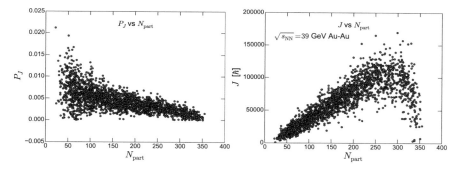

Fig. 8.12 Left: global Λ polarization at mid-rapidity as a function of the number of participating nucleons N_{part} in the initial state. Each point represents one hydrodynamic configuration in an ensemble of 2000 event-by-event calculations for 0–50% central Au-Au collisions at $\sqrt{s_{\text{NN}}} = 39$ GeV. Right: out-of-plane component of initial angular momentum versus number of participating nucleons N_{part} in the same calculation

total energy, J/E, which is shown in the bottom panel of the same figure. Yet, one can see that the J/E shows only a mild decrease as collision energy increases.

In Fig. 8.12, we show the distribution of the global polarization of Λ as a function of centrality (i.e. N_{part}), where each point corresponds to a hydrodynamic evolution with a given fluctuating initial condition characterized by N_{part}; in the right panel, one can see the corresponding distribution of total angular momentum **J**. We observe that the total angular momentum distribution has a maximum at a certain range of N_{part} and drops to zero for the most central events (where the impact parameter is zero) and most peripheral ones (where the system becomes small). In contrast to that, the polarization shows a steadily increasing trend towards peripheral collisions, where it starts to fluctuate largely from event to event because of the smallness of the fireball, a situation where the initial state fluctuations start to dominate in the hydrodynamic stage.

The overall trend of the impact parameter (and centrality) dependence of the global polarization is confirmed in the PICR calculation [11], see Figure 8.13. This

Fig. 8.13 The scaled impact parameter $b_0 = b/(2R)$ dependence of global polarization in PICR model for $\sqrt{s_{NN}}$ =11.5, 27.0 and 62.4 GeV

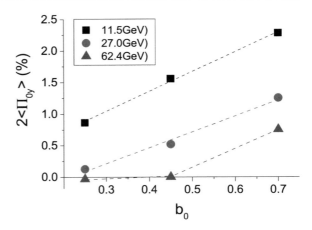

Fig. 8.14 Same as Fig. 8.10 but with bands added, which correspond to variations of the model parameters

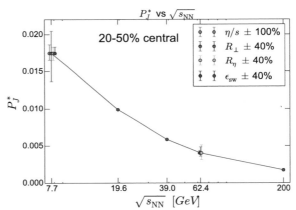

figure shows the global polarization in Au+Au collisions as a function of ratio of impact parameter b to Au's nuclear radius R, i.e. $b_0 = b/2R$. One could see that the polarization at different energies indeed approximately takes a linear increase with the increase of impact parameter, except for 62.4GeV due to the vanishing polarization signals at relatively central collisions. This linear dependence clearly indicates that the polarization in our model arises from the initial angular momentum. However, the polarization's linear dependence on b is somewhat different from the angular momentum's quadratic dependence on b. This is because the angular momentum L is an extensive quantity dependent on the system's mass, while the polarization Π is an intensive quantity (Fig. 8.14).

Finally, the energy dependence of Λ polarization in the 3FD calculation is similar to the other two hydrodynamic models, see Fig. 8.16, top panel. Contrary to the decrease of the Λ polarization at mid-rapidity with collision energy—a trend which will be discussed in the next subsection—the polarization of all Λ actually grows with the energy, as can be seen on the bottom panel of Fig. 8.16. The latter is explained in the 3FD calculation by the vorticity being pushed out to the fragmentation regions.

Fig. 8.15 Λ polarization and angular momentum within space–time rapidity $|\eta| < 0.5$ slab, the slope of directed flow dv_1/dy and elliptic flow v_2 in 3FD calculations of Au-Au collisions with fixed impact parameter $b = 8$ fm. The plot is taken from [15]

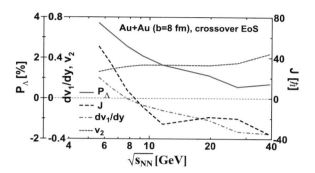

Fig. 8.16 Collision energy dependence of the components of polarization vector of Λ, calculated in its rest frame, in 3-fluid dynamics for 20–50% central Au-Au collisions

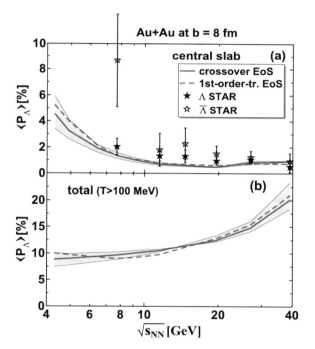

The 3FD calculation further demonstrates that the Λ polarization at the mid-space–time rapidity slab of matter $|\eta| < 0.5$ does not correlate with the total angular momentum of the slab, as shown in Fig. 8.15. In this calculation, the momentum of the slab changes sign around collision energy $\sqrt{s_{NN}} \approx 9$ GeV, whereas the polarization remains positive. Also, the polarization follows correlates neither with the slope of directed flow (which also changes sign around the same energy as the angular momentum) nor with the elliptic flow v_2.

Parameter dependence. As it has been mentioned above, the parameters of the model are set to monotonically depend on collision energy in order to approach the experimental data for basic hadronic observables. The question may arise whether the collision energy dependence of P^y is the result of an interplay of collision energy

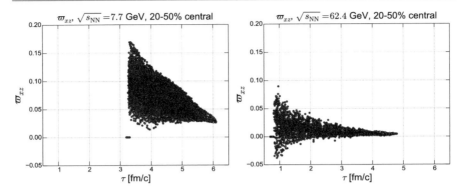

Fig. 8.17 Evolution of ϖ_{xz} in mid-rapidity ($|y| < 0.3$) slice of particlization surface, projected onto time axis (right column). The hydrodynamic evolutions start from averaged initial state corresponding to 20–50% central Au-Au collisions at $\sqrt{s_{NN}} = 7.7$ (top row) and 62.4 GeV (bottom row)

dependencies of the parameters. The UrQMD+vHLLE calculation argues that it is not the case: in Fig. 8.14 one can see how the p_T integrated polarization component P^y varies at two selected collision energies, $\sqrt{s_{NN}} = 7.7$ and 62.4 GeV, when the granularity of the initial state controlled by $R_\perp$, R_η parameters, shear viscosity to entropy ratio of the fluid medium η/s and particlization energy density ϵ_{sw} change. It turns out that a variation of $R_\perp$ within $\pm40\%$ changes P^y by $\pm20\%$, and a variation of R_η by $\pm40\%$ changes P^y by $\pm25\%$ at $\sqrt{s_{NN}} = 62.4$ GeV only. The variations of the remaining parameters affect P^y much less. We thus conclude that the observed trend in p_T integrated polarization is robust with respect to variations of parameters of the model.

Discussion on the Energy Dependence

Now we have to understand the excitation function of the p_T integrated P^y which is calculated in the hydrodynamic models. As it has been mentioned, P^y at low momentum (which contributes most to the p_T integrated polarization) has a dominant contribution proportional to $\varpi_{xz} p_0$. It turns out that the pattern and magnitude of ϖ_{xz} over the particlization hypersurface do change with collision energy.

The latter is demonstrated in Fig. 8.17, for two selected collision energies. For this purpose, two single hydrodynamic calculations were performed with averaged initial conditions from 100 initial state UrQMD simulations each. At $\sqrt{s_{NN}} = 62.4$ GeV, because of the baryon transparency effect,[3] the x, z components of beta vector at mid-rapidity are small and do not have a regular pattern, therefore the distribution of ϖ_{xz} in the hydrodynamic cells close to particlization energy density includes both positive and negative parts, as it is seen on the corresponding plot in the right column.

---

[3]The phenomenon of baryon transparency describes transporting the baryon charge of the colliding nuclei to the forward and backward rapidities. Opposite to that, baryon stopping implies that the baryon charge from the colliding nuclei is *stopped* around mi-rapidity.

At $\sqrt{s_{NN}} = 7.7$ GeV, baryon stopping results in a shear flow structure, which leads to the same (positive) sign of the ϖ_{xz}.

In the right column of Fig. 8.17, we plot the corresponding ϖ_{xz} distributions over the particlization hypersurfaces projected on the proper time axis. Generally speaking, hydrodynamic evolution tends to dilute the initial vorticities. One can see that longer hydrodynamic evolution at $\sqrt{s_{NN}} = 62.4$ GeV in combination with the smaller absolute value of average initial vorticity results in factor 4–5 smaller average absolute vorticities at late times for $\sqrt{s_{NN}} = 62.4$ GeV than for $\sqrt{s_{NN}} = 7.7$ GeV. This results in a corresponding difference in the momentum integrated polarization at these two energies, that is mostly determined by low-p_T Λ which are preferentially produced from the Cooper–Frye hypersurface at late times.

The explanation of the collision energy dependence of the Λ polarization from 3FD is similar to the one above. The dominant effect is that the central-slab vorticity

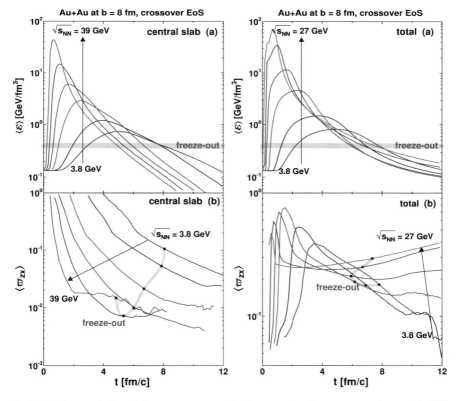

Fig. 8.18 Time evolution of the average energy density ε (top) and average ϖ_{zx} (bottom) in 3FD model [12], in the central slab of the fireball corresponding to the mid-rapidity (left) and in the whole system (right). The different curves represent the 3-fluid dynamic evolutions for Au-Au collisions at different collision energies from $\sqrt{s_{NN}} = 3.8$ GeV and up to $\sqrt{s_{NN}} = 39$ GeV. The calculation is performed with a fixed impact parameter $b = 8$ fm, which approximately corresponds to 30% centrality

at the freeze-out decreases with increasing collision energy because the vortical field is pushed out to the fragmentation regions. The evolution of the vortical field with collision energy and time at the central slab of the system is displayed in Fig.8.18, bottom left. In addition to that, similarly for the UrQMD+vHLLE calculations, the ϖ_{xz} decreases with time at any given collision energy, and the freeze-out time in 3FD actually decreases with collision energy, in the range $\sqrt{s_{NN}} = 3.8 \ldots 27$ GeV.

To summarize: in hydrodynamic models, the effect of hyperon polarization emerges in non-central collisions, where the angular momentum of the fireball is finite. However, the polarization does not (linearly) scale with the angular momentum of the system. The collision energy dependence in the hydrodynamic calculations is consistent with the experimental measurements by STAR collaboration in the Beam Energy Scan program at RHIC collider.

8.5   Hydrodynamic Calculations at $\sqrt{s_{NN}} = 200$ and 2760 GeV

As it was shown in the previous subsection, the mean, i.e. momentum averaged, polarization of Λ hyperons at mid-rapidity decreases with increasing collision energy. In $\sqrt{s_{NN}} = 200$ GeV Au-Au collisions at RHIC, the mean polarization is less than 0.3% [23], and at the LHC energies, it is presumably smaller, below the accuracy of the experimental measurement. Left panel of Fig. 8.20 demonstrated that, with the same assumptions about the initial state for the hydrodynamic expansion, the spin polarization tends to further decrease between the top RHIC and LHC energies.

However, hydrodynamic results from the previous subsection established not only the global ($\mathbf{p_T}$-integrated) polarization, but also patterns in local ($\mathbf{p_T}$-differential) polarization. The magnitude of the longitudinal component $P_\parallel$ on Fig. 8.7 was reaching 4% at high p_x and p_y, which is a few times larger than the mean polarization, aligned with the total orbital momentum of the fireball.

Indeed, as it has been mentioned in a subsection above, one of the earliest hydrodynamic calculations of Λ polarization [1] has already demonstrated the existence of local polarization, see Fig. 8.6.

A study [24] took it one step further and computed the local polarization of Λ hyperons in a hydrodynamic model using averaged initial state from Monte Carlo Glauber model with its parameters set as in [25]. With such an initial state, the mid-rapidity slice of the fireball has a small angular momentum. Nevertheless, the quadrupole patterns in the longitudinal polarization persisted at both $\sqrt{s_{NN}} = 200$ GeV RHIC and $\sqrt{s_{NN}} = 2760$ GeV LHC energies. The resulting transverse momentum dependence of P^{*z} is shown in Figure 8.19 for 20–50% central Au-Au collisions at $\sqrt{s_{NN}} = 200$ (RHIC) and 20–50% Pb-Pb collisions at $\sqrt{s_{NN}} = 2760$ GeV (LHC).

Particularly, the rotation–reflection symmetries imply that S^z has a Fourier decomposition involving only the sine of even multiples of the azimuthal angle φ:

$$S^z(\mathbf{p}_T, Y = 0) = \frac{1}{2} \sum_{k=1}^{\infty} f_{2k}(p_T) \sin 2k\varphi. \qquad (8.32)$$

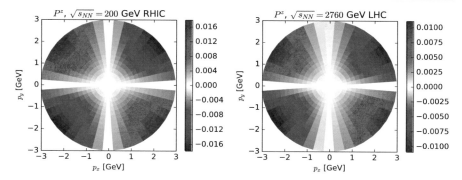

Fig. 8.19 Map of longitudinal component of polarization of mid-rapidity Λ from a hydrodynamic calculation corresponding to 20–50% central Au-Au collisions at $\sqrt{s_{NN}} = 200$ GeV (left) and 20–50% central Pb-Pb collisions at $\sqrt{s_{NN}} = 2760$ GeV (right)

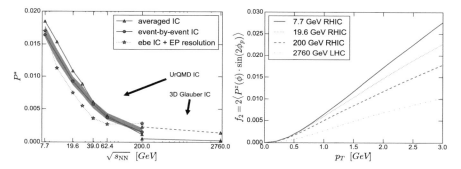

Fig. 8.20 Left panel: Global polarization of Λ hyperons in 20–50% central Au-Au (Pb-Pb) collisions at 7.7...200 GeV RHIC (2760 GeV LHC) energies. For the calculations with 3D Glauber IC (initial conditions), the solid one corresponding to longitudinally boost invariant initial flow, and the dashed one corresponding to a small amount of initial shear longitudinal flow as described in [1]. The lines connect the points to guide the eye. Right panel: Second-order Fourier harmonic coefficient of polarization component along the beam direction, calculated as a function of p_T for different collision energies; 200 and 2760 GeV points correspond to Monte Carlo Glauber IS

The corresponding second harmonic coefficients f_2 are displayed in fig. 8.20 for four different collision energies: 7.7, 19.6 GeV (calculated with initial state from the UrQMD cascade [10]), 200 and 2760 GeV (using averaged initial state from Monte Carlo Glauber model). It is worth noting that, while the P^y component, along the angular momentum, decreases by about a factor 10 between $\sqrt{s_{NN}} = 7.7$ and 200 GeV, f_2 decreases by only 35%. We also find that the mean p_T integrated value of f_2 stays around 0.2% at all collision energies, owing to two compensating effects: decreasing p_T differential $f_2(p_T)$ and increasing mean p_T with increasing collision energy. The P^y component in the UrQMD+vHLLE calculations is produced in the non-central collisions due to anisotropic transverse expansion (elliptic flow) driven by the global geometry of the fireball, whereas in central collisions, the initial state fluctuations dominate as shown in [26].

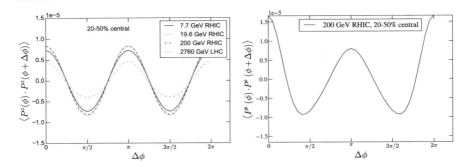

Fig. 8.21 Correlation of polarizations of two Λ hyperons as a function of their opening angle in the transverse plane. Left panel: average initial state, right panel: event-by-event hydrodynamic calculations with a Monte Carlo Glauber initial state

It is a quick exercise to show that from $P^z = 2S^z(\mathbf{p_T}) = f_2(p_T) \sin 2\phi$ it follows that

$$\langle P^z(\phi) P^z(\phi + \Delta\phi) \rangle = \frac{1}{2} f_2^2(p_T) \cos 2\Delta\phi, \tag{8.33}$$

which means that the correlation function of longitudinal polarization of two Λ hyperons, separated by the angle $\Delta\phi$ in the transverse momentum space, behaves as $\cos 2\Delta\phi$. Such behaviour one can see in Fig. 8.21 left. The right panel of Fig. 8.21 shows the correlation function from an ensemble of event-by-event hydrodynamic evolutions. In the latter case, the shape of the correlation function deviates from $\cos 2\Delta\phi$ because the underlying azimuthal angle dependence of the P^z after each hydrodynamic evolution in the event-by-event ensemble is randomly fluctuating with respect to the average $\sin 2\phi$ shape.

In fact, such correlation function of longitudinal, as well as transverse components of Λ polarization, has been reported in [26], see Fig. 8.22.

Connection Between Quadrupole Longitudinal Polarization and Elliptic Flow

Indeed, it can be shown that this component does not vanish even in the exact boost invariant scenario with no initial state fluctuations and that it decreases slowly with increasing center-of-mass energy. For the sake of simplicity, let us demonstrate that with an explicit calculation by assuming that the fluid is ideal, uncharged and that the *initial* transverse velocities u^x, u^y vanish. Accumulated evidence in relativistic heavy-ion collisions indicates that these are reasonable approximations at very high energy. Under such assumptions, it is known that the T-vorticity

$$\Omega_{\mu\nu} = \partial_\mu(T u_\nu) - \partial_\nu(T u_\mu) \tag{8.34}$$

vanishes at all times [1, 27], as a consequence of the equations of motion. In this case, the thermal vorticity reduces to [1]:

$$\varpi_{\mu\nu} = \frac{1}{T}\left(A_\mu u_\nu - A_\nu u_\mu\right) \tag{8.35}$$

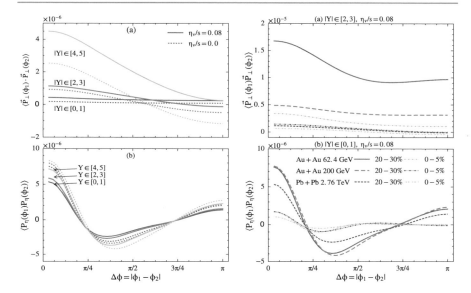

Fig. 8.22 Correlation of the transverse (top panel) and longitudinal (bottom panel) components of polarizations of two Λ hyperons as a function of their opening angle in the transverse plane. Left panel: correlation functions in different rapidity regions from the hydrodynamic calculation for $\sqrt{s_{NN}} = 2.76$ TeV. Right panel: correlation functions within one rapidity interval but at different collision energies. The plot is taken from [26]

A being the four-acceleration field. This form of the thermal vorticity shows its entirely relativistic nature, its spatial part being proportional to $(\mathbf{a} \times \mathbf{v})/c^2$ in the classical units. If we now substitute Eq. (8.35) in Eq. (8.18), we get:

$$S^\mu(p) = -\frac{1}{4m}\epsilon^{\mu\rho\sigma\tau} p_\tau \frac{\int_\Sigma d\Sigma_\lambda p^\lambda A_\rho \beta_\sigma n_F(1 - n_F)}{\int_\Sigma d\Sigma_\lambda p^\lambda n_F}, \qquad (8.36)$$

which shows that $S^z(p)$ can get contributions from the vector product of fields and momenta in the transverse plane, where they are expected to significantly develop even in the case of longitudinal boost invariance. From this equation on, we use a shortcut $n_F \equiv f(x, p)$. The uncharged perfect fluid equations of motion can be written as

$$A_\rho = \frac{1}{T}\nabla_\rho T = \frac{1}{T}\left(\partial_\rho T - u^\rho u \cdot \partial T\right).$$

If we plug the above acceleration expression in Eq. (8.37), only the first term with $\partial_\rho T$ gives a finite contribution as the second term vanishes owing to the presence of $\beta_\sigma u_\rho$ factor and the Levi-Civita tensor. Furthermore, since

$$\frac{\partial}{\partial p^\sigma} n_F = -\beta_\sigma n_F(1 - n_F),$$

we can rewrite Eq. (8.36) as

$$S^\mu(p) = \frac{1}{4mT} \epsilon^{\mu\rho\sigma\tau} p_\tau \frac{\int_\Sigma d\Sigma_\lambda p^\lambda \frac{\partial n_F}{\partial p^\sigma} \partial_\rho T}{\int_\Sigma d\Sigma_\lambda p^\lambda n_F}. \tag{8.37}$$

We can now integrate by parts the numerator in the above equation:

$$\int_\Sigma d\Sigma_\lambda p^\lambda \frac{\partial n_F}{\partial p^\sigma} \partial_\rho T = \frac{\partial}{\partial p^\sigma} \int_\Sigma d\Sigma_\lambda p^\lambda n_F \partial_\rho T - \int_\Sigma d\Sigma_\sigma n_F \partial_\rho T.$$

Another very reasonable assumption is that the decoupling hypersurface at high energy is described by the equation $T = T_c$ where T_c is the QCD pseudo-critical temperature. This entails that the normal vector to the hypersurface is the gradient of temperature. Then the final expression of the mean spin vector is

$$S^\mu(p) = \frac{1}{4mT} \epsilon^{\mu\rho\sigma\tau} p_\tau \frac{\frac{\partial}{\partial p^\sigma} \int_\Sigma d\Sigma_\lambda p^\lambda n_F \partial_\rho T}{\int_\Sigma d\Sigma_\lambda p^\lambda n_F}. \tag{8.38}$$

The longitudinal component of the mean spin vector S^z thus depends on the value of the temperature gradient on the decoupling hypersurface and its measurement can provide information thereupon. A simple solution of the above integral appears under the assumption of isochronous decoupling hypersurface, with the temperature field only depending on the Bjorken time $\tau = \sqrt{t^2 - z^2}$. In this case, the parameters describing the hypersurface are x, y, η with $\tau = const.$, and the only contribution to the numerator of the (8.38) arises from $\rho = 0$:

$$\int d\Sigma_\lambda p^\lambda n_F \frac{dT}{d\tau} \cosh \eta.$$

At $Y = 0$, the factor $\cosh \eta$ can be approximated with 1 because of the exponential fall-off $\exp[-(m_T/T) \cosh \eta]$ involved in n_F, therefore,

$$S^z(\mathbf{p}_T, Y = 0)\hat{\mathbf{k}} \simeq -\frac{dT/d\tau}{4mT} \hat{\mathbf{k}} \frac{\partial}{\partial\varphi} \log \int_\Sigma d\Sigma_\lambda p^\lambda n_F,$$

where φ is the transverse momentum azimuthal angle, counting from the reaction plane. In the above equation, the longitudinal spin component is a function of the spectrum alone at $Y = 0$. By expanding it in Fourier series in φ and retaining only the elliptic flow term, one obtains

$$\begin{aligned} S^z(\mathbf{p}_T, Y = 0) &\simeq -\frac{dT/d\tau}{4mT} \frac{\partial}{\partial\varphi} 2v_2(p_T) \cos 2\varphi \\ &= \frac{dT}{d\tau} \frac{1}{mT} v_2(p_T) \sin 2\varphi \end{aligned} \tag{8.39}$$

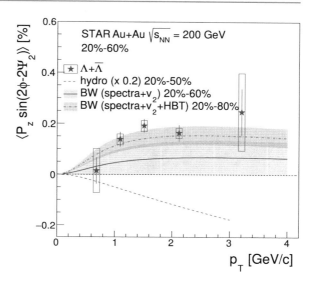

Fig. 8.23 The second-order Fourier sine coefficient of the longitudinal Λ and $\bar{\Lambda}$ polarizations as a function of p_T, measured by STAR for 20–60% Au-Au collisions at $\sqrt{s_{NN}} = 200$ GeV. The curves correspond to a Blast-Wave model calculation (unpublished) and the hydrodynamic calculation from [24]. The plot is taken from [28]

meaning, comparing this result to Eq. (8.32) that in this case:

$$f_2(p_T) = 2\frac{dT}{d\tau}\frac{1}{mT}v_2(p_T).$$

This simple formula only applies under special assumptions with regard to the hydrodynamic temperature evolution, but it clearly shows the salient features of the longitudinal polarization at mid-rapidity as a function of transverse momentum and how it can provide direct information on the temperature gradient at hadronization. It also shows, as has been mentioned, that it is driven by physical quantities related to transverse expansion and that it is independent of longitudinal expansion.

In 2019, STAR collaboration has published a measurement of the p_T and azimuthal angle dependence of the longitudinal polarization of Λ and $\bar{\Lambda}$ hyperons in Au-Au collisions at $\sqrt{s_{NN}} = 200$ GeV. The same quadrupole structure in the longitudinal polarization has been observed; however, its sign is the opposite to the hydrodynamic calculation in [24].

The same pattern but the opposite sign of the longitudinal Λ polarization, consistent with the measurement by STAR [28], has been reported in a hydrodynamic calculation in PICR model [29]. Since the origin of the polarization signal, as well as the basic ingredients of the PICR model appears to be similar to the other hydrodynamic calculations, it is not clear why the PICR calculation results in a different sign of the longitudinal polarization.

The discrepancy between the hydrodynamic calculations and the experimental measurement by STAR [28] remains an open question. Few ideas have been proposed to address the question. For example, in [30] it was found that when the thermal vorticity the formula for the spin polarization 8.15 is replaced by a projected thermal

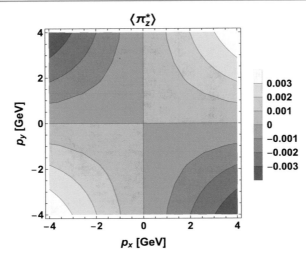

Fig. 8.24 Longitudinal component of Λ polarization, computed with projected thermal vorticity in Eq. 8.15 (for details see text). Note that the sign is compatible with the STAR result in this case. The plot is taken from [30]

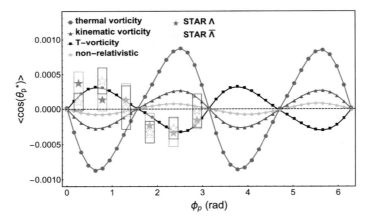

Fig. 8.25 Longitudinal component of Λ polarization, computed with different definitions of relativistic vorticity in Eq. 8.15. The plot is taken from [31]

vorticity[4]: $\varpi_{\text{proj}}^{\mu\nu} = \varpi_{\alpha\beta}\Delta_\alpha^\mu\Delta_\beta^\nu$, the resulting longitudinal component of polarization flips sign and becomes compatible with the experimental measurement.

Another recent study [31] suggests that swapping the thermal vorticity with the T-vorticity 8.6 in the same formula for the spin polarization 8.15 also results in the sign flip. However, as for now, it is not clear whether these suggestions will help

---

[4]The projection is made on a plane orthogonal to the direction of the collective flow velocity: $\varpi_{\text{proj}}^{\mu\nu}u_\nu = 0$.

solving the "sign problem" in the longitudinal polarization component between the hydrodynamic calculations and the experiment, since the basic derivation of the effect leads to the spin polarization to be proportional precisely to thermal vorticity and not to the projected thermal vorticity or the T-vorticity.

8.6   Acceleration, Grad T and Vorticity Contributions to Polarization

To gain insight into the physics of polarization in a relativistic fluid, it is very useful to decompose the gradients of the four-temperature vector in Eq. (8.18). We start off with the seperation of the gradients of the co-moving temperature and four-velocity field:

$$\partial_\mu \beta_v = \partial_\mu \left(\frac{1}{T} \right) + \frac{1}{T} \partial_\mu u_v .$$

Then, we can introduce the acceleration and the vorticity vector ω^μ with the usual definitions:

$$A^\mu = u \cdot \partial u^\mu$$
$$\omega^\mu = \frac{1}{2} \epsilon^{\mu\nu\rho\sigma} \partial_\nu u_\rho u_\sigma .$$

The antisymmetric part of the tensor $\partial_\mu u_\nu$ can then be expressed as a function of A and ω:

$$\frac{1}{2} \left(\partial_\nu u_\mu - \partial_\mu u_\nu \right) = \frac{1}{2} \left(A_\mu u_\nu - A_\nu u_\mu \right) + \epsilon_{\mu\nu\rho\sigma} \omega^\rho u^\sigma$$

therafter plugged into the (8.18) to give

$$S^\mu (x, p) = \frac{1}{8m} (1 - n_F) \epsilon^{\mu\nu\rho\sigma} p_\sigma \nabla_\nu (1/T) u_\rho \qquad (8.40)$$

$$+ \frac{1}{8m} (1 - n_F) \, 2 \, \frac{\omega^\mu u \cdot p - u^\mu \omega \cdot p}{T} \qquad (8.41)$$

$$- \frac{1}{8m} (1 - n_F) \frac{1}{T} \epsilon^{\mu\nu\rho\sigma} p_\sigma A_\nu u_\rho . \qquad (8.42)$$

Hence, polarization stems from three contributions: a term proportional to the gradient of temperature, a term proportional to the vorticity ω, and a term proportional to the acceleration. Further insight into the nature of these terms can be gained by choosing the particle rest frame, where $p = (m, \mathbf{0})$ and restoring the natural units. Equation (8.40) then certifies that the spin in the rest frame is proportional to the following combination:

$$\mathbf{S}^* (x, p) \propto \frac{\hbar}{K T^2} \gamma \mathbf{v} \times \nabla T + \frac{\hbar}{K T} \gamma (\boldsymbol{\omega} - (\boldsymbol{\omega} \cdot \mathbf{v}) \mathbf{v}/c^2) + \frac{\hbar}{K T} \gamma \mathbf{A} \times \mathbf{v}/c^2, \quad (8.43)$$

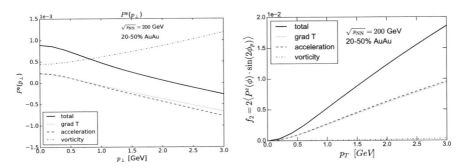

Fig. 8.26 Contributions to the global (left panel) and quadrupole longitudinal (right panel) components of Λ polarization stemming from gradients of temperature (dotted lines), acceleration (dashed lines) and vorticity (dash-dotted lines). Solid lines show the sums of all three contributions. The hydrodynamic calculation with vHLLE is performed with an averaged Monte Carlo Glauber IS corresponding to 20–50% central Au-Au collisions at $\sqrt{s_{NN}} = 200$ GeV RHIC energy

where $\gamma = 1/\sqrt{1 - v^2/c^2}$ and all three-vectors, including vorticity, acceleration and velocity, are observed in the particle rest frame.

The three independent contributions are now well discernible in Eq. (8.43). The second term scales like $\hbar\omega/KT$ and is the one already known from non-relativistic physics, proportional to the vorticity vector seen by the particle in its motion amid the fluid, with an additional term vanishing in the non-relativistic limit. The third term is a purely relativistic one and scales like $\hbar A/KTc^2$; it is usually overwhelmingly suppressed, except in heavy-ion collisions where the acceleration of the plasma is huge ($A \sim 10^{30} g$ at the outset of hydrodynamical stage). The first term, instead, is a new non-relativistic term [6] and applies to situations where the velocity field is not parallel to the temperature gradient. For ideal uncharged (thus relativistic) fluids, this term is related to the acceleration term because the equations of motion reduce to

$$\nabla_\mu T = T A_\mu/c^2. \tag{8.44}$$

Therefore, in the case of ideal uncharged fluid—which QGP is at a very high energy—the grad T and acceleration contributions will be exactly equal to each other.

Let's turn to the results from a realistic hydrodynamic calculation [32]. In Fig. 8.26, we plot the contributions to the global and quadrupole longitudinal polarization components from gradients of temperature, acceleration and vorticity individually, as well as their sum. One can see that the resulting p_T-integrated global polarization of Λ, which is dominated by its low-p_T contributions, has the largest contribution from the classical vorticity term. At the same time, f_2 has a negligible contribution from the vorticity term and virtually equal contributions from the grad T and acceleration terms. The latter result is expectable, as in hydrodynamics of ideal uncharged fluid, the temperature gradient and acceleration fields are related as follows:

$$A_\mu = \frac{1}{T}\Delta_{\mu\nu}\partial^\nu T. \tag{8.45}$$

Thus the small difference between the grad T and acceleration contributions seen in Fig. 8.26 shows that, even though the shear viscosity over entropy ratio in the calculations changes between $\eta/s = 0.08 \ldots 0.2$, the resulting hydrodynamic evolution is quantitatively not very different from the ideal one.

References

1. Becattini, F., Inghirami, G., Rolando, V., Beraudo, A., Del Zanna, L., De Pace, A., Nardi, M., Pagliara, G., Chandra, V.: A study of vorticity formation in high energy nuclear collisions. Eur. Phys. J. C **75**(9), 406 (2015)
2. Gourgoulhon, E., Publ, E.A.S.: Ser. **21**, 43 (2006)
3. Becattini, F., Csernai, L., Wang, D.J.: Phys. Rev. C **88**(3), 034905 (2013) [erratum: Phys. Rev. C **93**(6), 069901 (2016)] https://doi.org/10.1103/PhysRevC.88.034905 [arXiv:1304.4427 [nucl-th]]
4. Becattini, F., Bucciantini, L., Grossi, E., Tinti, L.: Eur. Phys. J. C **75**(5), 191 (2015)
5. Becattini, F.: Phys. Rev. Lett. **108**, (2012). https://doi.org/10.1103/PhysRevLett.108.244502. [arXiv:1201.5278 [gr-qc]]
6. Becattini, F., Chandra, V., Del Zanna, L., Grossi, E.: Relativistic distribution function for particles with spin at local thermodynamical equilibrium. Ann. Phys. **338**, 32 (2013)
7. Fang, R.H., Pang, L.G., Wang, Q., Wang, X.N.: Phys. Rev. C **94**(2) (2016)
8. Huovinen, P., Petersen, H.: Particlization in hybrid models. Eur. Phys. J. A **48**, 171 (2012)
9. Cooper, F., Frye, G.: Comment on the single particle distribution in the hydrodynamic and statistical thermodynamic models of multiparticle production. Phys. Rev. D **10**, 186 (1974)
10. Karpenko I., Huovinen P., Petersen H. and Bleicher M.: Estimation of the shear viscosity at finite net-baryon density from $A + A$ collision data at $\sqrt{s_{NN}} = 7.7 - 200$ GeV, Phys. Rev. C **91**(6), 064901 (2015)
11. Xie, Y., Wang, D., Csernai, L.P.: Global Λ polarization in high energy collisions. Phys. Rev. C **95**(3), 031901 (2017)
12. Ivanov, Y.B., Toneev, V.D., Soldatov, A.A.: Estimates of hyperon polarization in heavy-ion collisions at collision energies $\sqrt{s_{NN}} = 4$–40 GeV. Phys. Rev. C **100**(1), 014908 (2019)
13. Karpenko, I., Huovinen, P., Bleicher, M.: A 3+1 dimensional viscous hydrodynamic code for relativistic heavy ion collisions. Comput. Phys. Commun. **185**, 3016–3027 (2014)
14. Magas, V.K., Csernai, L.P., Strottman, D.D.: The Initial state of ultrarelativistic heavy ion collision. Phys. Rev. C **64** (2001)
15. Ivanov, Y.B., Soldatov, A.A.: Phys. Rev. C **102**(2), 024916 (2020). https://doi.org/10.1103/PhysRevC.102.024916, [arXiv:2004.05166 [nucl-th]]
16. Ivanov, Y.B., Russkikh, V.N., Toneev, V.D.: Phys. Rev. C **73** (2006). https://doi.org/10.1103/PhysRevC.73.044904 [arXiv:nucl-th/0503088 [nucl-th]]
17. Israel, W.: Nonstationary irreversible thermodynamics: A Causal relativistic theory. Annals Phys. **100**, 310 (1976); Israel, W., Stewart, J.M.: Transient relativistic thermodynamics and kinetic theory. Ann. Phys. **118**, 341 (1979)
18. Steinheimer, J., Schramm, S., Stocker, H.: J. Phys. G **38** (2011)
19. Becattini, F., Manninen, J., Gazdzicki, M.: Energy and system size dependence of chemical freeze-out in relativistic nuclear collisions. Phys. Rev. C **73** (2006)
20. Milekhin, G.A., Eksp, Zh: Teor. Fiz. **35**, 1185 (1958); Sov.Phys. JETP35, 829 (1959); Trudy FIAN16, 51 (1961)
21. Del Zanna, L., Chandra, V., Inghirami, G., Rolando, V., Beraudo, A., De Pace, A., Pagliara, G., Drago, A., Becattini, F.: Eur. Phys. J. C **73**, 2524 (2013). https://doi.org/10.1140/epjc/s10052-013-2524-5, [arXiv:1305.7052 [nucl-th]]
22. Karpenko I., Becattini F.: Study of Λ polarization in relativistic nuclear collisions at $\sqrt{s_{NN}} = 7.7$ –200 GeV. Eur. Phys. J. C **77**(4), 213 (2017)

23. Adam, J., et al.: [STAR], Global polarization of Λ hyperons in Au+Au collisions at $\sqrt{s_{NN}}$ = 200 GeV. Phys. Rev. C **98** (2018)
24. Becattini, F., Karpenko, I.: Collective longitudinal polarization in relativistic heavy-ion collisions at very high energy. Phys. Rev. Lett. **120**(1), 012302 (2018)
25. Bozek, P., Broniowski, W.: Transverse-momentum fluctuations in relativistic heavy-ion collisions from event-by-event viscous hydrodynamics. Phys. Rev. C **85** (2012)
26. Pang, L.G., Petersen, H., Wang, Q., Wang, X.N.: Phys. Rev. Lett. **117**(19), 192301 (2016) https://doi.org/10.1103/PhysRevLett.117.192301, [arXiv:1605.04024 [hep-ph]]
27. Deng, W.T., Huang, X.G.: Vorticity in heavy-ion collisions. Phys. Rev. C **93**(6), 064907 (2016)
28. Adam, J. et al.: [STAR], Polarization of Λ ($\bar{\Lambda}$) hyperons along the beam direction in Au+Au collisions at $\sqrt{s_{NN}}$ = 200 GeV. Phys. Rev. Lett. **123**(13), 132301 (2019)
29. Xie, Y., Wang, D., Csernai, L.P.: Eur. Phys. J. C **80**(1), 39 (2020) https://doi.org/10.1140/epjc/s10052-019-7576-8, [arXiv:1907.00773 [hep-ph]]
30. Florkowski, W., Kumar, A., Ryblewski, R., Mazeliauskas, A.: Phys. Rev. C **100**(5,)054907 (2019) https://doi.org/10.1103/PhysRevC.100.054907, [arXiv:1904.00002 [nucl-th]]
31. Wu, H.Z., Pang, L.G., Huang, X.G., Wang, Q.: Phys. Rev. Research. **1** (2019). https://doi.org/10.1103/PhysRevResearch.1.033058, [arXiv:1906.09385 [nucl-th]]
32. Karpenko, I., Becattini, F.: Lambda polarization in heavy ion collisions: from RHIC BES to LHC energies. Nucl. Phys. A **982**, 519–522 (2019)

Vorticity and Spin Polarization in Heavy Ion Collisions: Transport Models

9

Xu-Guang Huang, Jinfeng Liao, Qun Wang and Xiao-Liang Xia

Abstract

Heavy ion collisions generate strong fluid vorticity in the produced hot quark–gluon matter which could in turn induce measurable spin polarization of hadrons. We review recent progress on the vorticity formation and spin polarization in heavy ion collisions with transport models. We present an introduction to the fluid vorticity in non-relativistic and relativistic hydrodynamics and address various properties of the vorticity formed in heavy ion collisions. We discuss the spin polarization in a vortical fluid using the Wigner function formalism in which we derive the freeze-out formula for the spin polarization. Finally, we give a brief overview of recent theoretical results for both the global and local spin polarization of Λ and $\bar{\Lambda}$ hyperons.

X.-G. Huang (✉)
Physics Department, Center for Particle Physics and Field Theory, and Key Laboratory of Nuclear Physics and Ion-beam Application (MOE), Fudan University, Shanghai 200433, China
e-mail: huangxuguang@fudan.edu.cn

J. Liao
Physics Department and Center for Exploration of Energy and Matter, Indiana University, 2401 N Milo B. Sampson Lane, Bloomington, IN 47408, USA
e-mail: liaoji@indiana.edu

Q. Wang
Department of Modern Physics, University of Science and Technology of China, Hefei, Anhui 230026, China
e-mail: qunwang@ustc.edu.cn

X.-L. Xia
Physics Department and Center for Particle Physics and Field Theory, Fudan University, Shanghai 200433, China
e-mail: xiaxl@fudan.edu.cn

© Springer Nature Switzerland AG 2021
F. Becattini et al. (eds.), *Strongly Interacting Matter under Rotation*,
Lecture Notes in Physics 987,
https://doi.org/10.1007/978-3-030-71427-7_9

9.1   Introduction

Huge orbital angular momenta (OAM) are produced perpendicular to the reaction plane in non-central high-energy heavy ion collisions, and part of such huge OAM are transferred to the hot and dense matter created in collisions [1–7]. Due to the shear of the longitudinal flow, particles with spins can be polarized via the spin–orbit coupling in particle scatterings [1,6–8]. Such a type of spin polarization with respect to the reaction plane defined in the global laboratory frame of the collision is called the global polarization and is different from a particle's possible polarization with respect to its production plane which depends on the particle's momentum [1]. The global spin polarization of Λ and $\bar{\Lambda}$ has been measured by the STAR collaboration in Au+Au collisions over a wide range of beam energies, $\sqrt{s_{NN}} = 7.7 - 200\,\text{GeV}$ [9,10] and by ALICE collaboration in Pb+Pb collisions at 2.76 and 5.02 TeV [11]. The magnitude of the global spin polarization is about 2% at 7.7 GeV which decreases to be about 0.3% at 200 GeV and almost vanishes at LHC energies.

It has been shown that the spin–orbit coupling in microscopic particle scatterings can lead to the spin–vorticity coupling in a fluid when taking an ensemble average over random incoming momenta of colliding particles in a locally thermalized fluid [12]. In this way, the spin polarization is linked with the vorticity field in a fluid. To describe the STAR data on the global polarization of hyperons, the hydrodynamic and transport models have been used to calculate the vorticity field [13–24]. In hydrodynamic models, the velocity and in turn the vorticity fields in the fluid can be obtained naturally. In transport models, the phase space evolution of a multi-particle system is described by the Boltzmann transport equation with particle collisions, where the position and momentum of each particle in the system at any time are explicitly known. To extract the fluid velocity at one space–time point out of randomly distributed momenta in all events, a suitable coarse-graining method has to be used that can map the transport description into hydrodynamic information [18,19]. The vorticity field can then be computed based on the so-obtained fluid velocity. Once the vorticity field is obtained, the global polarization of hyperons can be calculated from an integral over the freeze-out hypersurface, which will be discussed in detail in Sects. 9.3 and 9.5. The calculations following the above procedure give results on the global polarization that agree with the data [22–28].

In this note, we will give a brief review of vorticity formation and spin polarization in heavy ion collisions with transport models. We use the Minkowskian metric $g_{\mu\nu} = \text{diag}(1, -1, -1, -1)$ and natural unit $k_B = c = \hbar = 1$ except for Sect. 9.3 in which $\hbar$ is kept explicitly.

9.2   Fluid Vorticity

9.2.1   Non-relativistic Case

In non-relativistic hydrodynamics, the (kinematic) vorticity is a (pseudo)vector field that describes the local angular velocity of a fluid cell. Mathematically, it is defined as

$$\boldsymbol{\omega}(\boldsymbol{x}, t) = \frac{1}{2}\nabla \times \boldsymbol{v}(\boldsymbol{x}, t),\tag{9.1}$$

where $\boldsymbol{v}$ is the flow velocity with its three components denoted as v_i ($i = 1, 2, 3$). Sometimes it is also defined without the pre-factor $1/2$ in Eq. (9.1). It can also be written in the tensorial form, $\omega_{ij} = (1/2)(\partial_i v_j - \partial_j v_i)$, so that $\omega_i = (1/2)\epsilon_{ijk}\omega_{jk}$, where ϵ_{ijk} is the three-dimensional anti-symmetric tensor. For an ideal fluid, the flow is governed by the Euler equation which can be written in terms of $\boldsymbol{\omega}$ as

$$\frac{\partial\boldsymbol{\omega}}{\partial t} = \nabla \times (\boldsymbol{v} \times \boldsymbol{\omega}).\tag{9.2}$$

This is called the vorticity equation. To arrive at Eq. (9.2), we have implicitly assumed the barotropic condition, $\nabla\rho \parallel \nabla P$, which is satisfied if the pressure P is a function of mass density ρ, $P = P(\rho)$. Equation (9.2) has interesting consequences. Let us define the circulation integral of the velocity field over a loop l co-moving with the fluid,

$$\Gamma = \oint_l \boldsymbol{v} \cdot d\boldsymbol{x} = 2\int_\Sigma \boldsymbol{\omega} \cdot d\boldsymbol{\sigma},\tag{9.3}$$

where Σ is a surface bounded by l with $d\boldsymbol{\sigma}$ being its infinitesimal area element. Note that the second equality in Eq. (9.3) follows from the Stokes theorem. It can be shown from Eq. (9.2),

$$\frac{d\Gamma}{d\tau} = 0,\tag{9.4}$$

with co-moving time derivative $d/d\tau$. This result is called Helmholtz–Kelvin theorem which states that the vortex lines move with the fluid. Physically, it is equivalent to the angular momentum conservation for a closed fluid filament in the absence of viscosity, as all forces acting on the filament would be normal to it and generate no torque. Another interesting consequence of Eq. (9.2) is the conservation of the flow helicity [29,30]

$$\mathcal{H}_f = \int d^3x\, \boldsymbol{\omega} \cdot \boldsymbol{v},\tag{9.5}$$

where the integral is over the whole space. Similar to energy, helicity is a quadratic invariant of the Euler equation of an ideal fluid although it is not positive definite. In the following, we will generalize the notion of vorticity to relativistic fluids and introduce the relativistic counterpart of the Helmholtz–Kelvin theorem and helicity conservation.

9.2.2   Relativistic Case

The generalization of vorticity to the relativistic case is not unique, and different definitions can be introduced for different purposes. Here we discuss four types of relativistic vorticity. The first one is called the *kinematic vorticity* defined as

$$\omega_K^\mu = \frac{1}{2}\epsilon^{\mu\nu\rho\sigma}u_\nu\partial_\rho u_\sigma, \tag{9.6}$$

which is a natural generalization of Eq. (9.1) as its spatial components recover Eq. (9.1) at non-relativistic limit. In the above, the four-velocity vector is defined by $u^\mu = \gamma(1, \mathbf{v})$ with $\gamma = 1/\sqrt{1 - v^2}$ the Lorentz factor. It is more convenient to define the kinematic vorticity tensor,

$$\omega_{\mu\nu}^K = -\frac{1}{2}(\partial_\mu u_\nu - \partial_\nu u_\mu), \tag{9.7}$$

so the kinematic vorticity vector is given by

$$\omega_K^\mu = -(1/2)\epsilon^{\mu\nu\rho\sigma}u_\nu\omega_{\rho\sigma}^K. \tag{9.8}$$

Note that the minus sign in Eqs. (9.7) and (9.8) is just a convention. The vorticity tensor and vector can also be defined without it. However, in either case (with or without the minus sign), the definition in Eq. (9.6) always holds. We note that the relationship between the vorticity tensor and vector in Eq. (9.8) also holds for the other types of vorticity definitions to be discussed below.

The second one is the *temperature vorticity* defined as

$$\omega_{\mu\nu}^T = -\frac{1}{2}[\partial_\mu(Tu_\nu) - \partial_\nu(Tu_\mu)], \tag{9.9}$$

where T is the temperature. The temperature vorticity for ideal neutral fluids is relevant to the relativistic version of Helmholtz–Kelvin theorem and helicity conservation [16, 19]. For an ideal neutral fluid, we can rewrite the Euler equation as

$$(\varepsilon + P)\frac{d}{d\tau}u^\mu = \nabla^\mu P, \tag{9.10}$$

with $d/d\tau = u^\mu\partial_\mu$ and $\nabla_\mu = \partial_\mu - u_\mu(d/d\tau)$. The Euler equation (9.10) can be put into the form of the Carter–Lichnerowicz equation with the help of the thermodynamic equation for a neutral fluid $dP = sdT$,

$$\omega_{\mu\nu}^T u^\nu = 0, \tag{9.11}$$

from which the relativistic Helmholtz–Kelvin theorem can be obtained immediately,

$$\frac{d}{d\tau}\oint Tu_\mu dx^\mu = 2\oint \omega_{\mu\nu}^T u^\mu dx^\nu = 0. \tag{9.12}$$

Using Eq. (9.11), we can also show that the temperature vorticity vector (multiplied by T) is conserved,

$$\partial_\mu(T\omega_T^\mu) = 4u^\mu\omega_{\mu\nu}^T\omega_T^\nu = 0, \tag{9.13}$$

where $\omega_T^\mu = -(1/2)\epsilon^{\mu\nu\rho\sigma}u_\nu\omega_{\rho\sigma}^T$. The conserved charge $\mathcal{H}_T = (1/2)\int d^3x\, T^2\gamma^2\boldsymbol{v} \cdot \boldsymbol{\nabla} \times \boldsymbol{v}$ is an extension of the helicity (9.5) to the relativistic case for an ideal neutral fluid.

The third type is the charged-fluid counterpart of the temperature vorticity which we call the *enthalpy vorticity*,

$$\omega_{\mu\nu}^W = -\frac{1}{2}[\partial_\mu(wu_\nu) - \partial_\nu(wu_\mu)], \tag{9.14}$$

where $w = (\varepsilon + P)/n$ is the enthalpy per particle and n is the charge density. In this case, the Euler equation (9.10) can be written in the following Carter–Lichnerowicz form:

$$u^\mu\omega_{\mu\nu}^W = \frac{1}{2}T\nabla_\nu(s/n). \tag{9.15}$$

If the flow is isentropic (s/n is a constant), we have $u^\mu\omega_{\mu\nu}^W = 0$, in the same form as Eq. (9.11). Therefore ,we have the conservation law for an ideal charged-fluid with the isentropic flow similar to Eq. (9.12),

$$\frac{d}{d\tau}\oint wu_\mu dx^\mu = 2\oint \omega_{\mu\nu}^W u^\mu dx^\nu = 0. \tag{9.16}$$

At the same time, the current $w\omega_w^\mu$ is conserved, $\partial_\mu(w\omega_w^\mu) = 0$, and the corresponding conserved charge is the enthalpy helicity, $\mathcal{H}_w = (1/2)\int d^3x\, w^2\gamma^2\boldsymbol{v} \cdot \boldsymbol{\nabla} \times \boldsymbol{v}$ [19].

The fourth vorticity is the *thermal vorticity*. It is defined as [16]

$$\omega_{\mu\nu}^\beta = -\frac{1}{2}[\partial_\mu(\beta u_\nu) - \partial_\nu(\beta u_\mu)]. \tag{9.17}$$

The thermal vorticity has an important property: for a fluid at global equilibrium, the four-vector $\beta_\mu = \beta u_\mu$ is a Killing vector and is given by $\beta_\mu = b_\mu + \omega_{\mu\nu}^\beta x^\nu$ with b_μ and $\omega_{\mu\nu}^\beta$ constant. Thus, the thermal vorticity characterizes the global equilibrium of the fluid. In addition, the thermal vorticity is responsible for the local spin polarization of particles in a fluid at global equilibrium which we will discuss in detail in the next section.

9.3   Spin Polarization in a Vortical Fluid

A semi-classical way to describe the space–time evolution of spin degrees of freedom
is through the spin-dependent distribution function. The quantum theory provides
a more rigorous description for the spin evolution through the Wigner function, a
quantum counterpart of the distribution function. For a relativistic spin-1/2 fermion,
one has to use the covariant Wigner function [31–34], which is a 4×4 matrix function
of position and momentum. Now the covariant Wigner function becomes a useful
tool to study the chiral magnetic and vortical effect and other related effects [35–42].
The Wigner function is equivalent to the quantum field and contains all information
that the quantum field does. Therefore the spin information in phase space is fully
encoded in the Wigner function from which one can obtain the quark polarization
from its axial-vector components.

The covariant Wigner function for spin-1/2 fermions in an external electromagnetic field is defined by [31–34]

$$
W_{\alpha\beta}(x, p) = \frac{1}{(2\pi)^4} \int d^4 y e^{-ip \cdot y} \left\langle \bar{\psi}_\beta \left(x + \frac{y}{2} \right) U \left(A; x + \frac{1}{2}y, x - \frac{1}{2}y \right) \psi_\alpha \left(x - \frac{y}{2} \right) \right\rangle,
\tag{9.18}
$$

where ψ_α and $\bar{\psi}_\beta$ are the fermionic field components ($\alpha, \beta = 1, 2, 3, 4$ are the spinor
indices), $U(A; x_2, x_1) = \exp\left[iQ \int_{x_1}^{x_2} dx^\mu A_\mu(x) \right]$ is the gauge link that makes gauge
invariance of the Wigner function with A_μ being the electromagnetic gauge potential,
and $\langle \hat{O} \rangle$ denotes the ensemble average of the operator $\hat{O}$ over thermal states. As a
4×4 complex matrix having 32 real variables, the Wigner function satisfies $W^\dagger = \gamma_0 W \gamma_0$, which reduces the number of independent variables to 16. Therefore, the
Wigner function can be expanded in terms of 16 generators of the Clifford algebra
$\{1, \gamma_5, \gamma^\mu, \gamma_5\gamma^\mu, \sigma^{\mu\nu}\}$ with $\gamma^5 \equiv i\gamma^0\gamma^1\gamma^2\gamma^3$ and $\sigma^{\mu\nu} \equiv \frac{i}{2}[\gamma^\mu, \gamma^\nu]$,

$$
W = \frac{1}{4} \left(\mathcal{F} + i\gamma^5 \mathcal{P} + \gamma^\mu \mathcal{V}_\mu + \gamma^5 \gamma^\mu \mathcal{A}_\mu + \frac{1}{2}\sigma^{\mu\nu} \mathcal{S}_{\mu\nu} \right),
\tag{9.19}
$$

where the coefficients are the scalar ($\mathcal{F}$), pseudoscalar ($\mathcal{P}$), vector ($\mathcal{V}_\mu$), axial-vector
($\mathcal{A}_\mu$), and tensor ($\mathcal{S}_{\mu\nu}$) components with 1, 1, 4, 4, and 6 independent variables,
respectively. Each component of W can be extracted by multiplying it with the
corresponding generator and taking a trace. These components are all real functions
of phase space coordinates and satisfy 32 real equations with 16 redundant equations.
For massless fermions, the equations for the vector and axial-vector component
are decoupled from the rest components. They can be linearly combined into the
right-handed and left-handed vector component, and both sectors satisfy the same
set of equations. By solving the set of equations, one can derive the right-handed
and left-handed currents which give the chiral magnetic and vortical effect in an
external electromagnetic field and a vorticity field [35–37,39,41–44]. For massive
fermions, the equations for the Wigner function components are all entangled and
hard to solve. Fortunately, there is a natural expansion parameter in these equations,

the Planck constant $\hbar$, which gives the order of quantum correction. The Wigner function components can thus be obtained by solving these questions order by order in $\hbar$, which is called semi-classical expansion [45–54].

The Wigner function components at the zeroth order in $\hbar$ are given by [45]

$$
\begin{aligned}
\mathcal{F}^{(0)}(x, p) &= m\delta(p^2 - m^2)V^{(0)}(x, p), \\
\mathcal{P}^{(0)}(x, p) &= 0, \\
\mathcal{V}^{(0)}_\mu(x, p) &= p_\mu \delta(p^2 - m^2)V^{(0)}(x, p), \\
\mathcal{A}^{(0)}_\mu(x, p) &= mn^{(0)}_\mu(x, p)\delta(p^2 - m^2)A^{(0)}(x, p), \\
\mathcal{S}^{(0)}_{\mu\nu}(x, p) &= m\Sigma^{(0)}_{\mu\nu}(x, p)\delta(p^2 - m^2)A^{(0)}(x, p),
\end{aligned}
\tag{9.20}
$$

with

$$
\begin{aligned}
V^{(0)}(x, p) &\equiv \frac{2}{(2\pi\hbar)^3}\sum_{e,s=\pm}\theta(ep^0)f_s^{(0)e}(x, ep), \\
A^{(0)}(x, p) &\equiv \frac{2}{(2\pi\hbar)^3}\sum_{e,s=\pm}s\theta(ep^0)f_s^{(0)e}(x, ep), \\
n^{(0)\mu}(x, p) &\equiv \theta(p^0)n^{+\mu}(x, p) - \theta(-p^0)n^{-\mu}(x, p), \\
\Sigma^{(0)}_{\mu\nu}(x, p) &= -\frac{1}{m}\epsilon_{\mu\nu\alpha\beta}p^\alpha n^{(0)\beta},
\end{aligned}
\tag{9.21}
$$

where $e = \pm$ denotes particle/antiparticle, $s = \pm$ denotes spin up/down, and $f_s^{(0)e}$ are the distribution functions. In Eq. (9.21), $n^\mu(p, n)$ is the spin four-vector and $n^{\pm\mu}(x, p)$ are spin four-vector for particle/antiparticle given by

$$
\begin{aligned}
n^{+\mu}(x, p) &= \left(\frac{n^+ \cdot p}{m}, n^+ + \frac{n^+ \cdot p}{m(m + E_p)}p\right), \\
n^{-\mu}(x, p) &= \left(\frac{n^- \cdot p}{m}, -n^- - \frac{n^- \cdot p}{m(m + E_p)}p\right),
\end{aligned}
\tag{9.22}
$$

where $n^\pm$ are spin quantization directions for particle/antiparticle in the particle's rest frame. In general, n^+ can be different from n^-. We note that $n^{+\mu}(x, p)$ can be expressed by a Lorentz boost from the the particle's rest frame to the lab frame in which the particle has the momentum p

$$
n^{+\mu}(x, p) = \Lambda^\mu_\nu(-v_p)n^{+\nu}(0, n^+).
\tag{9.23}
$$

Here $\Lambda^\mu_\nu(-v_p)$ is the Lorentz transformation for $v_p = p/E_p$ and $n^{+\nu}(0, n^+) = (0, n^+)$ is the four-vector of the spin quantization direction in the particle's rest frame. One can check that $n^{+\mu}(x, p)$ satisfies $n^+_\mu n^\mu_+ = -1$ and $n^+ \cdot p = 0$. Similarly $n^{-\mu}(x, p)$ for the antiparticle can be expressed by

$$
n^{-\mu}(x, p) = \Lambda^\mu_\nu(v_p)n^{-\nu}(0, n^-),
\tag{9.24}
$$

where $n^{-\nu}(\mathbf{0}, \boldsymbol{n}^-) = (0, -\boldsymbol{n}^-)$.

We see in Eqs. (9.20) and (9.21) that the axial-vector component corresponds to the spin four-vector. We can rewrite the last line of Eq. (9.21) in another form [45]:

$$n_\mu^{(0)} = -\frac{1}{2m}\epsilon_{\mu\nu\alpha\beta}p^\nu\Sigma^{(0)\alpha\beta}, \tag{9.25}$$

where $n_\mu^{(0)}$ is the Pauli–Lubanski pseudovector and $\Sigma^{(0)\alpha\beta}$ plays the role of a spin angular momentum tensor.

At the first order in $\hbar$, the axial-vector component is [38,45]

$$\mathscr{A}_\mu^{(1)} = m\bar{n}_\mu^{(1)}\delta(p^2 - m^2) + \tilde{F}_{\mu\nu}p^\nu V^{(0)}\delta'(p^2 - m^2), \tag{9.26}$$

where $\tilde{F}_{\mu\nu} = (1/2)\epsilon_{\mu\nu\alpha\beta}F^{\alpha\beta}$ and

$$\bar{n}_\mu^{(1)} \equiv -\frac{1}{2m}\epsilon_{\mu\nu\alpha\beta}p^\nu\bar{\Sigma}^{(1)\alpha\beta}, \tag{9.27}$$

is the first-order on-shell correction to $n_\mu^{(0)}A^{(0)}$. In Eq. (9.27), $\bar{\Sigma}^{(1)\alpha\beta}$ can be decomposed as

$$\bar{\Sigma}^{(1)\alpha\beta} = \frac{1}{2}\chi^{\alpha\beta} + \Xi^{\alpha\beta}, \tag{9.28}$$

where the tensor $\Xi^{\alpha\beta}$ is symmetric and satisfies $p_\alpha\Xi^{\alpha\beta} = 0$. The evolution equations for $\chi^{\alpha\beta}$ and for $\Xi^{\alpha\beta}$ are [45]

$$
\begin{aligned}
p \cdot \nabla^{(0)}\chi_{\mu\nu} &= 0, \\
p \cdot \nabla^{(0)}\Xi_{\mu\nu} &= F^\alpha_{\ \mu}\Xi_{\nu\alpha} - F^\alpha_{\ \nu}\Xi_{\mu\alpha},
\end{aligned}
\tag{9.29}
$$

where $\nabla^{(0)\mu} \equiv \partial_x^\mu - F^{\mu\nu}\partial_{p\nu}$. The component $\chi_{\mu\nu}$ satisfies the constraint

$$p^\nu\chi_{\mu\nu} = \nabla_\mu^{(0)}V^{(0)}. \tag{9.30}$$

In global equilibrium, a special choice of $\chi_{\mu\nu}$ is

$$\chi_{\mu\nu} = -\omega_{\mu\nu}^\beta\frac{\partial V^{(0)}}{\partial(\beta p_0)}, \tag{9.31}$$

where $\omega_{\mu\nu}^\beta$ is the thermal vorticity tensor (9.17) and

$$V^{(0)} \equiv \frac{2}{(2\pi\hbar)^3}\sum_s\left[\theta(u \cdot p)f_s^{(0)+} + \theta(-u \cdot p)f_s^{(0)-}\right],$$

$$f_s^{(0)\pm} = \frac{1}{\exp(\beta u \cdot p \mp \beta\mu_s) + 1}. \tag{9.32}$$

Here u^μ is the flow velocity and $\omega_{\mu\nu}$ is the vorticity tensor. Therefore, the vorticity-dependent part of the axial-vector component in Eq. (9.26) reads [38,45]

$$\mathcal{A}_\mu^{(1)} = \frac{1}{4}\epsilon_{\mu\nu\rho\sigma} p^\nu \omega_\beta^{\rho\sigma} \frac{\partial V^{(0)}}{\partial(\beta u \cdot p)} \delta(p^2 - m^2). \qquad (9.33)$$

We can integrate $\mathcal{A}_\mu^{(1)}$ over p_0 to make the momentum of the particle/antiparticle to be on the mass shell. The average spin per particle (with an additional factor $1/2$ from the particle's spin) is given by

$$S_\mu^\pm = -\frac{1}{8(u \cdot p)}\epsilon_{\mu\nu\rho\sigma} p^\nu \omega_\beta^{\rho\sigma} (1 - f_{\mathrm{FD}}^\pm), \qquad (9.34)$$

where $f_{\mathrm{FD}}^\pm$ is the on-shell Fermi–Dirac distribution function with p_0 replaced by $\pm E_p$ ($E_p \equiv \sqrt{m^2 + \boldsymbol{p}^2}$) in $f_s^{(0)\pm}$ for a particle/antiparticle, respectively. We can generalize the above equilibrium formula to a hydrodynamic process at a freeze-out hypersurface σ_μ [38,49,55], and in this case, the average spin per particle is given by

$$S^\mu(p) = -\frac{1}{8m}\epsilon^{\mu\nu\rho\sigma} p_\nu \frac{\int d\sigma_\lambda p^\lambda \omega_{\rho\sigma}^\beta (u \cdot p)^{-1} f_{\mathrm{FD}}(1 - f_{\mathrm{FD}})}{\int d\sigma_\lambda p^\lambda f_{\mathrm{FD}}}, \qquad (9.35)$$

where we have suppressed the index $\pm$ for the particle/antiparticle since the above formula is valid for both particles and antiparticles. If the momentum is not large compared with the particle mass, we have $u \cdot p \approx m$ and Eq. (9.35) recovers the result in Refs. [38,49,55] which is widely used in calculating the hadron polarization in heavy ion collisions.

9.4 Vorticity in Heavy Ion Collisions

There are multiple sources of vorticity in heavy ion collisions. One source is the global orbital angular momentum (OAM) of the two colliding nuclei in non-central collisions. Geometrically, this OAM is perpendicular to the reaction plane.[1] After the collision, a fraction of the total OAM is retained in the produced quark–gluon matter and induces vorticity. As we will discuss later in this section, in the mid-rapidity region for $\sqrt{s}$ larger than about $10\,\mathrm{GeV}$, such a generated vorticity decreases with the increasing beam energy, consistent with the measured global spin polarization of Λ and $\bar{\Lambda}$ hyperons. The second source of the vorticity is the jet-like fluctuation in the fireball which can induce a smoke-loop type vortex around the fast-moving particle [4]. The direction of such vorticity is not correlated to the reaction plane and thus does not contribute to the global Λ polarization. Instead, on an event-by-event

---

[1]Strictly speaking, this is true only after taking the average over many collision events, as the collision geometry itself (and thus the direction of the OAM) suffers from event-by-event fluctuations.

basis, it generates a near-side longitudinal spin–spin correlation [56]. The third source of the vorticity is the inhomogeneous expansion of the fire ball [18,23,56–58]. In particular, anisotropic flows in the transverse plane can produce a quadrupole pattern of the longitudinal vorticity along the beam direction while the inhomogeneous transverse expansion can produce transverse vorticity circling the longitudinal axis. There may be other sources of vorticity, e.g., the strong magnetic field created by fast-moving spectators may magnetize the quark–gluon matter and potentially lead to vorticity along the direction of the magnetic field through the so-called Einstein–de Haas effect.

Vorticity formation in high-energy nuclear collisions has been extensively studied in relativistic hydrodynamic models, such as ECHO-QGP [16], PICR [14,15], and CLVisc [56] in (3+1) dimensions. Using the ECHO-QGP code [59], different vorticities in relativistic hydrodynamics are studied in the context of directed flow in non-central collisions [16]. The evolution of the kinematic vorticity has been calculated using the PICR hydrodynamic code [14]. Using CLVisc [60,61] with event-by-event fluctuating initial conditions, the vorticity distributions have been calculated. A structure of vortex-pairing in the transverse plane due to the convective flow of hot spots in the radial direction is found to possibly form in high-energy heavy ion collisions.

In this section, we will focus on the kinematic and thermal vorticity based on transport models such as the AMPT model, but the discussion will also involve other types of vorticity. Before we go into the details, let us first discuss the setup of numerical simulations for extracting vorticity structures from the AMPT model [18] as well as the HIJING model [19] with partons as basic degrees of freedom.

9.4.1 Setup of Computation in Transport Models

According to the definitions in Sect. 9.2, in order to calculate the kinematic and thermal vorticity, we first need to obtain the velocity field u^μ (with normalization $u^\mu u_\mu = 1$) and the temperature field T. A natural way to achieve this is by using the energy-momentum tensor $T^{\mu\nu}$ through which we can define the velocity field and the energy density ε as the eigenvector and eigenvalue of $T^{\mu\nu}$, respectively,

$$T^{\mu\nu} u_\nu = \varepsilon u^\mu. \tag{9.36}$$

The temperature T can be determined from ε as a function of T by assuming a local equilibrium. In transport models such as HIJING, AMPT, or UrQMD, the position and momentum of each particle is known at any moment. A simple way to determine $T^{\mu\nu}$ as a function of space–time is by the coarse-grained method. This is done by splitting the whole space–time volume into grid cells and calculating an event average of $\sum_i p_i^\mu p_i^\nu / p_i^0$ inside each space–time cell,

$$T^{\mu\nu}(x) = \frac{1}{\Delta x \Delta y \Delta z} \left\langle \sum_i \frac{p_i^\mu p_i^\nu}{p_i^0} \right\rangle, \tag{9.37}$$

where i labels a particle inside the cell. The event average is taken to cancel the random or thermal motion of particles in each space–time cell, and finally the collective motion is kept.

Another way is to introduce a function $\Phi(x, x_i)$ to smear a physical quantity (such as the momentum) of the ith particle at x_i in an event. In such a way, we can construct a continuous function of that physical quantity [19,23]. Physically, function $\Phi(x, x_i)$ reflects the quantum nature of the particle as a wave-packet. With $\Phi(x, x_i)$, the phase space distribution can be obtained as

$$f(x, p) = \frac{1}{N} \sum_i (2\pi)^3 \delta^{(3)}[p - p_i(t)]\Phi[x, x_i(t)], \tag{9.38}$$

where $N = \int d^3x \Phi(x, x_i)$ is a normalization factor. Then the energy-momentum tensor is given by

$$T^{\mu\nu}(x) = \int \frac{d^3p}{(2\pi)^3} \frac{p^\mu p^\nu}{p^0} f(x, p) = \frac{1}{N} \sum_i \frac{p_i^\mu p_i^\nu}{p_i^0} \Phi(x, x_i). \tag{9.39}$$

The choice of the smearing function is important. Here we give two examples.

(a) The Δ smearing. This is given by generalizing the δ function $\delta^{(3)}[x - x_i(t)]$ (corresponding to a zero smearing) to

$$\Phi_\Delta[x, x_i(t)] = \delta_\Delta^{(3)}[x - x_i(t)], \tag{9.40}$$

which is 1 if $|x - x_i(t)| < \Delta x$, $|y - y_i(t)| < \Delta y$, $|z - z_i(t)| < \Delta z$, and is 0 otherwise. This is actually the coarse-grained method as we have discussed earlier in this subsection.

(b) The Gaussian smearing [19,60,62]. This is given by

$$\Phi_G[x, x_i(\tau)] = K \exp\left[-\frac{(x - x_i)^2}{2\sigma_x^2} - \frac{(y - y_i)^2}{2\sigma_y^2} - \frac{(\eta - \eta_i)^2}{2\sigma_\eta^2}\right], \tag{9.41}$$

where we have adopted the Milne coordinate (τ, x, y, η) with $\eta = (1/2)\ln[(t + z)/(t - z)]$ being the space–time rapidity and $\tau = \sqrt{t^2 - z^2}$ being the proper time instead of the Minkowski coordinate, and K and $\sigma_{x,y,z}$ are parameters that can be determined by fitting to experimental data. As a convention for the coordinate system, the z-axis is along the beam direction of the projectile, the x-axis is along the impact parameter from the target to the projectile nucleus, and the y-axis is along $\hat{z} \times \hat{x}$, see Fig. 9.1.

Fig. 9.1 The coordinate system of a heavy ion collision. Here, 'T' is for target and 'P' is for projectile

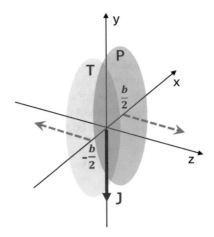

9.4.2 Results for Kinematic Vorticity

The kinematic vorticity (9.6) is a natural extension of the non-relativistic vorticity (9.1) which is a direct measure of the angular velocity of the fluid cell. We will discuss a series of features of the kinematic vorticity (including the non-relativistic one).

Centrality dependence. It is expected that for a given collision energy, the total angular momentum of the two colliding nuclei with respect to the collision center increases with the centrality or equivalently impact parameter. As a result, the vorticity is expected to increase with the centrality too. This is indeed the case as shown in Fig. 9.2 in which the average non-relativistic and relativistic vorticity in y-direction $\langle \bar{\omega}_y \rangle$ at initial time ($\tau_0 = 0.4$ fm for $\sqrt{s} = 200$ GeV and $\tau_0 = 0.2$ fm for $\sqrt{s} = 2.76$ TeV) and mid-rapidity are plotted as functions of the impact parameter b. The average is over both the transverse overlapping region (indicated by an overline of ω_y) and the collision events (indicated by $\langle \cdots \rangle$), see Ref. [19] for details. We see that the magnitude of the kinematic vorticity is big, for example, $|\omega_y|$ is about 10^{20} s^{-1} at $b = 10$ fm and $\sqrt{s} = 200$ GeV, a value surpassing the vorticity of any

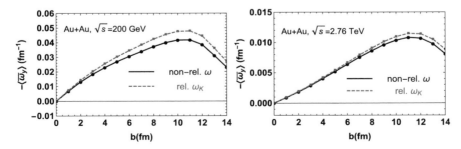

Fig. 9.2 The y-components of the non-relativistic vorticity in Eq. (9.1) and the relativistic kinematic vorticity in Eq. (9.6) at $\tau = \tau_0$ and $\eta = 0$ in 200 GeV Au + Au and 2.76 TeV Pb + Pb collisions [19]

Fig. 9.3 The collision energy dependence of the kinematic vorticity at mid-rapidity [19]

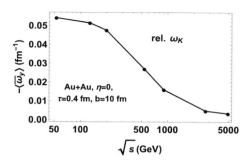

other known fluids. We also notice that the kinematic vorticity begins to decrease with b when $b \gtrsim 2R_A$ with R_A being the nucleus radius, reflecting the fact that the two colliding nuclei begin to separate.

Energy dependence. It is obvious that the total angular momentum of the two colliding nuclei grows with collision energy $\sqrt{s}$ at a fixed impact parameter. Naively one would then expect a similar energy dependence of the vorticity. However, as shown in Fig. 9.3 (and also in Fig. 9.6), the y-component of the kinematic vorticity at mid-rapidity decreases as $\sqrt{s}$ increases. Such a behavior features the relativistic effect in the mid-rapidity region: as $\sqrt{s}$ increases, two nuclei become more transparent to each other and leave the mid-rapidity region more boost invariant which supports lower vorticity. To put it in another way: while the total angular momentum of the colliding system increases with the beam energy, the fraction of that angular momentum carried by the fireball at mid-rapidity decreases rapidly with the beam energy [18]. At low energy, the relativistic effect becomes less important and the fireball acquires a considerably more fraction of the system's angular momentum, leading to a much increased vorticity [18,19]. At very low energy, however, the total angular momentum would be small and the vorticity becomes inevitably small again [21].

Correlation to the participant plane. Geometrically, it is expected that the direction of the vorticity should be perpendicular to the reaction plane. However, this is true only at the optical limit or after event average. In reality, the nucleons in the nucleus are not static but always move from time to time, leading to the event-by-event fluctuation at the moment of collisions. Such event-by-event fluctuations can smear the direction of the vorticity from being perfectly perpendicular to the reaction plane. To quantify this effect, one can study the azimuthal-angle correlation between the vorticity and the participant plane (which can describe the overlapping region more accurately than the reaction plane), $\langle \cos[2(\psi_\omega - \psi_2)] \rangle$, where ψ_ω and ψ_2 denote the azimuthal angle of the vorticity and the participant plane of the second order, respectively. The result is shown in Fig. 9.4. We see that the correlation is significantly suppressed in the most central (due to the strong fluctuation in ψ_ω) and most peripheral (due to the strong fluctuation in ψ_2) collisions. We note that a similar feature can also be observed in magnetic fields [63,64].

Spatial distribution. The vorticity is inhomogeneous in the transverse plane (the x-y plane in Fig. 9.1). As seen in Fig. 9.5 (left panel), the non-relativistic vorticity

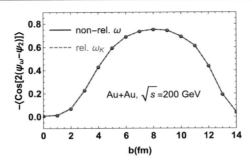

Fig. 9.4 The correlation between the direction of the vorticity ψ_ω and the second-order participant plane ψ_2 [19]

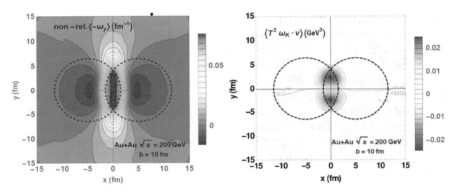

Fig. 9.5 The spatial distribution of the non-relativistic vorticity (left panel) and the helicity (right panel) in the transverse plane at $\eta = 0$ for RHIC Au + Au collisions at $\sqrt{s} = 200\,\text{GeV}$ [19]. See also discussion around Eq. (9.13)

varies more steeply along the x direction in accordance with the elliptic shape of the overlapping region. The event average of the helicity field $h_T^0 \equiv (1/2)T^2 \boldsymbol{v} \cdot \boldsymbol{\nabla} \times \boldsymbol{v}$ as defined below in Eq. (9.13) is depicted in Fig. 9.5 (right panel). Clearly, the reaction plane separates the region with the positive helicity from that with the negative helicity, due simply to the fact that $\langle v_y \rangle$ changes its sign across the reaction plane while $\langle \omega_y \rangle$ does not change sign. We note that a similar feature also exists for the electromagnetic helicity $\langle \boldsymbol{E} \cdot \boldsymbol{B} \rangle$ in heavy ion collisions [65].

Time evolution. In the hot quark–gluon medium, the fluid velocity evolves in time, so does the vorticity. Understanding the time evolution of the vorticity is also important for understanding vorticity-driven effects such as spin polarization. The results for the non-relativistic vorticity as functions of time in an AMPT simulation are presented in Fig. 9.6. We see that at a very early stage $-\langle \bar{\omega}_y \rangle$ (in Ref. [18], the spatial average is weighted by the inertia moment) briefly increases with time which is probably due to a decrease of inertia moment by parton scatterings before the transverse radial expansion is developed. After reaching a maximum value at ~ 1 fm, $-\langle \bar{\omega}_y \rangle$ follows a steady decrease with time because of the system's expansion.

Fig. 9.6 The time evolution of the non-relativistic vorticity for different collision energies [18]

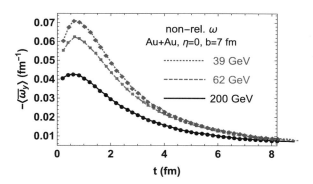

To understand how the system's expansion brings the vorticity down, we can consider the dissipation equation for the non-relativistic vorticity,

$$\frac{\partial \boldsymbol{\omega}}{\partial t} = \nabla \times (\boldsymbol{v} \times \boldsymbol{\omega}) + \nu \nabla^2 \boldsymbol{\omega}, \tag{9.42}$$

where $\nu = \eta/(\varepsilon + P) = \eta/(sT)$ is the kinematic shear viscosity with η being the shear viscosity and s the entropy density. Thus, the change of the vorticity can be driven by either the fluid flow (the first term on the right-hand side) or by the viscous damping (the second term on the the right-hand side). The ratio of the two terms can be characterized by the Reynolds number $Re = UL/\nu$ with U and L being the characteristic velocity and system size, respectively. If $Re \ll 1$, the second term dominates and the vorticity is damped by the shear viscosity with a time scale $t_\omega \sim L^2/(4\nu)$. If $Re \gg 1$, the first term in Eq. (9.42) dominates and the vortex flux is nearly frozen in the fluid (see the discussion in Sect. 9.2.1 about the Helmholtz–Kelvin theorem). In this case, the vorticity decreases due to the system's expansion. Considering Au + Au collisions at $\sqrt{s} = 200$ GeV as an example. Typically, we can assume $U \sim 0.1 - 1$, $L \sim 5$ fm, $T \sim 300$ MeV, and $\eta/s \sim 1/(4\pi)$ for the strongly coupled QGP, then we have $Re \sim 10 - 100$. Thus, the vorticity decays as shown in Fig. 9.6 mainly due to the system's expansion, see Refs. [18, 19] for more discussions.

9.4.3   Results for Thermal Vorticity

The thermal vorticity (9.17) can be decomposed into the part proportional to the kinematic vorticity and the part related to temperature gradients,

$$\varpi_{\mu\nu} \equiv \omega^\beta_{\mu\nu} = \beta \omega^K_{\mu\nu} + u_{[\nu} \partial_{\mu]} \beta, \tag{9.43}$$

where $[\cdots]$ means anti-symmetrization of indices. Note that in this subsection, we will use ϖ to denote the thermal vorticity in order to be consistent with the traditional notation widely used in literature. Thus, in many aspects, the thermal vorticity behaves similarly to the kinematic vorticity. But the difference between the two vorticities becomes significant when the temperature gradient is large.

Fig. 9.7 The time evolution
of the zx-component of the
thermal vorticity at
space–time rapidity $\eta = 0$
and impact parameter $b = 9$
fm for different collision
energies. The figure is taken
from Ref. [23]

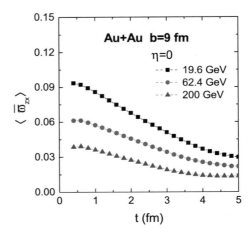

Time evolution. In Fig. 9.7, we show the zx-component of the thermal vorticity
in Au + Au collisions at $\eta = 0$, $b = 9$ fm and $\sqrt{s} = 19.6, 62.4, 200$ GeV. Here,
the thermal vorticity is averaged over the transverse plane first (weighted by the
energy density and indicated by an overline) and then over collision events (indicated
by $\langle \cdots \rangle$). Comparing with Fig. 9.6, except for a very short early time, the time
evolution of the thermal vorticity is similar to the kinematic vorticity, so is the energy
dependence: both the thermal and kinematic vorticity decrease with $\sqrt{s}$. This can be
understood from the fact that at higher collision energies, both terms in Eq. (9.43)
become smaller at $\eta = 0$ as two colliding nuclei become more transparent to each
other and make the mid-rapidity region more boost invariant.

Spatial distribution. In the left panel of Fig. 9.8, we show the spatial distribution
of the event-averaged thermal vorticity $\varpi_\perp = (\varpi_{yz}, \varpi_{zx})$ on the transverse plane
at $\eta = 0$. We take $t = 0.6$ fm for the Au + Au collisions at $\sqrt{s} = 19.6$ GeV as an
example. The arrows represent $\langle \varpi_\perp \rangle$ and colors represent the magnitude of $\langle \varpi_{zx} \rangle$.
We see two vorticity loops associated with the motion of the participant nucleons
in the projectile and target nucleus, respectively. The right panel shows the radial
component of $\langle \varpi_\perp \rangle$, and a clear sign separation by the reaction plane is observed.

As we have already discussed, the source of vorticity is multifold. The inhomogeneous expansion of the fireball serves as a good generator of the vorticity. To see
this more clearly, let us consider a non-central collision and parameterize its velocity
profile at a given moment as

$$
\begin{aligned}
v_r &\sim \bar{v}_r(r, z) \left[1 + 2c_r \cos(2\phi)\right], \\
v_z &\sim \bar{v}_z(r, z) \left[1 + 2c_z \cos(2\phi)\right], \\
v_\phi &\sim 2c_\phi \bar{v}_\phi(r, z) \sin(2\phi),
\end{aligned}
\tag{9.44}
$$

where r, z, and ϕ are the radial, longitudinal, and azimuthal coordinates respectively, and c_r, c_z, and c_ϕ characterize the eccentricity in v_r, v_z, and v_ϕ, respectively.
For high-energy collisions, the expansion respects approximately a $z \to -z$ reflec-
tion symmetry which requires that $\bar{v}_r(r, z) = \bar{v}_r(r, -z)$, $\bar{v}_z(r, z) = -\bar{v}_z(r, -z)$, and

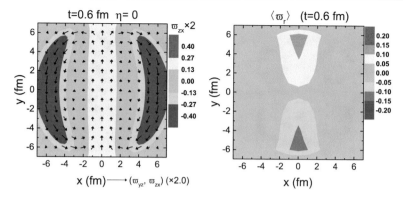

Fig. 9.8 The distribution of the event-averaged thermal vorticity on the transverse plane at $t = 0.6$ fm, $\eta = 0$, and $\sqrt{s} = 19.6\,\mathrm{GeV}$ for Au + Au collisions, averaged over the centrality range 20–50%. Left panel: arrows represent $\langle\varpi_\perp\rangle = (\langle\varpi_{yz}\rangle, \langle\varpi_{zx}\rangle)$ and colors represent the magnitude of $\langle\varpi_{zx}\rangle$. Right panel: the radial thermal vorticity $\langle\varpi_r\rangle = \hat{r}\cdot\langle\varpi_\perp\rangle$. The figures are taken from Ref. [23]

$\bar{v}_\phi(r, z) = \bar{v}_\phi(r, -z)$. Thus we find very interesting features in the non-relativistic kinematic vorticity field, $\boldsymbol{\omega} = (1/2)\nabla \times \boldsymbol{v}$, from the velocity profile (9.44).

First, at mid-rapidity $\eta = 0$ or $z = 0$ in a non-central collision, the longitudinal non-relativistic kinematic vorticity ω_z can be nonzero while the transverse component ω_r and ω_ϕ vanish. In particular, we have $\omega_z \sim \sin(2\phi)$ at mid-rapidity, featuring a quadrupole distribution as illustrated in the left panel of Fig. 9.9.[2] Such a quadrupole structure in the non-relativistic vorticity field is a result of the positive elliptic flow v_2. Quite similarly, the longitudinal component of the thermal vorticity also shows a quadrupole structure in the transverse plane in the right panel of Fig. 9.9, in which the results of ϖ_{xy} in the transverse plane of Au + Au collisions at $t = 0.6$ fm, $\eta = 0$, and $\sqrt{s} = 19.6\,\mathrm{GeV}$ are presented. Surprisingly, in each quadrant, the thermal vorticity ϖ_{xy} has an opposite sign compared to the non-relativistic vorticity ω_z. This means that the contributions from acceleration and temperature gradient to the thermal vorticity are large and outperform that from the velocity gradient.

Second, at finite rapidity, all three components of $\boldsymbol{\omega}$ can be finite and the transverse vorticity is dominated by the ϕ component. The origin of this ϕ-directed vortex is similar to the onset of the smoke-loop vortex as illustrated in Fig. 9.10 (upper-left). More precisely, $\omega_\phi \sim (1/2)[\partial\bar{v}_r/\partial z - \partial\bar{v}_z/\partial r]$ changes sign under the reflection transformation $z \to -z$ or $\eta \to -\eta$, such a behavior exists in non-central as well as central collisions. In the positive rapidity region $\eta > 0$, the first term in ω_ϕ is usually negative while the second term is positive, so the direction of the ϕ-directed vortex depends on the relative strength of two terms. A similar smoke-loop pattern for the thermal vorticity also exists, see the lower panels of Fig. 9.10. The projection to the

---

[2]We note that the left panel of Fig. 9.9 is just for illustrative purpose, the real velocity profile is much more complicated including components that can contribute a positive v_2 but an opposite vortical structure to the one shown in the figure.

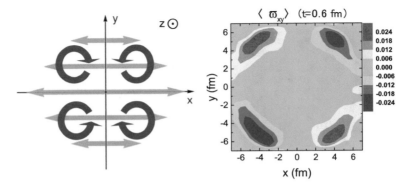

Fig. 9.9 Left panel: illustration of an anisotropic expansion of the fireball in the transverse plane in non-central collisions. Such a flow profile represents a positive elliptic flow v_2 and a quadrupolar distribution of the longitudinal kinematic vorticity Eq. (9.1) in the transverse plane. Right panel: the longitudinal component of the thermal vorticity distributed in the transverse plane at $t = 0.6$ fm, $\eta = 0$, and $\sqrt{s} = 19.6\,$GeV in Au + Au collisions. The results are obtained by averaging over events in 20–50% centrality. A remarkable difference between the left and right panels is the sign difference of the longitudinal vorticity in each quadrant. The figure is taken from Ref. [23]

reaction plane forms a quadrupole structure for ϖ_{zx} as shown in the upper-right panel of Fig. 9.10. We will discuss how this intricate local vortical structure can be reflected in the spin polarization of Λ hyperons in the next section.

9.5 Λ Polarization in Heavy Ion Collisions

An important consequence of the vorticity field is that particles with spin can be polarized. The detailed mechanism for such a spin polarization has been discussed in Sect. 9.3. In this section, we review the numerical simulation based on transport models for the spin polarization of one specific hyperon, Λ, and its antiparticle, $\bar{\Lambda}$. The reason why the Λ hyperon is chosen is that its weak decay $\Lambda \rightarrow p + \pi^-$ which violates the parity symmetry, so the daughter proton emits preferentially along the spin direction of Λ in its rest frame. More precisely, if $\boldsymbol{P}_\Lambda^*$ is the spin polarization of Λ in its rest frame (hereafter, we will use an asterisk to indicate Λ's rest frame), the angular distribution of the daughter protons is given by

$$\frac{1}{N_p}\frac{dN_p}{d\Omega^*} = \frac{1}{4\pi}\left(1 + \alpha \hat{\boldsymbol{p}}^* \cdot \boldsymbol{P}_\Lambda^*\right), \qquad (9.45)$$

where $\boldsymbol{p}^*$ is the momentum of the proton in the rest frame of Λ (a hat over a vector denotes its unit vector), Ω^* is the solid angle of $\boldsymbol{p}^*$, and $\alpha \approx 0.642 \pm 0.013$ is the decay constant. Thus, experimentally, one can extract P_Λ^* by measuring $dN_p/d\Omega^*$ [9,66,67]. The above discussion applies equally well to $\bar{\Lambda}$ but with a negative decay constant $-\alpha$. Our purpose is to discuss the current theoretical understanding of $\boldsymbol{P}_\Lambda^*$ induced by the vorticity in heavy ion collisions. We note that the vorticity induced spin polarization can also lead to other interesting consequences like the spin alignment of

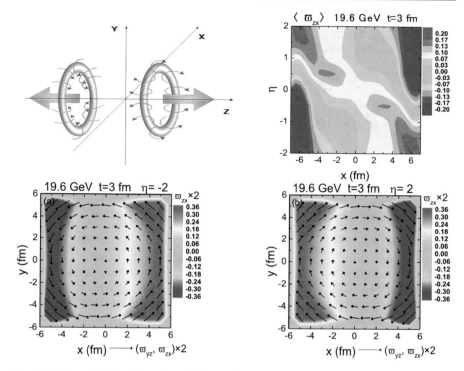

Fig. 9.10 Upper-left: the illustration of the smoke-loop type vortices due to the fast longitudinal expansion. Note that the radial expansion inhomogeneous in z or η direction results in similar vortices. Upper-right: the distribution of event-averaged thermal vorticity in the reaction plane (the $x - \eta$ plane) for Au + Au collisions at $\sqrt{s} = 19.6$ GeV. Lower-left and lower-right: the vector plot for the thermal vorticity projected to the transverse plane at space–time rapidity $\eta = -2, 2$ for Au + Au collisions at $\sqrt{s} = 19.6$ GeV averaged over events in 20–50% centrality range. The background color represents the magnitude and sign of ϖ_{zx}. The figures are from Ref. [23]

vector mesons, [2,68–70], enhancement of the yield of hadrons with higher spin [71], spin–spin correlation [56], and polarization of emitted photons [72], which, however, will not be discussed.

The basic assumption that enables us to link the vorticity and Λ spin polarization is the local equilibrium of the spin degree of freedom leading to the formula for a spin-s fermion with mass m and momentum p^μ produced at point x [38,49,55,73],

$$S^\mu(x, p) = -\frac{s(s + 1)}{6m}(1 - n_F)\epsilon^{\mu\nu\rho\sigma} p_\nu \varpi_{\rho\sigma}(x) + O(\varpi)^2, \qquad (9.46)$$

where $n_F(p_0)$ is the Fermi–Dirac distribution function with $p_0 = \sqrt{\mathbf{p}^2 + m^2}$ being the energy of the fermion. We should note that this formula can be shown to be hold at global equilibrium, as we derived in Sect. 9.3, but here we assume that it holds also at local equilibrium. For Λ and $\bar{\Lambda}$, we have $s = 1/2$. If the fermion mass is much larger than the temperature as in the case of Λ and $\bar{\Lambda}$ produced in heavy ion collisions at RHIC and LHC energies, we can approximate $1 - n_F \approx 1$. Using $S^{*\mu} = (0, \boldsymbol{S}^*)$ to denote the spin vector in Λ's rest frame, the Lorentz transformation from the laboratory frame gives

$$\boldsymbol{S}^* = \boldsymbol{S} - \frac{\boldsymbol{p} \cdot \boldsymbol{S}}{p_0(p_0 + m)} \boldsymbol{p}. \tag{9.47}$$

Finally, the spin polarization of Λ in the direction $\boldsymbol{n}$ is given by

$$P_n^* = \frac{1}{s} \boldsymbol{S}^* \cdot \boldsymbol{n}. \tag{9.48}$$

In the following, for simplicity, we will use P_n to denote P_n^* if there is no confusion. In a transport model like AMPT, Eqs. (9.46)–(9.48) are used to calculate the Λ polarization.

The global polarization. In the last few years, transport models such as AMPT have been widely used in the study of the Λ polarization. The results of various groups are consistent with each other to a large extent. Here, we mainly show the results of Refs. [22–24]. In Fig. 9.11, theoretical results for the spin polarization of Λ and $\bar{\Lambda}$ are compared with experimental data. The simulations are done for Au + Au collisions in the centrality range 20−50% and rapidity range $|Y| < 1$. We see very good agreement between numerical results and data, which gives strong support for the vorticity interpretation of the measured Λ polarization. We have three comments. (1) The simulations include only the polarization caused by the vorticity, so there is no difference between Λ and $\bar{\Lambda}$ in the calculation. The data shows a difference between Λ and $\bar{\Lambda}$ although the errors are large. This is not fully understood. A possible source for such a difference might be the magnetic field because Λ and $\bar{\Lambda}$ have an opposite magnetic moment. (2) The simulation given in Fig. 9.11 counts only the Λ and $\bar{\Lambda}$ coming from the hadronization of quarks in the AMPT model (called the primary or primordial Λ and $\bar{\Lambda}$). However, a big fraction (∼80%) of the measured Λ and $\bar{\Lambda}$ hyperons are from the decay of higher-lying hyperons such as Σ^0, Σ^*, Ξ, and Ξ^*. However, such a feed-down contribution to the Λ polarization is small: it can reduce about 10−20% of the spin polarization of primordial Λ's [74,75]. (3) Recently, HADES collaboration reported the measurement of the Λ polarization at $\sqrt{s} = 2.4$ GeV which shows a nearly vanishing P_y [76]. This means that the energy dependence of P_y at low energies is not monotonous. The AMPT model is not applicable in such a low-energy region, and it is necessary to use other transport models, such as UrQMD and IQMD, to calculate the Λ polarization at very low energy [21].

Local polarization and polarization harmonics. The above analysis is for the integrated spin polarization over the azimuthal angle and rapidity and p_T region, so is called the global polarization. As we have shown in Sect. 9.4.3, the thermal vorticity has a nontrivial distribution in coordinate space, especially the quadrupole structure shown in Figs. 9.9 and 9.10, leading to a nontrivial spin-polarization distribution in momentum space following Eq. (9.35).

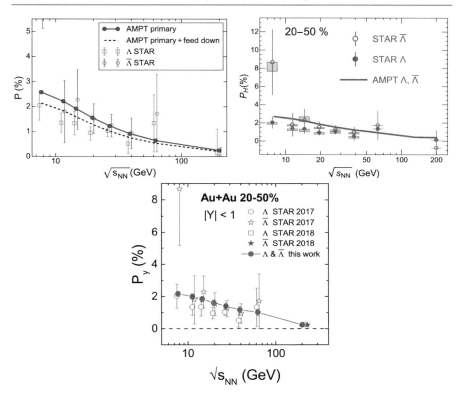

Fig. 9.11 The global Λ and $\bar{\Lambda}$ spin polarization simulated in AMPT model with comparison to the experimental data. Shown are the polarization along y direction in 20–50% centrality range of Au+Au collisions from different working groups which are consistent to each other [22–24]

Here, we show the results from Ref. [58]. In Fig. 9.12, we present the Λ spin polarization as functions of Λs' momentum azimuthal angle ϕ_p for Au + Au collisions at 200 GeV (left) and Pb + Pb collisions at 2760 GeV (right). As illustrated in Fig. 9.10, the inhomogeneous expansion of the fireball can generate transverse vorticity loops, the directions of which are clockwise and counterclockwise in positive and negative rapidity regions, respectively. As a consequence, the transverse Λ spin polarization (P_x, P_y) should have a similar structure. To extract this effect, P_x and P_y are weighted by the sign of rapidity and then averaged in local azimuthal-angle bins. The results shown in the upper two panels in Fig. 9.12 present good harmonic behaviors $P_x \mathrm{sgn}(Y) \sim \sin(\phi_p)$ and $P_y \mathrm{sgn}(Y) \sim -\cos(\phi_p)$, which agree with the direction of the transverse vorticity loop. On the other hand, the longitudinal vorticity has a quadrupole structure on the transverse plane shown in Fig. 9.9. Correspondingly, the longitudinal spin polarization P_z shows a $-\sin(2\phi_p)$ behavior in the lowest panels in Fig. 9.12.

Figure 9.13 shows another way to present the local polarization [23]. In the left panel, we present the distribution of the transverse spin polarization P_y on the $\phi - Y$ plane where ϕ is the momentum azimuthal angle and Y is the rapidity. Clearly, as

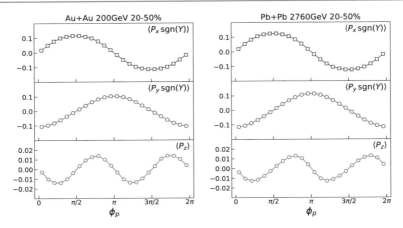

Fig. 9.12 The Λ spin polarization as functions of Λs' momentum azimuthal angle ϕ_p in 20–50% central Au + Au collisions at 200 GeV (left) and Pb + Pb collisions at 2760 GeV (right) [58]

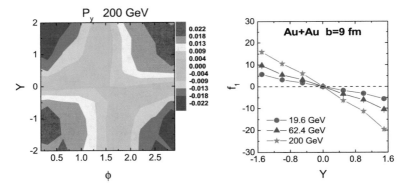

Fig. 9.13 Left: the polarization P_y on $\phi - Y$ plane which is the spin-polarization response to the upper-right panel of Fig. 9.10 up to the linear order in thermal vorticity according to Eq. (9.46). Right: the directed spin flow f_1 defined in Eq. (9.49) versus rapidity Y [23]

a spin-polarization response to the quadrupole structure of the vorticity field shown in the upper-right panel of Fig. 9.10, $P_y(Y, \phi)$ also shows a quadrupole structure. To characterize such a nontrivial ϕ dependence of $P_y(\phi)$ at a given rapidity Y, we can decompose $P_y(Y, \phi)$ into a harmonic series,

$$P_y(Y, \phi) = \frac{1}{2\pi} \frac{dP_y}{dY} \left\{ 1 + 2 \sum_{n=1}^{\infty} f_n \cos[n(\phi - \Phi_n)] \right\}, \qquad (9.49)$$

where Φ_n defines the nth harmonic plane for spin with the corresponding harmonic coefficient f_n. The first harmonic coefficient, f_1, shown in the right panel of Fig. 9.13), is induced by the vorticity from collective expansion, which is odd in rapidity and peaks at finite rapidity in accordance with Fig. 9.10). The measurement

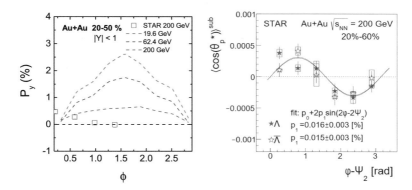

Fig. 9.14 Left: the polarization P_y as a function of the azimuthal angle ϕ. The red squares are experimental data [10]. Right: the experimental results of longitudinal polarization $P_z(\phi)$ from STAR Collaboration [81]. Note that $P_z \sim \langle \cos\theta_p^* \rangle$

of such a directed flow of spin polarization would be the indicator of the quadrupole structure in the vorticity field due to inhomogeneous expansion of the fireball.

The "sign problem". Although the global polarization P_y can be described well by simulations based on the thermal vorticity following Eq. (9.35), such relation fails in describing the azimuthal dependence of P_y at mid-rapidity. In fact, the theoretical calculations, including both the transport model and hydrodynamic model calculations, found that $P_y(\phi)$ at mid-rapidity grows from $\phi = 0$ (i.e., the in-plane direction) to $\phi = \pi/2$ (i.e., the out-of-plane direction) while the experimental data shows an opposite trend, see the left panel of Fig. 9.14. In addition, the longitudinal polarization $P_z(\phi)$ at mid-rapidity also has a similar "sign problem": the theoretical calculations predicted that $P_z(\phi) \sim -\sin(2\phi)$ as shown in Fig. 9.12. This can also be seen in Fig. 9.9 as the spin polarization is roughly proportional to the thermal vorticity following Eq. (9.35). However, the data shows an opposite sign, see the right panel of Fig. 9.14. These sign problems challenge the thermal vorticity interpretation of the measured Λ polarization and are puzzles at the moment.

Here we have several comments about them. (1) As we have already discussed, the feed-down decays of other strange baryons constitute a major contribution to the total yield of Λ and $\bar{\Lambda}$. Thus, to bridge the measured spin polarization and the vorticity, we must take into account the feed-down contributions. In addition, Λ hyperon produced from the feed-down decay can have opposite spin polarization compared to its parent particle in some decay channels, e.g., $\Sigma^0 \to \Lambda + \gamma$. Recently, the feed-down effects have been carefully studied in Refs. [74,75]. Although the feed-down contribution suppresses the polarization of primordial Λ, it is not strong enough to resolve the sign problem. (2) At the moment, most of the theoretical studies are based on Eq. (9.35) which assumes global equilibrium for the spin degree of freedom. This might not be the case for realistic heavy ion collisions. In a non-equilibrium state or even near-equilibrium state, the spin polarization is not determined by the thermal vorticity and should be treated as an independent dynamic variable. This requires new theoretical frameworks like the spin hydrodynamics; see, e.g., [77,78]. Recently, there have been progresses in developing these new frameworks and hopefully, the numerical simulations based on them can give a more accurate description of the Λ polarization and insight into the sign problem. (3) There have been theoretical explanations of the sign problem based on chiral kinetic theory [79,80], blast-wave model [81], and hydrodynamics [82,83]. But they introduce new assumptions such as the presence of net chirality [79], kinematic vorticity, or T-vorticity dominance of the polarization [81–83]) which need further examinations; see Refs. [84–88] for recent reviews.

The magnetic polarization. Finally, let us discuss one intriguing aspect of the measured global polarization: there is a visible difference between hyperons and anti-hyperons, especially in the low beam energy region. While the error bars are still too large to unambiguously identify a splitting between $P_{\bar{\Lambda}}$ and P_{Λ}, the difference shown by these data is significant enough to warrant a serious investigation into the probable causes. One natural and plausible explanation could be the magnetic polarization effect (which distinguishes particles from antiparticles) in addition to the rotational polarization (which is "blind" to particle/antiparticle identities) [73,89–91]. Indeed the hyperon Λ and anti-hyperon $\bar{\Lambda}$ have negative and positive magnetic moments, respectively. When subject to an external magnetic field, $\bar{\Lambda}$ spin would be more aligned along the field direction while $\bar{\Lambda}$ spin would be more aligned against the field direction. This could indeed qualitatively explain the observed splitting with $P_{\bar{\Lambda}} > P_{\Lambda}$, provided that the magnetic field in heavy ion collisions is indeed approximately parallel to average vorticity and possibly survive long enough till the freeze-out time.

To examine whether this idea may work, quantitative simulations have been carried out recently within the AMPT framework [90]. Under the presence of both vorticity and magnetic field, the polarization given in Eq. (9.46) should be modified to include both effects as follows:

$$S^{\mu}(x, p) = -\frac{s(s+1)}{6m}(1 - n_F)\epsilon^{\mu\nu\rho\sigma} p_{\nu} \left[\varpi_{\rho\sigma}(x) \mp 2(eF_{\rho\sigma})\, \mu_{\Lambda}/T_f \right] , \quad (9.50)$$

where the $\mp$ sign is for Λ and $\bar{\Lambda}$ while $\mu_{\Lambda} = 0.613/(2m_N)$ is the absolute value of the hyperon/anti-hyperon magnetic moment, with $m_N = 938$MeV being the nucleon mass. T_f is the local temperature upon the particle's formation. Here, we focus on the electromagnetic field component that is most relevant to the global polarization effect, namely $eB_y = eF_{31} = -eF_{13}$ along the out-of-plane direction. By adopting a certain parameterization of the time dependence for the magnetic field with a lifetime parameter t_B, one could then investigate how the polarization splitting depends on the B field lifetime. The left plan of Fig. 9.15 shows how the magnetic field lifetime t_B would quantitatively influence the polarization of Λ and $\bar{\Lambda}$. As one can see, with increasing magnetic field lifetime (which means stronger magnetic field at late time in the collisions), the $P_{\bar{\Lambda}}$ steadily increases while the P_{Λ} decreases at all collision energies. With long enough t_B, eventually the $P_{\bar{\Lambda}}$ always becomes larger than P_{Λ}. By comparing with experimentally measured polarization splitting, one could actually extract the optimal value (or a constraint) on the magnetic field lifetime. Such analysis is shown in the right panel of Fig. 9.15. A number of different parameterizations

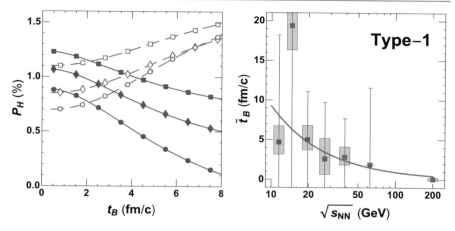

Fig. 9.15 Left: The dependence on magnetic field lifetime parameter t_B of the global polarization signals P_H for hyperons ($H \rightarrow \Lambda$, blue solid curves with filled symbols) and anti-hyperons ($H \rightarrow \bar{\Lambda}$, red dashed curves with open symbols) at beam energy $\sqrt{s_{NN}} =$19.6 (square), 27 (diamond), and 39 (circle) GeV, respectively [90]. Right: The optimal value of magnetic field lifetime parameter $\tilde{t}_B$ extracted from polarization splitting ΔP data for a range of collision beam energy $\sqrt{s_{NN}}$. The results in this plot use the parameterization $eB(t) = eB(0)/[1 + (t - t_0)^2/t_B^2]$ (see [90] for details). The solid curve is from fitting analysis with a formula $\tilde{t}_B = \frac{A}{\sqrt{s_{NN}}}$. The error bars are converted from the corresponding errors of experimental data in [9, 10]

were studied in [90] and the overall analysis suggests an empirical formula for possible magnetic field lifetime: $t_B = A/\sqrt{s_{NN}}$ with $A = 115 \pm 16$ GeV $\cdot$ fm/c. Interestingly, this is considerably longer than the expected vacuum magnetic field lifetime without any medium effect, which could be estimated by $t_{vac} \simeq 2R_A/\gamma \simeq 26$ GeV $\cdot$ fm $\cdot$ c$^{-1}/\sqrt{s_{NN}}$. Such extended magnetic field lifetime, as indicated by polarization difference, may imply a considerable role of the medium-generated dynamical magnetic field especially at low beam energy [91].

9.6 Summary

The non-vanishing global spin polarization of Λ and $\bar{\Lambda}$ has been measured by STAR collaboration in Au+Au collisions at $\sqrt{s_{NN}} = 7.7 - 200$ GeV. Microscopically, such a global polarization originates from the spin–orbit coupling of particle scatterings in a fluid with local vorticity. It has been shown that the spin–orbit coupling can lead to a spin–vorticity coupling when taking an ensemble average over random incoming momenta of scattering particles in a locally thermalized fluid. With the spin–vorticity coupling, the local spin polarization can be obtained from the vorticity field in the fluid. The global polarization is an integration of the local one in the whole phase space. In this note, we review recent progress on the vorticity formation and spin polarization in heavy ion collisions with transport models. We present an introduction of the fluid vorticity in non-relativistic and relativistic hydrodynamics. We discuss the spin polarization in a vortical fluid in the Wigner function formalism for massive spin-1/2 fermions, in which we derive the freeze-out formula for the spin polarization in heavy ion collisions. Then we show results for various properties of the kinematic and thermal vorticity with transport models, including: the evolution in time and space, the correlation to the participant plane, the collision energy dependence, etc. Finally, we give a brief overview of recent theoretical results for the spin polarization of Λ and $\bar{\Lambda}$ including the global and local polarization, the polarization harmonics in azimuthal angles, the sign problem in the longitudinal polarization as well as the polarization difference between particles and antiparticles.

Acknowledgements We thank Wei-Tian Deng, Hui Li, Yu-Chen Liu, Yin Jiang, Zi-Wei Lin, Long-Gang Pang, Shuzhe Shi, De-Xian Wei, and Xin-Nian Wang for collaborations and discussions. This work is supported in part by the NSFC Grants No. 11535012, No. 11675041, and No. 11890713, as well as by the NSF Grant No. PHY-1913729 and by the U.S. Department of Energy, Office of Science, Office of Nuclear Physics, within the framework of the Beam Energy Scan Theory (BEST) Topical Collaboration.

References

1. Liang, Z.T., Wang, X.N.: Phys. Rev. Lett. **94** (2005). arXiv:nucl-th/0410079 [nucl-th]
2. Liang, Z.T., Wang, X.N.: Phys. Lett. B **629**, 20–26 (2005). arXiv:nucl-th/0411101 [nucl-th]
3. Voloshin, S.A.: arXiv:nucl-th/0410089 [nucl-th]
4. Betz, B., Gyulassy, M., Torrieri, G.: Phys. Rev. C **76** (2007). arXiv:0708.0035 [nucl-th]
5. Becattini, F., Piccinini, F., Rizzo, J.: Phys. Rev. C **77** (2008). arXiv:0711.1253 [nucl-th]
6. Gao, J.H., Chen, S.W., Deng, W.t., Liang, Z.T., Wang, Q., Wang, X.N.: Phys. Rev. C **77**, 044902 (2008). arXiv:0710.2943 [nucl-th]
7. Huang, X.G., Huovinen, P., Wang, X.N.: Phys. Rev. C **84** (2011). arXiv:1108.5649 [nucl-th]
8. Chen, S.W., Deng, J., Gao, J.H., Wang, Q.: Front. Phys. China **4**, 509–16 (2009) arXiv:0801.2296 [hep-ph]
9. Adamczyk, L., et al.: STAR. Nature **548**, 62–65 (2017). arXiv:1701.06657 [nucl-ex]
10. Adam, J., et al.: STAR. Phys. Rev. C **98** (2018). arXiv:1805.04400 [nucl-ex]
11. Acharya, S., et al.: ALICE. Phys. Rev. C **101** (2020). arXiv:1909.01281 [nucl-ex]
12. Zhang, J.J., Fang, R.H., Wang, Q., Wang, X.N.: Phys. Rev. C **100**, 064904 (2019). arXiv:1904.09152 [nucl-th]
13. Baznat, M., Gudima, K., Sorin, A., Teryaev, O.: Phys. Rev. C **88** (2013). arXiv:1301.7003 [nucl-th]
14. Csernai, L.P., Magas, V.K., Wang, D.J.: Phys. Rev. C **87** (2013). arXiv:1302.5310 [nucl-th]
15. Csernai, L.P., Wang, D.J., Bleicher, M., Stöcker, H.: Phys. Rev. C **90** (2014)
16. Becattini, F., Inghirami, G., Rolando, V., Beraudo, A., Del Zanna, L., De Pace, A., Nardi, M., Pagliara, G., Chandra, V.: Eur. Phys. J. C **75**, 406 (2015). arXiv:1501.04468 [nucl-th]
17. Teryaev, O., Usubov, R.: Phys. Rev. C **92** (2015)
18. Jiang, Y., Lin, Z.W., Liao, J.: Phys. Rev. C **94** (2016). arXiv:1602.06580 [hep-ph]
19. Deng, W.T., Huang, X.G.: Phys. Rev. C **93** (2016). arXiv:1603.06117 [nucl-th]
20. Ivanov, Y.B., Soldatov, A.A.: Phys. Rev. C **95** (2017). arXiv:1701.01319 [nucl-th]
21. Deng, X.G., Huang, X.G., Ma, Y.G., Zhang, S.: Phys. Rev. C **101** (2020). arXiv:2001.01371 [nucl-th]
22. Li, H., Pang, L.G., Wang, Q., Xia, X.L.: Phys. Rev. C **96** (2017). arXiv:1704.01507 [nucl-th]
23. Wei, D.X., Deng, W.T., Huang, X.G.: Phys. Rev. C **99** (2019). arXiv:1810.00151 [nucl-th]
24. Shi, S., Li, K., Liao, J.: Phys. Lett. B **788**, 409–413 (2019). arXiv:1712.00878 [nucl-th]
25. Karpenko, I., Becattini, F.: Eur. Phys. J. C **77**, 213 (2017). arXiv:1610.04717 [nucl-th]
26. Xie, Y., Wang, D., Csernai, L.P.: Phys. Rev. C **95** (2017). arXiv:1703.03770 [nucl-th]

27. Sun, Y., Ko, C.M.: Phys. Rev. C **96** (2017). arXiv:1706.09467 [nucl-th]
28. Xie, Y.L., Bleicher, M., Stöcker, H., Wang, D.J., Csernai, L.P.: Phys. Rev. C **94** (2016). arXiv:1610.08678 [nucl-th]
29. Moffatt, H.K.: J. Fluid Mech. **35**, 117 (1969)
30. Moreau, J.J., Acad, C.R.: Sci. Paris **252**, 2810 (1961)
31. Heinz, U.W.: Phys. Rev. Lett. **51**, 351 (1983)
32. Elze, H.T., Gyulassy, M., Vasak, D.: Nucl. Phys. B **276**, 706–728 (1986)
33. Vasak, D., Gyulassy, M., Elze, H.T.: Annals Phys. **173**, 462–492 (1987)
34. Zhuang, P., Heinz, U.W.: Annals Phys. **245**, 311–338 (1996). arXiv:nucl-th/9502034 [nucl-th]
35. Gao, J.H., Liang, Z.T., Pu, S., Wang, Q., Wang, X.N.: Phys. Rev. Lett. **109** (2012). arXiv:1203.0725 [hep-ph]
36. Chen, J.W., Pu, S., Wang, Q., Wang, X.N.: Phys. Rev. Lett. **110** (2013). arXiv:1210.8312 [hep-th]
37. Gao, J.H., Wang, Q.: Phys. Lett. B **749** 542–546 (2015). arXiv:1504.07334 [nucl-th]
38. Fang, R.H., Pang, L.G., Wang, Q., Wang, X.N.: Phys. Rev. C **94**, 024904 (2016). arXiv:1604.04036 [nucl-th]
39. Hidaka, Y., Pu, S., Yang, D.L.: Phys. Rev. D **95** (2017). arXiv:1612.04630 [hep-th]
40. Mueller, N., Venugopalan, R.: Phys. Rev. D **97** (2018). arXiv:1701.03331 [hep-ph]
41. Huang, A., Shi, S., Jiang, Y., Liao, J., Zhuang, P.: Phys. Rev. D **98** (2018). arXiv:1801.03640 [hep-th]
42. Liu, Y.C., Gao, L.L., Mameda, K., Huang, X.G.: Phys. Rev. D **99** (2019). arXiv:1812.10127 [hep-th]
43. Gao, J.H., Pu, S., Wang, Q.: Phys. Rev. D **96**, 016002 (2017). arXiv:1704.00244 [nucl-th]
44. Gao, J.H., Liang, Z.T., Wang, Q., Wang, X.N.: Phys. Rev. D **98** (2018). arXiv:1802.06216 [hep-ph]
45. Weickgenannt, N., Sheng, X.L., Speranza, E., Wang, Q., Rischke, D.H.: Phys. Rev. D **100** (2019). arXiv:1902.06513 [hep-ph]
46. Gao, J.H., Liang, Z.T.: Phys. Rev. D **100** (2019). arXiv:1902.06510 [hep-ph]
47. Hattori, K., Hidaka, Y., Yang, D.L.: Phys. Rev. D **100** (2019). arXiv:1903.01653 [hep-ph]
48. Wang, Z., Guo, X., Shi, S., Zhuang, P.: Phys. Rev. D **100** (2019). arXiv:1903.03461 [hep-ph]
49. Liu, Y.C., Mameda, K., Huang, X.G.: Chin. Phys. C **44** (2020). arXiv:2002.03753 [hep-ph]
50. Sheng, X.L., Wang, Q., Huang, X.G.: Phys. Rev. D **102**(2), 025019 (2020). https://doi.org/10.1103/PhysRevD.102.025019, [arXiv:2005.00204 [hep-ph]]
51. Florkowski, W., Kumar, A., Ryblewski, R.: Phys. Rev. C **98**(4), 044906 (2018). https://doi.org/10.1103/PhysRevC.98.044906, [arXiv:1806.02616 [hep-ph]]
52. Yang, D.L., Hattori, K., Hidaka, Y.: JHEP **20**, 070 (2020). https://doi.org/10.1007/JHEP07(2020)070. arXiv:2002.02612 [hep-ph]
53. Weickgenannt, N., Speranza, E., Sheng, X.l., Wang, Q., Rischke, D.H.: arXiv:2005.01506 [hep-ph]
54. Wang, Z., Guo, X., Zhuang, P.: arXiv:2009.10930 [hep-th]
55. Becattini, F., Chandra, V., Del Zanna, L., Grossi, E.: Annals Phys. **338**, 32–49 (2013). arXiv:1303.3431 [nucl-th]
56. Pang, L.G., Petersen, H., Wang, Q., Wang, X.N.: Phys. Rev. Lett. **117** (2016). arXiv:1605.04024 [hep-ph]
57. Becattini, F., Karpenko, I.: Phys. Rev. Lett. **120** (2018). arXiv:1707.07984 [nucl-th]
58. Xia, X.L., Li, H., Tang, Z.B., Wang, Q.: Phys. Rev. C **98** (2018). arXiv:1803.00867 [nucl-th]
59. Del Zanna, L., Chandra, V., Inghirami, G., Rolando, V., Beraudo, A., De Pace, A., Pagliara, G., Drago, A., Becattini, F.: Eur. Phys. J. C **73**, 2524 (2013). arXiv:1305.7052 [nucl-th]
60. Pang, L., Wang, Q., Wang, X.N.: Phys. Rev. C **86** (2012). arXiv:1205.5019 [nucl-th]
61. Pang, L.G., Petersen, H., Wang, X.N.: Phys. Rev. C **97** (2018). arXiv:1802.04449 [nucl-th]
62. Hirano, T., Huovinen, P., Murase, K., Nara, Y.: Prog. Part. Nucl. Phys. **70** 108–158 (2013). arXiv:1204.5814 [nucl-th]
63. Bloczynski, J., Huang, X.G., Zhang, X., Liao, J.: Phys. Lett. B **718**, 1529–1535 (2013). arXiv:1209.6594 [nucl-th]

64. Bloczynski, J., Huang, X.G., Zhang, X., Liao, J.: Nucl. Phys. A **939**, 85–100 (2015). arXiv:1311.5451 [nucl-th]
65. Deng, W.T., Huang, X.G.: Phys. Rev. C **85** (2012). arXiv:1201.5108 [nucl-th]
66. Abelev, B.I., et al.: STAR. Phys. Rev. C **76** (2007). arXiv:0705.1691 [nucl-ex]
67. Siddique, I., Liang, Z.T., Lisa, M.A., Wang, Q., Xu, Z.B.: Chin. Phys. C **43** (2019). arXiv:1710.00134 [nucl-th]
68. Sheng, X.L., Oliva, L., Wang, Q.: Phys. Rev. D **101** (2020). arXiv:1910.13684 [nucl-th]
69. Sheng, X.L., Wang, Q., Wang, X.N.: Phys. Rev. D **102** (2020). arXiv:2007.05106 [nucl-th]
70. Xia, X.L., Li, H., Huang, X.G., Huang, H.Z.: arXiv:2010.01474 [nucl-th]
71. Taya, H., et al.: [ExHIC-P], Phys. Rev. C **102**, 021901 (2020). arXiv:2002.10082 [nucl-th]
72. Ipp, A., Di Piazza, A., Evers, J., Keitel, C.H.: Phys. Lett. B **666**, 315–319 (2008). arXiv:0710.5700 [hep-ph]
73. Becattini, F., Karpenko, I., Lisa, M., Upsal, I., Voloshin, S.: Phys. Rev. C **95** (2017). arXiv:1610.02506 [nucl-th]
74. Xia, X.L., Li, H., Huang, X.G., Huang, H.Z.: Phys. Rev. C **100** (2019). arXiv:1905.03120 [nucl-th]
75. Becattini, F., Cao, G., Speranza, E.: Eur. Phys. J. C **79**, 741 (2019). arXiv:1905.03123 [nucl-th]
76. Kornas, F. for HADES Collaboration, Talk given at Strange Quark Matter 2019, Bali, Italy, June 11–15, 2019
77. Florkowski, W., Friman, B., Jaiswal, A., Speranza, E.: Phys. Rev. C **97** (2018). arXiv:1705.00587 [nucl-th]
78. Hattori, K., Hongo, M., Huang, X.G., Matsuo, M., Taya, H.: Phys. Lett. B **795**, 100–106 (2019). arXiv:1901.06615 [hep-th]
79. Sun, Y., Ko, C.M.: Phys. Rev. C **99** (2019). arXiv:1810.10359 [nucl-th]
80. Liu, S.Y.F., Sun, Y., Ko, C.M.: Phys. Rev. Lett. **125** (2020). arXiv:1910.06774 [nucl-th]
81. Adam, J., et al.: STAR. Phys. Rev. Lett. **123** (2019). arXiv:1905.11917 [nucl-ex]
82. Florkowski, W., Kumar, A., Ryblewski, R., Mazeliauskas, A.: Phys. Rev. C **100** (2019). arXiv:1904.00002 [nucl-th]
83. Wu, H.Z., Pang, L.G., Huang, X.G., Wang, Q.: Phys. Rev. Res. **1** (2019). arXiv:1906.09385 [nucl-th]
84. Wang, Q.: Nucl. Phys. A **967**, 225–232 (2017) arXiv:1704.04022 [nucl-th]
85. Liu, Y.C., Huang, X.G.: Nucl. Sci. Tech. **31**, 56 (2020). arXiv:2003.12482 [nucl-th]
86. Huang, X.G.: arXiv:2002.07549 [nucl-th]
87. Becattini, F., Lisa, M.A.: arXiv:2003.03640 [nucl-ex]
88. Gao, J.H., Ma, G.L., Pu, S., Wang, Q.: Nucl. Sci. Tech. **31**, 90 (2020). arXiv:2005.10432 [hep-ph]
89. Muller, B., Schaefer, A.: Phys. Rev. D **98** (2018). arXiv:1806.10907 [hep-ph]
90. Guo, Y., Shi, S., Feng, S., Liao, J.: Phys. Lett. B **798** (2019). arXiv:1905.12613 [nucl-th]
91. Guo, X., Liao, J., Wang, E.: Sci. Rep. **10**, 2196 (2020). arXiv:1904.04704 [hep-ph]

Connecting Theory to Heavy Ion Experiment

10

Gaoqing Cao and Iurii Karpenko

Abstract

Only a fraction of all Λ and $\bar{\Lambda}$ hyperons detected in heavy ion collisions are produced from the hot and dense matter directly at the hadronization. These hyperons are called the *primary* hyperons. The rest of the hyperons are products of the decays of heavier hyperon states, which in turn are produced at the hadronization. As such, the polarization of only primary hyperons can be described with the formulae introduced in Sect. 8. For the rest of the hyperons, the polarization transfer in the decays has to be computed, and convoluted with the polarization of the mother hyperon. In this chapter, a derivation of the polarization transfer coefficients, as well as the computation of the mean polarization of all Λ hyperons detected in the experiment, is presented. The chapter is concluded with the calculation of the resonance contributions to the global and local Λ polarizations.

10.1 Introduction

Only a fraction of all Λ and $\bar{\Lambda}$ hyperons detected in heavy ion collisions are produced directly at the hadronization stage and thus are called *primary*. Indeed, a large fraction thereof stems from decays of heavier hyperons and one should account for the feed-down from higher lying resonances when trying to extract information about the vorticity from measurements of Λ polarizations. Particularly, the most important feed-down channels involve the strong decay $\Sigma^* \to \Lambda + \pi$, the electromagnetic (EM) decay $\Sigma^0 \to \Lambda + \gamma$, and the weak decay $\Xi \to \Lambda + \pi$ [1]. Of course, there are

G. Cao (✉)
Sun Yat-sen University, Guangzhou 510275, China
e-mail: caogaoqing@mail.sysu.edu.cn

I. Karpenko
Czech Technical University in Prague, Břehová 7, 11519 Prague, Czech Republic
e-mail: yu.karpenko@gmail.com

© Springer Nature Switzerland AG 2021
F. Becattini et al. (eds.), *Strongly Interacting Matter under Rotation*,
Lecture Notes in Physics 987,
https://doi.org/10.1007/978-3-030-71427-7_10

also many heavier resonances which decay to either Σ^0 or Λ. As a matter of fact, in the heavy ion collisions of RHIC at energy $\sqrt{s_{NN}} = 200$ GeV, the primary Λ hyperons were predicted to contribute only a quarter to all that measured [1]. Therefore, the non-primary Λ contributions from heavier hyperon decays dominate the final yield and may alter the polarization features of primary Λ. When polarized particles decay, their daughters are themselves polarized because of angular momentum conservation. In general, the fractions of polarization, which are inherited by the daughters or transferred from the mother to the daughters, depend on the momenta of the daughters in the rest frame of the mother.

Even though the theoretical predictions [1] and experimental measurements are consistent with each other on the global polarization of Λ, that is, along the direction of the total angular momentum of fireball, they contradict with each other for the sign of either the transverse local polarizations (TLPs) [with the global one excluded] or longitudinal local polarization (LLP); see the theoretical calculations with primary Λ [2–5] and recent experimental measurements at STAR [6,7]. Before any further great efforts are devoted to solving the "sign puzzles", one has to firstly check the simplest possibility: the feed-down effect from higher lying hyperon decays on the Λ polarization. In this section, we're going to explore the feed-down effects on Λ polarizations, especially the local ones. The theoretical derivations are mainly based on [1,8,9] and the relevant experimental measurements were reported in [6,7]. All the secondary contributions to Λ are two-body decays, thus we can generally denote the decay as $H \rightarrow \Lambda + X$, where H and X refer to the mother particle (heavier hyperon) and byproduct daughter particle (usually pions π or photon γ), respectively.

In the following, H, Λ and X will be simply called *Mother, Daughter* and *Byproduct*, so that the derivations and discussions can be generally applied to any other two-body fermion decays. Local thermodynamic equilibrium will be assumed for both the kinematics and spin dynamics of the system, and the small scattering interactions between hadrons will be neglected after the hadronization stage. Generally, three reference frames are involved in the study: the QGP frame (QGPF) which is the center of mass of the colliding nuclei in a collider experiment and where the laboratory observations base, the Mother's rest frame (MRF) and the Daughter's rest frame (DRF). We assign different notations for the physical quantities in these frames: regular in the QGPF, subscript "$*$" in the MRF, and subscript "o" in the DRF.

This chapter is arranged as follows. As a first trial, in Sect. 10.2, we derive the proportional coefficients for global polarization transfers from Mothers to Daughters by following the simple momentum-integrated formalism. The following sections mainly focus on the complicated local polarization transfers: In Sect. 10.3, spin density matrix and derivation of momentum-dependent polarization are presented for the Mothers. Based on that, in Sect. 10.4, we further use a formalism of reduced spin density matrix to calculate the local polarization of the Daughter in the decays, which is then weighted over the momentum distributions of the Mothers in Sect. 10.5. Finally, we compare our numerical calculations with the experimental measurements in Sect. 10.6.

10.2   Global Polarization Transfer to the Daughter

As long as one is interested in the momentum-integrated mean spin vector (MSV) of the Daughter in *its* own rest frame, we will show that a simple linear rule applies with respect to that of the Mother, that is,

$$\mathbf{S}_{\Lambda o} = C_S \, \mathbf{S}_{H*}, \tag{10.1}$$

where C_S is a coefficient which may or may not depend on the dynamical matrix elements. The proportionality between these two MSVs should be expected as, once the momentum integrations are carried out, the only special direction for the MSV of the Daughter is parallel to that of the MSV of the Mother. In many two-body decays, the conservation laws constrain the final state to such an extent that the coefficient C_S is *independent* of the dynamical matrix elements. This happens, e.g., in the strong decay $\Sigma^*(1385) \to \Lambda\pi$ and the EM decay $\Sigma^0 \to \Lambda\gamma$, but not in the weak decay $\Xi \to \Lambda\pi$. Thus, this section will be devoted to determining the exact expressions of the coefficient C_S for both strong and EM decays. For the exploration of global polarization transfer, the summations over all angular momentum components of the Daughter and Byproduct, λ_Λ, λ'_Λ and λ_X, should be understood for both the MSVs and the normalization factors. For brevity, we will suppress the summation symbol "$\sum$" over these indices in this section.

We will work out the exact relativistic results. In the relativistic framework, the use of the helicity basis is very convenient for complete descriptions of the helicity and alternative spin formalisms; we refer the readers to [10–13]. For the Mother, with spin j and the z component m in its rest frame,[1] decaying into two particles Λ and X, the final state $|\psi\rangle$ can be written as a superposition of states with definite momenta and helicities:

$$|\psi\rangle \propto \int d\Omega_* \, D^j(\phi_*, \theta_*, 0)^{m*}_\lambda \, |\mathbf{p}_*, \lambda_\Lambda, \lambda_X\rangle \, T^j(\lambda_\Lambda, \lambda_X). \tag{10.2}$$

Here, $\lambda = \lambda_\Lambda - \lambda_X$ with λ_Λ and λ_X the helicities of the daughters in the MRF, $\mathbf{p}_*$ is the three-momentum of the Daughter Λ with θ_* and ϕ_* its spherical coordinates and $d\Omega_* = \sin\theta_* d\theta_* d\phi_*$ the corresponding infinitesimal solid angle, D^j is the Wigner rotation matrix in the representation of spin j, and $T^j(\lambda_\Lambda, \lambda_X)$ are the reduced dynamical amplitudes depending only on the final helicities.

Then, the relativistic MSV of the Daughter is given by

$$S^\mu_{\Lambda*} = \langle\psi| \, \widehat{S}^\mu_{\Lambda*} \, |\psi\rangle$$

---

[1] In the rest frame, helicity coincides with the eigenvalue of the spin operator $\widehat{S}$, conventionally $\widehat{S}_3$; see the textbooks.

with $\langle \psi | \psi \rangle = 1$, hence we have explicitly

$$
\begin{aligned}
S_{\Lambda *}^{\mu} &= \frac{\int d\Omega_* D^j (\phi_*, \theta_*, 0)_{\lambda}^{m*} D^j (\phi_*, \theta_*, 0)_{\lambda'}^{m} \langle \lambda_{\Lambda}' | \widehat{S}_{\Lambda *}^{\mu} | \lambda_{\Lambda} \rangle T^j (\lambda_{\Lambda}, \lambda_X) T^j (\lambda_{\Lambda}', \lambda_X)^*}{\int d\Omega_* |D^j (\phi_*, \theta_*, 0)_{\lambda}^{m*}|^2 |T^j (\lambda_{\Lambda}, \lambda_X)|^2} \\
&= \frac{\int d\Omega_* D^j (\phi_*, \theta_*, 0)_{\lambda}^{m*} D^j (\phi_*, \theta_*, 0)_{\lambda'}^{m} \langle \lambda_{\Lambda}' | \widehat{S}_{\Lambda *}^{\mu} | \lambda_{\Lambda} \rangle T^j (\lambda_{\Lambda}, \lambda_X) T^j (\lambda_{\Lambda}', \lambda_X)^*}{\frac{4\pi}{2j+1} |T^j (\lambda_{\Lambda}, \lambda_X)|^2}. \quad (10.3)
\end{aligned}
$$

According to our conventions, we emphasize that the numerator should sum over λ_{Λ}, λ_{Λ}', and λ_X and the denominator over λ_{Λ} and λ_X, separately. To derive this expression, we have used the known results for the integrals of the Wigner D matrices and the fact that the operator $\widehat{S}_{\Lambda *}$ does not change the momentum eigenvalue as well as the helicity of the Byproduct X.

Now, the most important term in (10.3) is the transition amplitude of the spin operator: $\langle \lambda_{\Lambda}' | \widehat{S}_{\Lambda *}^{\mu} | \lambda_{\Lambda} \rangle$. To evaluate it, we decompose the spin operator as the following:

$$
\widehat{S}_{\Lambda *} = \sum_i \widehat{S}_{\Lambda *}^i n_i (p_*)
$$

with $n_i (p_*)$ the three space-like unit vectors orthogonal to the four-momentum p_*. Their expression can be obtained by applying the so-called *standard Lorentz transformation* $[p_*]$, which turns the unit time vector $\hat{t}$ into the direction of the four-momentum p_* [10], to the three spatial axis vectors $\mathbf{e}_i$, namely,

$$
n_i (p_*) = [p_*](\mathbf{e}_i),
$$

so that

$$
\widehat{S}_{\Lambda *} = [p_*] \left(\sum_i \widehat{S}_{\Lambda *}^i \mathbf{e}_i \right) \quad (10.4)
$$

by taking advantage of the linearity of $[p_*]$. It is more convenient to rewrite the sum in the argument of $[p_*]$ along the spherical vector basis:

$$
\mathbf{e}_{\pm} = \mp \frac{1}{\sqrt{2}} (\mathbf{e}_1 \pm i \mathbf{e}_2), \quad \mathbf{e}_0 = \mathbf{e}_3,
$$

upon which the D^j matrix elements are defined. We have

$$
\sum_i \widehat{S}_{\Lambda *}^i \mathbf{e}_i = -\frac{1}{\sqrt{2}} \widehat{S}_{\Lambda *}^- \mathbf{e}_+ + \frac{1}{\sqrt{2}} \widehat{S}_{\Lambda *}^+ \mathbf{e}_- + \widehat{S}_{\Lambda *}^0 \mathbf{e}_0 = \sum_{n=-1}^{1} a_n \widehat{S}_{\Lambda *}^{-n} \mathbf{e}_n, \quad (10.5)
$$

where $\widehat{S}_{\Lambda *}^{\pm} = \widehat{S}_{\Lambda *}^1 \pm i \widehat{S}_{\Lambda *}^2$ are the familiar spin ladder operators and $a_n = -n/\sqrt{2} + \delta_{n,0}$. The actions of these new operators onto the helicity ket $|\lambda_{\Lambda}\rangle$ are precisely the well-known ones onto the eigenstate of the z component of angular momentum operator with eigenvalue λ_{Λ}, e.g.,

$$
\langle \lambda_{\Lambda}' | \widehat{S}_{\Lambda *}^0 | \lambda_{\Lambda} \rangle = \lambda_{\Lambda} \delta_{\lambda_{\Lambda}, \lambda_{\Lambda}'}.
$$

Then, we can utilize (10.4) and (10.5) to rewrite the transition amplitude in a more explicit form as

$$\langle \lambda'_\Lambda | \widehat{S}_{\Lambda *} | \lambda_\Lambda \rangle = \sum_{n=-1}^{1} a_n \langle \lambda'_\Lambda | \widehat{S}_{\Lambda *}^{-n} | \lambda_\Lambda \rangle \, [p_*](\mathbf{e}_n). \tag{10.6}$$

In order to work out (10.6), we need to find an explicit expression for the standard transformation $[p_*]$. In principle, it can be chosen freely but our choice treats λ_Λ the Daughter's helicity [11,12], then the expression is

$$[p_*] = R_z(\phi_*) R_y(\theta_*) L_z(\xi). \tag{10.7}$$

This is just a Lorentz boost along the z-axis with a hyperbolic angle ξ such that $\sinh \xi = \|\mathbf{p}_*\|/m_\Lambda$, followed by a rotation around the y-axis with an angle θ_* and another one around the z-axis with an angle ϕ_*. Then,

$$[p_*](\mathbf{e}_\pm) = R_z(\phi_*) R_y(\theta_*)(\mathbf{e}_\pm) = \sum_{l=-1}^{1} D^1(\phi_*, \theta_*, 0)^l_{\pm 1} \mathbf{e}_l$$

as $\mathbf{e}_\pm$ are Lorentz invariant under the boost along the z-axis. Conversely, $\mathbf{e}_0$ is not invariant under the Lorentz boost and transforms as

$$[p_*](\mathbf{e}_0) = \cosh \xi R_z(\phi_*) R_y(\theta_*)(\mathbf{e}_0) + \sinh \xi R_z(\phi_*) R_y(\theta_*)(\hat{t})$$

$$= \sum_{l=-1}^{1} \frac{\varepsilon_{\Lambda *}}{m_\Lambda} D^1(\phi_*, \theta_*, 0)^l_0 \mathbf{e}_l + \frac{\mathrm{p}_*}{m_\Lambda} \hat{t},$$

where $\mathrm{p}_* = \|\mathbf{p}_*\|$ and the energy $\varepsilon_{\Lambda *} = \sqrt{\mathrm{p}_*^2 + m_\Lambda^2}$. By substituting these transformations into (10.6), we eventually get the most explicit form

$$\langle \lambda'_\Lambda | \widehat{S}_{\Lambda *} | \lambda_\Lambda \rangle = \sum_{l,n} b_n D^1(\phi_*, \theta_*, 0)^l_n \langle \lambda'_\Lambda | \widehat{S}_{\Lambda *}^{-n} | \lambda_\Lambda \rangle \, \mathbf{e}_l + \lambda_\Lambda \delta_{\lambda_\Lambda, \lambda'_\Lambda} \frac{\mathrm{p}_*}{m_\Lambda} \hat{t}, \tag{10.8}$$

where $b_n = -n/\sqrt{2} + \gamma_{\Lambda *} \delta_{n,0}$ with $\gamma_{\Lambda *} = \varepsilon_{\Lambda *}/m_\Lambda$ the Lorentz factor of the Daughter.

We can now write down the fully expanded expression of the MSV $S_{\Lambda *}$ following (10.3). The time component is especially simple: By using (10.8), one has

$$S_{\Lambda *}^0 = \frac{\mathrm{p}_*}{m_\Lambda} \frac{\lambda_\Lambda \int d\Omega_* \, |D^j(\phi_*, \theta_*, 0)^{m*}_\lambda|^2 |T^j(\lambda_\Lambda, \lambda_X)|^2}{\frac{4\pi}{2j+1} |T^j(\lambda_\Lambda, \lambda_X)|^2}, \tag{10.9}$$

which then reduces to

$$S_{\Lambda *}^0 = \frac{\mathrm{p}_*}{m_\Lambda} \frac{\lambda_\Lambda |T^j(\lambda_\Lambda, \lambda_X)|^2}{|T^j(\lambda_\Lambda, \lambda_X)|^2} \tag{10.10}$$

after carrying out the integral over Ω_* in the numerator. Similarly, the spatial components read

$$
\mathbf{S}_{\Lambda*} = \frac{T^j(\lambda_\Lambda, \lambda_X) T^j(\lambda'_\Lambda, \lambda_X)^*}{\frac{4\pi}{2j+1}|T^j(\lambda_\Lambda, \lambda_X)|^2} \sum_{n,l} \langle \lambda'_\Lambda | \widehat{S}^{-n}_{\Lambda*} | \lambda_\Lambda \rangle
$$
$$
\times b_n \int d\Omega_* \, D^j(\phi_*, \theta_*, 0)^{m*}_\lambda D^j(\phi_*, \theta_*, 0)^m_{\lambda'} D^1(\phi_*, \theta_*, 0)^l_n \mathbf{e}_l. \quad (10.11)
$$

Since the analytic results for the angular integrals in (10.11) are well known, the expression can be greatly simplified in terms of Clebsch-Gordan coefficients:

$$
\mathbf{S}_{\Lambda*} = \frac{T^j(\lambda_\Lambda, \lambda_X) T^j(\lambda'_\Lambda, \lambda_X)^* \sum_{n,l} b_n \langle \lambda'_\Lambda | \widehat{S}^{-n}_{\Lambda*} | \lambda_\Lambda \rangle \langle jm | j1 | ml \rangle \langle j\lambda | j1 | \lambda'n \rangle \mathbf{e}_l}{T^j(\lambda_\Lambda, \lambda_X)|^2}
$$
$$
= \frac{T^j(\lambda_\Lambda, \lambda_X) T^j(\lambda'_\Lambda, \lambda_X)^* \sum_n b_n \langle \lambda'_\Lambda | \widehat{S}^{-n}_{\Lambda*} | \lambda_\Lambda \rangle \langle jm | j1 | m0 \rangle \langle j\lambda | j1 | \lambda'n \rangle \mathbf{e}_0}{T^j(\lambda_\Lambda, \lambda_X)|^2} \quad (10.12)
$$

Note that the only non-vanishing spatial component of the MSV $\mathbf{S}_{\Lambda*}$ is the one along the z-axis or proportional to $\mathbf{e}_0 = \mathbf{e}_3$. As the Mother is polarized along the z direction by construction, this is a consequence of rotational invariance. We'd like to mention that the integrands in both (10.9) and (10.11), as functions of the angular variables θ_* and ϕ_*, are proportional to the MSV $S_{\Lambda*}(\mathbf{p}_*)$ at some given momentum $\mathbf{p}_*$; see more details in the following sections.

So far, what we have calculated is the MSV of the Daughter in the Mother's rest frame. However, one is more interested in the MSV in the Daughter's rest frame. For a given momentum $\mathbf{p}_*$, this can be obtained by means of Lorentz boost:

$$
\mathbf{S}_{\Lambda o}(\mathbf{p}_*) = \mathbf{S}_{\Lambda*}(\mathbf{p}_*) - \frac{\mathbf{p}_*}{\varepsilon_{\Lambda*}(\varepsilon_{\Lambda*} + m_\Lambda)} \mathbf{S}_{\Lambda*}(\mathbf{p}_*) \cdot \mathbf{p}_*.
$$

As $\mathbf{S}_{\Lambda*}$ is a four-vector orthogonal to p_* and hence $\mathbf{S}_{\Lambda*}(\mathbf{p}_*) \cdot \mathbf{p}_* = S^0_{\Lambda*}(\mathbf{p}_*)\varepsilon_{\Lambda*}$, we can evaluate the momentum-integrated MSV in DRF by the following:

$$
\mathbf{S}_{\Lambda o} \equiv \langle \mathbf{S}_{\Lambda o}(\mathbf{p}_*) \rangle = \langle \mathbf{S}_{\Lambda*}(\mathbf{p}_*) \rangle - \frac{\langle \mathbf{p}_* S^0_{\Lambda*}(\mathbf{p}_*) \rangle}{\varepsilon_{\Lambda*} + m_\Lambda} = \mathbf{S}_{\Lambda*} - \frac{\langle \mathbf{p}_* S^0_{\Lambda*}(\mathbf{p}_*) \rangle}{\varepsilon_{\Lambda*} + m_\Lambda}. \quad (10.13)
$$

The first term on the right-hand side is just the MSV shown in (10.12). While by adopting (10.9) and an alternative presentation of the momentum

$$
\mathbf{p}_* = \mathrm{p}_* \sum_{l=-1}^1 D^1(\phi_*, \theta_*, 0)^l_0 \mathbf{e}_l,
$$

the second term can be evaluated according to

$$
\begin{aligned}
\langle \mathbf{p}_* S^0_{\Lambda*}(\mathbf{p}_*) \rangle &= \frac{\mathbf{p}_*^2}{m_\Lambda} \frac{\lambda_\Lambda |T^j(\lambda_\Lambda, \lambda_X)|^2 \sum_{l=-1}^{1} \mathbf{e}_l \int d\Omega_* \, |D^j(\phi_*, \theta_*, 0)^{m*}_\lambda|^2 D^1(\phi_*, \theta_*, 0)^l_0}{\frac{4\pi}{2j+1} |T^j(\lambda_\Lambda, \lambda_X)|^2} \\
&= \frac{\mathbf{p}_*^2}{m_\Lambda} \frac{\lambda_\Lambda |T^j(\lambda_\Lambda, \lambda_X)|^2 \sum_{l=-1}^{1} \mathbf{e}_l \, \langle jm| \, j1 \, |ml\rangle \, \langle j\lambda| \, j1 \, |\lambda 0\rangle}{|T^j(\lambda_\Lambda, \lambda_X)|^2} \\
&= \frac{\mathbf{p}_*^2}{m_\Lambda} \frac{\lambda_\Lambda |T^j(\lambda_\Lambda, \lambda_X)|^2 \, \langle jm| \, j1 \, |m0\rangle \, \langle j\lambda| \, j1 \, |\lambda 0\rangle}{|T^j(\lambda_\Lambda, \lambda_X)|^2} \mathbf{e}_0.
\end{aligned}
\tag{10.14}
$$

Then, by collecting both (10.12) and (10.14) in (10.13), one finally gets

$$
S_{\Lambda o} = \frac{T^j(\lambda_\Lambda, \lambda_X) T^j(\lambda'_\Lambda, \lambda_X)^* \sum_n c_n \langle \lambda'_\Lambda| \, \widehat{S}^{-n}_{\Lambda*} |\lambda_\Lambda\rangle \, \langle jm| \, j1 \, |m0\rangle \, \langle j\lambda| \, j1 \, |\lambda'n\rangle}{|T^j(\lambda_\Lambda, \lambda_X)|^2} \mathbf{e}_0
\tag{10.15}
$$

with the parameter

$$
c_n = -\frac{n}{\sqrt{2}} + \left(\gamma_{\Lambda*} - \frac{\beta^2_{\Lambda*} \gamma^2_{\Lambda*}}{\gamma_{\Lambda*}+1} \right) \delta_{n,0} = -\frac{n}{\sqrt{2}} + \delta_{n,0}
\tag{10.16}
$$

the same as a_n. Note the disappearance of any dependence on the energy of the Daughter or the masses involved in the decay, once the MSV of the Daughter is boosted back to its rest frame; see also (10.1) and (10.18).

The MSV in (10.15) pertains to the Mother in state $|jm\rangle$, which is a pure eigenstate of its spin operator $\widehat{S}_z$ in its rest frame. For a mixed state, the MSV should be weighted over the probabilities P_m of different eigenstates.[2] Since it is known that

$$
\langle jm| \, j1 \, |m0\rangle = \frac{m}{\sqrt{j(j+1)}},
$$

the weighted average turns out to be

$$
S_{\Lambda o} = \sum_{m=-j}^{j} m P_m \mathbf{e}_0 \frac{T^j(\lambda_\Lambda, \lambda_X) T^j(\lambda'_\Lambda, \lambda_X)^* \sum_{n=-1}^{1} c_n \langle \lambda'_\Lambda| \, \widehat{S}^{-n}_{\Lambda*} |\lambda_\Lambda\rangle \, \langle j\lambda| \, j1 \, |\lambda'n\rangle}{\sqrt{j(j+1)} |T^j(\lambda_\Lambda, \lambda_X)|^2}.
\tag{10.17}
$$

It is easy to identify that $\sum_{m=-j}^{j} m P_m \mathbf{e}_0$ is just the MSV of the Mother $\mathbf{S}_{H*}$, so we finally verify that (10.1) holds, that is, the MSV of the Daughter in DRF is proportional to that of the Mother in MRF. And the explicit form of the proportional coefficient is now clear,

$$
C_S = \frac{T^j(\lambda_\Lambda, \lambda_X) T^j(\lambda'_\Lambda, \lambda_X)^* \sum_{n=-1}^{1} c_n \langle \lambda'_\Lambda| \, \widehat{S}^{-n}_{\Lambda*} |\lambda_\Lambda\rangle \, \langle j\lambda| \, j1 \, |\lambda'n\rangle}{\sqrt{j(j+1)} |T^j(\lambda_\Lambda, \lambda_X)|^2}.
\tag{10.18}
$$

---

[2] In the non-polarized case, P_m is the same for any $m \in [-j, \ldots, j]$.

According to the group theory, the Clebsch-Gordan coefficients involved in the estimation of (10.18) can be given directly as

$$\langle j\lambda | j1 | \lambda 0 \rangle = \frac{\lambda}{\sqrt{j(j+1)}}, \quad \langle j\lambda | j1 | (\lambda \mp 1) \pm 1 \rangle = \mp \sqrt{\frac{(j \mp \lambda + 1)(j \pm \lambda)}{2j(j+1)}}. \tag{10.19}$$

As has been mentioned before, the somewhat surprising feature of (10.18) is that C_S doesn't explicitly depend on the masses involved in the decay as c_n is independent of them. There of course might be an implicit dependence on the masses through the dynamical amplitudes T^j, but this actually cancels out due to the normalization in several important instances.

If the decay is driven by parity-conserving interaction, such as the strong decay $\Sigma^* \to \Lambda\pi$ and EM decay $\Sigma^0 \to \Lambda\gamma$, there is a known relation between the parity partners for the dynamical amplitudes [13]:

$$T^j(-\lambda_\Lambda, -\lambda_X) = \eta_H \eta_\Lambda \eta_X (-1)^{j - S_\Lambda - S_X} T^j(\lambda_\Lambda, \lambda_X). \tag{10.20}$$

Here, η_H, η_Λ and η_X are the intrinsic parities of the Mother, Daughter, and Byproduct, and j, S_Λ and S_X are their spins, respectively. Note that the helicity is constrained to $\lambda_X = \pm S_X$ in (10.20) if the Byproduct is massless [12]. In all these cases, one has

$$|T^j(-\lambda_\Lambda, -\lambda_X)|^2 = |T^j(\lambda_\Lambda, \lambda_X)|^2. \tag{10.21}$$

The (10.20) and (10.21) have interesting consequences: First of all, it can be readily realized that the time component of the MSV (10.10) vanishes. Secondly, if only one dynamical amplitude is independent in (10.18) because of the constraint from (10.20), the coefficient C_S can be finally reduced to a constant that is determined only by the conservation laws. We will see below that this is precisely the case for the decays $\Sigma^* \to \Lambda\pi$ and $\Sigma^0 \to \Lambda\gamma$.

A. Strong Decay $\Sigma^* \to \Lambda\pi$

In this case, $\lambda_X = 0$, $\lambda = \lambda_\Lambda$, $j = 3/2$ and $T^j(\lambda)$ is proportional to $T^j(-\lambda)$ through a phase factor, which turns out to be 1 according to (10.20). As $\lambda_\Lambda = \pm 1/2$, there is only one independent reduced helicity amplitude; thus, the coefficient C_S simplifies to

$$C_S = \sum_{n=-1}^{1} \sum_{\lambda, \lambda'} \langle \lambda' | \widehat{S}_{\Lambda^*}^{-n} | \lambda \rangle \frac{c_n}{\sqrt{j(j+1)}} \frac{\langle j\lambda | j1 | \lambda' n \rangle}{2S_\Lambda + 1}. \tag{10.22}$$

We now evaluate the three terms in the above summation over n one by one. For $n = 0$, one obtains

$$\sum_{\lambda = \pm 1/2} \frac{1}{2} \lambda^2 \frac{1}{j(j+1)} = \frac{1}{15},$$

where the first equation in (10.19) has been used. For $n = 1$, the corresponding ladder operator in (10.22) is $\widehat{S}_{\Lambda^*}^{-}$, which selects the term with $\lambda' = -1/2$ and $\lambda = 1/2$ as the only non-vanishing contribution. Similarly, for $n = -1$, the corresponding ladder operator $\widehat{S}_{\Lambda *}^{+}$ in (10.22) selects the opposite combination: $\lambda' = 1/2$ and $\lambda = -1/2$. According to the second equation in (10.19), the corresponding Clebsch-Gordan coefficients are opposite to each other for $n = \pm 1$. Then, by inserting (10.16), their contributions turn out to be the same, that is,

$$\frac{1}{2}\sqrt{\frac{8}{15}}\frac{1}{\sqrt{2}}\frac{1}{\sqrt{j(j+1)}} = \frac{2}{15}.$$

Therefore, the coefficient C_S is just

$$C_S = \frac{1}{15} + 2\frac{2}{15} = \frac{1}{3}, \tag{10.23}$$

which indicates that the MSV of the Daughter is along that of the Mother.

B. Electromagnetic Decay $\Sigma^0 \to \Lambda\gamma$

This case is fully relativistic as the Byproduct is a photon, then the helicity basis is compelling with $\lambda_X = \pm 1$. Now $j = 1/2$, and (10.3) indicates that

$$|\lambda| = |\lambda_\Lambda - \lambda_X| = 1/2,$$

thus only two choices are possible:

$$\lambda_X = 1 \implies \lambda_\Lambda = 1/2 \implies \lambda = -1/2 ,$$
$$\lambda_X = -1 \implies \lambda_\Lambda = -1/2 \implies \lambda = 1/2 ,$$

from which we can generally identify $\lambda_X = 2\lambda_\Lambda$ and $\lambda = -\lambda_\Lambda$ in (10.18). The same argument applies to $\lambda' = \lambda'_\Lambda - \lambda_X$, so we also have $\lambda_X = 2\lambda'_\Lambda$, whence $\lambda'_\Lambda = \lambda_\Lambda$ and $\lambda' = \lambda$. This in turn implies that only the term with $n = 0$ contributes in (10.18), which then reads

$$C_S = \frac{\lambda_\Lambda |T^j(\lambda_\Lambda, 2\lambda_\Lambda)|^2 \langle j - \lambda_\Lambda j 1 | -\lambda_\Lambda 0 \rangle}{\sqrt{j(j+1)}|T^j(\lambda_\Lambda, 2\lambda_\Lambda)|^2}. \tag{10.24}$$

Like the previous case, there is only one independent dynamical amplitude because of the constraint from (10.21), so (10.24) becomes

$$C_S = \sum_{\lambda_\Lambda = \pm 1/2} \lambda_\Lambda \frac{(-\lambda_\Lambda)}{j(j+1)} \frac{1}{2S_\Lambda + 1} \tag{10.25}$$

by inserting the first equation in (10.19). With the spins $j = S_\Lambda = 1/2$, we eventually recover the known result [14, 15]:

$$C_S = -\frac{1}{3},$$

which indicates that the MSV of the Daughter is along the opposite direction to that of the Mother.

10.3   Spin Density Matrix for the Mother and Its Polarization

In general, the Mother's eigenstates do not have definite spins in a local thermodynamic equilibrium (LTE) system with angular momentum-vorticity coupling, which
means that the Mother's spin can be altered. To account for the spin transition, the
spin density matrix (SDM) can be defined as follows: For the Mother with four-
momentum p_H in QGPF, the form is given by

$$\Theta(p_H)_{\sigma\sigma'} = \frac{\mathrm{tr}(\widehat{\rho}a^\dagger(p_H)_{\sigma'}a(p_H)_\sigma)}{\sum_\sigma \mathrm{tr}(\widehat{\rho}a^\dagger(p_H)_\sigma a(p_H)_\sigma)}, \tag{10.26}$$

where $a^\dagger(p_H)_\sigma$ and $a(p_H)_\sigma$ are creation and annihilation operators of the Mother
in the spin state σ, respectively. As mentioned before, the meaning of σ depends on
the choice of the *standard Lorentz transformation* $[p_H]$ which transforms $\hat{t}$ to the
direction of p_H [10]. For convenience and consistency, we adopt the choice that σ
stands for the particle's helicity [12] in the following. Similar to that for the Daughter
(10.7), the transformation is explicitly given by

$$[p_H] = \mathsf{R}(\phi,\theta,0)\mathsf{L}_z(\xi) = \mathsf{R}_z(\phi)\mathsf{R}_y(\theta)\mathsf{L}_z(\xi), \tag{10.27}$$

where the functions have the same meanings as those in (10.7) but is for the Mother
in QGPF here.

Then, by operating the transformation over the space-like orthonormal vector
basis $\mathbf{e}_i$ as $n_i(p_H) \equiv [p_H](\mathbf{e}_i)$ [1,10], one can readily determine the MSV of the
Mother from the SDM (10.26) as

$$S_H^\mu(p_H) = \sum_{i=1}^{3} \mathrm{tr}\left[D^j(\mathsf{J}^i)\Theta(p_H)\right]n_i(p_H)^\mu = \sum_{i=1}^{3}[p_H]_i^\mu\mathrm{tr}\left[D^j(\mathsf{J}^i)\Theta(p_H)\right], \tag{10.28}$$

where J^i are the angular momentum generators of the Mother and $D^j(\mathsf{J}^i)$ their irreducible representation matrices with total spin S. It should be stressed that, in spite
of the appearance of the Lorentz transformation $[p_H]$, the MSV is independent of its
particular choice as should be for any observables. Actually, the SDM (10.26) also
depends on the convention of $[p_H]$ implicitly through the definition of the spin variable σ, which just compensates the explicit dependence. By adopting the covariant
form of the irreducible representation matrix

$$D^j(\mathsf{J}^\lambda) = -\frac{1}{2}\epsilon^{\lambda\mu\nu\rho}D^j(\mathsf{J}_{\mu\nu})\hat{t}_\rho, \tag{10.29}$$

which indicates $D^j(\mathsf{J}^0) = 0$ for the unit time vector $\hat{t} = (1,0,0,0)$, the MSV (10.28)
can be conveniently rewritten with the full Lorentz covariant indices as

$$S_H^\mu(p_H) = [p_H]_\nu^\mu\mathrm{tr}\left[D^j(\mathsf{J}^\nu)\Theta(p_H)\right]. \tag{10.30}$$

Now, the most important mission is to evaluate the SDM for a general spin S, which is not an easy task in quantum field theory (QFT): Even for the simplest non-trivial case with the density operator involving the angular momentum-vorticity coupling, an exact solution is unknown. However, it is possible to find an explicit exact solution for a single species of relativistic quantum particles by neglecting quantum statistic (or quantum field) effects. In this case, the general density operator $\widehat{\rho}$ for a system in equilibrium is given by

$$\widehat{\rho} = \frac{1}{Z} \exp\left[-b \cdot \widehat{P} + \frac{1}{2}\varpi : \widehat{J}\right],$$

where b is a time-like constant four-vector, ϖ an anti-symmetric constant tensor, and $\widehat{P}$ and $\widehat{J}$ are the conserved total four-momentum and total angular momentum operators, respectively. As the scattering effects are neglected in our study, the system can be viewed as a set of non-interacting distinguishable particles. Then we can write

$$\widehat{P} = \sum_i \widehat{P}_i, \qquad \widehat{J} = \sum_i \widehat{J}_i,$$

and consequently $\widehat{\rho} = \otimes_i \widehat{\rho}_i$ with the density operator for a single particle species

$$\widehat{\rho}_i = \frac{1}{Z_i} \exp\left[-b \cdot \widehat{P}_i + \frac{1}{2}\varpi : \widehat{J}_i\right].$$

By following the Poincaré group algebra for the generators of translations $\widehat{P}_\mu$ and Lorentz transformations $\widehat{J}_{\mu\nu}$ [11,12]

$$[\widehat{P}_\mu, \widehat{P}_\nu] = 0, \qquad [\widehat{P}_\tau, \widehat{J}_{\mu\nu}] = -i(\widehat{P}_\mu \eta_{\nu\tau} - \widehat{P}_\nu \eta_{\mu\tau}), \qquad (10.31)$$

it can be shown that

$$\widehat{F}_{kj} \equiv \left[\left[(-b \cdot \widehat{P}_i), \left(\frac{1}{2}\varpi : \widehat{J}_i\right)^{(k)}\right], (-b \cdot \widehat{P}_i)^{(j)}\right]$$

$$= -(-i)^k \widehat{P}_i^\mu \underbrace{\left(\varpi_{\mu\nu_1}\varpi^{\nu_1\nu_2} \dots \varpi_{\nu_{k-1}\nu_k}\right)}_{k \text{ times}} b^{\nu_k} \delta_{j0}, \qquad (10.32)$$

where $[\widehat{A}, \widehat{B}^{(k)}] = [[\cdots[\widehat{A}, \widehat{B}], \cdots], \widehat{B}]$ with k times of nesting commutations and $\widehat{F}_{kj}$ commute with each other for any k and j. In the most general case with arbitrary operators $\widehat{A}$ and $\widehat{B}$, an identity has been derived:

$$e^{\widehat{A}+\widehat{B}} = e^{\widehat{A}} e^{\widehat{B}} e^{-\frac{1}{2}[\widehat{A},\widehat{B}]} e^{\frac{1}{6}[\widehat{A}^{(2)},\widehat{B}] - \frac{1}{3}[\widehat{A},\widehat{B}^{(2)}]} \cdots,$$

where the higher level commutation exponents in "$\cdots$" rely on the lower ones through some recursion relations [16]. In our present case, all the non-vanishing commutation exponents must be functions of $\widehat{P}_\mu$ according to (10.31) and commute with each other. Thus, a general identity can be applied to the density operator, and we have [16]

$$\widehat{\rho}_i = \frac{1}{Z_i} \exp\left\{\sum_{k=1}^\infty \frac{(-1)^k \widehat{F}_{k0}}{(k+1)!}\right\} \exp[-b \cdot \widehat{P}_i] \exp\left[\frac{1}{2}\varpi : \widehat{J}_i\right]. \quad (10.33)$$

Then, it can be rewritten in a very simple factorized form as

$$\widehat{\rho}_i = \frac{1}{Z_i} \exp[-\tilde{b} \cdot \widehat{P}_i] \exp\left[\frac{1}{2}\varpi : \widehat{J}_i\right] \quad (10.34)$$

by defining a ϖ-dependent effective four-vector

$$\tilde{b}_\mu = \sum_{k=0}^\infty \frac{i^k}{(k+1)!} \underbrace{\left(\varpi_{\mu\nu_1}\varpi^{\nu_1\nu_2}\ldots\varpi_{\nu_{k-1}\nu_k}\right)}_{k \text{ times}} b^{\nu_k}.$$

As the non-commutative operators $\widehat{P}_i$ and $\widehat{J}_i$ are completely separated from each other into two independent multiplying exponential functions in (10.34), the SDM for the Mother can be reduced to a simple form:

$$\Theta(p_H)_{\sigma\sigma'} = \frac{\langle p_H, \sigma | \widehat{\rho}_H | p_H, \sigma' \rangle}{\sum_\sigma \langle p_H, \sigma | \widehat{\rho}_H | p_H, \sigma \rangle} = \frac{\langle p_H, \sigma | \exp\left[\frac{1}{2}\varpi : \widehat{J}_H\right] | p_H, \sigma' \rangle}{\sum_\sigma \langle p_H, \sigma | \exp\left[\frac{1}{2}\varpi : \widehat{J}_H\right] | p_H, \sigma \rangle}. \quad (10.35)$$

It is now completely determined by its single particle density operator $\widehat{\rho}_H$ or more precisely the angular momentum-dependent part.

To derive the explicit form for (10.35), we use a convenient analytic continuation technique: We first derive $\Theta(p_H)$ for imaginary ϖ and then continue the result back to real value. In the former case, $\widehat{\Lambda} \equiv \exp[\varpi : \widehat{J}_H/2]$ is just a unitary representation of Lorentz transformation, then the well-known relations in group theory can be used to obtain

$$\Theta(p_H)_{\sigma\sigma'} = \frac{\langle p_H, \sigma | \widehat{\Lambda} | p_H, \sigma' \rangle}{\sum_\sigma \langle p_H, \sigma | \widehat{\Lambda} | p_H, \sigma \rangle} = \frac{W(p_H)_{\sigma\sigma'} 2\varepsilon_H \delta^3(\mathbf{p}_H - \Lambda(\mathbf{p}_H))}{W(p_H)_{\sigma\sigma} 2\varepsilon_H \delta^3(\mathbf{p}_H - \Lambda(\mathbf{p}_H))}. \quad (10.36)$$

Here, $\Lambda(\mathbf{p}_H)$ stands for the spatial part of the four-vector $\Lambda(p_H)$, and the covariant normalization scheme is used for the Mother eigenstates, that is,

$$\langle p_H, \sigma | p'_H, \sigma' \rangle = 2\varepsilon_H \delta^3(\mathbf{p}_H - \mathbf{p}'_H)\delta_{\sigma\sigma'}.$$

In (10.36), the matrix $W(p_H)$ is the so-called Wigner rotation matrix:

$$W(p_H) = D^j([\Lambda p_H]^{-1}\Lambda[p_H]),$$

where D^j is the $(2S + 1)$-dimensional representation, the so-called $(0, 2S + 1)$ [12], of the SO(1,3)-SL(2,C) matrices in the argument [8]. Altogether, the SDM for the Mother is simply

$$\Theta(p_H)_{\sigma\sigma'} = \frac{D^j([p_H]^{-1}\Lambda[p_H])_{\sigma\sigma'}}{\text{tr}\left[D^j(\Lambda)\right]}, \tag{10.37}$$

which seems appropriate to be analytically continued to real ϖ.

However, it is not satisfactory yet as the analytic continuation of (10.37) to real ϖ, that is,

$$D^j(\Lambda) \rightarrow \exp\left[\frac{1}{2}\varpi : J_H\right], \tag{10.38}$$

does not give rise to a Hermitian matrix for $\Theta(p_H)$ as it should. This problem can be fixed by taking into account the fact that $W(p_H)$ is the representation of a rotation hence unitary. We thus replace $W(p_H)$ with $(W(p_H) + W(p_H)^{-1\dagger})/2$ in (10.36) and obtain, by using the transparency to the adjoint operation property of SL(2,C) representations,

$$\Theta(p_H) = \frac{D^j([p_H]^{-1}\Lambda[p_H]) + D^j([p_H]^\dagger \Lambda^{-1\dagger}[p_H]^{-1\dagger})}{\text{tr}\left[D^j(\Lambda) + D^j(\Lambda)^{-1\dagger}\right]}.$$

As the analytic continuation of $\Lambda^{-1\dagger}$-related part reads

$$D^j(\Lambda^{-1\dagger}) \rightarrow \exp\left[\frac{1}{2}\varpi : D^{j\dagger}(J)\right], \tag{10.39}$$

the final expression of the SDM in a rotational system is

$$\Theta(p_H) = \frac{\sum_{O=1,\dagger}\left[D^j([p_H]^{-1}\exp[\varpi : D^j(J)/2][p_H])\right]^O}{\text{tr}\left[\exp[\varpi : D^j(J)/2] + \exp[\varpi : D^{j\dagger}(J)/2]\right]}, \tag{10.40}$$

which is manifestly Hermitian.

The expression can be further simplified: By taking the involved matrices as SO(1,3) transformations and using known relations in group theory, we have

$$[p_H]^{-1}\exp\left[\frac{1}{2}\varpi : J\right][p_H] = \exp\left[\frac{1}{2}\varpi^{\mu\nu}[p_H]^{-1}J_{\mu\nu}[p_H]\right] = \exp\left[\frac{1}{2}\varpi_*^{\alpha\beta}(p_H)J_{\alpha\beta}\right],$$

where the effective anti-symmetric tensor ϖ_* is defined as

$$\varpi_*^{\alpha\beta}(p_H) \equiv \varpi^{\mu\nu}[p_H]_\mu^{-1\alpha}[p_H]_\nu^{-1\beta}. \tag{10.41}$$

Actually, $\varpi_*^{\alpha\beta}$ have physical meanings themselves, that is, the components of thermal vorticity tensor in the MRF. They are obtained from the ones in QGPF by taking the inverse transformation of $[p_H]$. Finally, (10.40) becomes

$$\Theta(p_H) = \frac{D^j(\exp[\varpi_*(p_H) : D^j(J)/2]) + D^j(\exp[\varpi_*(p_H) : D^{j\dagger}(J)/2])}{\mathrm{tr}(\exp[\varpi : D^j(J)/2]) + \exp[\varpi : D^{j\dagger}(J)/2])}.$$

$$(10.42)$$

In many cases, such as in peripheral heavy ion collisions, the thermal vorticity ϖ is usually $\ll 1$ due to the relatively large proper temperature [3], so the SDM can be expanded in power series around $\varpi = 0$. Taking into account the fact that the generators of Lorentz transformation are traceless, that is $\mathrm{tr}(J_H) = 0$, we have

$$\Theta(p_H)^{\sigma}_{\sigma'} \simeq \frac{\delta^{\sigma}_{\sigma'}}{2j+1} + \frac{1}{4(2j+1)}\varpi_*^{\mu\nu}(p_H)\left(D^j(J_{\mu\nu}) + D^{j\dagger}(J_{\mu\nu})\right)^{\sigma}_{\sigma'}$$

to the order $o(\varpi)$. The representation $D^j(J_{\mu\nu})$ can be decomposed as the following:

$$D^j(J_{\mu\nu}) = \epsilon_{\mu\nu\rho\tau}D^j(J^\rho)\hat{t}^\tau + D^j(K_\nu)\hat{t}_\mu - D^j(K_\mu)\hat{t}_\nu \qquad (10.43)$$

with $D^j(J^i)$ Hermitian and $D^j(K^i)$ anti-Hermitian matrices, respectively. Then, we find that the SDM is only rotation relevant:

$$\Theta(p_H)^{\sigma}_{\sigma'} \simeq \frac{\delta^{\sigma}_{\sigma'}}{2j+1} + \frac{1}{2(2j+1)}\varpi_*^{\mu\nu}(p_H)\epsilon_{\mu\nu\rho\tau}D^j(J^\rho)^{\sigma}_{\sigma'}\hat{t}^\tau. \qquad (10.44)$$

Note that the number of the generators $D^S(J)$ is more than three for $S > 1/2$, but only three is involved in (10.44) with the others functioning through higher order terms of ϖ. By substituting (10.44) into (10.30), only the second term of (10.44) contributes

$$S_H^\mu(p_H) = [p_H]_\kappa^\mu \frac{1}{2(2j+1)}\varpi_*^{\alpha\beta}(p_H)\epsilon_{\alpha\beta\rho\tau}\mathrm{tr}\left(D^j(J^\rho)D^j(J^\kappa)\right)\hat{t}^\tau$$

$$= -\frac{j(j+1)}{6}[p_H]_\rho^\mu \varpi_{*\alpha\beta}(p_H)\epsilon^{\alpha\beta\rho\tau}\hat{t}_\tau = -\frac{j(j+1)}{6m_H}\epsilon^{\alpha\beta\mu\tau}\varpi_{\alpha\beta}p_{H\tau},$$

$$(10.45)$$

where we have transformed back to the QGPF by inserting (10.41) in the last equality. As we will see in next section, the MSV can be boosted to the MRF to give the true spin observables $\mathbf{S}_{H*}$, and then the polarization of the Mother is defined as $\mathbf{P}_H = \mathbf{S}_{H*}/j$.

10.4 Local Polarization Transfer to the Daughter

Now, with the Mother's local polarization determined in the previous section, it's the right time to study the polarization transfer to the Daughter from the feed-down effect of the Mother in two-body decays. Concretely, the most important mission is to derive the reduced spin density matrix for the Daughter in the MRF by tracing over the quantum states of the Byproduct, which thus indicates that this reduced SDM should be mixed rather than pure in general. In the MRF, the magnitude of the three-momentum of the Daughter is fixed due to energy-momentum conservation, that is,

$$p_* = p_{\Lambda *} \equiv \frac{1}{2m_H} \prod_{s,t=\pm} (m_H + s\, m_\Lambda + t\, m_X)^{1/2}. \tag{10.46}$$

As mentioned in Sect. 10.2, as long as the decay hasn't been observed, contribution of the Mother state to the quantum superposition of the Daughter and Byproduct reads in the helicity basis as [10,12,13]

$$|p_* j m \lambda_\Lambda \lambda_X\rangle \propto T^j(\lambda_\Lambda, \lambda_X) \int d\Omega_* \, D^j(\phi_*, \theta_*, 0)_\lambda^{m\,*} \, |\boldsymbol{p}_* \lambda_\Lambda \lambda_X\rangle. \tag{10.47}$$

Once a measurement is made for the momentum of either final particle and hence $\boldsymbol{p}_*$ is fixed down, we can define the non-integrated form of the two-body spin density operator as[3]

$$\widehat{\rho}(\boldsymbol{p}_*) = \frac{T^j(\lambda_\Lambda, \lambda_X) T^j(\lambda'_\Lambda, \lambda'_X)^* D^j(\phi_*, \theta_*, 0)_\lambda^{m\,*} D^j(\phi_*, \theta_*, 0)_{\lambda'}^m \, |\boldsymbol{p}_* \lambda_\Lambda \lambda_X\rangle \langle \boldsymbol{p}_* \lambda'_\Lambda \lambda'_X|}{|T^j(\lambda_\Lambda, \lambda_X)|^2 |D^j(\phi, \theta, 0)_\lambda^m|^2 \langle \boldsymbol{p}_* \lambda_\Lambda \lambda_X | \, \boldsymbol{p}_* \lambda_\Lambda \lambda_X\rangle} \tag{10.48}$$

for a given state of the Mother with z component of spin m.

However, as we've illuminated in the previous section, the spin state of the Mother can be shifted according to (10.42) in a rotational system. In this case, the two-body density operator should be a mixing of different spin states of the Mother:

$$\sum_{m,n=-j}^{j} \Theta(p_H)_n^m \, |p_* j m \lambda_\Lambda \lambda_X\rangle \langle p_* j n \lambda'_\Lambda \lambda'_X|, \tag{10.49}$$

rather than the pure one with $\Theta(p_H)_n^m \to \delta_n^m$. To be consistent with the setup, the involved matrices $D^j(J)$ in (10.42) are now also defined in the MRF, which then allows us to apply the usual matrix algebra in later explicit evaluations. Following

---

[3]For brevity, the summation convention is assumed: If an angular momentum component index (only for superscripts and subscripts) shows more than once in the formula, the index should be summed over. For example, we should sum over m in the numerator of (10.48) and over m, λ_Λ and λ_X in the denominator as $|D^j(\phi, \theta, 0)_\lambda^m|^2 = D^j(\phi, \theta, 0)_{\lambda_\Lambda - \lambda_X}^{m\,*} D^j(\phi, \theta, 0)_{\lambda_\Lambda - \lambda_X}^m$.

(10.49), a more general density operator for the daughters with fixed momentum $\mathbf{p}_*$ reads

$$\widehat{\rho}(\mathbf{p}_*) \propto T^j(\lambda_\Lambda, \lambda_X) T^j(\lambda'_\Lambda, \lambda'_X)^* D^j(\phi_*, \theta_*, 0)^{m*}_\lambda \Theta(p_H)^m_n D^j(\phi_*, \theta_*, 0)^n_{\lambda'}$$
$$|\mathbf{p}_* \lambda_\Lambda \lambda_X\rangle\langle\mathbf{p}_* \lambda'_\Lambda \lambda'_X|. \tag{10.50}$$

Then, the normalized two-body spin density matrix follows directly:

$$\Theta(\phi_*, \theta_*)^{\lambda_\Lambda \lambda_X}_{\lambda'_\Lambda \lambda'_X} = \frac{T^j(\lambda_\Lambda, \lambda_X) T^j(\lambda'_\Lambda, \lambda'_X)^* D^j(\phi_*, \theta_*, 0)^{m*}_\lambda \Theta(p_H)^m_n D^j(\phi_*, \theta_*, 0)^n_{\lambda'}}{|T^j(\lambda_\Lambda, \lambda_X)|^2 D^j(\phi_*, \theta_*, 0)^{m*}_\lambda \Theta(p_H)^m_n D^j(\phi_*, \theta_*, 0)^n_\lambda}. \tag{10.51}$$

Combining (10.30) and (10.51), the MSV of the Daughter can be obtained from (10.30) as

$$S^\mu_{\Lambda*}(\mathbf{p}_*) = [p_*]^\mu_\nu D^{S_\Lambda}(J^\nu)^{\lambda'_\Lambda}_{\lambda_\Lambda} \Theta(\phi_*, \theta_*)^{\lambda_\Lambda \lambda_X}_{\lambda'_\Lambda \lambda_X}, \tag{10.52}$$

where the summation over λ_X reduces the two-body SDM to the single-particle one for the Daughter. In general, the MSV of the Daughter depends on $(2j+1)^2 - 1$ real parameters through the $(2j+1)$-dimensional and trace 1 Hermitian SDM of the Mother. This means the MSV of the Daughter cannot be definitely determined by the MSV of the Mother, which only involves 3 real parameters, except for $j = 1/2$. Indeed, this was well known in the literature [17,18] and was illuminated explicitly in [9]. Nevertheless, the SDM of the primary Mother can be well approximated by the first-order expansion form (10.44), which surprisingly implies that the MSV of the Daughter can be definitely determined by that of the Mother now, as we will see in (10.55).

By applying the approximation (10.44) for $\Theta(p_H)^m_n$ to the (10.51), we explore the feed-down effect to first order in thermal vorticity $\varpi_*(p_H)$. The first term in (10.44) is proportional to the identity matrix and selects $m = n$ in (10.51), then one is left with

$$D^j(\phi_*, \theta_*, 0)^{m*}_\lambda D^j(\phi_*, \theta_*, 0)^m_{\lambda'} = \delta^\lambda_{\lambda'}$$

due to the unitary of D^j's. On the other hand, the second term gives rise to the three D-matrices multiplying term:

$$D^j(\phi_*, \theta_*, 0)^{m*}_\lambda D^j(J^\rho)^m_n D^j(\phi_*, \theta_*, 0)^n_{\lambda'} = D^{j(-1)}(\phi_*, \theta_*, 0)^\lambda_m D^j(J^\rho)^m_n D^j(\phi_*, \theta_*, 0)^n_{\lambda'},$$

which, according to a well-known relation in group representation theory [12], equals

$$\mathsf{R}(\phi_*, \theta_*, 0)^\rho_\tau D^j(J^\tau)^\lambda_{\lambda'}, \tag{10.53}$$

where the rotation R transforms the z-axis unit vector $\mathbf{e}_3$ into the $\mathbf{p}_*$ direction. Altogether, we get the explicit form of (10.51) as

$$\Theta(\phi_*, \theta_*)^{\lambda_\Lambda \lambda_X}_{\lambda'_\Lambda \lambda'_X} \simeq \frac{\delta^\lambda_{\lambda'} + \frac{1}{2}\varpi_*(p_H)^{\alpha\beta} \epsilon_{\alpha\beta\rho\nu} D^j(J^\tau)^\lambda_{\lambda'} \mathsf{R}(\phi_*, \theta_*, 0)^\rho_\tau \hat{i}^\nu}{\left[T^j(\lambda_\Lambda, \lambda_X) T^j(\lambda'_\Lambda, \lambda'_X)^*\right]^{-1} \sum^{\lambda_X}_{\lambda_\Lambda} |T^j(\lambda_\Lambda, \lambda_X)|^2}, \tag{10.54}$$

where the denominator gives the normalization factor solely determined by the dynamical amplitudes. For parity conservative decays with the property (10.21), by substituting (10.54) into (10.52), we find that the first term of (10.54) does not contribute as $\mathrm{tr}\, D^{S_\Lambda}(\mathsf{J}^\nu) = 0$ and the MSV of the Daughter is proportional to the thermal vorticity $\varpi_*(p_H)$ in the MRF:

$$
\begin{aligned}
S^\mu_{\Lambda*}(\mathbf{p}_*) &= \frac{1}{2}\varpi_*(p_H)^{\alpha\beta}\epsilon_{\alpha\beta\rho\nu}\hat{t}^\nu \sum_{\lambda_X,\lambda'_X} \frac{\delta_{\lambda_X\lambda'_X}[p_*]^\mu_\kappa D^{S_\Lambda}(\mathsf{J}^\kappa)^{\lambda'_\Lambda}_{\lambda_\Lambda} D^j(\mathsf{J}^\tau)^{\lambda}_{\lambda'_\Lambda} \mathsf{R}(\phi_*,\theta_*,0)^\rho_\tau}{[T^j(\lambda_\Lambda,\lambda_X)T^j(\lambda'_\Lambda,\lambda'_X)^*]^{-1} \sum^{\lambda_X}_{\lambda_\Lambda} |T^j(\lambda_\Lambda,\lambda_X)|^2} \\
&= -\frac{3S_{H*\rho}(p_H)}{j(j+1)} \sum_{\lambda_X,\lambda'_X} \frac{\delta_{\lambda_X\lambda'_X}[p_*]^\mu_\kappa D^{S_\Lambda}(\mathsf{J}^\kappa)^{\lambda'_\Lambda}_{\lambda_\Lambda} D^j(\mathsf{J}^\tau)^{\lambda}_{\lambda'_\Lambda} \mathsf{R}(\phi_*,\theta_*,0)^\rho_\tau}{[T^j(\lambda_\Lambda,\lambda_X)T^j(\lambda'_\Lambda,\lambda'_X)^*]^{-1} \sum^{\lambda_X}_{\lambda_\Lambda} |T^j(\lambda_\Lambda,\lambda_X)|^2}. \quad (10.55)
\end{aligned}
$$

In the last step, (10.45) has been used to reexpress the formula in terms of the MSV of the Mother in its rest frame, $S_{H*}(p_H)$.

As the first step, we would like to apply (10.55) to the simplest parity conservative decays: strong decays with the Byproduct $X = \pi$ and EM decays with $X = \gamma$. As mentioned in Sect. 10.2, the dynamical amplitude has a definite sign under parity inversion in these cases; see (10.20). After that, the parity violating weak decays will be discussed in more detail with the Byproduct $X = \pi$.

A. Strong Decays

For the strong decay $H \to \Lambda + \pi$, the spin-parity structure is explicitly $j^{\eta_H} \to 1/2^+ + 0^-$. Hence, (10.20) becomes

$$
T^j(-\lambda_\Lambda, 0) = P_S T^j(\lambda_\Lambda, 0), \qquad P_S \equiv \eta_H(-1)^{j+\frac{1}{2}},
$$

and (10.55) can be reduced to

$$
\begin{aligned}
S^\mu_{\Lambda*}(\mathbf{p}_*) &= -\frac{3S_{H*\rho}(p_H)}{j(j+1)} \frac{[p_*]^\mu_\kappa D^{1/2}(\mathsf{J}^\kappa)^{\lambda'_\Lambda}_{\lambda_\Lambda} D^j(\mathsf{J}^\tau)^{\lambda_\Lambda}_{\lambda'_\Lambda} \mathsf{R}(\phi_*,\theta_*,0)^\rho_\tau}{[T^j(\lambda_\Lambda,0)T^j(\lambda'_\Lambda,0)^*]^{-1} \sum_{\lambda_\Lambda=\pm\frac{1}{2}} |T^j(\lambda_\Lambda,0)|^2} \\
&= -\frac{3S_{H*\rho}(p_H)}{j(j+1)} \frac{P_S + (1-P_S)\delta_{\lambda_\Lambda\lambda'_\Lambda}}{2} [p_*]^\mu_\kappa D^{1/2}(\mathsf{J}^\kappa)^{\lambda'_\Lambda}_{\lambda_\Lambda} D^j(\mathsf{J}^\tau)^{\lambda_\Lambda}_{\lambda'_\Lambda} \mathsf{R}(\phi_*,\theta_*,0)^\rho_\tau \\
&= \frac{3S_{H*\rho}(p_H)}{j(j+1)} \left\{ P_S C^j_\tau [p_*]^\mu_\tau \mathsf{R}(\phi_*,\theta_*,0)^{\rho\tau} + (1-P_S)C^j_3 [p_*]^\mu_3 \mathsf{R}(\phi_*,\theta_*,0)^{\rho 3} \right\} \\
&= \frac{3S_{H*\rho}(p_H)}{j(j+1)} \left\{ P_S C^j [p_*]^\mu_\tau \mathsf{R}(\phi_*,\theta_*,0)^{\rho\tau} - (P_S C^j - C^j_3)[p_*]^\mu_3 \mathsf{R}(\phi_*,\theta_*,0)^{\rho 3} \right\}.
\end{aligned}
$$
$$(10.56)$$

In the derivation, the following conventions of $D^j(J)$ matrices are used for the Mother and Daughter:

$$
\begin{cases}
D^{1/2}(\mathsf{J}^1) = \frac{\sigma_1}{2}, & D^{1/2}(\mathsf{J}^2) = \frac{\sigma_2}{2}, & D^{1/2}(\mathsf{J}^3) = \frac{\sigma_3}{2}, \\
D^{3/2}_{\mathrm{r}}(\mathsf{J}^1) = \sigma_1, & D^{3/2}_{\mathrm{r}}(\mathsf{J}^2) = \sigma_2, & D^{3/2}_{\mathrm{r}}(\mathsf{J}^3) = \frac{\sigma_3}{2},
\end{cases} \quad (10.57)
$$

where $D_r^{3/2}$ are the 2×2 matrices for the Mother with spin 3/2, reduced due to the restrictions of the indices $\lambda_\Lambda, \lambda'_\Lambda = \pm 1/2$. One can easily check that

$$C_\tau^{1/2} = C^{1/2}, \qquad C_\tau^{3/2} = \left(\frac{1}{2}, \frac{1}{2}, \frac{1}{2}, \frac{1}{4}\right) = C^{3/2} - \frac{1}{4}\delta_\tau^3$$

with $C^{1/2} = 1/4$ and $C^{3/2} = 1/2$, where the irrelevant coefficients C_0^j, as $R^{\rho 0} = \eta^{\rho 0}$, are introduced for the brevity of presentations.

In the helicity scheme, the matrix $[p_*]$ can be expanded according to (10.7), so we take advantage of the orthogonality of rotations R to obtain

$$S_{\Lambda *}^\mu(\mathbf{p}_*) = \frac{3}{j(j+1)} \Big\{ P_S C^j L_{\hat{\mathbf{p}}_*}(\xi)_\rho^\mu S_{H*}^\rho(p_H) - (P_S C^j - C_3^j) S_{H*\rho}(P) R(\phi_*, \theta_*, 0)_\nu^\mu$$
$$L_z(\xi)_3^\nu R(\phi_*, \theta_*, 0)^{\rho 3} \Big\}, \tag{10.58}$$

where $L_{\hat{\mathbf{p}}_*}(\xi) = R(\phi_*, \theta_*, 0) L_z(\xi) R^{-1}(\phi_*, \theta_*, 0)$ is the pure Lorentz boost transforming $\hat{t}$ into the $\mathbf{p}_*$ direction in the MRF. The Lorentz transformation involved in the second term of (10.58) can be expressed explicitly as a function of $\hat{\mathbf{p}}_*$, that is,

$$R(\phi_*, \theta_*, 0)_\nu^\mu L_z(\xi)_3^\nu R(\phi_*, \theta_*, 0)^{\rho 3}$$
$$= R(\phi_*, \theta_*, 0)_3^\mu L_z(\xi)_3^3 R(\phi_*, \theta_*, 0)^{\rho 3} + R(\phi_*, \theta_*, 0)_0^\mu L_z(\xi)_3^0 R(\phi_*, \theta_*, 0)^{\rho 3}$$
$$= -\cosh\xi\, \hat{\mathbf{p}}_*^\mu \hat{\mathbf{p}}_*^\rho - \sinh\xi\, \hat{\mathbf{p}}_*^\rho \delta_0^\mu = -\frac{\varepsilon_{\Lambda *}}{m_\Lambda} \hat{\mathbf{p}}_*^\rho \hat{\mathbf{p}}_*^\mu - \frac{P_*}{m_\Lambda} \hat{\mathbf{p}}_*^\rho \delta_0^\mu, \tag{10.59}$$

then (10.58) becomes

$$S_{\Lambda *}^\mu(\mathbf{p}_*) = \frac{3}{j(j+1)} \Big[P_S C^j L_{\hat{\mathbf{p}}_*}(\xi)_\rho^\mu S_{H*}^\rho(p_H) - (P_S C^j - C_3^j)\Big(\frac{\varepsilon_{\Lambda *}}{m_\Lambda} S_{H*}(p_H) \cdot \hat{\mathbf{p}}_* \hat{\mathbf{p}}_*^\mu$$
$$+ \frac{P_*}{m_\Lambda} S_{H*}(p_H) \cdot \hat{\mathbf{p}}_* \delta_0^\mu \Big) \Big]. \tag{10.60}$$

So with the help of the well-known formulae for the pure Lorentz boost:

$$L_{\hat{\mathbf{p}}_*}(\xi)_\rho^0 = \frac{\varepsilon_{\Lambda *}}{m_\Lambda} \eta_\rho^0 - \frac{P_{*\rho}}{m_\Lambda}, \qquad L_{\hat{\mathbf{p}}_*}(\xi)_\rho^i = \eta_\rho^i - \frac{\mathbf{p}_*^i P_{*\rho}}{m_\Lambda(\varepsilon_{\Lambda *} + m_\Lambda)} - \frac{\mathbf{p}_*^i}{m_\Lambda}\eta_{\rho 0}, \tag{10.61}$$

we get the explicit forms for the time and spatial components of the MSV of the Daughter in the MRF as

$$S_{\Lambda *}^0(\mathbf{p}_*) = \frac{3C_3^j}{j(j+1)} \frac{1}{m_\Lambda} S_{H*}(p_H) \cdot \mathbf{p}_*,$$

$$\mathbf{S}_{\Lambda *}(\mathbf{p}_*) = \frac{3}{j(j+1)} \Big[P_S C^j \mathbf{S}_{H*}(p_H) - \Big(P_S C^j - \frac{\varepsilon_{\Lambda *}}{m_\Lambda} C_3^j\Big) \mathbf{S}_{H*}(p_H) \cdot \hat{\mathbf{p}}_* \hat{\mathbf{p}}_* \Big].$$

Finally, we boost the MSV to the DRF as that is the one measured in experiments and find

$$S_{\Lambda o}(p_*) = S_{\Lambda *}(p_*) - S_{\Lambda *}^0(p_*)\frac{p_*}{\varepsilon_{\Lambda *} + m_\Lambda}$$

$$= \frac{3}{j(j+1)}\left[P_S C^j S_{H*}(p_H) - (P_S C^j - C_3^j)S_{H*}(p_H) \cdot \hat{p}_* \hat{p}_*\right]. \quad (10.62)$$

The average over the whole solid angle Ω_* gives

$$\langle S_{\Lambda o}(p_*)\rangle = \frac{3}{j(j+1)}\left[P_S C^j - \frac{1}{2}(P_S C^j - C_3^j)\int_0^\pi d\theta_* \ \sin\theta_* \cos^2\theta_*\right]\langle S_{H*}(p_H)\rangle$$

$$= \frac{2P_S C^j + C_3^j}{j(j+1)}\langle S_{H*}(p_H)\rangle \quad (10.63)$$

for a given $S_{H*}(p_H)$ independent of $\hat{p}_*$. The result is consistent with that found in Sect. 10.2. In Sect. 10.5, we will see that $S_{H*}(p_H)$ does depend on $\hat{p}_*$ for a given momentum of the Daughter in the QGPF, thus the application of (10.63) should be taken cautiously.

B. Electromagnetic Decays

For the EM decay $H \rightarrow \Lambda + \gamma$, the spin-parity structure is explicitly $j^{\eta_H} \rightarrow 1/2^+ + 1^-$. Hence, (10.20) becomes

$$T^j(-\lambda_\Lambda, -\lambda_X) = P_{EM}T^j(\lambda_\Lambda, \lambda_X), \qquad P_{EM} \equiv \eta_H(-1)^{j-\frac{1}{2}}$$

with $P_{EM} = -P_s$ for the same η_H and j. In this case, (10.55) can be reduced to

$$S_{\Lambda *}^\mu(p_*) = -\frac{3S_{H*\rho}(p_H)}{j(j+1)}\sum_{\lambda_X=\pm1}^{\lambda_X'=\pm1} \frac{\delta_{\lambda_X \lambda_X'}[p_*]_\kappa^\mu D^{1/2}(J^\kappa)_{\lambda_\Lambda}^{\lambda_\Lambda'} D^j(J^\tau)_{\lambda'}^\lambda R(\phi_*, \theta_*, 0)_\tau^\rho}{[T^j(\lambda_\Lambda, \lambda_X)T^j(\lambda_\Lambda', \lambda_X')^*]^{-1}\sum_{\lambda_X=\pm1}^{\lambda_\Lambda=\pm1/2}|T^j(\lambda_\Lambda, \lambda_X)|^2}. \quad (10.64)$$

Keeping in mind the parity inversion properties of the irreducible representation matrices:

$$D^j(J^\tau)_{-\lambda'}^{-\lambda} = (-1)^{\delta_{\tau 1}+1}D^j(J^\tau)_{\lambda'}^\lambda$$

with the component indices $|\lambda|, |\lambda'| \leq j$, it is easy to show that

$$S_{\Lambda *}^\mu(p_*) = -\frac{3S_{H*\rho}(p_H)}{2j(j+1)}\sum_{\lambda_X=\pm1}^{\lambda_X'=\pm1} \frac{\delta_{\lambda_X \lambda_X'}[p_*]_\kappa^\mu \sum_{s=\pm}\left(D^{1/2}(J^\kappa)_{s\lambda_\Lambda}^{s\lambda_\Lambda'} D^j(J^\tau)_{s\lambda'}^{s\lambda}\right)R(\phi_*, \theta_*, 0)_\tau^\rho}{[T^j(\lambda_\Lambda, \lambda_X)T^j(\lambda_\Lambda', \lambda_X')^*]^{-1}\sum_{\lambda_X=\pm1}^{\lambda_\Lambda=\pm1/2}|T^j(\lambda_\Lambda, \lambda_X)|^2}$$

$$= -\frac{3S_{H*\rho}(p_H)}{j(j+1)}\sum_{\lambda_X=\pm1}^{\lambda_X'=\pm1} \frac{\delta_{\lambda_X \lambda_X'}[p_*]_\kappa^\mu D^{1/2}(J^\kappa)_{\lambda_\Lambda}^{\lambda_\Lambda'} D^j(J^\kappa)_{\lambda'}^\lambda R(\phi_*, \theta_*, 0)_\kappa^\rho}{[T^j(\lambda_\Lambda, \lambda_X)T^j(\lambda_\Lambda', \lambda_X')^*]^{-1}\sum_{\lambda_X=\pm1}^{\lambda_\Lambda=\pm1/2}|T^j(\lambda_\Lambda, \lambda_X)|^2}$$

$$= -\frac{3S_{H*\rho}(p_H)}{j(j+1)}\left(N^j - (N^j - N_3^j)\delta^{\kappa 3}\right)[p_*]_\kappa^\mu R(\phi_*, \theta_*, 0)_\kappa^\rho, \quad (10.65)$$

where the $D^j(J^\kappa)$ and $T^j(\lambda_\Lambda, \lambda_X)$ relevant normalization factors are

$$N^j = \frac{\sqrt{3}}{2} \frac{\sum_{\lambda_\Lambda = \pm 1/2} T^j(\lambda_\Lambda, 1) T^j(-\lambda_\Lambda, 1)^*}{\sum_{\lambda_X = \pm 1}^{\lambda_\Lambda = \pm 1/2} |T^j(\lambda_\Lambda, \lambda_X)|^2}, \quad N_3^j = \sum_{\lambda_\Lambda = \pm 1/2} \frac{(-1)^{\lambda_\Lambda + \frac{1}{2}}(1 - \lambda_\Lambda)|T^j(\lambda_\Lambda, 1)|^2}{\sum_{\lambda_X = \pm 1}^{\lambda_\Lambda = \pm 1/2} |T^j(\lambda_\Lambda, \lambda_X)|^2}.$$

(10.66)

Now, we can immediately identify the similarity between (10.56) and (10.65). So the final results for the MSV of the Daughter and the averaged one in the DRF can be given directly as

$$\mathbf{S}_{\Lambda o}(\mathbf{p}_*) = \frac{3}{j(j+1)}\left[N^j \mathbf{S}_{H*}(p_H) - (N^j - N_3^j)\mathbf{S}_{H*}(p_H) \cdot \hat{\mathbf{p}}_* \hat{\mathbf{p}}_*\right], \quad (10.67)$$

$$\langle \mathbf{S}_{\Lambda o}(\mathbf{p}_*)\rangle = \frac{2N^j + N_3^j}{j(j+1)}\langle \mathbf{S}_{H*}(p_H)\rangle \quad (10.68)$$

by changing the coefficients

$$P_S C^j \to N^j, \qquad C_3^j \to N_3^j$$

from (10.62) and (10.63).

For $j = 3/2$, the formula (10.67) cannot be simplified further in general, as the transition amplitudes in (10.66) cannot be canceled out. However, fortunately for the study of Λ polarization, only the EM decay $\Sigma^0 \to \Lambda\gamma$ is relevant and $j^{\eta_{\Sigma^0}} = 1/2^+$. Then, we immediately find that $N^{1/2} = 0$ and $N_3^{1/2} = -1/4$ due to the restrictions $|\lambda|, |\lambda'| \leq 1/2$ [1], so the MSV and the averaged one are explicitly

$$\mathbf{S}_{\Lambda o}(\mathbf{p}_*) = -\mathbf{S}_{H*}(p_H) \cdot \hat{\mathbf{p}}_* \hat{\mathbf{p}}_*, \quad (10.69)$$

$$\langle \mathbf{S}_{\Lambda o}(\mathbf{p}_*)\rangle = -\frac{1}{3}\langle \mathbf{S}_{H*}(p_H)\rangle. \quad (10.70)$$

The result is also consistent with that found in Sect. 10.2.

C. Weak Decays

For weak decays, it is well known that the dynamical transition amplitude is a mixture of parity even and odd modes. Only one kind of weak decay channel is relevant to Λ polarization, that is, $\Xi \to \Lambda + \pi$, so we stick to the simple case with $j^{\eta_\Xi} = 1/2^+$. First of all, due to parity violation, the first term of (10.54) will give rise to a finite contribution to the MSV of the Daughter [9]. Assuming the dynamical amplitude in the following form:

$$T_w^{1/2}(\pm 1/2, 0) = T_e \pm T_o,$$

this contribution is simply

$$S'_{\Lambda*}{}^\mu(\mathbf{p}_*) = \frac{\alpha_w}{2m_\Lambda}(p_*\eta^{\mu 0} + \varepsilon_{\Lambda*}\hat{\mathbf{p}}_*^\mu), \qquad \alpha_w = \frac{2\mathrm{Re}(T_e^* T_o)}{|T_e|^2 + |T_o|^2}, \quad (10.71)$$

and the corresponding MSV in the DRF is proportional to the three-momentum unit vector:

$$S'_{\Lambda o}(\mathbf{p}_*) = \frac{\alpha_w}{2}\hat{\mathbf{p}}_*. \tag{10.72}$$

Next, the polarization transfer effect from the Mother can be deduced from (10.55) as

$$
\begin{aligned}
S''_{\Lambda *}{}^{\mu}(\mathbf{p}_*) &= -4S_{H*\rho}(p_H)\frac{[p_*]^{\mu}_{\kappa}D^{1/2}(\mathsf{J}^{\kappa})^{\lambda'_{\Lambda}}_{\lambda_{\Lambda}}D^{1/2}(\mathsf{J}^{\tau})^{\lambda_{\Lambda}}_{\lambda'_{\Lambda}}\mathsf{R}(\phi_*,\theta_*,0)^{\rho}_{\tau}}{\left[T^j(\lambda_{\Lambda},0)T^j(\lambda'_{\Lambda},0)^*\right]^{-1}\sum_{\lambda_{\Lambda}=\pm\frac{1}{2}}|T^j(\lambda_{\Lambda},0)|^2} \\
&= -S_{H*\rho}(p_H)[p_*]^{\mu}_{\kappa}\left((1-\gamma_w)\delta^{\kappa 3}\delta^{\tau 3}-\gamma_w\eta^{\kappa\tau}+\epsilon^{\kappa\tau 3}\beta_w\right)\mathsf{R}(\phi_*,\theta_*,0)^{\rho}_{\tau}
\end{aligned}
\tag{10.73}
$$

with the dynamical parameters

$$\beta_w = \frac{2\mathrm{Im}(T_e^*T_o)}{|T_e|^2+|T_o|^2}, \qquad \gamma_w = \frac{|T_e|^2-|T_o|^2}{|T_e|^2+|T_o|^2}. \tag{10.74}$$

Again, we can immediately recognize the similarity between the first two terms of (10.73) and those in the strong decay (10.56), hence simple alternations of the coefficients will give the final results. By noticing $L_z(\xi)^{\nu}_{\tau}=\delta^{\nu}_{\tau}$ for $\tau=1,2$ in (10.27), the Lorentz transformation in the last term of (10.73) can be evaluated as

$$\epsilon^{\kappa\tau 3}[p_*]^{\mu}_{\kappa}\mathsf{R}(\phi_*,\theta_*,0)^{\rho}_{\tau}=\epsilon^{\kappa\tau 3}\mathsf{R}(\phi_*,\theta_*,0)^{\mu}_{\kappa}\mathsf{R}(\phi_*,\theta_*,0)^{\rho}_{\tau}.$$

As we already know the explicit forms of the involved rotations:

$$
\begin{aligned}
\mathsf{R}(\phi_*,\theta_*,0)^{\mu}_1 &= (\cos\phi_*\cos\theta_*,\ \sin\phi_*\cos\theta_*,\ -\sin\theta_*), \\
\mathsf{R}(\phi_*,\theta_*,0)^{\mu}_2 &= (-\sin\phi_*,\ \cos\phi_*,\ 0),
\end{aligned}
$$

the transformation can be shown to be simply

$$\epsilon^{\kappa\tau 3}\mathsf{R}(\phi_*,\theta_*,0)^{\mu}_{\kappa}\mathsf{R}(\phi_*,\theta_*,0)^{\rho}_{\tau}=\epsilon^{\mu\nu\rho}\hat{p}_{*\nu}. \tag{10.75}$$

So, gathering (10.59), (10.61), and (10.75) all in (10.73), we will find

$$
\begin{aligned}
S''_{\Lambda *}{}^{0}(\mathbf{p}_*) &= \frac{1}{m_{\Lambda}}S_{H*}(p_H)\cdot\mathbf{p}_*, \\
S''_{\Lambda *}(\mathbf{p}_*) &= \gamma_w S_{H*}(p_H)+\frac{\varepsilon_{\Lambda *}-\gamma_w m_{\Lambda}}{m_{\Lambda}}S_{H*}(p_H)\cdot\hat{\mathbf{p}}_*\hat{\mathbf{p}}_*+\beta_w S_{H*}(p_H)\times\hat{\mathbf{p}}_*,
\end{aligned}
$$

and the MSV in the DRF is

$$
\begin{aligned}
S''_{\Lambda o}(\mathbf{p}_*) &= \gamma_w S_{H*}(p_H)+(1-\gamma_w)S_{H*}(p_H)\cdot\hat{\mathbf{p}}_*\hat{\mathbf{p}}_*+\beta_w S_{H*}(p_H)\times\hat{\mathbf{p}}_* \\
&= S_{H*}(p_H)\cdot\hat{\mathbf{p}}_*\hat{\mathbf{p}}_*+\beta_w S_{H*}(p_H)\times\hat{\mathbf{p}}_*+\gamma_w\hat{\mathbf{p}}_*\times(S_{H*}(p_H)\times\hat{\mathbf{p}}_*). \tag{10.76}
\end{aligned}
$$

Table 10.1 Polarization transfer formulae for the decay $H \rightarrow \Lambda + X$ in the Mother's rest frame

Decay channels	Local polarization $\mathbf{P}_\Lambda$	$\langle \mathbf{P}_\Lambda \rangle / \langle \mathbf{P}_{H*} \rangle$
A. Strong[a]	$\frac{6}{(j+1)}\left[P_S C^j \mathbf{P}_{H*} - (P_S C^j - C_3^j)\mathbf{P}_{H*} \cdot \hat{\mathbf{p}}_* \hat{\mathbf{p}}_* \right]$	$\frac{2(2P_S C^j + C_3^j)}{(j+1)}$
$1/2^+ \rightarrow 1/2^+ 0^-$	$-\mathbf{P}_{H*} + 2\mathbf{P}_{H*} \cdot \hat{\mathbf{p}}_* \hat{\mathbf{p}}_*$	$-1/3$
$1/2^- \rightarrow 1/2^+ 0^-$	$\mathbf{P}_{H*}$	1
$3/2^+ \rightarrow 1/2^+ 0^-$	$\frac{3}{5}\left[2\mathbf{P}_{H*} - \mathbf{P}_{H*} \cdot \hat{\mathbf{p}}_* \hat{\mathbf{p}}_* \right]$	1
$3/2^- \rightarrow 1/2^+ 0^-$	$\frac{3}{5}\left[-2\mathbf{P}_{H*} + 3\mathbf{P}_{H*} \cdot \hat{\mathbf{p}}_* \hat{\mathbf{p}}_* \right]$	$-3/5$
B. Electromagnetic[b]	$\frac{6}{(j+1)}\left[N^j \mathbf{P}_{H*} - (N^j - N_3^j)\mathbf{P}_{H*} \cdot \hat{\mathbf{p}}_* \hat{\mathbf{p}}_* \right]$	$\frac{2(2N^j + N_3^j)}{(j+1)}$
$1/2^\pm \rightarrow 1/2^+ 1^-$	$-\mathbf{P}_{H*} \cdot \hat{\mathbf{p}}_* \hat{\mathbf{p}}_*$	$-1/3$
C. Weak[c]		
$1/2^+ \rightarrow 1/2^+ 0^-$	$(\alpha_w + \mathbf{P}_{H*} \cdot \hat{\mathbf{p}}_*)\hat{\mathbf{p}}_* + \beta_w \mathbf{P}_{H*} \times \hat{\mathbf{p}}_* + \gamma_w \hat{\mathbf{p}}_* \times (\mathbf{P}_{H*} \times \hat{\mathbf{p}}_*)$	$\frac{1+2\gamma_w}{3}$

[a] $P_S \equiv \eta_H (-1)^{j+\frac{1}{2}}$, $C^{1/2} = 1/4$, $C^{3/2} = 1/2$, and $C_3^{1/2} = C_3^{3/2} = 1/4$
[b] See (10.66) for the definitions of N^j and N_3^j.
[c] See (10.71) and (10.74) for the definitions of α_w, β_w, and γ_w

Finally, the total MSV of the Daughter is

$$\mathbf{S}_{\Lambda o}(\mathbf{p}_*) = \mathbf{S}'_{\Lambda o}(\mathbf{p}_*) + \mathbf{S}''_{\Lambda o}(\mathbf{p}_*),$$

and the average over the whole solid angle Ω_* gives

$$\langle \mathbf{S}_{\Lambda o}(\mathbf{p}_*) \rangle = \left[\gamma_w + (1 - \gamma_w)\frac{1}{2}\int_0^\pi d\theta_* \, \sin\theta_* \cos^2\theta_* \right] \langle \mathbf{S}_{H*}(p_H) \rangle$$
$$= \frac{1+2\gamma_w}{3} \langle \mathbf{S}_{H*}(p_H) \rangle. \tag{10.77}$$

As expected from the arguments in Sect. 10.2, the spontaneous local polarization $\mathbf{S}'_{\Lambda o}(\mathbf{p}_*)$ doesn't contribute to the global one.

For the convenience of future use, we summarize all the polarization transfers from the decays of the Mother with polarization vector $\mathbf{P}_{H*} = \mathbf{S}_{H*}(p_H)/j$ to the Daughter with polarization vector $\mathbf{P}_\Lambda = 2\mathbf{S}_{\Lambda o}(\mathbf{p}_*)$ in Table 10.1, where explicit decay channels are also listed. The results are completely consistent with those given in Ref. [9] to a linear order of the thermal vorticity. We notice that $P_S = -1$ in the strong decay $1/2^+ \rightarrow 1/2^+ 0^-$, that is, only the dynamical amplitude with odd parity is involved. Thus, the polarization transfer result can be alternatively derived from the more general formula of the weak decay $1/2^+ \rightarrow 1/2^+ 0^-$ by setting $\alpha_w = \beta_w = 0$ and $\gamma_w = -1$ as $T_e = 0$. The results are truly consistent with each other according to Table 10.1.

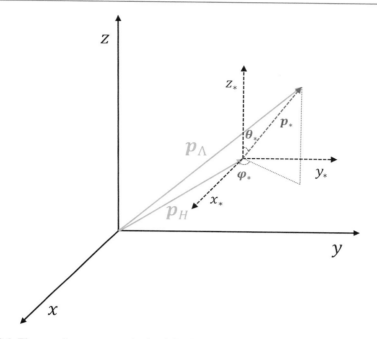

Fig. 10.1 The coordinate systems in the QGP frame, with solid axis and vector lines, and the Mother's rest frame, with dashed axis and vector lines. $\mathbf{p}_H$ and $\mathbf{p}_\Lambda$ are the momenta of the Mother and the Daughter in the QGP frame, respectively. $\mathbf{p}_*$ is the momentum of the Daughter in the Mother's rest frame with the azimuthal angle ϕ_* and polar angle θ_*

10.5 Average Over the Momentum of the Mother

In the previous section, we have established the formulae for the polarization transfer in two-body decays, where the momentum of the Daughter is given in the Mother's rest frame. However, we are more interested in the polarization inherited by the Daughter as a function of its momentum $\mathbf{p}_\Lambda$ in the QGP frame. In the QGPF, the Mother is in a momentum distribution which has to be averaged over before useful results are obtained to compare with experimental measurements. So first of all in this section, we establish the Mother's momentum averaged formula for the mean spin vector of the Daughter with a given momentum in the QGPF. The coordinate systems are parallel to each other in the QGPF and the MRF; see Fig. 10.1, where the momenta of the Mother $\mathbf{p}_H$ and the Daughter $\mathbf{p}_\Lambda$ in the QGPF and the momentum of the Daughter $\mathbf{p}_*$ in the MRF are also illuminated. Note that these momenta are related to each other through the Lorentz boost from the QGPF to the MRF, rather than the simple triangle algebra for vectors in a single coordinate system; see Appendix 10.6.

Let $n(\mathbf{p}_H)$ be the un-normalized momentum distribution of the Mother in the QGPF such that $\int \mathrm{d}^3 p_H \, n(\mathbf{p}_H)$ yields the total number of the Mother; one would then define the MSV of the Daughter fed-down from a specific decay as

$$\mathbf{S}_{\Lambda o}(\mathbf{p}_\Lambda) = \frac{\int \mathrm{d}^3 \mathrm{p}_H \, n(\mathbf{p}_H) \, \mathbf{S}_{\Lambda o}(\mathbf{p}_*)}{\int \mathrm{d}^3 \mathrm{p}_H \, n(\mathbf{p}_H)},$$

where the MSV of the Daughter in the MRF $\mathbf{S}_{\Lambda o}(\mathbf{p}_*)$ is listed in the second column of Table 10.1. Since the magnitude of $\mathbf{p}_*$ is fixed in two-body decay, see (10.46), the three components of $\mathbf{p}_H$ are not completely independent for a given $\mathbf{p}_\Lambda$; see the Lorentz boost relation:

$$\varepsilon_{\Lambda *} = \frac{\varepsilon_H}{m_H}\varepsilon_\Lambda - \frac{1}{m_H}\mathbf{p}_H \cdot \mathbf{p}_\Lambda = \sqrt{\mathrm{p}_{\Lambda *}^2 + m_\Lambda^2}$$

with the energy $\varepsilon_{H/\Lambda} = \sqrt{\mathbf{p}_{H/\Lambda}^2 + m_{H/\Lambda}^2}$ in the QGPF. Taking into account this fact, one should redefine the MSV of the Daughter with momentum $\mathbf{p}_\Lambda$ in the QGPF by multiplying the integrands by a delta function, that is,

$$\mathbf{S}_{\Lambda o}(\mathbf{p}_\Lambda) = \frac{\int \mathrm{d}^3 \mathrm{p}_H \, n(\mathbf{p}_H) \, \mathbf{S}_{\Lambda o}(\mathbf{p}_*) \delta(p_* - p_{\Lambda *})}{\int \mathrm{d}^3 \mathrm{p}_H \, n(\mathbf{p}_H) \delta(p_* - p_{\Lambda *})}. \tag{10.78}$$

By altering the integration variable from $\mathbf{p}_H$ to $\mathbf{p}_*$ through the Lorentz boost relation (see Appendix.1),

$$\mathbf{p}_H = \frac{2m_H(\varepsilon_{\Lambda *} + \varepsilon_\Lambda)(\mathbf{p}_\Lambda - \mathbf{p}_*)}{(\varepsilon_{\Lambda *} + \varepsilon_\Lambda)^2 - (\mathbf{p}_\Lambda - \mathbf{p}_*)^2} = \frac{m_H(\varepsilon_{\Lambda *} + \varepsilon_\Lambda)(\mathbf{p}_\Lambda - \mathbf{p}_*)}{m_\Lambda^2 + \varepsilon_\Lambda \varepsilon_{\Lambda *} + \mathbf{p}_\Lambda \cdot \mathbf{p}_*} \Longrightarrow \hat{\mathbf{p}}_H = \frac{\mathbf{p}_\Lambda - \mathbf{p}_*}{|\mathbf{p}_\Lambda - \mathbf{p}_*|} \tag{10.79}$$

and completing the integrations over the magnitude p_*, only solid angle integrations are left over:

$$\mathbf{S}_{\Lambda o}(\mathbf{p}_\Lambda) = \frac{\int \mathrm{d}\Omega_* \, n(\mathbf{p}_H) \left\| \frac{\partial \mathbf{p}_H}{\partial \mathbf{p}_*} \right\| \mathbf{S}_{\Lambda o}(\mathbf{p}_*)}{\int \mathrm{d}\Omega_* n(\mathbf{p}_H) \left\| \frac{\partial P}{\partial \mathbf{p}_*} \right\|}. \tag{10.80}$$

Here and in the following, one should keep in mind that p_* is fixed to $p_{\Lambda *}$ and the absolute value of the determinant of the Jacobian (AVDJ) reads (see Appendix.1)

$$\left\| \frac{\partial \mathbf{p}_H}{\partial \mathbf{p}_*} \right\| = \frac{m_H^3 (\varepsilon_{\Lambda *} + \varepsilon_\Lambda)^2 \left[(\varepsilon_{\Lambda *} + \varepsilon_\Lambda)^2 - (\varepsilon_\Lambda \varepsilon_{\Lambda *} + \mathbf{p}_\Lambda \cdot \mathbf{p}_* + m_\Lambda^2) \right]}{\varepsilon_{\Lambda *}(\varepsilon_\Lambda \varepsilon_{\Lambda *} + \mathbf{p}_\Lambda \cdot \mathbf{p}_* + m_\Lambda^2)^3}. \tag{10.81}$$

The most involved thing in the evaluation of (10.80) is that $\mathbf{S}_{\Lambda o}(\mathbf{p}_*)$ implicitly depends on ϕ_* and θ_* through $\mathbf{S}_{H*}(p_H)$, besides explicitly through $\hat{\mathbf{p}}_*$. The features of $\mathbf{S}_{H*}(p_H)$ have been well studied by following the symmetries, associated with the parity inversion and rotation around the total angular momentum axis, of the fireball produced in peripheral heavy ion collisions. So the three components of

$S_{H*}(p_H)$ can be expanded as Fourier series of the momentum azimuthal angle ϕ_H to the second-order harmonics [3,4,21]:

$$S_{H*x} \simeq \frac{2j(j+1)}{3} \left[h_1(p_H^T, Y_H) \sin\phi_H + h_2(p_H^T, Y_H) \sin 2\phi_H \right],$$

$$S_{H*y} \simeq \frac{2j(j+1)}{3} \left[g_0(p_H^T, Y_H) + g_1(p_H^T, Y_H) \cos\phi_H + g_2(p_H^T, Y_H) \cos 2\phi_H \right],$$

$$S_{H*z} \simeq \frac{2j(j+1)}{3} f_2(p_H^T, Y_H) \sin 2\phi_H, \tag{10.82}$$

where p_H^T and Y_H are the magnitudes of the transverse momentum and the rapidity of the Mother, respectively. According to the (10.45), the prefactor $2j(j+1)/3$ is extracted out from all the functions f, g, and h so that they don't depend on the total spin j any more. The aforementioned symmetries imply that h_1 and g_1 are odd functions of Y_H whereas g_0, f_2, g_2, and h_2 are even. Furthermore, in a right-handed reference frame with x-axis on the reaction plane and y-axis in the direction opposite to the total angular momentum, both the hydrodynamic model [2] and the AMPT model [4] predicted the magnitudes of all the coefficient functions and particularly their signs to be

$$h_1(p_H^T, Y_H > 0) > 0, \quad h_2(p_H^T, Y_H) < 0, \quad g_0(p_H^T, Y_H) < 0,$$
$$g_1(p_H^T, Y_H > 0) < 0, \quad g_2(p_H^T, Y_H) > 0, \quad f_2(p_H^T, Y_H) < 0. \tag{10.83}$$

For the study of Λ polarization, the Mother's masses are at most 24% larger than that of Λ, so we can assume f, g, and h to be the same as those for the primary Λ according to (10.41).

It's more convenient to represent $\mathbf{p}_H$ and $\mathbf{p}_\Lambda$ with cylindrical coordinates and $\mathbf{p}_*$ with the spherical ones as

$$\mathbf{p}_H = p_H^T \cos\phi_H \mathbf{e}_1 + p_H^T \sin\phi_H \mathbf{e}_2 + p_{Hz}\mathbf{e}_3,$$
$$\mathbf{p}_\Lambda = p_\Lambda^T \cos\phi_\Lambda \mathbf{e}_1 + p_\Lambda^T \sin\phi_\Lambda \mathbf{e}_2 + p_{\Lambda z}\mathbf{e}_3,$$
$$\mathbf{p}_* = p_* \sin\theta_* \cos\phi_* \mathbf{e}_1 + p_* \sin\theta_* \sin\phi_* \mathbf{e}_2 + p_* \cos\theta_* \mathbf{e}_3. \tag{10.84}$$

In the following, we stick to the simplest case of midrapidity Λ with $p_{\Lambda z} = 0$. Then, the rightmost equality in (10.79) can be used to express the trigonometric functions of the Mother in terms of the spherical coordinates of the Daughter as

$$\sin 2\phi_H = \frac{p_*^2 \sin^2\theta_* \sin 2\phi_* + p_\Lambda^{T\,2} \sin 2\phi_\Lambda - 2p_* p_\Lambda^T \sin\theta_* \sin(\phi_* + \phi_\Lambda)}{p_*^2 \sin^2\theta_* + p_\Lambda^{T\,2} - 2p_* p_\Lambda^T \sin\theta_* \cos(\phi_* - \phi_\Lambda)}$$
$$= \mathcal{A}(\theta_*, \psi) \sin 2\phi_\Lambda + \mathcal{B}(\theta_*, \psi) \cos 2\phi_\Lambda,$$

$$\cos 2\phi_H = \frac{p_*^2 \sin^2\theta_* \cos 2\phi_* + p_\Lambda^{T\,2} \cos 2\phi_\Lambda - 2p_* p_\Lambda^T \sin\theta_* \cos(\phi_* + \phi_\Lambda)}{p_*^2 \sin^2\theta_* + p_\Lambda^{T\,2} - 2p_* p_\Lambda^T \sin\theta_* \cos(\phi_* - \phi_\Lambda)}$$
$$= \mathcal{A}(\theta_*, \psi) \cos 2\phi_\Lambda - \mathcal{B}(\theta_*, \psi) \sin 2\phi_\Lambda,$$

$$\sin \phi_H = \frac{p_\Lambda^T \sin \phi_\Lambda - p_* \sin \theta_* \sin \phi_*}{\sqrt{p_*^2 \sin^2 \theta_* + p_\Lambda^{T\,2} - 2p_* p_\Lambda^T \sin \theta_* \cos(\phi_* - \phi_\Lambda)}}$$
$$= \mathcal{C}(\theta_*, \psi) \sin \phi_\Lambda + \mathcal{D}(\theta_*, \psi) \cos \phi_\Lambda,$$
$$\cos \phi_H = \frac{p_\Lambda^T \cos \phi_\Lambda - p_* \sin \theta_* \cos \phi_*}{\sqrt{p_*^2 \sin^2 \theta_* + p_\Lambda^{T\,2} - 2p_* p_\Lambda^T \sin \theta_* \cos(\phi_* - \phi_\Lambda)}}$$
$$= \mathcal{C}(\theta_*, \psi) \cos \phi_\Lambda - \mathcal{D}(\theta_*, \psi) \sin \phi_\Lambda \qquad (10.85)$$

with the introduced variable $\psi = \phi_* - \phi_\Lambda$ and the auxiliary functions:

$$\mathcal{A}(\theta_*, \psi) = \frac{p_*^2 \sin^2 \theta_* \cos 2\psi - 2p_* p_\Lambda^T \sin \theta_* \cos \psi + p_\Lambda^{T\,2}}{p_*^2 \sin^2 \theta_* + p_\Lambda^{T\,2} - 2p_* p_\Lambda^T \sin \theta_* \cos \psi},$$
$$\mathcal{B}(\theta_*, \psi) = \frac{p_*^2 \sin^2 \theta_* \sin 2\psi - 2p_* p_\Lambda^T \sin \theta_* \sin \psi}{p_*^2 \sin^2 \theta_* + p_\Lambda^{T\,2} - 2p_* p_\Lambda^T \sin \theta_* \cos \psi},$$
$$\mathcal{C}(\theta_*, \psi) = \frac{p_\Lambda^T - p_* \sin \theta_* \cos \psi}{\sqrt{p_*^2 \sin^2 \theta_* + p_\Lambda^{T\,2} - 2p_* p_\Lambda^T \sin \theta_* \cos \psi}},$$
$$\mathcal{D}(\theta_*, \psi) = \frac{-p_* \sin \theta_* \sin \psi}{\sqrt{p_*^2 \sin^2 \theta_* + p_\Lambda^{T\,2} - 2p_* p_\Lambda^T \sin \theta_* \cos \psi}}. \qquad (10.86)$$

One can easily verify the even-odd and normalization features of the auxiliary functions, that is,

$$\mathcal{A}(\theta_*, -\psi) = \mathcal{A}(\theta_*, \psi), \qquad \mathcal{B}(\theta_*, -\psi) = -\mathcal{B}(\theta_*, \psi),$$
$$\mathcal{C}(\theta_*, -\psi) = \mathcal{C}(\theta_*, \psi), \qquad \mathcal{D}(\theta_*, -\psi), = -\mathcal{D}(\theta_*, \psi)$$
$$\mathcal{A}^2(\theta_*, \psi) + \mathcal{B}^2(\theta_*, \psi) = 1, \qquad \mathcal{C}^2(\theta_*, \psi) + \mathcal{D}^2(\theta_*, \psi) = 1. \qquad (10.87)$$

With the variable transformation $\phi_* \to \psi$, the integration over the solid angle $d\Omega_*$ in (10.80) can be replaced by another one:

$$\int d\Omega_* = \int_0^\pi d\theta_* \sin \theta_* \int_{-\phi_\Lambda}^{2\pi - \phi_\Lambda} d\psi = \int_0^\pi d\theta_* \sin \theta_* \int_{-\pi}^\pi d\psi,$$

where the last step is owing to the 2π-periodic in ψ of all the functions in the integrands.

The spectrum function $n(\mathbf{p}_H)$ depends on the specific model of the collision, but it must be *even* in "$\cos \theta_H$" because of the symmetries of the colliding system and isotropic in the transverse plane when the usually small elliptic flow is neglected. So, $n(\mathbf{p}_H)$ can be assumed to only depend on the magnitudes of its longitudinal and transverse momenta p_H^L and p_H^T to a very good approximation. In this case, p_H^L and

Table 10.2 The even ("+") and odd ("−") properties of relevant functions in (10.80) (top row) with respect to the variables (first column).

Variables	h_1, g_1	h_2, g_0, g_2, f_2	$\mathcal{A}, \mathcal{C}$	$\mathcal{B}, \mathcal{D}$	$n(\mathbf{p}_H)$	$\left\|\frac{\partial \mathbf{p}_H}{\partial \mathbf{p}_*}\right\|$
$\cos\theta_*$	−	+	+	+	+	+
ψ	+	+	+	−	+	+

p_H^T can be given explicitly with the variables for the Daughter by following (10.79) as

$$\mathrm{p}_H^L = m_H \frac{(\varepsilon_{\Lambda*} + \varepsilon_\Lambda)|\mathrm{p}_* \cos\theta_*|}{m_\Lambda^2 + \varepsilon\varepsilon_{\Lambda*} + \mathrm{p}_\Lambda^T \mathrm{p}_* \sin\theta_* \cos\psi}, \qquad (10.88)$$

$$\mathrm{p}_H^T = m_H \frac{(\varepsilon_{\Lambda*} + \varepsilon_\Lambda)\sqrt{\mathrm{p}_*^2 \sin^2\theta_* + \mathrm{p}_\Lambda^{T\,2} - 2\mathrm{p}_\Lambda^T \mathrm{p}_* \sin\theta_* \cos\psi}}{m_\Lambda^2 + \varepsilon\varepsilon_{\Lambda*} + \mathrm{p}_\Lambda^T \mathrm{p}_* \sin\theta_* \cos\psi}, \qquad (10.89)$$

which imply that the Mother's spectrum function is even in both $\cos\theta_*$ and ψ. Then, as the polar angle of the Mother is given by

$$\cos\theta_H \equiv \hat{\mathbf{p}}_H \cdot \mathbf{e}_3 = -\frac{\mathrm{p}_* \cos\theta_*}{|\mathbf{p}_\Lambda - \mathbf{p}_*|} = -\frac{\mathrm{p}_* \cos\theta_*}{\sqrt{\mathrm{p}_*^2 + \mathrm{p}_\Lambda^{T\,2} - 2\mathrm{p}_\Lambda^T \mathrm{p}_* \sin\theta_* \cos\psi}}, \qquad (10.90)$$

the rapidity Y_H is found to be an odd function of $\cos\theta_*$ and an even function of ψ, that is,

$$m_H^T \sinh Y_H = \mathbf{p}_H \cdot \mathbf{e}_3 = \mathrm{p}_H^T \tan\theta_H,$$

where the transverse mass $m_H^T = \sqrt{(\mathrm{p}_H^T)^2 + m_H^2}$. For the convenience of future discussions, we summarize the even-oddness of all the functions relevant to the evaluation of (10.80) in Table 10.2.

Now, we can well understand the advantage of introducing the new variable "ψ": In this way, p_H^T and p_H^L, thus $g, f, h, n(\mathbf{p}_H)$ and $\left\|\frac{\partial \mathbf{p}_H}{\partial \mathbf{p}_*}\right\|$, are all independent of the observable ϕ_Λ and the integrations over the new solid angles ψ and θ_* can be numerically carried out easily.

We now pay attention to the parity-conservative strong and EM decays first, which share very similar expressions for the MSV of the Daughter; see Table 10.1. For brevity, the following general formula will be used:

$$\mathbf{S}_{\Lambda o}^{PC}(\mathbf{p}_*) = \frac{3}{j(j+1)} \left[A\, \mathbf{S}_{H*} + B\, \mathbf{S}_{H*} \cdot \hat{\mathbf{p}}_* \hat{\mathbf{p}}_* \right], \qquad (10.91)$$

where the strong (EM) decay coefficients $A = P_S C^j\ (N^j)$ and $B = C_3^j - P_S C^j$ $(N_3^j - N^j)$ are constants solely determined by the helicity properties of the transition amplitudes. Then, the integrands for the transverse and longitudinal components of the MSV of the Daughter can be given with spherical coordinates as

$$S_{\Lambda x}^{PC}(\mathbf{p}_*) = \frac{3}{2j(j+1)}\Big[2S_{H*x}\Big(A + B\cos^2\phi_*\sin^2\theta_*\Big) + B\Big(S_{H*z}\cos\phi_*\sin 2\theta_*$$
$$+ S_{H*y}\sin 2\phi_*\sin^2\theta_*\Big)\Big],$$

$$S_{\Lambda y}^{PC}(\mathbf{p}_*) = \frac{3}{2j(j+1)}\Big[2S_{H*y}\Big(A + B\sin^2\phi_*\sin^2\theta_*\Big) + B\Big(S_{H*z}\sin\phi_*\sin 2\theta_*$$
$$+ + S_{H*x}\sin 2\phi_*\sin^2\theta_*\Big)\Big],$$

$$S_{\Lambda z}^{PC}(\mathbf{p}_*) = \frac{3}{2j(j+1)}\Big[2S_{H*z}\Big(A + B\cos^2\theta_*\Big) + B\Big(S_{H*x}\cos\phi_*\sin 2\theta_*$$
$$+ + S_{H*y}\sin\phi_*\sin 2\theta_*\Big)\Big]. \tag{10.92}$$

Inserting (10.82) into these integrands with the help of the trigonometric function relations (10.85) and using the even-odd properties listed in Table 10.2, only the following terms are non-vanishing when the integrations over the solid angle are taken into account (see Appendix.2):

$$S_{\Lambda x}^{PC}(\mathbf{p}_*) = [h_2 F\mathcal{A} + Bg_0\sin^2\theta_*\cos 2\psi]\sin 2\phi_\Lambda + \frac{B}{2}l_2^+\mathcal{F}_2^-\sin^2\theta_*\sin 4\phi_\Lambda,$$

$$S_{\Lambda y}^{PC}(\mathbf{p}_*) = \Big[g_0 F + \frac{B}{2}l_2^-\mathcal{F}_2^+\sin^2\theta_*\Big] + \Big[g_2 F\mathcal{A} - Bg_0\sin^2\theta_*\cos 2\psi\Big]\cos 2\phi_\Lambda$$
$$- \frac{B}{2}l_2^+\mathcal{F}_2^-\sin^2\theta_*\cos 4\phi_\Lambda,$$

$$S_{\Lambda z}^{PC}(\mathbf{p}_*) = \Big[2f_2\Big(A + B\cos^2\theta_*\Big)\mathcal{A} + \frac{B}{2}l_1^+(\mathcal{C}\cos\psi - \mathcal{D}\sin\psi)\sin 2\theta_*\Big]\sin 2\phi_\Lambda, \tag{10.93}$$

where we define the auxiliary functions as

$$l_n^\pm = h_n \pm g_n, \qquad F = 2A + B\sin^2\theta_*, \qquad \mathcal{F}_n^\pm = \mathcal{A}\cos n\psi \pm \mathcal{B}\sin n\psi.$$

So, both the single-ϕ_H harmonics in the transverse components of the MSV of the Mother contribute to the LLP $S_{\Lambda z}^{PC}(\mathbf{p}_*)$, while its local feature $\sim \sin 2\phi_H$ is well inherited by the Daughter with the polarization $\sim \sin 2\phi_\Lambda$. The decays also give rise to higher mode of harmonics to the TLPs of the Daughter, that is, $\sin 4\phi_\Lambda$ and $\cos 4\phi_\Lambda$, even though the primary ones of the Mother are only to $2\phi_H$ harmonics. As both h_1 and g_1 vanish for primary Λ with $p_{z*} = 0$, we arrive at a conclusion: only even-time harmonics of ϕ_Λ are relevant to midrapidity Λ polarizations, even after collecting the feed-downs from the strong and EM decays of the primary Mothers.

In weak decay, the previous polarization transfer pattern (10.91) remains important with the coefficients defined as $A = \gamma_w/4$ and $B = (1 - \gamma_w)/4$ now. However, more terms are involved in weak decay, that is, the α_w- and β_w-dependent terms listed in Table 10.1. The α_w term is irrelevant to the initial polarization of the Mother, and the contributions to the transverse and longitudinal components of the MSV of the Daughter can be given directly as

$$S_{\Lambda x}^{\alpha_w}(\mathbf{p}_*) = \frac{\alpha_w}{2} \sin\theta_* \cos\psi \cos\phi_\Lambda,$$
$$S_{\Lambda y}^{\alpha_w}(\mathbf{p}_*) = \frac{\alpha_w}{2} \sin\theta_* \cos\psi \sin\phi_\Lambda,$$
$$S_{\Lambda z}^{\alpha_w}(\mathbf{p}_*) = 0. \tag{10.94}$$

The explicit forms for the corresponding contributions from the β_w term are

$$S_{\Lambda x}^{\beta_w}(\mathbf{p}_*) = \beta_w \Big(S_{H*y} \cos\theta_* - S_{H*z} \sin\phi_* \sin\theta_* \Big),$$
$$S_{\Lambda y}^{\beta_w}(\mathbf{p}_*) = \beta_w \Big(S_{H*z} \cos\phi_* \sin\theta_* - S_{H*x} \cos\theta_* \Big),$$
$$S_{\Lambda z}^{\beta_w}(\mathbf{p}_*) = \beta_w \Big(S_{H*x} \sin\phi_* \sin\theta_* - S_{H*y} \cos\phi_* \sin\theta_* \Big). \tag{10.95}$$

Then, by following a similar procedure as that for the strong and EM decays, the terms giving rise to finite contributions are just

$$S_{\Lambda x}^{\beta_w}(\mathbf{p}_*) = \frac{\beta_w}{4} \Big[(-f_2 \mathcal{F}_1^+ \sin\theta_* + g_1 C \cos\theta_*) \cos\phi_\Lambda + f_2 \mathcal{F}_1^- \sin\theta_* \cos 3\phi_\Lambda \Big],$$
$$S_{\Lambda y}^{\beta_w}(\mathbf{p}_*) = \frac{\beta_w}{4} \Big[(f_2 \mathcal{F}_1^+ \sin\theta_* - h_1 C \cos\theta_*) \sin\phi_\Lambda + f_2 \mathcal{F}_1^- \sin\theta_* \sin 3\phi_\Lambda \Big],$$
$$S_{\Lambda z}^{\beta_w}(\mathbf{p}_*) = \frac{\beta_w}{4} \sin\theta_* (l_2^- \mathcal{F}_1^+ \cos\phi_\Lambda - l_2^+ \mathcal{F}_1^- \cos 3\phi_\Lambda). \tag{10.96}$$

Notice that α_w and β_w terms only give rise to odd-time harmonics of ϕ_Λ, contrary to the even ones in strong and EM decays.

Thus, by gathering (10.93),(10.94), and (10.96), the most general integrands for the transverse and longitudinal components of the MSV of the Daughter fed-down from a single decay in HICs are

$$S_{\Lambda i}(\mathbf{p}_*) = S_{\Lambda i}^{PC}(\mathbf{p}_*) + S_{\Lambda i}^{\alpha_w}(\mathbf{p}_*) + S_{\Lambda i}^{\beta_w}(\mathbf{p}_*), \qquad i = x, y, z. \tag{10.97}$$

For the transverse components of the Λ polarization, the β_w contributions are much less important than the α_w ones because of the smallness of the polarization coefficients h, g, and f, thus, they are suppressed in the following discussions. For the longitudinal component, the β_w term from weak decay breaks the pure $\sin 2\phi_\Lambda$ polarization structure of the Daughter inherited from the strong and EM decays in principle. However, for the specific case with Λ polarization, the relevant decay parameters for Ξ^0 and Ξ^- are [20]:

$$\alpha_w^{\Xi^0} = -0.347, \quad \beta_w^{\Xi^0} = \tan(0.366 \pm 0.209)\gamma_w^{\Xi^0}, \quad \gamma_w^{\Xi^0} = 0.85,$$
$$\alpha_w^{\Xi^-} = -0.392, \quad \beta_w^{\Xi^-} = \tan(0.037 \pm 0.014)\gamma_w^{\Xi^-}, \quad \gamma_w^{\Xi^-} = 0.89.$$

So $\beta_w^{\Xi^-}/\gamma_w^{\Xi^-} = 0.037 \pm 0.014$ is very small and the breaking effect can be safely neglected for the weak decay of Ξ^-; but $\beta_w^{\Xi^0}/\gamma_w^{\Xi^0} = 0.158 - 0.648$, the breaking effect might be large for that of Ξ^0. If we assume $|h_2|, g_2 \lesssim |f_2|/2$ which is always true in HICs [2], the magnitudes of the integrated coefficients in front of $\cos\phi_\Lambda$ and $\cos 3\phi_\Lambda$ are at least one order smaller than that of $\sin 2\phi_\Lambda$ for the largest ratio: $\beta_w^{\Xi^0}/\gamma_w^{\Xi^0} = 0.648$. The reason can be well understood by comparing the prefactors in the integrands: Keeping only the dominant $\mathcal{A}$-related term in $\mathcal{F}_1^\pm$, the ratios between the prefactors are roughly

$$\frac{\beta_w^{\Xi^0}}{\gamma_w^{\Xi^0} + (1 - \gamma_w^{\Xi^0})\cos^2\theta_*} \frac{h_2 \mp g_2}{2f_2} \sin\theta_* \cos\psi.$$

Then, they are double trigonometric functions suppressed especially by $\cos\psi$ when carrying out the integrations, besides the initial suppression by $\beta_w^{\Xi^0}/\gamma_w^{\Xi^0}$. Similar comparisons can also be applied to the TLPs, thus, the contributions from β_w term will be neglected for Λ polarization in the following.

At sufficiently high energy, because of the approximate longitudinal boost invariance, we expect all the functions g, h, and f in (10.82) to be very weakly dependent on the rapidity Y_H. As a consequence, compared to the other rapidity-even functions, the rapidity-odd functions h_1 and g_1 can be safely neglected as they vanish at midrapidity $Y_H = 0$. Finally, by inserting (10.93) and (10.94) into (10.80), the total transverse and longitudinal components of the MSV of Λ can be put in simple forms as

$$\mathbf{S}_{\Lambda x}(\mathbf{p}_\Lambda^T) = \frac{1}{2}\left(H_1^{\text{Tot}}(\mathbf{p}_\Lambda^T)\cos\phi_\Lambda + H_2^{\text{Tot}}(\mathbf{p}_\Lambda^T)\sin 2\phi_\Lambda + H_4^{\text{Tot}}(\mathbf{p}_\Lambda^T)\sin 4\phi_\Lambda\right)$$

$$= \frac{1}{2}\sum_{M=\Lambda}^{H}\left(H_1^M(\mathbf{p}_\Lambda^T)\cos\phi_\Lambda + H_2^M(\mathbf{p}_\Lambda^T)\sin 2\phi_\Lambda + H_4^M(\mathbf{p}_\Lambda^T)\sin 4\phi_\Lambda\right)$$

$$\equiv \frac{1}{2}\left[R_{\Lambda p}h_2\sin 2\phi_\Lambda + \sum_H R_H\left(h_1^H\cos\phi_\Lambda + h_2^H\sin 2\phi_\Lambda + h_4^H\sin 4\phi_\Lambda\right)\right],$$

$$h_1^H(\mathbf{p}_\Lambda^T) = \frac{\alpha_w^H}{N_H}\int d\Omega_* n(\mathbf{p}_H)\left\|\frac{\partial\mathbf{p}_H}{\partial\mathbf{p}_*}\right\|\sin\theta_*\cos\psi,$$

$$h_2^H(\mathbf{p}_\Lambda^T) = \frac{2}{N_H}\int d\Omega_* n(\mathbf{p}_H)\left\|\frac{\partial\mathbf{p}_H}{\partial\mathbf{p}_*}\right\|\left[h_2(\mathbf{p}_H^T)F^H\mathcal{A} + B^H g_0(\mathbf{p}_H^T)\sin^2\theta_*\cos 2\psi\right],$$

$$h_4^H(\mathbf{p}_\Lambda^T) = \frac{B^H}{N_H}\int d\Omega_* n(\mathbf{p}_H)\left\|\frac{\partial\mathbf{p}_H}{\partial\mathbf{p}_*}\right\|l_2^+(\mathbf{p}_H^T)\mathcal{F}_2^-\sin^2\theta_*, \tag{10.98}$$

$$\mathbf{S}_{\Lambda y}(\mathbf{p}_\Lambda^T) = \frac{1}{2}\left(G_0^{\text{Tot}}(\mathbf{p}_\Lambda^T) + G_1^{\text{Tot}}(\mathbf{p}_\Lambda^T)\sin\phi_\Lambda + G_2^{\text{Tot}}(\mathbf{p}_\Lambda^T)\cos 2\phi_\Lambda + G_4^{\text{Tot}}(\mathbf{p}_\Lambda^T)\cos 4\phi_\Lambda\right)$$

$$= \frac{1}{2}\sum_{M=\Lambda}^{H}\left(G_0^M(\mathbf{p}_\Lambda^T) + G_1^M(\mathbf{p}_\Lambda^T)\sin\phi_\Lambda + G_2^M(\mathbf{p}_\Lambda^T)\cos 2\phi_\Lambda + G_4^M(\mathbf{p}_\Lambda^T)\cos 4\phi_\Lambda\right)$$

$$\equiv \frac{1}{2}\left[R_{\Lambda p}(g_0 + g_2\cos 2\phi_\Lambda) + \sum_H R_H\left(g_0^H + h_1^H\sin\phi_\Lambda + g_2^H\cos 2\phi_\Lambda\right.\right.$$

$$\left.\left. - h_4^H\cos 4\phi_\Lambda\right)\right],$$

$$g_0^H(\mathbf{p}_\Lambda^T) = \frac{1}{N_H}\int d\Omega_* n(\mathbf{p}_H)\left\|\frac{\partial\mathbf{p}_H}{\partial\mathbf{p}_*}\right\|\left[2g_0(\mathbf{p}_H^T)F^H + B^H l_2^-(\mathbf{p}_H^T)\mathcal{F}_2^+\sin^2\theta_*\right],$$

$$g_2^H(\mathbf{p}_\Lambda^T) = \frac{2}{N_H}\int d\Omega_* n(\mathbf{p}_H)\left\|\frac{\partial\mathbf{p}_H}{\partial\mathbf{p}_*}\right\|\left[g_2(\mathbf{p}_H^T)F^H\mathcal{A} - B^H g_0(\mathbf{p}_H^T)\sin^2\theta_*\cos 2\psi\right]. \tag{10.99}$$

$$\mathbf{S}_{\Lambda z}(\mathbf{p}_\Lambda^T) = \frac{1}{2} F_2^{\text{Tot}}(\mathbf{p}_\Lambda^T) \sin 2\phi_\Lambda = \frac{1}{2} \left[\sum_{M=\Lambda}^{H} F_2^M(\mathbf{p}_\Lambda^T) \right] \sin 2\phi_\Lambda$$

$$\equiv \frac{1}{2} \left[R_{\Lambda p} f_2(\mathbf{p}_\Lambda^T) + \sum_H R_H f_2^H(\mathbf{p}_\Lambda^T) \right] \sin 2\phi_\Lambda,$$

$$f_2^H(\mathbf{p}_\Lambda^T) = \frac{4}{\mathcal{N}_H} \int d\Omega_* \, n(\mathbf{p}_H) \left\| \frac{\partial \mathbf{p}_H}{\partial \mathbf{p}_*} \right\| f_2(\mathbf{p}_H^T) \left(A^H + B^H \cos^2 \theta_* \right) \mathcal{A} \qquad (10.100)$$

with the normalization

$$\mathcal{N}_H = \int d\Omega_* n(\mathbf{p}_H) \left\| \frac{\partial \mathbf{p}_H}{\partial \mathbf{p}_*} \right\|.$$

Here, $h^H(\mathbf{p}_\Lambda^T)$, $g^H(\mathbf{p}_\Lambda^T)$ and $f^H(\mathbf{p}_\Lambda^T)$ are the polarization transfer coefficients from the Mother and R's are the Λ number fractions from different contribution channels: $R_{\Lambda p}$ primary and R_H secondary. Due to the 2π-periodicity of all the components with respect to ϕ_Λ, we can fold the transverse ones $\mathbf{S}_{\Lambda x}(\mathbf{p}_\Lambda^T)$ once over the region $\phi_\Lambda \in (-\pi/2, 3\pi/2)$ to $(-\pi/2, \pi/2)$ and $\mathbf{S}_{\Lambda y}(\mathbf{p}_\Lambda^T)$ twice over the region $\phi_\Lambda \in (0, 2\pi)$ to $(0, \pi/2)$, respectively. Then, all the trivial harmonics of ϕ_Λ contributed from α_w^H terms will be removed from (10.98) and (10.99), and the even-time harmonics of ϕ_Λ can be explored in advance. For cascade decays, the evaluations of the MSV of the last Daughter should be done step by step, that is, iterating (10.98), (10.99), and (10.100) over and over until the Daughter we're interested in. Take the EM decay $\Sigma^0 \to \Lambda\gamma$, for example; we should first obtain the total polarization coefficients for Σ^0 including both the primary contributions and feed-downs from higher lying resonances. Then, these total polarization coefficients, instead of the primary ones, are used to evaluate the contribution of Σ^0 decay to Λ polarization; see [1,9] for numerical calculations.

10.6 Theoretical Predictions and Sign Puzzles

In this section, we perform numerical calculations by adopting (10.98), (10.99), and (10.100), and compare the results with experimental measurements if available. In [8], we just focused on the most important feed-down effects on the LLP of the Λ, that is, from the strong and EM decay channels with the Mother $H = \Sigma^*$ and $H = \Sigma^0$, respectively. A more complete study of all decay channels had been performed in [9] and the conclusion remains the same for LLP. As mentioned before, these two parity conservative channels correspond to the decay types $3/2^+ \to 1/2^+ 0^-$ and $1/2^+ \to 1/2^+ 1^-$, and the decay coefficients are, respectively,

$$A^{\Sigma^*} = 1/2, \ B^{\Sigma^*} = -1/4; \qquad A^{\Sigma^0} = 0, \ B^{\Sigma^0} = -1/4.$$

The fractions of primary and secondary Λ can be estimated by means of the statistical hadronization model. At the hadronization temperature $T = 164$ MeV and baryon

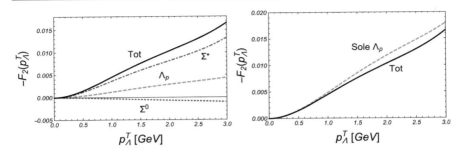

Fig. 10.2 Left panel: longitudinal polarization coefficients $F_2(\mathrm{p}_\Lambda^T)$ of the Λ. Primary (Λ_p) and secondary ($H = \Sigma^*, \Sigma^0$) components, weighted with the production fractions are shown together with the resulting sum $F_2^{\mathrm{Tot}}(\mathrm{p}_\Lambda^T)$ (solid line). Right panel: comparison between the total polarization coefficient $F_2^{\mathrm{Tot}}(\mathrm{p}_\Lambda^T)$ of the Λ and the one $f_2(\mathrm{p}_\Lambda^T)$ of only primary Λ [3]

chemical potential of 30 MeV for $\sqrt{s_{\mathrm{NN}}} = 200$ GeV Au+Au collisions, they turn out to be [19]:

$$R_{\Lambda_p} = 0.243, \quad R_{\Sigma^*} = 0.359, \quad R_{\Sigma^0} = 0.275 * 60\%, \tag{10.101}$$

where 60% is the contribution fraction from primary Σ^0 and the left from higher lying resonance decays is assumed to cancel out for simplicity. At this hadronization temperature, the quantum statistics effects are negligible for all these particles, so the Boltzmann distinguishable particle assumption adopted in Sect. 10.3 is an excellent approximation.

To perform numerical evaluations for the longitudinal component of the MSV of the Daughter (10.100), two ingredients are still unknown: the primary LLP prefactor $f_2(\mathrm{p}^T)$ and the momentum spectrum $n(\mathbf{p}_H)$. A precise fit to the data obtained in [1] for $f_2(\mathrm{p}_\Lambda^T)$ of the primary Λ yields

$$f_2(\mathrm{p}_\Lambda^T) = \left[-7.71 \left(\mathrm{p}_\Lambda^T \right)^2 + 3.32 \left(\mathrm{p}_\Lambda^T \right)^3 - 0.471 \left(\mathrm{p}_\Lambda^T \right)^4 \right] \times 10^{-3}$$

with p_Λ^T's unit "GeV". As far as $n(\mathbf{p}_H)$ is concerned, it is plausible that the dependence on its form is very mild, because it shows in both the numerator and denominator of $f_2^H(\mathrm{p}_\Lambda^T)$. For the purpose of approximate calculations, we have assumed a spectrum of the following form [8]:

$$n(\mathbf{p}_H) \propto \frac{1}{\cosh Y_H} \mathrm{e}^{-m_H^T/T_s} = \frac{m_H^T}{\varepsilon_H} \mathrm{e}^{-m_H^T/T_s}, \tag{10.102}$$

where T_s is a phenomenological parameter describing the slope of the transverse momentum spectrum. It had been checked that the final results are almost independent of T_s within a realistic range: $T_s = 0.2 - 0.8$ GeV.

The relevant polarization prefactors $F_2^M(\mathrm{p}_\Lambda^T)$ for primary and secondary decay components and the total $F_2^{\mathrm{Tot}}(\mathrm{p}_\Lambda^T)$ are shown together in Fig. 10.2, and the associated LLP features are illuminated in Fig. 10.3 where we choose $\mathrm{p}_\Lambda^T = 2$ GeV as an

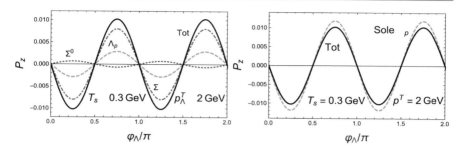

Fig. 10.3 Left panel: the azimuthal angle dependence of the longitudinal polarization $P_z = S_{\Lambda z}(p_\Lambda^T)/S = 2S_{\Lambda z}(p_\Lambda^T)$ of the Λ. Primary (Λ_p) and secondary ($H = \Sigma^*, \Sigma^0$) components, weighted with the production fractions are shown together with the resulting sum (solid line) at fixed transverse momentum $p_\Lambda^T = 2$ GeV and slope parameter $T_s = 0.3$ GeV. Right panel: comparison between the total polarization profile of the Λ and that of only primary Λ [3]

example. As expected from the polarization transfer coefficients list in Table 10.1 and the fractions in (10.101), the strong and EM decays give large positive and small negative feedbacks to the primary Λ polarization, respectively; see the left panel in Fig. 10.2. It happens that $F_2^{Tot}(p_\Lambda^T)$ is close to the primary $f_2(p_\Lambda^T)$ and only slightly suppressed in large p_Λ^T region; see the right panel in Fig. 10.2. In principle, there are also feedbacks from EM decay of secondary Σ^0 and weak decays of Ξ's (positive), but their weights in Λ productions are quite limited and definitely not able to flip the sign of $f_2(p_\Lambda^T)$ in Fig. 10.2; see the results presented in [9]. Compared to the theoretical predictions for the LLP profiles in Fig. 10.3, the experimental measurements nicely verified the $\sin 2\phi_\Lambda$ feature but with an opposite sign [6,7]; see Fig. 10.4 for both Λ and $\bar\Lambda$ polarizations. We'd like to point out that this contradiction is not due to different conventions of the coordinate system in the theoretical and experimental studies. It is a real *sign puzzle* because the experimental measurements follow the same sign as that given by the differential of elliptic flow: $-\partial_\phi v_2(\phi)$ [6,7] but the theoretical predictions give opposite sign due to the negative prefactor $dT/d\tau$ [3]. Of course, this statement bases on the fact that the model calculations could well reproduce the elliptic flows measured in HICs; see hydrodynamic [23] and AMPT [24] simulations for example.

For the TLPs, radial component P_r was discovered mainly due to the parity-violating effect from weak decays [9]. According to (10.98) and (10.99), the radial polarization of Λ should be approximately proportional to $\alpha_w^\Xi R_\Xi$ with $R_\Xi \sim 15\%$ [9]. The results are shown in Fig. 10.5 for $p_\Lambda^T = 2$ GeV, where the $-\cos\phi_\Lambda$ and $-\sin\phi_\Lambda$ features are just inherited from the sign of α_w; see [9] for more realistic calculations. Now getting rid of the spontaneous radial polarization, we focus on the folded TLPs with the feed-down effect of the form (10.93). First of all, the folded results for $S_{\Lambda x}^{PC}(p_\Lambda^T)$ are studied and shown in Fig. 10.6, where very nice $2\phi_\Lambda$ harmonics can be identified. The higher harmonic $\sim \sin 4\phi_\Lambda$ vanishes here because the chosen parameters satisfy $h_2 + g_2 = 0$ and we've checked that this contribution is very weak even for $g_2 = h_2 = -f_2/4$.

Fig. 10.4 The experimental
measurements of the
longitudinal local
polarizations of Λ and $\bar{\Lambda}$
hyperons as functions of the
azimuthal angle ϕ relative to
the second-order event plane
Ψ_2 for $20\% - 60\%$ centrality
bin in $\sqrt{s_{NN}} = 200$GeV
Au+Au collisions [6,7].
Solid lines show the fit with
the function $\sin(2(\phi - \Psi_2))$

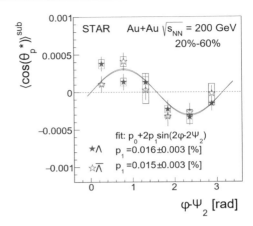

Fig. 10.5 The radial
polarizations P_r^x (red dashed
line) and P_r^y (blue dotted
line) of the Λ as functions of
the azimuthal angle at fixed
transverse momentum
$p_\Lambda^T = 2$ GeV. Only the
dominate α_w term is adopted
for illumination

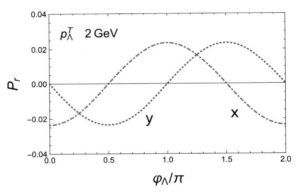

For the more involved TLP $\mathbf{S}_{\Lambda y}^{PC}(p_\Lambda^T)$, the comparison between theoretical predictions for $p_\Lambda^T = 2$ GeV and experimental measurements is illuminated in Fig. 10.7. Due to different conventions for the y-axis, the polarization $-P_y$ predicted in the theoretical study corresponds to P_H measured in experiments. Then, we immediately find that the signs of the azimuthal angle averaged $-P_y$ and the global P_H are consistent with each other, which just follow that of the total angular momentum. However, the relative magnitudes between the in-plane ($\phi_\Lambda = 0$) and out-plane ($\phi_\Lambda = \pi/2$) polarizations are opposite in the theoretical and experimental studies. The theoretical profile originates from the opposite signs between g_0 and g_2 as discussed in (10.83) and the feed-down effect from the Mothers would not change that; see also [9]. So, this is another *sign puzzle* in Λ polarization and definitely rules out the naive guess that the contradictions between theoretical and experimental results are only due to different conventions of the coordinate system.

As indicated in (10.93), secondary decays can give rise to $4\phi_\Lambda$ harmonic of Λ polarization along the total angular momentum even though only up to $2\phi_H$ harmonics of the primary Mother polarizations are considered. Similar to $\mathbf{S}_{\Lambda x}^{PC}(p_\Lambda^T)$, this higher harmonic vanishes in the left panel of Fig. 10.7 because of the choice $h_2 + g_2 = 0$ and this contribution is still very weak even for $g_2 = h_2 = -f_2/4$. We

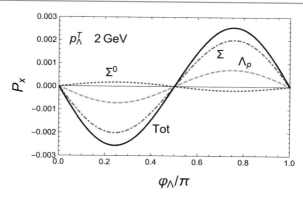

Fig. 10.6 The azimuthal angle dependence of the folded transverse polarization $P_x = 2S_{\Lambda x}(p_\Lambda^T)$ of the Λ. The parameters and denotations are the same as Fig. 10.7

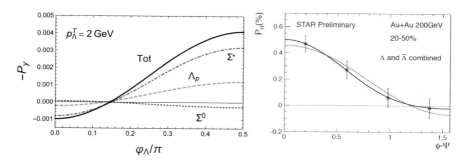

Fig. 10.7 Left panel: the azimuthal angle dependence of the folded polarization along total angular momentum $P_y = 2S_{\Lambda z}(p_\Lambda^T)$ of the Λ. The parameters and denotations are the same as Fig. 10.3, and we choose $g_0 = -0.004$ and $g_2 = -h_2 = -f_2/4$ according to the simulations in [2]. Right panel: the experimental measurements of the polarizations of Λ and $\bar\Lambda$ hyperons as functions of the azimuthal angle ϕ relative to the first-order event plane Ψ for $20\% - -50\%$ centrality bin in $\sqrt{s_{NN}} = 200$GeV Au+Au collisions [6]. Solid and dotted lines show the fits with even cosine harmonics up to quadruple and double angles, respectively

give the best fits to the experimental data in the right panel of Fig. 10.7: Though the fit with up to $4\phi_\Lambda$ harmonics has more advantage to reproduce the central values, the fit with up to $2\phi_\Lambda$ harmonics is also consistent with the data within error bars.

We conclude that while the theoretical predictions and the experimental measurements are consistent with each other for the global polarization of Λ, the azimuthal angle dependences for either the longitudinal and transverse polarizations give opposite signs. Though the component $S_{\Lambda x}^{PC}(p_\Lambda^T)$ has not been measured in experiments, we expect the sign to be also opposite to the theoretical one, which then shares the same origin as the previous *sign puzzles*. Taking into account the feed-down effect of higher lying hyperon decays [8,9], the final amplitudes of the $2\phi_\Lambda$ harmonics are almost the same as that given by the primary Λ. Thus, sign flips are still impossible even after taking into account the contributions from resonance decays. Compared to the global polarization, the local polarizations always involve the thermal vorticity with time component (TVWTC) [2,3], so the answers to the *sign puzzles* might be closely related to this component. Actually, several definitions of vorticity [21] including "thermal", "kinematic", and "temperature" ones are compared in [22]: The kinematic one gives the same signs as the thermal one which indicates the overwhelming role of VWTC, while the temperature one gives the correct signs as the experiments because its dependence on temperature is inverse to that of thermal one. Besides, getting rid of the TVWTC, the sign was found to be consistent with experimental measurements for the LLP of Λ [5,7]. In this models, the opposite effects seem to be simply originated from the opposite contributions of ϖ_{01}, ϖ_{02} (< 0), and ϖ_{12} (> 0) to the MSV of the hyperons in the last equality of (10.45). However, even the hydrodynamic simulations, following the non-relativistic definition of vorticity, don't give the same sign as the experimental measurement. Thus, the reason is not so trivial. We have a better proposal: It might be the higher order derivative corrections to the commonly adopted thermal vorticity that change the whole features.

Appendix 1 Lorentz Boost and Jacobian Determinant

In this Appendix, we demonstrate details to derive (10.79) and (10.81) shown in Sect. 10.5. As mentioned in the context, $p_\Lambda^\mu = (\varepsilon_\Lambda, \mathbf{p}_\Lambda)$ and $p_*^\mu = (\varepsilon_{\Lambda*}, \mathbf{p}_*)$ are the four-momenta of Λ in the QGP frame and Mother's rest frame, respectively, and $p_H^\mu = (\varepsilon_H, \mathbf{p}_H)$ the four-momentum of the Mother in QGPF. The pure Lorentz boost transforming the momentum of Λ from QGPF to MRF reads

$$\varepsilon_{\Lambda*} = \gamma_H(\varepsilon_\Lambda - \mathbf{v}_H \cdot \mathbf{p}_\Lambda), \tag{10.103}$$

$$\mathbf{p}_* = \mathbf{p}_\Lambda + \left(\frac{\gamma_H - 1}{\mathbf{v}_H^2}\mathbf{v}_H \cdot \mathbf{p}_\Lambda - \gamma_H \varepsilon_\Lambda\right)\mathbf{v}_H, \tag{10.104}$$

where $\mathbf{v}_H = \mathbf{p}_H/\varepsilon_H$ is the velocity of the Mother and $\gamma_H = \varepsilon_H/m_H$ the corresponding Lorentz factor. Hence, the explicit forms of (10.103) and (10.104) are

$$\varepsilon_{\Lambda*} = \frac{1}{m_H}(\varepsilon_H\varepsilon_\Lambda - \mathbf{p}_H \cdot \mathbf{p}_\Lambda), \tag{10.105}$$

$$\mathbf{p}_* = \mathbf{p}_\Lambda + \left[\frac{\mathbf{p}_H \cdot \mathbf{p}_\Lambda}{m_H(\varepsilon_H + m_H)} - \frac{\varepsilon_\Lambda}{m_H}\right]\mathbf{p}_H, \tag{10.106}$$

then the expression of $\mathbf{p}_H \cdot \mathbf{p}_\Lambda$ from (10.105) can be substituted into (10.106) to get

$$\mathbf{p}_* = \mathbf{p}_\Lambda + \left[\frac{\varepsilon_H\varepsilon_\Lambda - m_H\varepsilon_{\Lambda*}}{m_H(\varepsilon_H + m_H)} - \frac{\varepsilon_\Lambda}{m_H}\right]\mathbf{p}_H = \mathbf{p}_\Lambda - \frac{\varepsilon_{\Lambda*} + \varepsilon_\Lambda}{\varepsilon_H + m_H}\mathbf{p}_H. \tag{10.107}$$

Moving $\mathbf{p}_\Lambda$ to the left-hand side of (10.107) and take square of both sides, we have

$$(\mathbf{p}_* - \mathbf{p}_\Lambda)^2 = \frac{(\varepsilon_{\Lambda*} + \varepsilon_\Lambda)^2}{(\varepsilon_H + m_H)^2}\mathbf{p}_H^2 = \frac{\varepsilon_H - m_H}{\varepsilon_H + m_H}(\varepsilon_{\Lambda*} + \varepsilon_\Lambda)^2, \tag{10.108}$$

which then gives the energy of the Mother in terms of the energy-momenta of the Daughter as

$$\varepsilon_H = m_H \frac{(\varepsilon_{\Lambda*} + \varepsilon_\Lambda)^2 + (\mathbf{p}_* - \mathbf{p}_\Lambda)^2}{(\varepsilon_{\Lambda*} + \varepsilon_\Lambda)^2 - (\mathbf{p}_* - \mathbf{p}_\Lambda)^2}. \tag{10.109}$$

By substituting (10.109) back into (10.107), the final expression for the momentum of the Mother follows directly

$$\mathbf{p}_H = 2m_H \frac{\varepsilon_+ \mathbf{p}_-}{\varepsilon_+^2 - \mathbf{p}_-^2} \quad \text{with} \quad \varepsilon_+ = \varepsilon_\Lambda + \varepsilon_{\Lambda*}, \quad \mathbf{p}_- = \mathbf{p}_\Lambda - \mathbf{p}_*. \tag{10.110}$$

Now, the above equation (10.110) can be easily adopted to alter the integration variable involved in (10.78) from $\mathbf{p}_H$ to $\mathbf{p}_*$ by fixing $\mathbf{p}_\Lambda$. The Jacobian matrix of the transformation can be evaluated as

$$\frac{\partial \mathrm{p}_{Hi}}{\partial \mathrm{p}_{*j}} = \frac{2m_H}{\varepsilon_+^2 - \mathbf{p}_-^2} \left\{ \left[\mathrm{p}_{-,i} \frac{\mathrm{p}_{*j}}{\varepsilon_{\Lambda*}} - \varepsilon_+ \delta_{ij} \right] - \frac{2\varepsilon_+ \mathrm{p}_{-,i}}{\varepsilon_+^2 - \mathbf{p}_-^2} \left[\varepsilon_+ \frac{\mathrm{p}_{*j}}{\varepsilon_{\Lambda*}} + \mathrm{p}_{-,j} \right] \right\} \tag{10.111}$$

for $i, j = x, y, z$, and the determinant follows directly after some algebraic manipulations:

$$\left| \frac{\partial \mathbf{p}_H}{\partial \mathbf{p}_*} \right| = \frac{4m_H^3 \varepsilon_+^2 (\varepsilon_+^2 + \mathbf{p}_-^2)}{\varepsilon_{\Lambda*}(\varepsilon_+^2 - \mathbf{p}_-^2)^3}. \tag{10.112}$$

Appendix 2 Integrands for the Transverse and Longitudinal Polarizations

Herein, we work out the integrands for the evaluations of the transverse and longitudinal components of the mean spin vector, fed down from the strong and EM decays. Taking the most complicated component $S_{\Lambda y}^{PC}(\mathbf{p}_*)$, along the total angular momentum, for example, inserting (10.82) into the second equation of (10.92) gives

$$S_{\Lambda y}^{PC}(\mathbf{p}_*) = 2(g_0 + g_1 \cos\phi_H + g_2 \cos 2\phi_H)\left(A + B \sin^2\phi_* \sin^2\theta_*\right) + B\left(f_2 \sin 2\phi_H\right.$$

$$\left. \sin\phi_* \sin 2\theta_* + (h_1 \sin\phi_H + h_2 \sin 2\phi_H) \sin 2\phi_* \sin^2\theta_*\right). \tag{10.113}$$

Because $h_1(P_T, Y_H)$ and $g_1(P_T, Y_H)$ are odd functions of Y_H thus also of "$\cos\theta_*$" and all the trigonometric functions of the Mother in (10.85) are even functions of "$\cos\theta_*$", the terms proportional to h_1 and g_1 do not contribute at all after integrating over θ_*. Likewise, the term proportional to $f_2(P_T, Y_H)$, which is an even function of "$\cos\theta_*$", vanishes upon integration over θ_* because the function $\sin 2\theta_*$ is odd. So we are left with

$$S_{\Lambda y}^{PC}(\mathbf{p}_*) = (g_0 + g_2 \cos 2\phi_H)\left(F - B \cos 2\phi_* \sin^2\theta_*\right) + B h_2 \sin 2\phi_H \sin 2\phi_* \sin^2\theta_*, \tag{10.114}$$

where $F = 2A + B \sin^2\theta_*$.

Inserting (10.85) and replacing ϕ_* by $\phi_\Lambda + \psi$, (10.114) becomes explicitly

$$[g_0 + g_2(\mathcal{A}\cos 2\phi_\Lambda - \mathcal{B}\sin 2\phi_\Lambda)]\left[F - B(\cos 2\phi_\Lambda \cos 2\psi - \sin 2\phi_\Lambda \sin 2\psi)\sin^2 \theta_*\right]$$
$$+ B h_2(\mathcal{A}\sin 2\phi_\Lambda + \mathcal{B}\cos 2\phi_\Lambda)(\cos 2\phi_\Lambda \sin 2\psi + \sin 2\phi_\Lambda \cos 2\psi)\sin^2 \theta_*. \tag{10.115}$$

Remember that any terms that are odd functions of "$\cos \theta_*$" or ψ vanish after solid angle integrations. Thus, by taking into account the even-oddness of the relevant functions listed in Table 10.2, the following terms are left:

$$(g_0 + g_2\mathcal{A}\cos 2\phi_\Lambda)(F - B\cos 2\phi_\Lambda \cos 2\psi \sin^2 \theta_*) - g_2\mathcal{B}B\sin^2 2\phi_\Lambda \sin 2\psi \sin^2 \theta_*$$
$$+ B h_2(\mathcal{A}\sin^2 2\phi_\Lambda \cos 2\psi + \mathcal{B}\cos^2 2\phi_\Lambda \sin 2\psi)\sin^2 \theta_*. \tag{10.116}$$

Finally, we adopt the double-angle relationships for the trigonometric functions:

$$\cos^2 x = \frac{1}{2}(\cos 2x + 1), \qquad \sin^2 x = \frac{1}{2}(-\cos 2x + 1)$$

to put the result (10.116) in harmonics of ϕ_Λ:

$$S^{PC}_{\Lambda y}(\mathbf{p}_*) = \left[g_0 F + \frac{B}{2}(h_2 - g_2)(\mathcal{A}\cos 2\psi + \mathcal{B}\sin 2\psi)\sin^2 \theta_*\right]$$
$$- (g_0 B\cos 2\psi \sin^2 \theta_* - g_2 F\mathcal{A})\cos 2\phi_\Lambda$$
$$- \frac{B}{2}(h_2 + g_2)(\mathcal{A}\cos 2\psi - \mathcal{B}\sin 2\psi)\sin^2 \theta_* \cos 4\phi_\Lambda. \tag{10.117}$$

One finds that h_2 and g_2 terms give rise to contributions to both global and $4\phi_\Lambda$ harmonic modes for the TLP P_y.

Similarly, h_1, g_1 and f_2 do not contribute to the TLP P_x because the relevant terms in the integrand $S^{PC}_{\Lambda x}(\mathbf{p}_*)$ are also odd functions of "$\cos \theta_*$". So by combining (10.82) and (10.85) with the first equation in (10.92), the integrand is explicitly

$$S^{PC}_{\Lambda x}(\mathbf{p}_*) = h_2(\mathcal{A}\sin 2\phi_\Lambda + \mathcal{B}\cos 2\phi_\Lambda)\left(F + B\cos 2\phi_* \sin^2 \theta_*\right)$$
$$+ B[g_0 + g_2(\mathcal{A}\cos 2\phi_\Lambda - \mathcal{B}\sin 2\phi_\Lambda)]\sin 2\phi_* \sin^2 \theta_*, \tag{10.118}$$

which becomes

$$h_2\left[\mathcal{A}\sin 2\phi_\Lambda\left(F + B\cos 2\psi \cos 2\phi_\Lambda \sin^2 \theta_*\right) - \frac{B}{2}\mathcal{B}\sin 2\psi \sin 4\phi_\Lambda \sin^2 \theta_*\right]$$
$$+ B\left[(g_0 + g_2\mathcal{A}\cos 2\phi_\Lambda)\cos 2\psi \sin 2\phi_\Lambda - g_2\frac{B}{2}\sin 4\phi_\Lambda \sin 2\psi\right]\sin^2 \theta_* \tag{10.119}$$

after replacing ϕ_* by $\phi_\Lambda + \psi$. And the double-angle relationships give

$$S^{PC}_{\Lambda x}(\mathbf{p}_*) = (h_2 F\mathcal{A} + g_0 B\cos 2\psi \sin^2 \theta_*)\sin 2\phi_\Lambda$$
$$+ \frac{B}{2}(h_2 + g_2)(\mathcal{A}\cos 2\psi - \mathcal{B}\sin 2\psi)\sin^2 \theta_* \sin 4\phi_\Lambda. \tag{10.120}$$

where we recognize that the coefficient of the $4\phi_\Lambda$ harmonic is opposite to that of $S_{\Lambda y}^{PC}(\mathbf{p}_*)$.

For the longitudinal component, g_0, g_2, and h_2 do not contribute because the relevant terms in the integrand $S_{\Lambda z}^{PC}(\mathbf{p}_*)$ are also odd functions of "$\cos\theta_*$". So by combining (10.82) and (10.85) with the third equation in (10.92), the integrand is explicitly

$$S_{\Lambda z}^{PC}(\mathbf{p}_*) = 2f_2(\mathcal{A}\sin 2\phi_\Lambda + \mathcal{B}\cos 2\phi_\Lambda)\left(A + B\cos^2\theta_*\right) + B[h_1(C\sin\phi_\Lambda + \mathcal{D}\cos\phi_\Lambda)$$
$$\cos\phi_* + g_1(C\cos\phi_\Lambda - \mathcal{D}\sin\phi_\Lambda)\sin\phi_*]\sin 2\theta_*, \tag{10.121}$$

which becomes

$$S_{\Lambda z}^{PC}(\mathbf{p}_*) = \left[2f_2\mathcal{A}\left(A + B\cos^2\theta_*\right) + \frac{B}{2}(h_1 + g_1)(C\cos\psi - \mathcal{D}\sin\psi)\sin 2\theta_*\right]\sin 2\phi_\Lambda \tag{10.122}$$

after replacing ϕ_* by $\phi_\Lambda + \psi$. Note that the LLP keeps the same harmonic as the primary one without any other mixing, that is, $\sim \sin 2\phi_\Lambda$.

References

1. Becattini, F., Karpenko, I., Lisa, M., Upsal, I., Voloshin, S.: Global hyperon polarization at local thermodynamic equilibrium with vorticity, magnetic field and feed-down. Phys. Rev. C **95**(5), 054902 (2017)
2. Karpenko I., Becattini F.: Study of Λ polarization in relativistic nuclear collisions at $\sqrt{s_{NN}} = 7.7$ -200 GeV. Eur. Phys. J. C **77**(4), 213 (2017)
3. Becattini, F., Karpenko, I.: Collective longitudinal polarization in relativistic heavy-ion collisions at very high energy. Phys. Rev. Lett. **120**(1), 012302 (2018)
4. Xia, X.L., Li, H., Tang, Z.B., Wang, Q.: Probing vorticity structure in heavy-ion collisions by local Λ polarization. Phys. Rev. C **98**, (2018)
5. Florkowski, W., Kumar, A., Ryblewski, R., Mazeliauskas, A.: Longitudinal spin polarization in a thermal model. Phys. Rev. C **100**, 054907 (2019)
6. Niida, T.: [STAR Collaboration]: Global and local polarization of Λ hyperons in Au+Au collisions at 200 GeV from STAR. Nucl. Phys. A **982**, 511 (2019)
7. Adam, J., et al.: [STAR Collaboration]: Polarization of Λ ($\bar{\Lambda}$) hyperons along the beam direction in Au+Au collisions at $\sqrt{s_{NN}} = 200$ GeV. Phys. Rev. Lett. **123**, 132301 (2019)
8. Becattini, F., Cao, G., Speranza, E.: Polarization transfer in hyperon decays and its effect in relativistic nuclear collisions. Eur. Phys. J. C **79**(9), 741 (2019)
9. Xia, X.L., Li, H., Huang, X.G., Huang, H.Z: Feed-down effect on Λ spin polarization. Phys. Rev. C **100**, 014913 (2019)
10. Moussa, P., Stora, R.: Angular analysis of elementary particle reactions. In: Proceedings of the 1966 International School on Elementary Particles, Hercegnovi. Gordon and Breach, New York/London (1968)
11. Weinberg, S.: The Quantum Theory of Fields, vol. I. Cambridge University Press, Cambridge (1995)
12. Tung, W.K.: Group Theory in Physics. World Scientific, Singapore (1985)
13. Chung, S.U.: Spin Formalisms. BNL preprint Report No. BNLQGS- 02-0900. Brookhaven National Laboratory, Upton, 2008. Updated version of CERN 71-8
14. Cha, M.H., Sucher, J.: Phys. Rev. **140**, B668 (1965)
15. Armenteros, R., et al.: Nucl. Phys. B **21**, 15 (1970)
16. Lin, Qg, Ka, X.L.: On the correction to an operator formula. Coll. Phys. **21**(12) (2002)

17. Leader, E.: Spin in Particle Physics. Cambridge University Press, Cambridge (2001)
18. Kim, J., Lee, J., Shim, J.S., Song, H.S.: Polarization effects in spin 3/2 hyperon decay. Phys. Rev. D **46**, 1060 (1992)
19. Becattini, F., Steinheimer, J., Stock, R., Bleicher, M.: Hadronization conditions in relativistic nuclear collisions and the QCD pseudo-critical line. Phys. Lett. B **764**, 241 (2017)
20. Tanabashi, M., et al.: [Particle Data Group]: Review of particle physics. Phys. Rev. D **98**(3), 030001 (2018)
21. Becattini F., et al.: A study of vorticity formation in high energy nuclear collisions. Eur. Phys. J. C **75**(9), 406 (2015)
22. Wu, H.Z., Pang, L.G., Huang, X.G., Wang, Q.: Local spin polarization in high energy heavy ion collisions. Phys. Rev. Res. **1**, 033058 (2019)
23. Kolb, P.F., Huovinen, P., Heinz, U.W., Heiselberg, H.: Elliptic flow at SPS and RHIC: from kinetic transport to hydrodynamics. Phys. Lett. B **500**, 232 (2001)
24. Lin, Zw, Ko, C.M.: Partonic effects on the elliptic flow at RHIC. Phys. Rev. C **65**, 034904 (2002)

QCD Phase Structure Under Rotation

11

Hao-Lei Chen, Xu-Guang Huang and Jinfeng Liao

Abstract

We give an introduction to the phase structure of QCD matter under rotation based on effective four-fermion models. The effects of the magnetic field on the rotating QCD matter are also explored. Recent developments along these directions are overviewed, with special emphasis on the chiral phase transition. The rotational effects on pion condensation and color superconductivity are also discussed.

11.1 Introduction

Exploration of the phase structure of quantum chromodynamics (QCD) is one of the most active researching frontiers of nuclear physics. In the past, most of the attention has been paid to the phase diagram in the plane of temperature T and baryon density or baryon chemical potential μ_B. At low temperature and low baryon chemical potential, the QCD matter is in a confined hadronic phase where the (approximate) chiral symmetry of QCD Lagrangian is spontaneously broken. With increasing temperature, QCD undergoes a transition from the hadronic matter to a deconfined and chirally symmetric state of quarks and gluons usually called the quark–gluon plasma

H.-L. Chen
Physics Department and Center for Particle Physics and Field Theory, Fudan University, Shanghai 200433, China
e-mail: hlchen15@fudan.edu.cn

X.-G. Huang (✉)
Physics Department, Center for Particle Physics and Field Theory, and Key Laboratory of Nuclear Physics and Ion-beam Application (MOE), Fudan University, Shanghai 200433, China
e-mail: huangxuguang@fudan.edu.cn

J. Liao
Physics Department and Center for Exploration of Energy and Matter, Indiana University, 2401 N Milo B. Sampson Lane, Bloomington, IN 47408, USA
e-mail: liaoji@indiana.edu

© Springer Nature Switzerland AG 2021   349
F. Becattini et al. (eds.), *Strongly Interacting Matter under Rotation*,
Lecture Notes in Physics 987,
https://doi.org/10.1007/978-3-030-71427-7_11

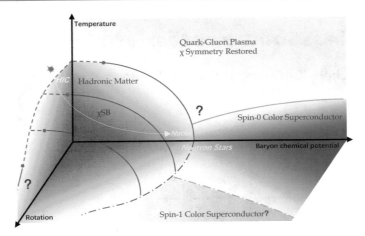

Fig. 11.1 A schematic phase diagram of QCD matter in the 3-dimensional parameter space spanned by the temperature−baryon chemical potential−rotation axes

(QGP). At zero μ_B, this transition is not a true phase transition (i.e. without thermodynamic singularity) but a rapid crossover with the crossover region determined roughly by the confinement scale $\Lambda_{QCD} \sim 200$ MeV. However, at finite μ_B, many model studies and theoretical arguments suggest that the restoration of chiral symmetry should occur via a first-order phase transition. The experimental search of the first-order phase transition at finite μ_B and its end point (i.e. the QCD critical end point) is one of the main goals of the RHIC beam energy scan program. At low T but very high μ_B, QCD is possibly in a color superconducting phase. This is confirmed at asymptotically high μ_B where the perturbative calculation shows that the ground state of QCD is a color–flavor-locking superconducting phase. At moderate μ_B, we lack a first-principle calculation and the model studies suggest a series of possible color superconducting phases as well as other exotic phases. See, e.g. Ref. [1] for review on QCD phase structure in the $T - \mu_B$ plane, Ref. [2] on color superconductivity, and Ref. [3] on the QCD critical end point and its experimental search. A schematic phase structure of QCD matter is shown in Fig. 11.1 where, in addition to the usual temperature- and baryon chemical-potential axes, a new dimension of finite global rotation is introduced. It is the influence of rotation on the QCD phase structures that is an emerging new direction of study and the main topic of our discussion.

The effects of the magnetic field on QCD phase structure have attracted a lot of interest in the past few decades. On the one hand, this is because the interplay between quantum electrodynamics (QED) and QCD proposes novel and interesting theoretical problems. On the other hand, this is also because strong magnetic fields do exist in a wide range of physical systems usually governed by QCD physics, e.g. the neutron stars and heavy-ion collision experiments. Recent theoretical studies have shown that the heavy-ion collisions generate the strongest magnetic fields (of the order of m_π^2) in the current universe; see reviews [4–6] and references therein. One remarkable consequence of the magnetic field is the magnetic catalysis of chiral condensate at zero temperature [7,8]. Namely, the chiral condensate $\langle \bar{\psi}\psi \rangle(B)$ is generally enhanced compared with $\langle \bar{\psi}\psi \rangle(0)$ at $T = 0$ where B is an external magnetic field. This is understood as a result of the dimensional reduction of charged-fermion dynamics in a strong magnetic field. Surprisingly, the lattice QCD simulations revealed that the critical temperature for chiral phase transition of QCD is not enhanced but suppressed by the magnetic field [9,10]. This abnormal phenomenon is dubbed by the inverse magnetic catalysis and is not fully understood yet. For reviews about the magnetic effects on QCD phase structure, see Refs. [11,12].

Recently, various properties of rotating QCD matter have become a new topic under active investigations. This was first inspired by the analogy between rotation and magnetic field. There is also a strong motivation from the discovery of extremely strong fluid vorticity structures (i.e. the local angular velocity or rotating frequency) in heavy-ion collisions where both experiments [13–15] and theories (see, e.g. Refs. [16–19]) find that the typical strength of the vorticity is about $10^{21} - 10^{22}$ s^{-1} (or tens of MeV) which may strongly influence the QCD matter. Furthermore, the rapidly rotating pulsars (neutron stars) provide another example of rotating QCD system, though the angular velocity is much smaller than the vorticity found in heavy-ion collisions. The rotation or fluid vorticity can polarize spin and thus induce a number of spin-related quantum phenomena. For example, it can induce a parity violating current called the chiral vortical effect (CVE) [20–23], which is analogous to an effect (chiral magnetic effect) induced by magnetic field [24,25]. It can also lead to detectable global polarization of hadrons with nonzero spin (e.g. Λ hyperons and vector mesons) in heavy-ion collisions [26–32].

It is quite natural to ask what are the influences of rotation on the QCD phase structure. This is, however, a hard question to answer, as the QCD phase transitions involve non-perturbative dynamics in general and the rotation further complicates the problem. (A lattice simulation of QCD in the rotating frame was given in Ref. [33], though the phase structure was not discussed.) In recent years, this question has been extensively studied by using effective models with quarks or hadrons as dynamic degrees of freedom [34–51]. It is found that in a uniformly rotating system at finite temperature, density, and magnetic field, angular velocity plays a role of an effective chemical potential for the angular momentum and its presence suppresses the spin-0 pairing of quarks [34,35,44]. It is also confirmed that such a uniformly rotational effect on thermodynamics is invisible at zero temperature and density [36,38,39]. For non-uniform rotation, even the ground state can be affected by the presence of rotation and exhibit a vortex structure under sufficiently rapid rotation [45].

Another very interesting question is the QCD phase structure when there are both rotation and magnetic field. Note that in both the heavy-ion collisions and the neutron stars, the strong rotation is accompanied by strong magnetic fields. Experimental measurements, showing a small difference between the hyperon and anti-hyperon spin polarization, are also indicative of the magnetic field and vorticity in a parallel configuration [52–54]. With a concurrent magnetic field, the rotation is found to create an even richer phase structure. For example, by using the Nambu–Jona-Lasinio (NJL) model, it was shown that under strong rotation the chiral condensate decreases with increasing magnetic field and eventually the chiral symmetry is restored. This phenomenon is named the "rotational magnetic inhibition" as it is an analogy to the magnetic inhibition phenomenon in the finite density system [34]. These studies suggest a phase diagram of QCD in temperature-baryon chemical potential-rotation space as illustrated in Fig. 11.1, which we shall explain in the subsequent sections.

In the following, we start with an introduction to the rotating frame and the thermodynamic potential of the NJL model in the rotating frame with or without the presence of a parallel magnetic field. We then discuss the effects of the rotation with or without a parallel magnetic field on chiral condensate as well as on other types of scalar condensates. We use the natural unit with $\hbar = c = k_B = 1$.

11.2   Rotating Frame

Let us consider a cylindrical system which is rigidly rotating with angular velocity Ω ($\Omega > 0$) in the inertial laboratory frame Σ' with the vector basis $(\hat{\partial}_{t'}, \hat{\partial}_{x'}, \hat{\partial}_{y'}, \hat{\partial}_{z'})$. The Minkowski metric is

$$\eta_{\mu\nu} = \eta^{\mu\nu} = \text{diag}(1, -1, -1, -1). \tag{11.1}$$

We can go to the non-inertial rotating frame Σ through the coordinate transformation (we assume the rotating axis is along the z-axis)

$$\begin{cases} x' = x \cos \Omega t - y \sin \Omega t \\ y' = x \sin \Omega t + y \cos \Omega t \\ z' = z \\ t' = t \end{cases} \tag{11.2}$$

From the coordinate transformation, it is easy to get the line element in the rotating frame

$$ds^2 = \eta_{\mu\nu}dx'^\mu dx'^\nu = g_{\mu\nu}dx^\mu dx^\nu = (1 - \Omega^2 r^2)dt^2 + 2\Omega y dx dt - 2\Omega x dy dt - dx^2 - dy^2 - dz^2, \tag{11.3}$$

where $r = \sqrt{x^2 + y^2}$. Thus the metric of the rotating frame is

$$g_{\mu\nu} = \begin{pmatrix} 1 - \Omega^2 r^2 & \Omega y & -\Omega x & 0 \\ \Omega y & -1 & 0 & 0 \\ -\Omega x & 0 & -1 & 0 \\ 0 & 0 & 0 & -1 \end{pmatrix}. \tag{11.4}$$

The Christoffel symbol is given by

$$\Gamma^\beta_{\mu\nu} = \frac{1}{2} g^{\beta\sigma} (\partial_\mu g_{\sigma\nu} + \partial_\nu g_{\sigma\mu} - \partial_\sigma g_{\mu\nu}), \tag{11.5}$$

and the nonzero components are

$$\Gamma^x_{tt} = -x\Omega^2, \quad \Gamma^x_{ty} = \Gamma^x_{yt} = -\Omega, \quad \Gamma^y_{tt} = -y\Omega^2, \quad \Gamma^y_{tx} = \Gamma^y_{xt} = \Omega. \tag{11.6}$$

The Riemann curvature is identically zero.

The Lagrangian density of a complex scalar field in curved spacetime[1] is

$$\mathcal{L} = \sqrt{-g}[|D_\mu \phi|^2 - (m^2 + \xi R)|\phi|^2] = -\phi^* \left[\frac{1}{\sqrt{-g}} D_\mu (\sqrt{-g} g^{\mu\nu} D_\nu \phi) + (m^2 + \xi R)\phi \right],$$

(11.7)

where m is the mass of the field, g is the determinant of $g_{\mu\nu}$, and ξ represents the coupling of the field with the Ricci curvature R which is 0 in the rotating frame. $D_\mu = \partial_\mu + iq A_\mu$ is the covariant derivative with q the charge and A_μ the $U(1)$ gauge field. Note that the vector field should transform as $A_\mu(x)dx^\mu = A'_\mu(x')dx'^\mu$ under coordinate transformation. The Klein–Gordon equation can be derived from the Lagrangian density by the Euler–Lagrangian equation (for $R = 0$),

$$\frac{1}{\sqrt{-g}} D_\mu (\sqrt{-g} g^{\mu\nu} D_\nu \phi) + m^2 \phi = 0.$$

(11.8)

In order to discuss the spinor field in curved spacetime, it is convenient to use the vierbein formalism [55,56]. The Dirac equation of free fermion in curved spacetime is

$$[i\gamma^\mu(\partial_\mu + iq A_\mu + \Gamma_\mu) - m]\psi(x) = 0,$$

(11.9)

where $\gamma^\mu = e_i^\mu \gamma^i$, e_i^μ is the vierbein, which satisfies $g_{\mu\nu} = e_\mu^i e_\nu^j \eta_{ij}$, and Γ_μ is the spin connection given by

$$\Gamma_\mu = -\frac{i}{4} \omega_{\mu ij} \sigma^{ij},$$

$$\omega_{\mu ij} = g_{\alpha\beta} e_i^\alpha (\partial_\mu e_j^\beta + \Gamma^\beta_{\mu\nu} e_j^\nu),$$

(11.10)

$$\sigma^{ij} = \frac{i}{2}[\gamma^i, \gamma^j].$$

The Greek and the Latin letters denote the indices in coordinate and tangent (local Miknowski) spaces, respectively.

In the following, we will adopt

$$e_0^t = e_1^x = e_2^y = e_3^z = 1 , \quad e_0^x = y\Omega , \quad e_0^y = -x\Omega ,$$

(11.11)

and other components are zero. From Eq. (11.10), it is straightforward to get the only nonzero spin connection

$$\Gamma_t = -i\Omega \sigma^{12}.$$

(11.12)

---

[1]To specify the terminology, any non-Minkowski metric with either zero or nonzero Riemann curvature is referred to as a "curved spacetime".

11.3 Nambu–Jona-Lasinio Model

The Nambu–Jona-Lasinio (NJL) model is a four-fermion model which is commonly used as a low-energy effective model of QCD [57–59]. In curved spacetime, the NJL Lagrangian is

$$\mathcal{L}_{\mathrm{NJL}} = \bar{\psi}(i\gamma^{\mu}\nabla_{\mu} - m_0)\psi + \frac{G}{2}[(\bar{\psi}\psi)^2 + (\bar{\psi}i\gamma^5\boldsymbol{\tau}\psi)^2], \qquad (11.13)$$

where ψ is the fermion field (representing the quarks), $\nabla_{\mu} = \partial_{\mu} + i\hat{Q}A_{\mu} + \Gamma_{\mu}$ is the covariant derivative, $\hat{Q}$ is the charge matrix in the flavor space, G is the coupling constant, and $\boldsymbol{\tau}$ is the generator of the flavor group. For one flavor NJL model, $\boldsymbol{\tau}$ is just 1. For the two-flavor NJL model, $\boldsymbol{\tau}$ should be the Pauli matrices of isospin $SU(2)$ group. In case that the quark current mass $m_0 = 0$, the NJL model has also chiral symmetry so that its symmetry group is $SU_R(2) \otimes SU_L(2) \otimes U_B(1)$, the same as that for QCD with two flavor quarks. When $m_0 \neq 0$ but small, the NJL model has an approximate chiral symmetry. Besides, the gauge color $SU(N_c)$ symmetry can be trivially assigned to Lagrangian (11.13) as an additional global symmetry. The generating functional of the NJL model (with vanishing sources) is

$$Z = \int \mathcal{D}[\bar{\psi}, \psi] \exp\left(i \int d^4x \sqrt{-g}\mathcal{L}_{\mathrm{NJL}}\right)$$
$$= \int \mathcal{D}[\bar{\psi}, \psi, \sigma, \pi] \exp\left\{i \int d^4x \sqrt{-g}\left[\bar{\psi}(i\gamma^{\mu}\nabla_{\mu} - m - i\gamma^5\pi \cdot \tau)\psi - \frac{\sigma^2 + \pi^2}{2G}\right]\right\},$$
$$(11.14)$$

where $m = m_0 + \sigma$. In the second line, we have performed the Hubbard–Stratonovich transformation by introducing an auxiliary scalar field σ and N_{π} pseudoscalar fields π (N_{π} is the number of the generators τ of the flavor group, e.g. $N_{\pi} = 3$ for the two-flavor case). In the rest of this chapter, unless otherwise stated, we will consider only the rotating frame so that the metric is given by Eq. (11.4) and $g = -1$. Under mean field approximation, the one-loop effective action is

$$\Gamma = \frac{1}{i}\ln Z = -\int d^4x \frac{\sigma^2 + \pi^2}{2G} + \frac{1}{i}\ln \det(i\gamma^{\mu}\nabla_{\mu} - m - i\gamma^5\pi \cdot \tau). \quad (11.15)$$

If we further assume that σ and π are constant, the second term can be evaluated as

$$\frac{1}{i}\ln \det(i\gamma^{\mu}\nabla_{\mu} - m - i\gamma^5\pi \cdot \tau) = \frac{1}{2i}\mathrm{Tr}\ln[-(i\partial_t)^2 + \hat{H}^2]$$
$$= \int dt \int \frac{dp_0}{2\pi} \sum_{\{\xi\}} \frac{1}{2i}\ln[-p_0^2 + \varepsilon_{\{\xi\}}^2]$$
$$(11.16)$$

where

$$\hat{H} = -i\gamma^0\gamma^x\nabla_x - i\gamma^0\gamma^y\nabla_y - i\gamma^0\gamma^z\nabla_z + m\gamma^0 + i\gamma^0\gamma^5\pi \cdot \tau + \hat{Q}A_t - i\Gamma_t,$$
$$(11.17)$$

which we assume to be time independent, i.e. $[\hat{H}, i\partial_t] = 0$, $\varepsilon_{\{\xi\}}$ is the eigenvalue of $\hat{H}$ with a set of quantum number $\{\xi\}$, and p_0 is the eigenvalue of $i\partial_t$. Then by employing Matsubara formalism

$$t \to -i\tau, \quad p_0 \to i\omega_n = i2\pi \frac{1}{\beta}\left(n + \frac{1}{2}\right), \quad \int \frac{dp_0}{2\pi} \to \frac{i}{\beta}\sum_n, \tag{11.18}$$

where $\beta = 1/T$ with T the temperature, we can obtain the thermodynamic potential

$$V_{\text{eff}} = -i\frac{1}{\beta V}\Gamma = \frac{\sigma^2 + \pi^2}{2G} - \sum_{\{\xi\}}\left[\frac{\varepsilon_{\{\xi\}}}{2} + \frac{1}{\beta}\ln(1 + e^{-\beta\varepsilon_{\{\xi\}}})\right]. \tag{11.19}$$

By minimizing the thermodynamic potential with respect to σ and π, we can get the gap equations

$$\frac{\partial V_{\text{eff}}}{\partial \sigma} = 0, \quad \frac{\partial V_{\text{eff}}}{\partial \pi} = 0. \tag{11.20}$$

The stable thermodynamic state is given by the solution of the gap equations associated with the global minimum of V_{eff}.

The above procedure can be easily modified to allow the condensates σ and π to be inhomogeneous in space. But in this case $\hat{H}$ is technically very hard to be diagonalized, so we have to adopt certain approximation or perturbative expansion during the calculation. One approximation scheme is the *local density approximation* which assumes that the derivative of the condensate is negligible compared to the condensate itself ($\partial\sigma \ll \sigma^2$ or more precisely $\partial^n\sigma \ll \sigma^{n+1}$ with $n > 0$; similarly for π) so that we can treat the condensates as constant in solving the eigenvalue problem. Then Eq. (11.16) is changed to

$$\frac{1}{2i}\int d^4x \int \frac{dp_0}{2\pi}\sum_{\{\xi\}}\ln[-(p_0)^2 + \varepsilon_{\{\xi\}}^2(x)]\Psi_{\{\xi\}}^\dagger \Psi_{\{\xi\}}, \tag{11.21}$$

where $\Psi_{\{\xi\}}$ is the eigenfunction of $\hat{H}$ under the local density approximation. The thermodynamic potential is accordingly changed to

$$V_{\text{eff}} = \frac{1}{\beta V}\int d^4x_E\left\{\frac{\sigma^2 + \pi^2}{2G} - \sum_{\{\xi\}}\left[\frac{\varepsilon_{\{\xi\}}}{2} + \frac{1}{\beta}\ln(1 + e^{-\beta\varepsilon_{\{\xi\}}})\right]\Psi_{\{\xi\}}^\dagger \Psi_{\{\xi\}}\right\}, \tag{11.22}$$

where x_E^μ is the Euclidean coordinate with compact time direction that is used in the Matsubara formalism. The gap equations are still given by Eq. (11.20).

11.4   Rotating Fermions Without Boundary

In this section, we will ignore the background gauge field and focus on the effect of rotation only. A uniformly rotating system must be finite so that the causality condition $\Omega R \leqslant 1$ (with R the transverse size of the system) is satisfied. This requires appropriate boundary conditions when solving the eigenvalue problem for $\hat{H}$. However, if we focus on the region far away from the boundary, we can ignore the influence of the boundary and take the $R \to \infty$ limit. We will consider this approximation in this section and examine the influence of the boundary in the next section. To simplify the discussion, we further assume that $\pi = 0$ and σ depends only on the transverse radius r.

The eigenvalue $\varepsilon_{\{\xi\}}$ of $\hat{H}$ is obtained by solving the Dirac equation under the local density approximation, which is given by (here $\{\xi\} = \{l, p_z, p_t, s\}$ with $s = \pm$ and we abbreviate $\varepsilon_{l,p_z,p_t,s}$ as $\varepsilon_{l,s}$) [35]

$$\varepsilon_{l,\pm} = \pm\sqrt{p_z^2 + p_t^2 + \sigma^2} - \Omega\left(l + \frac{1}{2}\right), \tag{11.23}$$

where p_z is the z-momentum, p_t is the transverse momentum magnitude, and $l = 0, \pm1, \ldots$ is the quantum number of the orbital angular momentum. We do not show the lengthy expression of the eigenfunction associated with $\varepsilon_{l,s}$; it is given in, e.g. Refs. [35,60,61]. From the dispersion relation (11.23), one can observe that rotation behaves very similar to a chemical potential, which has been noticed for a long time (see, e.g. Ref. [56]).

It is worthy to compare the effect of rotation with the magnetic field B on the dispersion relation. The latter gives the Landau levels: $\varepsilon_{n,\pm} = \pm\sqrt{p_z^2 + \sigma^2 + 2nqB}$. In a background magnetic field, the transverse motion is quantized while there is no transverse-motion quantization in the rotation without boundary. This is because the existence of the centrifugal force due to rotation prevents the formation of a quantum mechanical bound-state problem. In the background magnetic field, each Landau level is highly degenerate with degeneracy $N = \lfloor qBS/(2\pi)\rfloor$ with S the transverse area. As there is no transverse-motion quantization in the rotating case, we do not have such degeneracy. In fact, as we will show later, the Landau level degeneracy N counts the number of allowed angular-momentum modes accommodated at each Landau level, which is lifted when there is a rotation. Thus the rotation behaves quite differently from the magnetic field.

Following the procedure introduced in the previous section, one can get the thermodynamic potential as

$$V_{\text{eff}} = \int d^3r \left\{ \frac{\sigma^2}{2G} - \frac{2N_f N_c}{16\pi^2} \sum_l \int dp_t^2 \int dp_z \right.$$

$$\left. \times T \ln(1 + e^{\beta\varepsilon_l})(1 + e^{-\beta\varepsilon_l})[J_l(p_t r)^2 + J_{l+1}(p_t r)^2] \right\}, \tag{11.24}$$

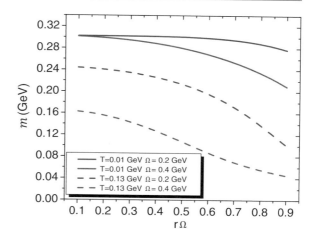

Fig. 11.2 The dependence of $m = \sigma + m_0$ on radial coordinate $r\Omega$ for several fixed values of Ω and T. (Taken from [35])

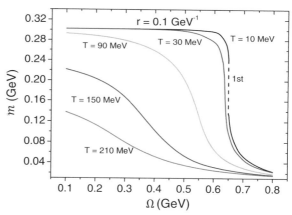

Fig. 11.3 Dependence of m at radius $r = 0.1\ \mathrm{GeV}^{-1}$ on Ω for various fixed values of T. (Taken from [35])

where $\varepsilon_l = \varepsilon_{l,+}$, $N_f = 2$ is the flavor number, $N_c = 3$ is the color number, and $J_l(x)$ is the Bessel function of the first kind. The condensate σ is obtained from the gap equation. The numerical results are shown in Figs. 11.2 and 11.3 with model parameters given in [35]. At all values of temperature, the mass gap m (and thus the chiral condensate σ) decreases with increasing values of Ω, which indicates the suppression effect of rotation on the chiral condensate. Furthermore, at low temperature, the chiral condensate experiences a first-order transition when Ω exceeds a critical value Ω_c, while at high temperature the chiral condensate vanishes with increasing Ω via a smooth crossover (it would be a second-order phase transition if $m_0 = 0$).

Figure 11.4 is the $T - \Omega$ phase diagram obtained in [35]. We can see that the chiral symmetry is broken at low temperature and slow rotation. Qualitatively speaking, the rotation will polarize the spin and orbital angular momentum of quarks along the direction of the angular velocity regardless of its charge, thus this polarization effect tends to destroy the pairing of the chiral condensate, which is total spin 0. If Ω is strong enough, the rotation would tend to forbid the formation of spin-0 pairing

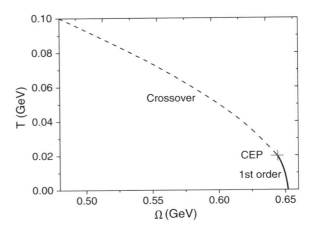

Fig. 11.4 The phase diagram on the $T - \Omega$ plane. (Taken from [35])

condensate and thus leads to a phase transition. For chiral condensate, this is very similar to the effect of baryon chemical potential μ_B which breaches the quark and anti-quark states and finally melts down the chiral condensate when μ_B is large enough. Hence, the chiral symmetry will be restored by increasing T and/or Ω. At high T and low Ω, there will be a smooth crossover. While at low T and high Ω, the transition is first order. The crossover and the first-order transition are separated by a critical end point. Figure 11.4 is very similar to the $T - \mu_B$ phase diagram, which could be understood by considering Ω as a sort of "chemical potential" for angular momentum.

Besides the chiral condensate, the authors of [35] also considered the diquark two-flavor superconducting (2SC) condensate at high density by adding the Lagrangian density (11.13) a chemical potential term and a diquark interaction

$$\mathcal{L}_d = G_d(i\psi^T C\gamma^5\psi)(i\psi^\dagger C\gamma^5\psi^*), \qquad (11.25)$$

where G_d is the diquark coupling constant and C is the charge conjugation operator. Following a similar procedure discussed in Sect. 11.3, one can get the gap equation for the diquark condensate $\Delta\epsilon^{\alpha\beta3}\epsilon_{ij} = -2G_d\langle i\psi_i^\alpha C\gamma^5\psi_j^\beta\rangle$ under the mean field approximation: $\partial V_{\text{eff}}/\partial\Delta = 0$. Their numerical result is shown in Fig. 11.5. We can see that the diquark condensate is also suppressed by the rotation simply because the 2SC pairing is also spin 0. Similarly, increasing Ω leads to a phase transition of melting of the diquark condensate, which is first order at low temperature while second order at high temperature.

Recently, the authors of [44] studied the influence of the rotation on chiral condensate with an additional vector channel interaction,

$$\mathcal{L}_V = -(G_V/2)[(\bar{\psi}\gamma^\mu\psi)^2 + (\bar{\psi}\gamma^\mu\gamma^5\psi)^2], \qquad (11.26)$$

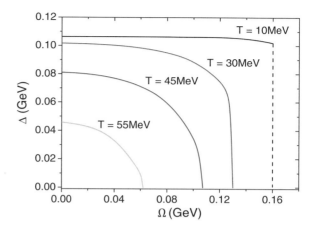

Fig. 11.5 The diquark condensate Δ at $r = 0.1$ GeV^{-1} as a function of Ω for several values of T and fixed value of $\mu = \mu_B/3 = 400$ MeV. (Taken from [35])

where G_V is the corresponding coupling constant. In the mean field approximation, the effective action becomes

$$\Gamma = -\int d^4x \left[\frac{\sigma^2}{2G} - \frac{(\mu - \tilde{\mu})^2}{2G_V} \right] + \frac{1}{i} \ln \det[(i\gamma^\mu \nabla_\mu - \sigma + \tilde{\mu}\gamma^0], \quad (11.27)$$

where the effective quark chemical potential is defined as $\tilde{\mu} = \mu - G_V \langle \psi^\dagger \psi \rangle$ with μ the quark chemical potential. The thermodynamic potential is $V_{\text{eff}} = -i\Gamma/(\beta V)$ and the two gap equations governing σ and $\tilde{\mu}$ are $\partial V_{\text{eff}}/\partial \sigma = \partial V_{\text{eff}}/\partial \tilde{\mu} = 0$. Figure 11.6 shows the $T - \Omega$ phase diagram with different chemical potentials. An interesting observation is that the increase of the chemical potential only shifts down the critical temperature T_C and does not change the critical angular velocity much. Similar behavior happens in $T - \mu$ diagram with different Ω, which again confirm the analogy between rotation and chemical potential.

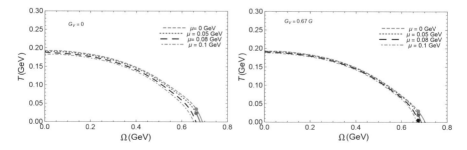

Fig. 11.6 The phase diagram in the $T - \Omega$ plane with different quark chemical potential μ for two different values of the vector coupling. (Taken from [44])

11.5 Boundary Conditions

As we stress in the last section, due to the requirement of causality, the absolute value of velocity $v = \Omega r$ should not exceed the speed of light. Thus, for uniform rotation, the system size should be limited by

$$\Omega R \leqslant 1. \tag{11.28}$$

In the previous section, we assume that the system is large enough and the boundary effects on the bulk condensates can be ignored. In this section, we will discuss the influence of the boundary by considering two kinds of the boundary conditions, the no-flux boundary condition [36,62] and the MIT bag boundary condition [62,63]; see Ref. [61] for more discussion on boundary conditions. Again, we consider a system with cylindrical symmetry so that the angular momentum is a good quantum number and we ignore the background gauge field. To make the discussions more transparent, we focus on σ condensate only and take $m_0 = 0$, $N_f = 1$, $N_c = 1$ in this section.

The no-flux boundary condition requires no total incoming flux at the spatial boundary to keep the total charge constant in the cylinder,

$$\int d\theta \, \bar{\psi} \gamma^r \psi \Big|_{r=R} = 0, \tag{11.29}$$

where $\gamma^r = \gamma^1 \cos\theta + \gamma^2 \sin\theta$. One important feature of the no-flux boundary condition is that it can guarantee the Dirac Hamiltonian to be Hermitian. Due to the boundary condition, the transverse momentum should take discrete values, so we will denote it as $p_{l,k}$. Since the condition Eq. (11.29) does not uniquely fix the solution of the Dirac equation, further requirement should be imposed [62]. For example,

$$p_{l,k} = \begin{cases} \xi_{l,k} R^{-1} & \text{for } l = 0, 1, \dots \\ \xi_{-l-1,k} R^{-1} & \text{for } l = -1, -2, \dots \end{cases} \tag{11.30}$$

where $\xi_{l,k}$ represents the kth zero of the Bessel function $J_l(x)$. By employing the local density approximation, the gap equation at $T = \mu = 0$ reads [36]

$$\sigma = \frac{\sigma}{4G} \int_{-\infty}^{+\infty} dp_z \sum_{l=-\infty}^{\infty} \sum_{k=1}^{\infty} \frac{2}{[J_{l+1}(p_{l,k}R)]^2 R^2} \frac{J_l(p_{l,k}r)^2 + J_{l+1}(p_{l,k}r)^2}{E} \theta(E - |\Omega j|), \tag{11.31}$$

where $E = \sqrt{p_{l,k}^2 + p_z^2 + \sigma^2}$ and $j = l + 1/2$. The expression of the gap equation is very similar to the finite density system; the effect of rotation appears only in the theta function with $|\Omega j|$ playing a role of an effective chemical potential here. Therefore, the rotational effect appears only when $E < |\Omega j|$ for some j. If we do not consider the boundary condition, the transverse momentum $p_{l,k}$ takes continuous value from 0 to $+\infty$. One can find a region of transverse momentum to satisfy $E < |\Omega j|$. However,

Fig. 11.7 Inhomogeneous chiral condensate σ as a function of the radial coordinate r. (Taken from [36])

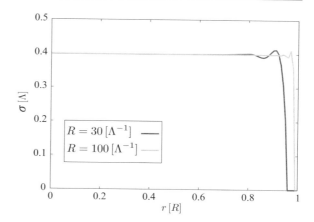

once we take the boundary condition (11.29) into account, there is no mode that satisfies $E < |\Omega j|$. In fact, E is minimized by setting $p_z = \sigma = 0$ and $k = 1$. By using an inequality for the zeros of the Bessel function [36,64]

$$\xi_{l,1} > l + 1.855757 l^{1/3} + 0.5 l^{-1/3} \quad \text{for} \quad l \geqslant 1,$$
$$\xi_{0,1} = 2.40493 > 1/2,$$
(11.32)

and the causality constraint $\Omega R \leqslant 1$, one has the inequality for $l \geqslant 0$

$$E - \Omega j \geqslant \frac{1}{R}(\xi_{l,1} - \Omega R j) \geqslant \frac{1}{R}(\xi_{l,1} - j) > 0.$$
(11.33)

In the same way, for $l < 0$, one can also prove that $E > |\Omega j|$. Thus uniform rotation has no effect in vacuum with the no-flux boundary condition. We can understand this fact by comparing it to the finite density system. In the finite density system, the effect of chemical potential will be visible only when it exceeds the mass threshold, which is known as the Silver Blaze problem in a finite density system. In a rotating system, the effective chemical potential $|\Omega j|$ can never exceed the threshold $p_{l,1}$, thus the uniform rotation cannot induce a visible effect in vacuum.

Although we do not have any rotational effect at $T = \mu = 0$, it is still worthwhile to see the finite-size effect with the no-flux boundary condition. By numerically solving the gap equation (11.31), one can get the result as shown in Fig. 11.7 [36]. One can observe that the local density approximation is invalid in the vicinity of the boundary. The oscillation comes from the cutoff Λ: As the NJL model is not renormalizable, an ultraviolet cutoff must be introduced. As R increases, the oscillation behavior becomes milder. The vanishing condensate at the boundary is a consequence of the condition (11.30).

In [38], the authors adopted another boundary condition, the MIT boundary condition

$$[i\gamma^\mu n_\mu(\theta) - 1]\psi \Big|_{r=R} = 0,$$
(11.34)

where $n_\mu(\theta) = (0, \cos\theta, -\sin\theta, 0)$ is a unit vector normal to the cylinder surface. It is easy to check that the MIT boundary condition Eq. (11.34) leads to

$$j^\mu n_\mu = 0 \quad \text{at} \quad r = R, \tag{11.35}$$

where $j^\mu = \bar{\psi}\gamma^\mu\psi$ is the current. This also leads to $\bar{\psi}\psi = 0$ at the boundary. Thus we have the current vanishing at any point on the surface of the cylinder, a condition that is stronger than the no-flux boundary condition (11.29).

By solving the Dirac equation (11.9) with $m = \sigma$ with the MIT boundary condition, the discrete transverse momentum $p_{l,k}$ is given by the kth positive root of

$$j_l^2(p_{l,k}R) + \frac{2\sigma}{p_{l,k}}j_l(p_{l,k}R) - 1 = 0, \tag{11.36}$$

where

$$j_l(x) = \frac{J_l(x)}{J_{l+1}(x)}. \tag{11.37}$$

The lowest energy spectrum $\tilde{\varepsilon}_l \equiv \varepsilon_{l,1,+}(p_z = 0)$ with $\varepsilon_{l,k,\pm} = \pm\sqrt{p_{l,k}^2 + p_z^2 + \sigma^2} - \Omega(l + 1/2)$ is shown in Fig. 11.8 [38]. It is easy to prove that the change of the orbital number $l \to -l - 1$ (i.e. $j \to -j$) will not affect the eigenvalue $p_{l,k}$ in Eq. (11.36):

$$p_{l,k} = p_{-l-1,k}, \tag{11.38}$$

which is a result of CP symmetry of the non-rotating system. Thus, we can see that energy spectrum is doubly degenerated in Fig. 11.8 at $\Omega = 0$. The spectrum will become asymmetric at $\Omega \neq 0$ because the rotation will explicitly break the CP symmetry. However, the spectrum is invariant under the simultaneous flips $j \to -j$ and $\Omega \to -\Omega$.

An interesting observation of Fig. 11.8 is that $\tilde{\varepsilon}_l$ is always positive indicating the inequality

$$E > |\Omega j| \tag{11.39}$$

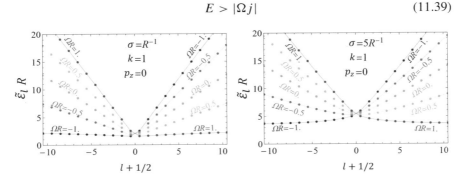

Fig. 11.8 Lowest energy eigenmodes with $k = 1$ and $p_z = 0$ versus the angular momentum $j = l + 1/2$ for various values of the rotation frequency Ω. (Taken from [38])

Fig. 11.9 The ground-state condensate σ in a non-rotating cylinder as a function of the coupling constant G at various fixed radii R. Here, Λ is the ultraviolet cutoff for the NJL model. (Taken from [38])

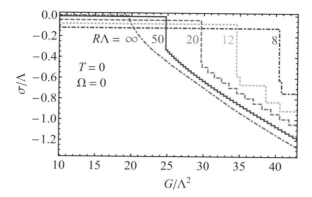

to hold also for the MIT boundary condition. In fact, this inequality can be proven similarly for the no-flux boundary condition. Thus, one again confirms that the uniform rotation has no effect on the chiral condensate in the vacuum. In other words, the cold vacuum does not rotate [38].

Then let's have a look at how the MIT boundary condition affects the chiral condensate at zero temperature and chemical potential. In [38], the authors assumed σ to be homogeneous and computed σ as a function of the coupling constant, as shown in Fig. 11.9. We can see that the negative condensate is favored which has multiple step-like discontinuities, similar to the Shubnikov–De Haas oscillation, due to the discretization of the excitation levels. If we take $R \to \infty$, the discontinuities will disappear. It should be mentioned that the MIT boundary condition explicitly breaks the chiral symmetry. Thus at small R, the boundary effect is strong and we have large σ. While at large R, the boundary effect becomes weak, and the explicit breaking of chiral symmetry can be ignored.

At finite temperature, the rotational effect becomes visible. The phase diagram in the $T - \Omega$ plane is obtained in [38] and shown in Fig. 11.10. The rotation in a finite cylinder tends to restore the chiral symmetry which is in agreement with the result in the previous section. However, the chiral phase transition is first order and is step-like due to the boundary condition. We note that in the restored phase, the

Fig. 11.10 The phase diagram of the rotating fermionic matter in the $T - \Omega$ plane with MIT boundary condition. (Taken from [38])

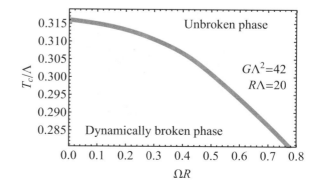

condensate is still nonzero but has a small value due to the explicit breaking of the chiral symmetry by the MIT boundary condition.

The MIT boundary condition can be generalized to [39]

$$[i\gamma^\mu n_\mu(\theta) - e^{-i\Theta\gamma^5}]\psi\Big|_{r=R} = 0, \tag{11.40}$$

where Θ is a chiral angle which parametrizes the chiral boundary condition. Then Eq. (11.36) becomes

$$j_l^2(p_{l,k}R) + \frac{2\sigma}{p_{l,k}} j_l(p_{l,k}R)\cos\Theta - 1 = 0. \tag{11.41}$$

The usual MIT boundary condition corresponds to $\Theta = 0$. One special choice of the chiral angle is $\Theta = \pi$, which only flip the sign of the mass term, $\sigma \to -\sigma$ in Eq. (11.36). Thus one can get all the previous results of the MIT boundary condition only with the sign flip in the condensate $\sigma \to -\sigma$. Another special choice is $\Theta = \pi/2$. In this case, the values of $p_{l,k}$ are independent of the mass. Since the spectrum is affected by the choice of the chiral angle Θ, one can expect that the phase diagram will exhibit a certain dependence on Θ. It is easy to prove that the thermodynamic potential has the following properties:

$$V_{\text{eff}}(\sigma, \Theta) = V_{\text{eff}}(-\sigma, \pi - \Theta) = V_{\text{eff}}(\sigma, 2\pi - \Theta). \tag{11.42}$$

Thus one only need to consider the interval $\Theta \in [0, \pi/2]$ while other values of Θ can be restored from Eq. (11.42). In Fig. 11.11 [39], we can see how the boundary condition (11.40) affects the chiral condensate in a non-rotating cylinder at finite temperature. At low values of the boundary angle Θ, the system resides in a phase with a dynamically unbroken chiral symmetry in which a weak, explicit, violation of the chiral symmetry occurs. The explicit breaking is caused by the fact that the boundary conditions are not invariant under the chiral transformation as mentioned above. At $\Theta = \Theta_c \approx 5\pi/24$, the chiral condensate suddenly changes from a small negative value to a larger negative value, i.e. a first-order transition occurs.

Fig. 11.11 The condensate σ in vacuum in the non-rotating cylinder as a function of the chiral angle Θ. The cylinder radius is $R = 20/\Lambda$ and the coupling $G = 42\Lambda^2$. (Taken from [39])

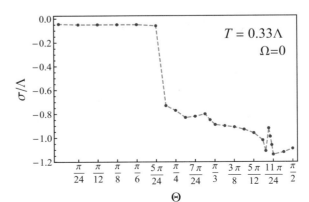

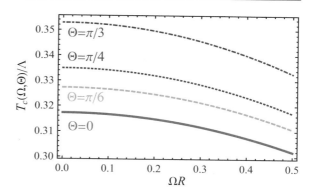

Fig. 11.12 Phase diagram of the rotating fermionic matter in (T, Ω) plane at different angles Θ. The symmetry breaking phase is at the lower part of the diagram. (Taken from [39])

The phase diagram in the temperature-rotation (T, Ω) plane for various angles of the boundary chiral angle Θ is shown in Fig. 11.12. From the similarity of all these phase lines, one can infer that the effect of rotation always tends to restore the chiral symmetry. On the other hand, the critical temperature of the chiral phase transition depends substantially on the boundary condition.

We note that besides the real solutions, Eq. (11.36) also has purely imaginary solutions [40]. These solutions are corresponding to the "edge states" because their wave functions are localized at the boundary.

11.6 Rotating Fermions with Background Magnetic Field

In this section, we will take into account the background magnetic field. And we will see that there will be some interesting effects caused by the combination of rotation and magnetic field. We consider the chiral limit $m_0 = 0$ in this section.

For simplicity, we introduce a constant magnetic field in the inertial lab frame Σ' along the z'-axis (which coincides with the z-axis in the rotating frame) and assume $q\mathbf{B} \cdot \mathbf{\Omega} > 0$. To preserve the rotational symmetry, we choose the symmetric gauge

$$A'_\mu = \left(0, \frac{1}{2}By', -\frac{1}{2}Bx', 0\right). \tag{11.43}$$

Making a coordinate transformation to the rotating frame, we obtain the gauge vector in the rotating frame:

$$A_\mu = \left(-\frac{1}{2}B\Omega r^2, \frac{1}{2}By, -\frac{1}{2}Bx, 0\right). \tag{11.44}$$

From the Klein–Gordon equation (11.8) and the Dirac equation (11.9), the dispersion relation is given by [34,65]

$$[\varepsilon + \Omega(l + s_z)]^2 = p_z^2 + (2\lambda + 1 - 2s_z)qB + m^2, \tag{11.45}$$

for both the bosons and fermions, where l and s_z are the quantum number of the orbit and spin angular momentum, respectively. For an unbounded system, λ is a non-negative integer and for a finite system λ depends on the boundary condition. Let us first focus on the unbounded case. The right-hand side of Eq. (11.45) is the well-known Landau-level quantization. Rotation enters the dispersion relation by a shift of the energy in the left-hand side, which is expected from the discussion in previous sections that rotation has a similar effect with the chemical potential.

At $\Omega = 0$, each Landau level is degenerate with the degeneracy factor

$$N = \left\lfloor \frac{qBS}{2\pi} \right\rfloor, \tag{11.46}$$

where S is the area of the xy-plane of the system. Thus for the λth Landau level, l takes integer values in

$$-\lambda \leqslant l \leqslant N - \lambda, \tag{11.47}$$

and labels the degenerate angular modes of the Landau level λ. At finite Ω, each Landau level is splitted to N non-degenerate levels separated by Ω. Precisely speaking, l should run up to $N - \lambda - 1$, but we consider sufficiently strong magnetic field or large S so that $N \gg 1$, thus we can approximate the upper bound to be $N - \lambda$.

Although here we don't impose any boundary condition on the wave function for the unbounded case, we still have to limit the system size by $R\Omega \leqslant 1$ to preserve the causality. On the other hand, in order to discuss the Landau quantization in the cylindrical system, the radius R should be larger than the magnetic length $1/\sqrt{qB}$. Therefore, our treatment here is legitimate if R is large enough to ignore the boundary effect on the Landau quantization, but not too large to maintain the causality. That is, the following condition should be imposed:

$$1/\sqrt{qB} \ll R \leqslant 1/\Omega. \tag{11.48}$$

For simplicity we first assume that the condensate is spatially homogeneous. Following the standard procedure, one can get the thermodynamic potential at zero temperature and chemical potential (for $m_0 = 0$) [34]

$$V_{\text{eff}} = \frac{\sigma^2}{2G} - \frac{qB}{2\pi} \sum_{\lambda=0}^{\infty} \alpha_\lambda \int_{-\infty}^{\infty} \frac{dp_z}{2\pi} \sqrt{p_z^2 + m_\lambda^2} + V_\Omega, \tag{11.49}$$

where the rotational contribution is

$$V_\Omega = -\frac{1}{S} \sum_{\lambda=0}^{\infty} \alpha_\lambda \sum_{l=-\lambda}^{N-\lambda} \theta(\Omega|j| - m_\lambda) \int_{-k_{\lambda j}}^{k_{\lambda j}} \frac{dp_z}{2\pi} \left[\Omega|j| - \sqrt{p_z^2 + m_\lambda^2} \right] \tag{11.50}$$

Fig. 11.13 Chiral condensate as a function of Ω and qB at weak coupling, where $\sigma_{\rm dyn}$ is the condensate at $qB = 0.2\Lambda^2$ and $\Omega = 0$. (Taken from [34])

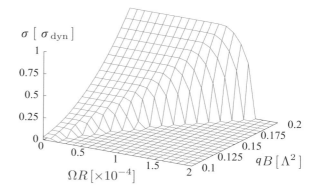

with $\alpha_\lambda = 2 - \delta_{\lambda 0}$, $m_\lambda^2 = 2\lambda q B + \sigma^2$, and $k_{\lambda j} = \sqrt{(\Omega j)^2 - m_\lambda^2}$. The gap equation reads

$$\frac{\sigma}{G} = \sigma \frac{qB}{2\pi} \sum_{\lambda=0}^{\infty} \alpha_\lambda \left[\int_{-\infty}^{\infty} \frac{dp_z}{2\pi} - \frac{1}{N} \sum_{l=-\lambda}^{N-\lambda} \theta(\Omega|j| - m_\lambda) \int_{-k_{\lambda j}}^{k_{\lambda j}} \frac{dp_z}{2\pi} \right] \frac{1}{\sqrt{p_z^2 + m_\lambda^2}}. \tag{11.51}$$

Since the integrand on the right-hand side (RHS) is positive, we observe that the presence of a nonzero Ω always gives a negative contribution to the RHS, and this requires a smaller σ (compared to the $\Omega = 0$ case) to balance the left-hand side (LHS). This is consistent with the results in previous sections that the rotation tends to suppress the chiral condensate. Another interesting observation is that the Ω-related terms in gap equation (11.51) are very similar to the gap equation at finite chemical potential, supporting the analogy between the rotational and density effects as we discussed before.

It is worthwhile to look at the pure magnetic-field effect on the chiral condensate. For simplicity, we consider the strong magnetic-field limit so that the lowest-Landau-level (LLL) approximation can apply. Under LLL approximation, the gap equation becomes

$$\frac{\sigma}{G} = \sigma \frac{qB}{2\pi} \int_{-\infty}^{\infty} \frac{dp_z}{2\pi} \frac{1}{\sqrt{p_z^2 + \sigma^2}} \approx \sigma \frac{qB}{2\pi^2} \ln\left(\frac{2\Lambda}{\sigma}\right) + O\left(\frac{\sigma}{\Lambda}\right)^0, \tag{11.52}$$

where Λ is a ultraviolet cutoff for p_z. The non-trivial solution to the above equation is $\sigma \approx 2\Lambda \exp[-G_c \Lambda^2/(GqB)]$ with $G_c = 2\pi^2/\Lambda^2$ the critical coupling for the onset of chiral condensate at $B = 0$. This shows that in the presence of a strong magnetic field, no matter how small the coupling G is, there is always a nonzero chiral condensate. This phenomenon is called the magnetic catalysis of chiral condensate [7, 8].

Let us discuss the following two cases separately: (A) $G < G_c$ and (B) $G > G_c$. In the weak coupling case ($G < G_c$), the 3-dimensional plot for the chiral conden-sate σ as a function of Ω and qB is shown in Fig. 11.13 [34] where $\sigma_{\rm dyn}$ is the

Fig. 11.14 Chiral
condensate as a function of
Ω and qB at strong coupling.
For large Ω, chiral symmetry
is restored by increasing qB,
which manifests the
rotational magnetic
inhibition. (Taken from [34])

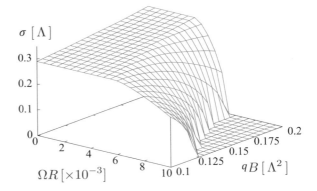

value of the condensate at $qB = 0.2\Lambda^2$ and $\Omega = 0$. From Fig. 11.13, we again see
that the rotation can restore the chiral symmetry. We note that the numerical results
in [34] are obtained using a smoothed cutoff function in regularizing the p_z inte-
gral [66]: $f(p_z, \Lambda) = \sinh(\Lambda/\delta\Lambda)/[\cosh(\sqrt{p_z^2 + 2\lambda qB}/\delta\Lambda) + \cosh(\Lambda/\delta\Lambda)]$ with
$\delta\Lambda = 0.05\Lambda$. In the strong coupling case ($G > G_c$), the 3-dimensional plot of σ on
Ω and qB is shown in Fig. 11.14. For small angular velocity, the chiral condensate is
almost independent of Ω and qB. With increasing Ω the chiral condensate is eventually suppressed by a larger magnetic field, i.e. a counterpart of the finite-density
inverse magnetic catalysis [67] is manifested. This phenomenon is named "rotational
magnetic inhibition" [34]. Similar observation was made also in Ref. [41].

The difference between the weak and strong coupling cases can be explained by
the contribution from the higher Landau levels. In the weak coupling case, only a
small number of the Landau levels contribute to the gap equation, while many more
Landau levels get involved as the coupling constant becomes larger. This is essential
for the realization of the rotational magnetic inhibition as well as of the inverse
magnetic catalysis at finite density.

One can go further by taking into account the boundary condition. In Ref. [68], the
authors adopt the no-flux boundary condition (11.29) and discuss the pure boundary
effect with magnetic field but without rotation. With the no-flux boundary condition,
λ cannot be an integer any more, and depends on l, which we will denote it as $\lambda_{l,k}$.
As discussed in Sect. 11.5, the no-flux boundary condition cannot uniquely fix the
solution, so we adopt a subsidiary condition [68],

$$\lambda_{l,k} = \begin{cases} \xi_{l,k} & \text{for } l = 0, 1, \dots \\ \xi_{-l-1,k} - l & \text{for } l = -1, -2, \dots \end{cases} \tag{11.53}$$

where $\xi_{l,k}$ denotes the kth zero of the confluent hypergeometric function $_1F_1(-\xi, l + 1, \alpha)$, and α is the dimensionless parameter defined by

$$\alpha \equiv \frac{1}{2}qBR^2. \tag{11.54}$$

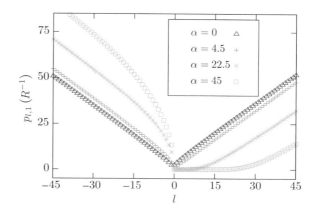

Fig. 11.15 Lowest transverse momentum $p_{l,1} = \sqrt{2qB\lambda_{l,1}}$ as a function of the angular momentum l for various α's. (Taken from [68])

Note that without the magnetic field, the system has charge-conjugation (C) and CP symmetries, and it is natural to adopt such subsidiary condition which preserves C and CP symmetries, while in the present case $B \neq 0$, C, and CP symmetries are explicitly broken by the magnetic field. Thus we choose Eq. (11.53) for the sake of convenience to connect to the $B = 0$ limit smoothly.

In Fig. 11.15 [68], we show the lowest transverse momentum $p_{l,1} = \sqrt{2qB\lambda_{l,1}}$ (this can be viewed as the finite-size modification to the LLL energy) as a function of the angular momentum l for various α. In the $B = 0$ case (purple triangular points), positive l modes and negative $(-l-1)$ modes have a degenerated $p_{l,1}$ due to the CP invariance which implies $j \leftrightarrow -j$ symmetry. At finite magnetic field, however, the momenta for the $l > 0$ branch are more suppressed than the $l < 0$ branch because the magnetic field violates CP symmetry and thus a particular direction of the angular momentum is favored. As α increases (i.e. either B or R increases), more $l > 0$ modes will be suppressed and finally, we recover the usual Landau zero modes ($\lambda_{0,1} = 0$ for $l \geqslant 0$) at $\alpha \to \infty$.

The wave functions with larger l peak at larger r due to the centrifugal force. In the presence of the magnetic field, the wave functions of spin-up and spin-down modes behave differently: with an increasing magnetic field, the spin-down modes are repelled outward further than the spin-up modes and are eventually accumulated at the boundary. (Note that in the unbounded system, the strong magnetic field would repel the spin-down-mode wave functions to infinity and leave the whole system spin-polarized; this is how the LLL dominates at a strong magnetic field.) In such a way, the low-energy phenomena closer to the boundary are more prominently affected by the strong magnetic field. One example is the chiral condensate which is a condensate of spin-aligned (but total angular momentum zero) quark–anti-quark pairs. In Fig. 11.16, the chiral condensate solved from the gap equation (Eq. (11.28) in Ref. [68]) in the local density approximation is shown. In a small magnetic field, $\alpha = 4.5$, we observe a situation similar to the zero-magnetic field case in comparison with Fig. 11.7. In contrast to $\alpha = 4.5$, the chiral condensate behavior for stronger magnetic fields ($\alpha = 22.5$ and 45 in Fig. 11.16) is qualitatively different. Away from the boundary, there is almost no difference. But near the boundary, the chiral con-

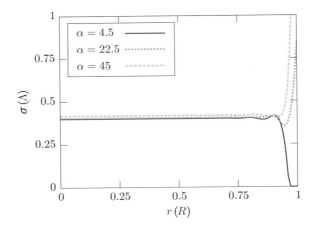

Fig. 11.16 Inhomogeneous chiral condensate as a function of the radial coordinate r for the choice of $R = 30\Lambda^{-1}$. (Taken from [68])

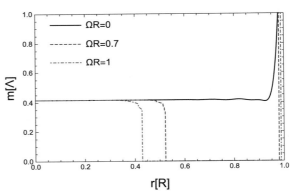

Fig. 11.17 Inhomogeneous chiral condensate as a function of the radial coordinate r for different angular velocity at $\alpha = 45$

densate is pushed up and its value increases with α. This abnormal enhancement is named "surface magnetic catalysis" in Ref. [68].

Then let us take rotation into account; the result is shown in Fig. 11.17. We can see that the rotational inhibition first occurs near the boundary. The inhibition effect gets closer to the center with the increase in the angular velocity. Due to the restriction of casuality, the angular velocity cannot be very large, thus the condensate at the center of the cylinder will not be inhibited. We can also observe that there is a competition between rotational magnetic inhibition and surface magnetic catalysis at the region very close to the boundary [47]. (Because the chiral condensate in the immediate vicinity of the boundary is strongly inhomogeneous, the local density approximation could break down. But the surface magnetic catalysis is expected to be qualitatively unchanged as it comes from a number of accumulating modes at $r \approx R$; see Refs. [47,68] for more discussions.)

Fig. 11.18 Chiral
condensate profile $\sigma(r)$
(solid lines) at different
temperatures under rotation
$\Omega = 0.06M_0[\exp(rM_0 -$
$10) + 1]^{-1}$ (dashed line).
Here M_0 is the chiral
condensate in infinite system
at $T = \Omega = 0$. (Taken from
[45])

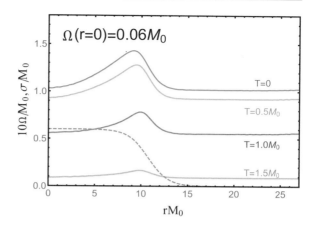

11.7 Inhomogeneity of Chiral Condensate: A BdG Treatment

During the above discussion, the local density approximation is adopted to deal with inhomogeneous condensate, which assumes that the condensate varies slowly with position, i.e. $\partial\sigma \ll \sigma^2$. As we can see from Figs. 11.7, 11.16, and 11.17, the local density approximation breaks down in the vicinity of the boundary. A more realistic way to handle the inhomogeneous condensate is to solve the Bogoliubov–de Gennes (BdG) equation, which has been utilized in [45,46] in the discussion of vortex in chiral condensate and also of the non-uniform rotation.

The BdG method amounts to solve the mean-field gap equations and the eigenvalue problem of the Hamiltonian self-consistently. Let us recall the Dirac Hamiltonian $\hat{H}[\sigma(x), \pi(x)]$ (see Eq. 11.17; the authors of [45,46] considered the case with one flavor and no background gauge field) and the thermodynamic potential $V_{\mathrm{eff}}[\sigma(x), \pi(x)]$ (see Eq. 11.22). The BdG equation is just the eigenequation of $\hat{H}$ with appropriate boundary conditions: $\hat{H}\Psi_{\{\xi\}}(x) = \varepsilon_{\{\xi\}}\Psi_{\{\xi\}}(x)$ where $\Psi_{\{\xi\}}(x)$ and $\varepsilon_{\{\xi\}}$ depend on the profile of $\sigma(x)$ and $\pi(x)$. Supplemented by the gap equations $\delta V_{\mathrm{eff}}/\delta\sigma(x) = \delta V_{\mathrm{eff}}/\delta\pi(x) = 0$, the BdG equation determines the condensates and the corresponding wave functions.

In Ref. [45], the authors studied the 2+1D NJL model rather than 3+1D, since the NJL model is renormalizable in 2+1D. The authors studied the chiral condensate under non-uniform rotation. Choosing a Woods–Saxon-shaped rotation profile $\Omega(r) = \Omega_0/[\exp(r - r_0) + 1]$, the BdG method gives the transverse profile of the chiral condensate as shown in Fig. 11.18. Since the angular velocity Ω_0 shown in Fig. 11.18 is small, the rotational suppression effect is not evident. Instead, the chiral condensate σ increases slowly in the range of the rotation plateau and falls back to $\sigma(\Omega = 0)$ rapidly beyond the rotational region. A bump structure appears around r_0 which reflects the large gradient of Ω_0 around r_0. Physically, the bump structure may be due to the rotational energy shift $\Delta E(r) = -\Omega(r)J_z$ which results in an angular-momentum-dependent radial force $F(x) = -\partial_r\Delta E(r) = \partial_r\Omega(r)J_z$ playing an analogous role as the centrifugal force.

Fig. 11.19 Chiral condensate with a vortex excitation of $\kappa = 1$ for different values of R and T at $\Omega = 0$. (Taken from [46])

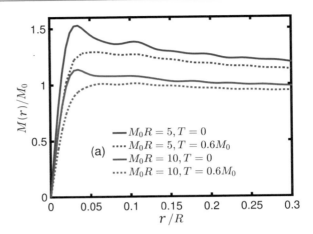

It is natural to consider the possible vortex structure induced by rotation by employing the BdG method. In Ref. [46], the authors studied the one flavor NJL model in 2+1D rotating spacetime with a constant angular velocity Ω. Due to the $U(1)$ chiral symmetry of the system, one can define a complex order parameter

$$\Delta(x) = \sigma(x) + i\pi(x) = M(x)e^{i\phi(x)}, \tag{11.55}$$

where M and ϕ are set to be real. The phase $\phi(x)$ is chosen to be

$$\phi = i\kappa\theta, \quad \kappa \in \mathbb{Z}, \tag{11.56}$$

with θ the azimuthal angle. Our previous discussions only cover the case with $\kappa = 0$. The case with $\kappa \neq 0$ corresponds to the quantized vortex state. Solving the BdG equation with the vortex ansatz can give us the vortex structure. If the condensate with such a vortex provides lower thermodynamic potential than the one without the vortex (i.e. for the case of $\kappa = 0$), it means that the vortex structure is thermodynamically favored.

The vortex core structure with $\kappa = 1$ is shown in Fig. 11.19 [46]. The structure is expected to be qualitatively similar for finite Ω. For the $\Omega = 0$ case, the vortex corresponds to an excited state and is not thermodynamically stable. But since the vortex carries a finite angular momentum, it is more favorable when the system is rotating. Indeed, one can examine the difference in thermodynamic potential, $\delta V_{\text{eff}} = V_{\kappa=0} - V_{\kappa=1}$ to determine the favored thermodynamical state, and when Ω exceeds a critical value Ω_c, δV_{eff} becomes positive meaning that the vortex state becomes more stable. For large system size, the critical angular velocity exceeds the causality bound (i.e. $\Omega_c R > 1$), thus it is impossible for a vortex state to stably exist [46].

11.8   Mesonic Superfluidity

We have focused on the chiral condensate in the previous sections. In this section, we discuss the condensation of other types of pairings, particularly the pions. Such condensate usually triggers a superfluidity at finite density, so in the following we will not distinguish the term "condensate" and the term "superfluid". As pions are $J = 0$ mesons, it is clear that the pion condensation would not be favored by uniform rotation. On the other hand, as $J = 1$ meson, the rho condensation would be favored by uniform rotation [43,51]. In Ref. [43], the authors took into account the isospin chemical potential in the NJL model and considered the rotational effect on the pion and rho condensates. The Lagrangian is the sum of two-flavor NJL Lagrangian (11.13) (with $A_\mu = 0$), $\mathcal{L}_I = (\mu_I/2)\bar{\psi}\gamma^0\tau_3\psi$, and $\mathcal{L}_\rho = -(G_\rho/2)(\bar{\psi}\gamma_\mu\boldsymbol{\tau}\psi)^2$. Considering the unbounded case and the local density approximation to the condensates $\sigma = -G\langle\bar{\psi}\psi\rangle$, $\pi = -G\langle\bar{\psi}i\gamma_5\tau_3\psi\rangle$, and $\rho = -G_\rho\langle\bar{\psi}i\gamma_0\tau_3\psi\rangle$, the thermodynamic potential can be derived in a similar manner as described in Sect. 11.4. The result is $V_{\text{eff}} = \int d^3r\mathcal{V}_{\text{eff}}(r)$ with

$$\mathcal{V}_{\text{eff}}(r) = \frac{\sigma^2 + \pi^2}{2G} - \frac{\rho^2}{2G_\rho} - \frac{N_f N_c}{16\pi^2}\sum_{a=\pm}\sum_l \int dp_t^2 \int dp_z$$
$$\times [J_l(p_t r)^2 + J_{l+1}(p_t r)^2]T\ln(1 + e^{\beta\varepsilon_l^a})(1 + e^{-\beta\varepsilon_l^a}), \tag{11.57}$$

where $\varepsilon_l^\pm = \sqrt{\pi^2 + (\sqrt{m^2 + p_t^2 + p_z^2} \pm \tilde{\mu}_I/2)^2} - \Omega(l + 1/2), m = m_0 + \sigma, \tilde{\mu}_I = \mu_I + G_\rho\rho$, and $N_f = 2$, $N_c = 3$. The thermodynamically equilibrium state is specified by the minimum of V_{eff}. As Ω and μ_I vary, the true equilibrium state would vary as well leading to phase transitions. The phase diagram so obtained is plotted in Fig. 11.20 (see Ref. [43] for the parameter setup) which is characterized by three distinctive regions: a vacuum-like sigma-dominated phase in the low isospin chemical potential and the slow rotation region, a pion-superfluid phase in the mid-to-high isospin density with moderate rotation, and a rho-superfluid phase in the high isospin and rapid rotation region. A second-order transition line separates the sigma-dominant and the pion-dominant regions while a first-order transition line separates the pion-dominant and the rho-dominant regions, with a tri-critical point (TCP) connecting them. Note that without electromagnetic interactions, there is no distinction among the isospin-1 triplet states $\pi^{0,\pm}$ or among the $\rho^{0,\pm}$ and one is free to choose the condensation to be in any direction of the isospin space.

In the above discussion we considered rotating isospin matter. Adding baryon charges into such isospin matter would induce intriguing phenomenon through chiral anomaly. The underlying mechanism can be understood through the delicate chiral vortical effect (CVE) for axial current $j_5^{a\mu} = \langle\bar{\psi}\gamma^\mu\gamma_5\tau^a\rangle$. Consider the case of $a = 3$ for two-flavor quarks at $T = 0$. The CVE current is given by

$$j_5^3 = \frac{\mu_B\mu_I}{\pi^2}\boldsymbol{\Omega}. \tag{11.58}$$

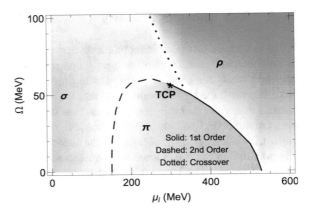

Fig. 11.20 Phase diagram on the $\Omega - \mu_I$ plane for mesonic superfluidity in isospin matter under rotation. (Taken from [43])

This current is protected by chiral anomaly and is thus exact regardless of the energy scale. Therefore, we can write down a low-energy effective Hamiltonian for the CVE current (11.58) in terms of pion field [37]:

$$\mathcal{H}_{\text{CVE}} = -\frac{\mu_B \mu_I}{2\pi^2 f_\pi} \nabla \pi_0 \cdot \boldsymbol{\Omega}, \tag{11.59}$$

where f_π is the pion decay constant. Adding the kinetic terms for π_0 field, we get the total effective Hamiltonian as $H = \int d^3x \mathcal{H}$ with $\mathcal{H}$ being (suppose $\boldsymbol{\Omega}$ to be along the z-axis)

$$\mathcal{H} = \frac{1}{2}\left[(\partial_r \pi_0)^2 + \frac{1-(\Omega r)^2}{r^2}(\partial_\theta \pi_0)^2 + (\partial_z \pi_0)^2\right] + m_\pi^2 f_\pi^2\left(1 - \cos\frac{\pi_0}{f_\pi}\right) - \frac{\mu_B \mu_I}{2\pi^2 f_\pi}\Omega\partial_z \pi_0, \tag{11.60}$$

where m_π is pion mass. The ground state must minimize H which is specified by the equations of motion for π_0: $\partial_r \pi_0 = \partial_\theta \pi_0 = 0$ and $\partial_z^2 \pi_0 = m_\pi^2 f_\pi \sin(\pi_0/f_\pi)$. The solution to the equations of motion is given by [37]

$$\cos\frac{\pi_0(\bar{z})}{2f_\pi} = \text{sn}(\bar{z}, k), \tag{11.61}$$

where $\text{sn}(\bar{z}, k)$ is the Jacobi elliptic function of modulus $k \in [0, 1]$ and $\bar{z} = m_\pi z/k$. This solution describes an interesting chiral soliton lattice structure of π_0 condensate along the $\boldsymbol{\Omega}$ direction. In each lattice cell, the angular momentum, baryon, and isospin charges are topological quantities (namely, independent of the shape of $\pi_0(z)$). Similar chiral-anomaly-induced neutral pion condensate occurs also in parallel electric and magnetic fields [69,70] and in external magnetic field with high baryon chemical potentials in which the chiral soliton lattice can also appear [71]. Also, similar chiral soliton lattice structure can also be realized in η' condensate in rotating baryonic matter [72]. We note that substituting solution (11.61) into H one can find that only when Ω exceeds a critical value $\Omega_c = 8\pi m_\pi f_\pi^2/(\mu_B|\mu_I|)$, the chiral soliton lattice corresponds to the lowest energy and is the true ground state of QCD. On the other hand, Ω must be much smaller that the QCD scale to guarantee the availability of the effective theory (which also guarantees that the rotational suppression of π_0 condensate is negligible); see more discussions in Ref. [37].

We have discussed the novel effects of rotation on the neutral pion condensate at finite isospin and baryon chemical potentials. In the following, we discuss the effect of a magnetic field on the rotating pionic matter which is related to charged pion superfluid. This was first considered in Refs. [41,42]. By solving the Klein–Gordon equation (11.8), one can get the dispersion relation for charged pion $\pi^{\pm}$

$$E = \sqrt{|qB|(2n+1) + p_z^2 + m_\pi^2} - \text{sgn}(q)\Omega l, \qquad (11.62)$$

where $q = e > 0$ for positively charged pions and $q = -e$ for negatively charged pions. It is easy to see that the degeneracy of each Landau level is lifted. In particular, the π^+ in the LLL splits down, and the π^- in the LLL splits up. Thus effectively the Ωl plays the role of an isospin chemical potential $\mu_I = \Omega l$ for π^+ and $-\mu_I = -\Omega l$ for π^-. Therefore, when $\mu_N = N\Omega$ ($N = \lfloor |qBS|/(2\pi) \rfloor$ is the Landau degeneracy) exceeds the effective mass of π^+ in the LLL, $m_0 = \sqrt{eB + m_\pi^2}$, but it is still below the π^+ effective mass in the first LL, the LLL π^+ may Bose-condense. As Ω increased, higher Landau-level π^+ may Bose-condense sequentially. Again, we emphasize that we are considering an unbounded system with the constraint $\Omega \leqslant 1/R \ll \sqrt{eB}$ implicitly assumed: the first inequality is due to the causality and the second one is to make the Landau quantization sensible.

Recently, such possible charged pion condensation was re-examined by using the NJL model [47,48]. This is important because π^+ comprises a u quark and a $\bar{d}$ anti-quark with their angular momenta (both the spin and the orbital one) antiparallel to each other. Thus, both the magnetic field and rotation would tend to suppress the π^+ condensate. In addition, the rotational magnetic inhibition discussed in Sect. 11.6 would take place which may also influence the charged pion condensate. So we need to re-consider the charged pion condensate from the quark-level dynamics. The Lagrangian is Eq. (11.13). Since the $SU(2)$ isospin symmetry is explicitly broken to $U(1)_{I_3}$ by the background magnetic field, it is very difficult to diagonalize the Hamiltonian Eq. (11.17) with nonzero charged $\pi^{\pm} = \pi^1 \mp \pi^2$ condensates. One can instead consider a Ginzburg–Landau approach for the $\pi^{\pm}$ fields but with the σ field taken into account self-consistently through gap equation. This amounts to examine the stability of the σ-condensed phase against the onset of charged pion condensate. For this purpose, we expand the thermodynamic potential in π,

$$V_{\text{eff}} = V_{\text{eff}}^{(0)} + V_{\text{eff}}^{(2)} + \cdots,$$

$$V_{\text{eff}}^{(0)} = \frac{1}{\beta V} \int d^4 x_E \frac{\sigma^2}{2G} - \frac{1}{\beta V} \text{Tr} \ln(i\slashed{\nabla} + \mu_B \gamma^0 - \sigma), \qquad (11.63)$$

$$V_{\text{eff}}^{(2)} = \frac{1}{\beta V} \int d^4 x_E \frac{\pi^2}{2G} - \frac{1}{2\beta V} \text{Tr}[(i\slashed{\nabla} + \mu_B \gamma^0 - \sigma)^{-1} \gamma^5 \boldsymbol{\pi} \cdot \boldsymbol{\tau}]^2,$$

where a baryon chemical potential μ_B is introduced. Note that we set $\pi^0 = 0$ since we only focus on the charged pion fields.

In Eq. (11.63), $V_{\text{eff}}^{(0)}$ has been discussed plentifully in previous sections. What we are interested in is $V_{\text{eff}}^{(2)}$ which is quadratic in π. If we write

$$V_{\text{eff}}^{(2)} = \int d^4x'_{\text{E}} d^4x_{\text{E}} \, \pi^+(x') \, C^{(2)}(x', x) \, \pi^-(x), \qquad (11.64)$$

where $C^{(2)}(x', x)$ is the inverse pion propagator at one-quark-loop level. The charged pion condensation would be favored if $C^{(2)}(x', x)$ is not semi-positive definite (as a matrix). In Refs. [47,48], the sign of $C^{(2)}(x', x)$ is examined by taking the following ansatz to simplify $V_{\text{eff}}^{(2)}$:

$$\pi^+(x')\pi^-(x) = e^{ie \int_{x'}^{x} A_\mu dz^\mu} \tilde{\pi}^+ \tilde{\pi}^-, \qquad (11.65)$$

where $\tilde{\pi}^+$ and $\tilde{\pi}^-$ are gauge-independent condensates which are assumed to be constants and the exponential factor is the Wilson line which connects x' and x. This Wilson line should be so chosen that the Schwinger phase in the quark propagator entering $C^{(2)}(x', x)$ is canceled and hence $V_{\text{eff}}^{(2)} = C^{(2)}\tilde{\pi}^+\tilde{\pi}^-$ is gauge invariant with the constant $C^{(2)} = \int d^4x'_{\text{E}} d^4x_{\text{E}} \, C^{(2)}(x', x)e^{ie \int_{x'}^{x} A_\mu dz^\mu}$ playing the role of the Ginzburg–Landau coefficient. The Schwinger phase $\Theta(x, x')$ is the gauge-dependent factor in the quark propagator, $S(x, x') = e^{i\Theta(x,x')} S_{\text{inv}}(x, x')$ so that $\Theta(x, x) = 0$ is satisfied and S_{inv} is gauge invariant. For constant electromagnetic field in rotating frame, it can be shown that the Schwinger phase is given by $\Theta(x, x') = -q \int_x^{x'} A_\mu(z)dz^\mu$ (q: the quark charge) along the geodesic between x and x'. The numerical study in Ref. [47] shows that under reasonable parameter choice, the Ginzburg–Landau coefficient is positive at $\mu_B = 0$ disfavoring the onset of charged pion condensate. But once a large negative μ_B is supplied, there can indeed be a region in Ω and B so that the charged pion condensation is favored. On the other hand, if one insists using another integral path away from the geodesic, one can indeed find charged pion condensate even at $\mu_B = 0$ [47,48]. Finally, we want to emphasize that even one can rule out the possibility of homogeneous charged pion condensates $\tilde{\pi}^{\pm}$ as given in the ansatz (11.65), it remains still the possibility of an inhomogeneous condensates to be favored. Besides, as the charged pion condensate triggers an electric superconductivity, the Meissner effect would repel the magnetic field from the bulk superconductor and only allow inhomogeneous magnetic vortices. These new possibilities demand future study.

11.9   Summary

In this article we have presented an overview of the recent progress on the understanding of the QCD phase structure under rotation. The rotation plays a role in polarizing the spin (and orbital motion) of underlying microscopic particles and in turn induces a number of novel effects on various condensates of quark–(anti-)quark pairs. An enriched phase diagram in the T-μ_B-Ω space is sketched in Fig. 11.1. At low T, μ_B, and Ω, we have the usual hadronic matter with the spontaneous chiral symmetry breaking. At high T, either for small or large Ω, the quarks and gluons are expected to be liberated from the hadrons and form a deconfined quark–gluon plasma phase with restored chiral symmetry. At small T and Ω but asymptotically high μ_B, QCD is in the color superconducting phase with merely spin-0 quark–quark pairings while at moderate μ_B the ground state of QCD is to a large content unknown. At high μ_B with increasing Ω, it is plausible to expect that QCD possibly undergoes a transition from spin-0 to spin-1 color superconductor. Now with a new dimension of rotation, the QCD phase diagram becomes even more interesting and rich. The rotation is found to considerably influence the chiral condensate. Adding a magnetic field or isospin chemical potential would lead to additional structures through condensations in various mesonic channels. We have discussed recent results on these effects in great detail by using effective models in a rotating frame. To summarize, the properties and phase structures of QCD matter under rotation is an interesting emerging direction with many new theoretical questions still to be explored and answered. It is also important to investigate potential implications and observable effects for the QCD system with large vorticity in heavy-ion collisions as well as the nuclear matter inside fast rotating compact stars. One would anticipate a lot of exciting progress to be made in the near future.

Acknowledgements We thank Kenji Fukushima, Defu Hou, Kazuya Mameda, Kentaro Nishimura, Yin Jiang, Naoki Yamamoti, and Hui Zhang for collaborations. This work is supported in part by the NSFC Grants No. 11535012 and No. 11675041, as well as by the NSF Grant No. PHY-1913729 and by the U.S. Department of Energy, Office of Science, Office of Nuclear Physics, within the framework of the Beam Energy Scan Theory (BEST) Topical Collaboration.

References

1. Fukushima, K., Hatsuda, T.: Rept. Prog. Phys. **74** (2011). arXiv:1005.4814 [hep-ph]
2. Alford, M.G., Schmitt, A., Rajagopal, K., Schafer, T.: Rev. Mod. Phys. **80**, 1455 (2008). arXiv:0709.4635 [hep-ph]
3. Bzdak, A., Esumi, S., Koch, V., Liao, J., Stephanov, M., Xu, N.: Phys. Rept. **853**, 1–87 (2020). arXiv:1906.00936 [nucl-th]
4. Hattori, K., Huang, X.G.: Nucl. Sci. Tech. **28**, 26 (2017). arXiv:1609.00747 [nucl-th]
5. Huang, X.G.: Rept. Prog. Phys. **79** (2016). arXiv:1509.04073 [nucl-th]
6. Kharzeev, D.E., Liao, J., Voloshin, S.A., Wang, G.: Prog. Part. Nucl. Phys. **88**, 1–28 (2016). https://doi.org/10.1016/j.ppnp.2016.01.001. arXiv:1511.04050 [hep-ph]
7. Gusynin, V.P., Miransky, V.A., Shovkovy, I.A.: Phys. Rev. Lett. **73**, 3499 (1994) Erratum: [Phys. Rev. Lett. **76**, 1005 (1996)] [hep-ph/9405262]
8. Gusynin, V.P., Miransky, V.A., Shovkovy, I.A.: Nucl. Phys. B **462**, 249 (1996). [hep-ph/9509320]
9. Bali, G.S., Bruckmann, F., Endrodi, G., Fodor, Z., Katz, S.D., Krieg, S., Schafer, A., Szabo, K.K.: JHEP **1202**, 044 (2012). arXiv:1111.4956 [hep-lat]
10. Bali, G.S., Bruckmann, F., Endrodi, G., Fodor, Z., Katz, S.D., Schafer, A.: Phys. Rev. D **86** (2012). arXiv:1206.4205 [hep-lat]
11. Andersen, J.O., Naylor, W.R., Tranberg, A.: Rev. Mod. Phys. **88** (2016). arXiv:1411.7176 [hep-ph]
12. Miransky, V.A., Shovkovy, I.A.: Phys. Rept. **576**, 1 (2015). arXiv:1503.00732 [hep-ph]

13. Adamczyk, L., et al.: STAR Collaboration. Nature **548**, 62 (2017). arXiv:1701.06657 [nucl-ex]
14. Adam, J., et al.: STAR Collaboration. Phys. Rev. C **98** (2018). arXiv:1805.04400 [nucl-ex]
15. Adam, J., et al., [STAR Collaboration]: Phys. Rev. Lett. **123**(13), 132301 (2019). arXiv:1905.11917 [nucl-ex]
16. Deng, W.T., Huang, X.G.: Phys. Rev. C **93** (2016). arXiv:1603.06117 [nucl-th]
17. Jiang, Y., Lin, Z.W., Liao, J.: Phys. Rev. C **94**, 044910 (2016) Erratum: [Phys. Rev. C **95**, no. 4, 049904 (2017)]. arXiv:1602.06580 [hep-ph]
18. Shi, S., Li, K., Liao, J.: Phys. Lett. B **788**, 409–413 (2019). arXiv:1712.00878 [nucl-th]
19. Deng, X.G., Huang, X.G., Ma, Y.G., Zhang, S.: Phys. Rev. C **101** (2020). arXiv:2001.01371 [nucl-th]
20. Vilenkin, A.: Phys. Rev. D **20**, 1807 (1979)
21. Erdmenger, J., Haack, M., Kaminski, M., Yarom, A.: JHEP **0901**, 055 (2009). arXiv:0809.2488 [hep-th]
22. Banerjee, N., Bhattacharya, J., Bhattacharyya, S., Dutta, S., Loganayagam, R., Surowka, P.: JHEP **1101**, 094 (2011). arXiv:0809.2596 [hep-th]
23. Son, D.T., Surowka, P.: Phys. Rev. Lett. **103** (2009). arXiv:0906.5044 [hep-th]
24. Kharzeev, D.E., McLerran, L.D., Warringa, H.J.: Nucl. Phys. A **803**, 227 (2008). arXiv:0711.0950 [hep-ph]
25. Fukushima, K., Kharzeev, D.E., Warringa, H.J.: Phys. Rev. D **78** (2008). arXiv:0808.3382 [hep-ph]
26. Liang, Z.T., Wang, X.N.: Phys. Rev. Lett. **94**, 102301 (2005). Erratum: [Phys. Rev. Lett. **96**, 039901 (2006)] [nucl-th/0410079]
27. Liang, Z.T., Wang, X.N.: Phys. Lett. B **629**, 20–26 (2005). arXiv:nucl-th/0411101
28. Huang, X.G., Liao, J., Wang, Q., Xia, X.L.: in this volume
29. Gao, J.H., Liang, Z.T., Wang, Q., Wang, X.N.: in this volume. arXiv:2009.04803 [nucl-th]
30. Becattini, F.: in this volume. arXiv:2004.04050 [hep-th]
31. Liu, Y.C., Huang, X.G.: Nucl. Sci. Tech. **31**, 56 (2020). arXiv:2003.12482 [nucl-th]
32. Becattini, F., Lisa, M.A.: (2020). arXiv:2003.03640 [nucl-ex]
33. Yamamoto, A., Hirono, Y.: Phys. Rev. Lett. **111** (2013). arXiv:1303.6292 [hep-lat]
34. Chen, H.L., Fukushima, K., Huang, X.G., Mameda, K.: Phys. Rev. D **93** (2016). arXiv:1512.08974 [hep-ph]
35. Jiang, Y., Liao, J.: Phys. Rev. Lett. **117** (2016). arXiv:1606.03808 [hep-ph]
36. Ebihara, S., Fukushima, K., Mameda, K.: Phys. Lett. B **764**, 94 (2017). arXiv:1608.00336 [hep-ph]
37. Huang, X.G., Nishimura, K., Yamamoto, N.: JHEP **1802**, 069 (2018). arXiv:1711.02190 [hep-ph]
38. Chernodub, M.N., Gongyo, S.: JHEP **1701**, 136 (2017). arXiv:1611.02598 [hep-th]
39. Chernodub, M.N., Gongyo, S.: Phys. Rev. D **95** (2017). arXiv:1702.08266 [hep-th]
40. Chernodub, M.N., Gongyo, S.: Phys. Rev. D **96** (2017). arXiv:1706.08448 [hep-th]
41. Liu, Y., Zahed, I.: Phys. Rev. D **98** (2018). arXiv:1710.02895 [hep-ph]
42. Liu, Y., Zahed, I.: Phys. Rev. Lett. **120** (2018). arXiv:1711.08354 [hep-ph]
43. Zhang, H., Hou, D., Liao, J.: (2020). arXiv:1812.11787 [hep-ph]
44. Wang, X., Wei, M., Li, Z., Huang, M.: Phys. Rev. D **99** (2019). arXiv:1808.01931 [hep-ph]
45. Wang, L., Jiang, Y., He, L., Zhuang, P.: Phys. Rev. C **100** (2019). arXiv:1901.00804 [nucl-th]
46. Wang, L., Jiang, Y., He, L., Zhuang, P.: Phys. Rev. D **100** (2019). arXiv:1901.04697 [nucl-th]
47. Chen, H.L., Huang, X.G., Mameda, K.: arXiv:1910.02700 [nucl-th]
48. Cao, G., He, L.: Phys. Rev. D **100** (2019). arXiv:1910.02728 [nucl-th]
49. Zhang, Z., Shi, C., Luo, X., Zong, H.S.: Phys. Rev. D **101** (2020). arXiv:2003.03765 [nucl-th]
50. Zhang, Z., Shi, C., Luo, X., Zong, H.S.: Phys. Rev. D **102** (2020). arXiv:2006.00677 [math-ph]
51. Cao, G.: arXiv:2008.08321 [nucl-th]
52. Mueller, B., Schaefer, A.: Phys. Rev. D **98** (2018). arXiv:1806.10907 [hep-ph]
53. Guo, Y., Shi, S., Feng, S., Liao, J.: Phys. Lett. B **798** (2019). arXiv:1905.12613 [nucl-th]
54. Guo, X., Liao, J., Wang, E.: Sci. Rep. **10**, 2196 (2020). arXiv:1904.04704 [hep-ph]

55. Parker, L., Toms, D.: Quantum Field Theory in Curved Spacetime: Quantized Fields and Gravity. Cambridge University Press, Cambridge (2009)
56. Birrell, N.D., Davies, P.C.W.: Quantum Fields in Curved Space. Cambridge University Press, Cambridge (1984)
57. Klevansky, S.P.: Rev. Mod. Phys. **64**, 649 (1992)
58. Hatsuda, T., Kunihiro, T.: Phys. Rept. **247**, 221 (1994). [hep-ph/9401310]
59. Buballa, M.: Phys. Rept. **407**, 205 (2005). [hep-ph/0402234]
60. Ambrus, V.E., Winstanley, E.: Phys. Lett. B **734**, 296 (2014). arXiv:1401.6388 [hep-th]
61. Ambrus, V.E., Winstanley, E.: in this volume. arXiv:1908.10244 [hep-th]
62. Ambrus, V.E., Winstanley, E.: Phys. Rev. D **93** (2016). arXiv:1512.05239 [hep-th]
63. Chodos, A., Jaffe, R.L., Johnson, K., Thorn, C.B., Weisskopf, V.F.: Phys. Rev. D **9**, 3471 (1974)
64. Giordano, C., Laforgia, A.: J. Comput. Appl. Math. **9**, 221 (1983)
65. Mameda, K., Yamamoto, A.: PTEP **2016**, 093B05 (2016). arXiv:1504.05826 [hep-th]
66. Gorbar, E.V., Miransky, V.A., Shovkovy, I.A.: Phys. Rev. D **83** (2011). arXiv:1101.4954 [hep-ph]
67. Preis, F., Rebhan, A., Schmitt, A.: Lect. Notes Phys. **871**, 51 (2013). arXiv:1208.0536 [hep-ph]
68. Chen, H.L., Fukushima, K., Huang, X.G., Mameda, K.: Phys. Rev. D **96** (2017). arXiv:1707.09130 [hep-ph]
69. Cao, G., Huang, X.G.: Phys. Lett. B **757**, 1 (2016). arXiv:1509.06222 [hep-ph]
70. Wang, L., Cao, G., Huang, X.G., Zhuang, P.: Phys. Lett. B **780**, 273 (2018). arXiv:1801.01682 [nucl-th]
71. Brauner, T., Yamamoto, N.: JHEP **1704**, 132 (2017). arXiv:1609.05213 [hep-ph]
72. Nishimura, K., Yamamoto, N.: JHEP **07**, 196 (2020). arXiv:2003.13945 [hep-ph]

Relativistic Decomposition of the Orbital and the Spin Angular Momentum in Chiral Physics and Feynman's Angular Momentum Paradox

12

Kenji Fukushima and Shi Pu

Abstract

Over recent years we have witnessed tremendous progress in our understanding of the angular momentum decomposition. In the context of the proton spin problem in high-energy processes, the angular momentum decomposition by Jaffe and Manohar, which is based on the canonical definition, and the alternative by Ji, which is based on the Belinfante improved one, have been revisited under light shed by Chen et al. leading to seminal works by Hatta, Wakamatsu, Leader, etc. In chiral physics as exemplified by the chiral vortical effect and applications to the relativistic nucleus–nucleus collisions, sometimes referred to as a relativistic extension of the Barnett and the Einstein–de Haas effects, such arguments of the angular momentum decomposition would be of crucial importance. We pay our special attention to the fermionic part in the canonical and the Belinfante conventions and discuss a difference between them, which is reminiscent of a classical example of Feynman's angular momentum paradox. We point out its possible relevance to early-time dynamics in the nucleus–nucleus collisions, resulting in excess by the electromagnetic angular momentum.

K. Fukushima (✉)
Department of Physics, The University of Tokyo, 7-3-1 Hongo, Bunkyo-ku, Tokyo 113-0033, Japan
e-mail: fuku@nt.phys.s.u-tokyo.ac.jp

S. Pu
Department of Modern Physics, University of Science and Technology of China, Hefei 230026, China
e-mail: shipu@ustc.edu.cn

© Springer Nature Switzerland AG 2021
F. Becattini et al. (eds.), *Strongly Interacting Matter under Rotation*,
Lecture Notes in Physics 987,
https://doi.org/10.1007/978-3-030-71427-7_12

12.1 Prologue

Some time ago we, Fukushima and Pu, together with our bright colleague, Zebin Qiu, published a paper [1] on a relativistic extension of the Barnett effect [2] in the context of chiral materials. Our results are beautiful and robust, we believe, but at the same time, we had to overcome many conceptual confusions. We are 100% sure about our calculations, results, and conclusions, but we were unable to find 100% unshakable justification for our spin identification. We could not remove theoretical uncertainty to extract the orbital angular momentum (OAM) and the spin angular momentum (SAM) out of the total angular momentum that is conserved. We adopted the most natural assumption, meanwhile, we studied many preceding works; for example, we found Ref. [3] that makes a surprising assertion of the existence of individually conserved OAM and SAM derived from the Dirac equation. The more we studied, the more confusion we were falling into. The present contribution is not an answer to controversies, but more like a note of what we have understood so far, and some of our own thoughts based on them. Actually, in Ref. [1] we posed an important question of how to represent the Barnett effect in chiral hydrodynamics, but in the present article we will not mention this. We will report our progress on hydrodynamics with OAM and SAM somewhere else hopefully soon, and the present article is focused on the field's theoretical descriptions.

12.2 Basics—Angular Momenta in an Abelian Gauge Theory

In non-relativistic and classical theories, the spin is not a dynamical variable; spin-up and spin-down electrons are treated as distinct species and the total spin is conserved unless interactions allow for spin unbalanced processes. Dirac successfully generalized an equation proposed by Pauli, who first postulated such internal doubling, into a fully relativistic formulation. Eventually, Majorana and other physicists realized the usage of Cartan's spinors. Today, even undergraduate students are familiar with tensors and spinors according to the representation theory of Lorentz symmetry. In contemporary physics, symmetries and associated conserved quantities play essential roles. This article mainly addresses the angular momentum and the spin. Readers interested in the history of the spin are invited to consult a very nice book, *The Story of Spin*, by Sin-itiro Tomonaga (see Ref. [4] for an English translated version).

To begin with, we shall summarize some textbook knowledge about various assignments of angular momenta. Lorentz symmetry is characterized by the following transformation:

$$x^\mu \;\to\; x'^\mu = \Lambda^\mu_{\;\nu} x^\nu = (\delta^\mu_\nu + \epsilon^\mu_{\;\nu}) x^\nu , \tag{12.1}$$

where $\Lambda_{\mu\nu}$ and infinitesimal $\epsilon_{\mu\nu}$ are antisymmetric tensors. Let us take a simple Abelian gauge theory defined by the following Lagrangian density:

$$\mathcal{L} = \bar\psi (i\gamma^\mu D_\mu - m)\psi - \frac{1}{4} F^{\mu\nu} F_{\mu\nu} \tag{12.2}$$

with the covariant derivative, $D_\mu \equiv \partial_\mu + ieA_\mu$, and the field strength tensor, $F^{\mu\nu} \equiv \partial^\mu A^\nu - \partial^\nu A^\mu$. This theory involves vector and spinor fields which transform together with Eq. (12.1) as

$$A^\mu(x) \rightarrow A'^\mu(x) = \Lambda^\mu_{\ \nu} A^\nu(\Lambda^{-1}x), \qquad (12.3)$$

$$\psi(x) \rightarrow \psi'(x) = \Lambda_{\frac{1}{2}} \psi(\Lambda^{-1}x), \qquad (12.4)$$

where $\Lambda_{\frac{1}{2}} = \mathbf{1} - \frac{i}{2}\epsilon_{\mu\nu}\Sigma^{\mu\nu}$ with $\Sigma^{\mu\nu} \equiv \frac{i}{4}[\gamma^\mu, \gamma^\nu]$. Thus, for an infinitesimal transformation, the fields change as $A^\alpha(x) \rightarrow A^\alpha(x) + \frac{1}{2}\epsilon_{\mu\nu}\Delta A^{\mu\nu\alpha}(x)$ and $\psi(x) \rightarrow \psi(x) + \frac{1}{2}\epsilon_{\mu\nu}\Delta\psi^{\mu\nu}(x)$ (where we put $\frac{1}{2}$ for antisymmetrization) with

$$\Delta A^{\mu\nu\alpha}(x) = \Big[(x^\mu\partial^\nu - x^\nu\partial^\mu)g^{\alpha\beta} + (g^{\mu\alpha}g^{\nu\beta} - g^{\nu\alpha}g^{\mu\beta})\Big]A_\beta(x), \qquad (12.5)$$

$$\Delta\psi^{\mu\nu}(x) = \big(x^\mu\partial^\nu - x^\nu\partial^\mu - i\Sigma^{\mu\nu}\big)\psi(x). \qquad (12.6)$$

Now we can compute the Nöther current. From the gauge part, we find

$$J_A^{\lambda\mu\nu} = \frac{\partial\mathcal{L}}{\partial(\partial_\lambda A^\alpha)}\Delta A^{\mu\nu\alpha} = -F^\lambda_{\ \alpha}(x^\mu\partial^\nu - x^\nu\partial^\mu)A^\alpha - F^{\lambda\mu}A^\nu + F^{\lambda\nu}A^\mu. \qquad (12.7)$$

In the same way, we go on to obtain the fermionic contribution,

$$J_\psi^{\lambda\mu\nu} = \frac{\partial\mathcal{L}}{\partial(\partial_\lambda\psi)}\Delta\psi^{\mu\nu} = \bar\psi i\gamma^\lambda\big(x^\mu\partial^\nu - x^\nu\partial^\mu - i\Sigma^{\mu\nu}\big)\psi. \qquad (12.8)$$

They satisfy $\partial_\lambda(J_A^{\lambda\mu\nu} + J_\psi^{\lambda\mu\nu}) = 0$, and the conserved charge (i.e., $\lambda = 0$ component) is the total angular momentum. From these expressions, it would be a natural choice for us to define the "canonical" OAM and SAM as follows:

$$L_{A,\text{can}}^{\mu\nu} \equiv -F^0_{\ \alpha}(x^\mu\partial^\nu - x^\nu\partial^\mu)A^\alpha, \quad S_{A,\text{can}}^{\mu\nu} \equiv -F^{0\mu}A^\nu + F^{0\nu}A^\mu. \qquad (12.9)$$

$$L_{\psi,\text{can}}^{\mu\nu} \equiv i\psi^\dagger(x^\mu\partial^\nu - x^\nu\partial^\mu)\psi, \quad S_{\psi,\text{can}}^{\mu\nu} \equiv \psi^\dagger\Sigma^{\mu\nu}\psi. \qquad (12.10)$$

This is simply our choice for the moment, and one may say that the spin can be identified as the remaining operator in the homogeneous limit where all spatial derivatives drop.[1] These are not separately conserved quantities but only the sums, the total angular momenta, are conserved. We point out that the above decomposition has been long known in the context of the proton spin problem (see Refs. [5,6] for reviews). In the language of quantum chromodynamics (QCD), if the gauge field is extended to the non-Abelian gluon field and the temporal index is changed to + in the light-cone

---

[1]The spin identification in such a frame to drop spatial derivatives is emphasized by Yoshimasa Hidaka. Another physical constraint is the commutation relation, and this prescription would always give the correct commutation relation of the spin.

coordinates, $S^{\mu\nu}_{\psi,\text{can}}$ and $S^{\mu\nu}_{A,\text{can}}$ correspond to $\frac{1}{2}\Delta\Sigma$ and ΔG, respectively, in what is called the Jaffe–Manohar decomposition.

Such expressions have been known by all QCD physicists; they look firmly founded, but not very undoubted yet, for they are obviously gauge dependent. Among quantum field theoreticians, a common folklore is that non-gauge-invariant objects may well be unphysical. This story would remind readers of a famous problem that the canonical energy–momentum tensor is not gauge invariant, while the symmetrized one is. Interestingly, rotation and translational shift are coupled together, so that the angular momenta and the energy–momentum tensor (EMT) are linked. The canonical EMT for the Abelian gauge theory is derived as

$$T^{\mu\nu}_{A,\text{can}} = \frac{\partial\mathcal{L}}{\partial(\partial_\mu A^\alpha)}\partial^\nu A^\alpha - g^{\mu\nu}\mathcal{L}_A = -F^\mu_\alpha \partial^\nu A^\alpha + \frac{1}{4}g^{\mu\nu}F^{\alpha\beta}F_{\alpha\beta} \quad (12.11)$$

for the gauge part, which is clearly gauge dependent, and

$$T^{\mu\nu}_{\psi,\text{can}} = \frac{\partial\mathcal{L}}{\partial(\partial_\mu\psi)}\partial^\nu\psi - g^{\mu\nu}\mathcal{L}_\psi = \bar{\psi}i\gamma^\mu\partial^\nu\psi - g^{\mu\nu}\bar{\psi}(i\gamma^\alpha D_\alpha - m)\psi \quad (12.12)$$

for the fermion part. From now on, we impose onshellness and utilize the equations of motion. We would recall that the derivation of Nöther's theorem already requires the equations of motion. Then, we can safely drop the last term in $T^{\mu\nu}_{\psi,\text{can}}$, thanks to the Dirac equation. Then, for spatial μ and ν (denoted by i and j), it is straightforward to confirm the relation between the OAM and the EMT,

$$L^{ij}_{A/\psi,\text{can}} = x^i T^{0j}_{A/\psi,\text{can}} - x^j T^{0i}_{A/\psi,\text{can}}. \quad (12.13)$$

So far, apart from the gauge invariance, all these relations perfectly fit in with our intuition.

Now, let us shift gears to discussions on the symmetrized version of the EMT. To consider the physical meaning of the symmetric and the antisymmetric parts of the EMT, the above relation (12.13) is quite useful. For the gauge and the fermion parts, generally, we immediately see that the following relation holds:

$$0 = \partial_\lambda J^{\lambda\mu\nu} = \partial_\lambda\left(x^\mu T^{\lambda\nu}_{\text{can}} - x^\nu T^{\lambda\mu}_{\text{can}} + S^{\lambda\mu\nu}_{\text{can}}\right) \Rightarrow T^{\mu\nu}_{\text{can}} - T^{\nu\mu}_{\text{can}} = -\partial_\lambda S^{\lambda\mu\nu}_{\text{can}}, \quad (12.14)$$

where $T^{\mu\nu}_{\text{can}} \equiv T^{\mu\nu}_{A,\text{can}} + T^{\mu\nu}_{\psi,\text{can}}$ and $S^{\lambda\mu\nu}_{\text{can}} \equiv S^{\lambda\mu\nu}_{A,\text{can}} + S^{\lambda\mu\nu}_{\psi,\text{can}}$. Therefore, the antisymmetric part of the canonical EMT is the source of the spin current. The EMT as conserved currents is not unique, but can be added by $\partial_\lambda K^{\lambda\mu\nu}$ satisfying $K^{\lambda\mu\nu} = -K^{\mu\lambda\nu}$, which would not change the conservation laws. One of the most interesting and important choices of $K^{\lambda\mu\nu}$ is

$$K^{\lambda\mu\nu}_{\text{Bel}} = \frac{1}{2}\left(S^{\lambda\mu\nu}_{\text{can}} - S^{\mu\lambda\nu}_{\text{can}} + S^{\nu\mu\lambda}_{\text{can}}\right)$$

$$= -F^{\lambda\mu}A^\nu + \frac{i}{4}\bar{\psi}\left(-i\varepsilon^{\lambda\mu\nu\rho}\gamma_5\gamma_\rho + 2g^{\mu\nu}\gamma^\lambda - 2g^{\lambda\nu}\gamma^\mu\right)\psi, \quad (12.15)$$

which gives the Belinfante–Rosenfeld form of the EMT, i.e., $T_{\text{Bel}}^{\mu\nu} \equiv T_{\text{can}}^{\mu\nu} + \partial_\lambda K_{\text{Bel}}^{\lambda\mu\nu}$. In the above, we used $\{\gamma^\lambda, \gamma^\mu\gamma^\nu\} = 2g^{\mu\nu}\gamma^\lambda - 2i\varepsilon^{\lambda\mu\nu\rho}\gamma_5\gamma_\rho$ to reach the second line (with the conventional definition of $\gamma_5 \equiv i\gamma^0\gamma^1\gamma^2\gamma^3$). We can show that if $T_{\text{Bel}}^{\mu\nu}$ is plugged into Eq. (12.14), the source is exactly canceled and $T_{\text{Bel}}^{\mu\nu} - T_{\text{Bel}}^{\nu\mu} = 0$ follows, which means that $T_{\text{Bel}}^{\mu\nu}$ is symmetric. (This is exactly the point where many people are puzzled especially when they want to formulate the spin hydrodynamics that seems to require antisymmetric components of the EMT, but in this article we will not go into this issue. Interested readers can consult a review [7].)

Now, we proceed to concrete expressions of the Belinfante EMT in the Abelian gauge theory. After several lines of calculations, one can find, for the gauge part,

$$\tilde{T}_{A,\text{Bel}}^{\mu\nu} = -F_\alpha^\mu F^{\nu\alpha} - \bar{\psi}\gamma^\mu e A^\nu \psi + \frac{1}{4}g^{\mu\nu}F^{\alpha\beta}F_{\alpha\beta}\,, \tag{12.16}$$

where the second term appears from the equations of motion, $\partial_\mu F^{\mu\nu} = \bar{\psi}i\gamma^\nu\psi$. The fermionic part needs a bit more labor to sort expressions out. From the definition, it is almost instant to get

$$\tilde{T}_{\psi,\text{Bel}}^{\mu\nu} = \bar{\psi}i\gamma^\mu \overset{\leftrightarrow}{\partial}{}^\nu \psi + \frac{1}{4}\varepsilon^{\mu\nu\lambda\rho}\partial_\lambda(\bar{\psi}\gamma_5\gamma_\rho\psi)\,. \tag{12.17}$$

It would be more appropriate to redefine these forms to move one term from $\tilde{T}_{A,\text{Bel}}^{\mu\nu}$ to $\tilde{T}_{\psi,\text{Bel}}^{\mu\nu}$ (which unchanges the sum, i.e., $\tilde{T}_{A,\text{Bel}}^{\mu\nu} + \tilde{T}_{\psi,\text{Bel}}^{\mu\nu} = T_{A,\text{Bel}}^{\mu\nu} + T_{\psi,\text{Bel}}^{\mu\nu}$), then the gauge invariance is manifested as

$$T_{A,\text{Bel}}^{\mu\nu} \equiv -F_\alpha^\mu F^{\nu\alpha} + \frac{1}{4}g^{\mu\nu}F^{\alpha\beta}F_{\alpha\beta}\,, \tag{12.18}$$

$$T_{\psi,\text{Bel}}^{\mu\nu} \equiv \bar{\psi}i\gamma^\mu \overset{\leftrightarrow}{D}{}^\nu \psi + \frac{1}{4}\varepsilon^{\mu\nu\lambda\rho}\partial_\lambda(\bar{\psi}\gamma_5\gamma_\rho\psi)\,. \tag{12.19}$$

These are very desirable expressions and all the terms are manifestly gauge invariant, thus corresponding to physical observables in principle. At this point, one might have thought that $T_{\psi,\text{Bel}}^{\mu\nu}$ does not look symmetric with respect to μ and ν. In a quite non-trivial way, one can prove that the above fermionic part is alternatively expressed as $T_{\psi,\text{Bel}}^{\mu\nu} = \bar{\psi}i\gamma^{(\mu}\overset{\leftrightarrow}{D}{}^{\nu)}\psi$, which is obviously symmetric.

Coming back to the angular momentum, we can introduce the Belinfante "improved" form for the angular momentum, i.e.,

$$J_{\text{Bel}}^{\lambda\mu\nu} \equiv J^{\lambda\mu\nu} + \partial_\rho\left(x^\mu K_{\text{Bel}}^{\rho\lambda\nu} - x^\nu K_{\text{Bel}}^{\rho\lambda\mu}\right)\,. \tag{12.20}$$

Because of the antisymmetric property of $K_{\text{Bel}}^{\rho\lambda\mu}$, obviously, $\partial_\lambda J_{\text{Bel}}^{\lambda\mu\nu} = 0$ follows as long as $\partial_\lambda J^{\lambda\mu\nu} = 0$ holds. Therefore, this newly defined $J_{\text{Bel}}^{\lambda\mu\nu}$ may well be qualified as a conserved physical observable. These definitions lead us to extremely interesting expressions, namely,

$$J_{A/\psi,\text{Bel}}^{\lambda\mu\nu} = x^\mu \tilde{T}_{A/\psi,\text{Bel}}^{\lambda\nu} - x^\nu \tilde{T}_{A/\psi,\text{Bel}}^{\lambda\mu}\,. \tag{12.21}$$

Such relations imply that the total angular momentum is given by something that looks like the OAM alone if we use the Belinfante improved forms. We sometimes hear people saying that the spin is identically vanishing in the Belinfante form, but this statement should be taken carefully. The spin part is simply unseen and the total angular momentum seemingly appears like the OAM even though the spin is already included. In the analogy to the QCD spin physics, the angular momentum identification as in Eq. (12.21) is known as the Ji decomposition.

12.3 Dirac Fermions and Physical and Pure Gauge Potentials

Discussions on the gauge part are a little cumbersome, and in this article we will mainly focus on the fermion part only, which, however, does not mean we drop the gauge fields. Let us reiterate basic definitions from the previous overview. In the canonical identification, in Eq. (12.10), the OAM and the SAM are given, respectively, by

$$L_{\psi,\text{can}} \equiv -i\psi^{\dagger} \boldsymbol{x} \times \nabla \psi \,, \qquad S_{\psi,\text{can}} \equiv -\frac{1}{2}\bar{\psi}\gamma_5 \boldsymbol{\gamma}\psi \,, \qquad (12.22)$$

where we defined $L^i \equiv \frac{1}{2}\varepsilon^{ijk}L^{jk}$ and $S^i \equiv \frac{1}{2}\varepsilon^{ijk}S^{jk}$. As we already discussed, $L_{\psi,\text{can}}$ is not gauge invariant, thus it cannot be a physical observable supposedly. Then, what about the Belinfante form? We can make a decomposition using Eq. (12.19). The latter term may well be called the spin part, with which we can compute $J^{\lambda\mu\nu}_{\psi,\text{Bel}}$ according to Eq. (12.21), and subtract added terms in Eq. (12.20). Some calculations yield

$$\tilde{S}_{\psi,\text{Bel}} = -\frac{1}{2}\bar{\psi}\gamma_5 \boldsymbol{\gamma}\psi - \frac{1}{2}i\boldsymbol{x} \times \nabla(\psi^{\dagger}\psi) \,. \qquad (12.23)$$

This expression is not gauge invariant, thus we shall redefine the spin to the same form as the canonical one which is manifestly gauge invariant and move unwanted terms to the orbital part. Thus, in this convention, we can reasonably adopt the following definitions:

$$L_{\psi,\text{Bel}} \equiv -i\psi^{\dagger} \boldsymbol{x} \times \boldsymbol{D}\psi \,, \qquad S_{\psi,\text{Bel}} \equiv S_{\psi,\text{can}} \,. \qquad (12.24)$$

In the high-energy physics context, the above identification is called Ji's orbital and spin angular momenta of quarks. Again, we make a caution remark; the Belinfante form has the total angular momentum that looks like the OAM, but this does not mean that the spin vanishes. Some people may say that the latter in Eq. (12.24) cannot be true since the Belinfante EMT has no antisymmetric part. This kind of criticism is meaningful when we need to construct the angular momentum in terms of the EMT, which is the case in the spin hydrodynamics, for example, [7,8].[2] See also

---

[2]K. F. thanks Wojciech Florkowski and Hidetoshi Taya for simulating conversations on this point which seem not to be very consistent to each other, and thus we just refer to their review and original literature here.

Table 12.1 Breakdown of the total angular momentum J from various contributions in the canonical (Jaffe–Manohar) decomposition (upper) and the Belinfante (Ji) decomposition (lower)

Canonical	$J = \underbrace{-\dfrac{1}{2}\bar{\psi}\gamma_5\boldsymbol{\gamma}\psi}_{\frac{1}{2}\Delta\Sigma} + \underbrace{\boldsymbol{E}\times\boldsymbol{A}}_{\Delta G}\underbrace{-i\psi^{\dagger}(\boldsymbol{x}\times\boldsymbol{\nabla})\psi}_{L^q_{\mathrm{can}}} + \underbrace{\boldsymbol{E}(\boldsymbol{x}\times\boldsymbol{\nabla})\boldsymbol{A}}_{L^g_{\mathrm{can}}}$
Belinfante	$J = \underbrace{-\dfrac{1}{2}\bar{\psi}\gamma_5\boldsymbol{\gamma}\psi}_{\frac{1}{2}\Delta\Sigma}\underbrace{-i\psi^{\dagger}(\boldsymbol{x}\times\boldsymbol{D})\psi}_{L^q_{\mathrm{Ji}}} + \underbrace{\boldsymbol{x}\times(\boldsymbol{E}\times\boldsymbol{B})}_{J^g_{\mathrm{Ji}}}$

Refs. [9, 10] for observable effects of different spin tensors, which may be significant especially in nonequilibrium [11]. Probably one way to define the spin part out from the Belinfante symmetrized form of the EMT is the Gordon decomposition (as Berry defined the gauge-invariant optical spin [12]) which is also applicable to massless theories. In any case, if we do not have to refer to the EMT, Eq. (12.24) is just a natural way of defining $\boldsymbol{S}_{\psi,\mathrm{Bel}}$, satisfying the correct commutation relation. Now we symbolically summarize the decomposition and the corresponding QCD terminology in Table 12.1.

Now, in this convention, the spin part has no ambiguity; it is gauge invariant as it should be, representing a physical observable for sure. The subtle (and thus interesting) point is the orbital part, and then one may be tempted to conclude that the canonical one makes no physical sense, and this conclusion seems to be unbreakable. An intriguing possibility has been suggested, however, in the high-energy physics context [13] inspired by QED studies and photon experiments (see, for example, Ref. [14] for very inspiring but a little mystical discussions including Lipkin's Zilch which is a "useless" conserved charge in QED), which invoked interesting theoretical discussions; see Ref. [15], for example. In fact, this canonical form can be promoted to be a gauge-invariant canonical (gic) one (using the terminology of Ref. [16]) as

$$L_{\psi,\mathrm{can}} \to L_{\psi,\mathrm{gic}} \equiv -i\psi^{\dagger}\boldsymbol{x}\times\boldsymbol{D}_{\mathrm{pure}}\psi, \qquad (12.25)$$

where $\boldsymbol{D}_{\mathrm{pure}} \equiv \boldsymbol{\nabla} - ie\boldsymbol{A}_{\mathrm{pure}}$. Here, the vector potential is decomposed into two pieces, namely, $\boldsymbol{A} = \boldsymbol{A}_{\mathrm{phys}} + \boldsymbol{A}_{\mathrm{pure}}$ with $\boldsymbol{A}_{\mathrm{phys}}$ extracted as a gauge invariant part and $\boldsymbol{A}_{\mathrm{pure}}$ makes the field strength tensor vanishing; $\boldsymbol{\nabla}\times\boldsymbol{A}_{\mathrm{pure}} = 0$. More specifically, under a gauge transformation, $\boldsymbol{A}$ is changed as $\boldsymbol{A} \to \boldsymbol{A} + \boldsymbol{\nabla}\alpha$, and then, by definition, $\boldsymbol{A}_{\mathrm{phys}} \to \boldsymbol{A}_{\mathrm{phys}}$ and $\boldsymbol{A}_{\mathrm{pure}} \to \boldsymbol{A}_{\mathrm{pure}} + \boldsymbol{\nabla}\alpha$. One simplest decomposition satisfying these requirements is obtained from the Helmholtz decomposition, i.e., any vector can be represented as a sum of divergence free (transverse) and rotation free (longitudinal) vectors. For a more concrete demonstration, let us write down an explicit form as

$$A_{\mathrm{phys}} = \boldsymbol{\nabla}\times\boldsymbol{a}, \qquad A_{\mathrm{pure}} = -\boldsymbol{\nabla}\varphi, \qquad (12.26)$$

where

$$a(x) = \frac{1}{4\pi} \int_V dx' \, \frac{\nabla' \times A(x')}{|x - x'|} - \frac{1}{4\pi} \int_S dS' \times \frac{A(x')}{|x - x'|} \,, \qquad (12.27)$$

$$\varphi(x) = \frac{1}{4\pi} \int_V dx' \, \frac{\nabla' \cdot A(x')}{|x - x'|} - \frac{1}{4\pi} \int_S dS' \cdot \frac{A(x')}{|x - x'|} \,. \qquad (12.28)$$

In principle, now, all the terms involving A can be made gauge invariant. Then, a finite difference between the canonical and the Belinfante OAM is also a gauge-invariant quantity, which is often called the "potential" orbital angular momentum, i.e.,

$$L_{\psi,\text{Bel}} = L_{\psi,\text{gic}} - e\psi^\dagger x \times A_{\text{phys}}\psi \,. \qquad (12.29)$$

Here, we make a comment which is not crucial in the present discussions but essential for phenomenological applications and particularly for measurability. Even though the Helmholtz decomposition is unique, such a gauge-invariant decomposition itself is not unique. As discussed in Ref. [17], for example, a different choice could be possible and even preferable in the high-energy processes.

We note that Eq. (12.28) is highly non-local in space, and such "physical" photon should have a space-like extension. For static electromagnetic background fields, for example, photons are virtual and off shell, so that space-like components are experimentally accessible (or even the vector potentials are controlled from the beginning). In contrast, in the parton model at high energy, the gauge particles are on shell and travel at the speed of light (or speed of "gluon" so to speak). Then, for such propagating modes along the light-cone, the space-like profiles as in Eq. (12.28) are not to be probed by scatterings. In this case of the light-cone propagation, as prescribed in Ref. [17], the light-cone decomposition would be more physical. In the Abelian gauge theory, the alternative decomposition is as simple as

$$A^i_{\text{phys}}(x^-) \equiv \frac{1}{\partial^+} F^{+i} = \int dy^- \, \mathcal{K}(x^- - y^-) \, F^{+i}(y^-) \,, \qquad (12.30)$$

where $\mathcal{K}(x^-)$ is chosen according to the boundary condition at $x^- = \pm\infty$ in the light-cone gauge $A^+ = 0$; it is $\theta(x^-)$ for the retarded boundary condition, $-\theta(-x^-)$ for the advanced one, and $\frac{1}{2}[\theta(x^-) - \theta(-x^-)]$ for the mixed boundary condition. We would point out that not only in high-energy physics but also in the laser optics the spatially non-local decomposition in Eq. (12.28) may not be appropriate if the propagating lights (such as the monochromatic waves) are concerned. The analogy between physical contents in high-energy physics and optics has been sometimes emphasized in the literature (see Ref. [16], for example), but this important question of what would be the "natural" choice is frequently missing. Along these lines of the natural choice, a mathematical argument in connection to the geodesic in tangent space is found in Ref. [18]. In this article, the existence of A_{phys} suffices for our discussions at present.

12.4   Potential Angular Momentum and Physical Interpretation

One might have a feeling that such classification of slightly different OAMs (while the SAM is common in our convention) may be an academic problem, but we recall that each term represents some physical observable and the lack of correct understanding would cause paradoxical confusions. For instance, if one is interested in the Einstein–de Haas effect and/or the Barnett effect within a relativistic framework, an interplay between the OAM provided by mechanical rotation and the spin polarization measured by the magnetization underlies observable phenomena. We had discussed this issue with knowledgeable researchers, some of whom told us that such a relativistic extension of these effects may not exist after all... such a conclusion is typically drawn based on the proper knowledge of knowledgeable researchers that the covariant derivative makes the theoretical formulation manifestly gauge invariant and the derivative and the vector potential are inseparable then. In the previous section, however, we have already seen that we can evade this problem by introducing D_{pure}. Now, in this section, we would like to address a difference between D and D_{pure}.

This question would be highly reminiscent of a more familiar and classic problem of the kinetic and the canonical momenta of a charged particle under electromagnetic background. That is, in our convention of the covariant derivative, $\partial_\mu + ieA_\mu$ (i.e., e is taken to be negative), the canonical momentum should be $p_{\mathrm{can}} = m\dot{x} + eA$, while the kinetic one is $p_{\mathrm{kin}} = m\dot{x} = p_{\mathrm{can}} - eA$ in a non-relativistic system. Since the canonical momentum should fullfil the commutation relation, we should identify $p_{\mathrm{can}} = -i\hbar\nabla$ in the x-representation and p_{kin} corresponds to the covariant derivative. For the gauge-invariant definition of p_{can}, we can replace ∇ with D_{pure}. In other words, the translational symmetry is generated by not the covariant derivative but the derivative, so that p_{can} is the momentum that can be conserved for the symmetry reason. The difference can be easily understood in the simplest physical example; if a charged particle is placed in a constant and homogeneous electric field, then the electric field accelerates the charged particle. Therefore, on the one hand, p_{kin} should increase by the impulse, eEt. On the other hand, the vector potential $A = -Et$ gives the electric field, and obviously, $p_{\mathrm{can}} = p_{\mathrm{kin}} + eA$ is time independent and conserved. In summary, it is important to note the following differences:

$$D \quad \leftrightarrow \quad p_{\mathrm{kin}} \quad (\text{non-conserved}) ,$$
$$D_{\mathrm{pure}} \quad \leftrightarrow \quad p_{\mathrm{can}} \quad (\text{conserved}) . \qquad (12.31)$$

It might be a little counter-intuitive that D whose definition involves the gauge potential corresponds to the momentum carried by the charged particle only and D_{pure} gives the total conserved momentum. Physically speaking, however, such a correspondence is quite reasonable. In most cases, only the particle's p_{kin} can be directly measured, and this readily measurable quantity just corresponds to the covariant derivative. In reality, sometimes, p_{can} does matter as well especially when the conservation law accounts for observable phenomena.

In exactly the same way as p_{kin} and p_{can} of the charged particle, we can classify two orbital angular momenta as

$$x \times D \quad \leftrightarrow \quad L_{kin} \sim L_{\psi,Bel} \quad \text{(non-conserved)},$$
$$x \times D_{pure} \quad \leftrightarrow \quad L_{can} \sim L_{\psi,gic} \quad \text{(conserved)}. \tag{12.32}$$

The difference between L_{kin} and L_{can} is often called the "potential" angular momentum (see Ref. [19] for a recent analysis of this difference). Unlike the above trivial example of p_{kin} and p_{can} with a constant E, it could be often very non-trivial to imagine what physically causes the potential angular momentum. To see this more, armed with these general basics, let us turn to a concrete problem now. We shall take a very instructive example of Ref. [20] which is entitled, "Is the Angular Momentum of an Electron Conserved in a Uniform Magnetic Field?" and this title already explains the contents by itself. The authors of Ref. [20] considered the time evolution of the radial width ρ of an electron motion in a uniform magnetic field B using the Schrödinger equation. The Hamiltonian of such a (non-relativistic) system is given by

$$H = -\frac{\hbar^2 \nabla^2}{2m} + \frac{1}{2} m \omega_L^2 \rho^2 - i \hbar \omega_L \frac{\partial}{\partial \varphi}, \tag{12.33}$$

where $\omega_L = |eB|/(2m)$ (i.e., the Larmor frequency). In classical physics, the charged particle with electric charge e and mass m receives the Lorentz force to make a circular rotation with the cyclotron frequency $\omega_c = 2\omega_L = |eB|/m$. It is easy to write down the Heisenberg equation of motion for $\langle \rho^2 \rangle$ to find that its time evolution solves as [20]

$$\langle \rho^2 \rangle(t) = \tilde{\rho}^2 + \left(\langle \rho^2 \rangle(0) - \tilde{\rho}^2 \right) \cos(\omega_c t). \tag{12.34}$$

Because the kinetic orbital angular momentum along the magnetic direction (which is taken to be the z axis, as is the convention in the following discussions too) depends on the moment of inertia, and the moment of inertia is a function of the radial width, they are related to each other as $\langle (L_{kin})_z \rangle = $ (conserved canonical OAM) + $m\omega_L \langle \rho^2 \rangle$. Thus, these calculations explicitly show that $\langle L_{kin} \rangle$ is not conserved but has time oscillatory behavior $\propto \langle \rho^2 \rangle$. This is an interesting observation that illustrates qualitative differences between the classical and the quantum motions of an electron, but not such an unexpected one; in a general case, it is not L_{kin} but L_{can} that is conserved. The question worth thinking is what kind of physics fills in this gap by $m\omega_L \langle \rho^2 \rangle$.

The answer is explicated in Ref. [20]—this gap turns out to be exactly the angular momentum of the electromagnetic field. As we listed up in Table 12.1, the electromagnetic angular momentum in the Belinfante form reads

$$J_z^{field} = \int d^3x \, [x \times (E \times B)]_z. \tag{12.35}$$

This is an integration of x times the electromagnetic momentum represented by the Poynting vector, which might have looked more like the OAM, but this is the total angular momentum as we derived in our previous discussions of this article. As argued in Ref. [20], if the electromagnetic fields are static and $\nabla \times E = 0$ holds, this electromagnetic angular momentum can be rewritten into a convenient form as

$$J_z^{\text{field}} = \int d^3x \, (\nabla \cdot E)(x \times A_{\text{phys}})_z \,. \tag{12.36}$$

Here, we note that the integration by parts with $B = \nabla \times A = \nabla \times A_{\text{phys}}$ in Eq. (12.35) would lead to an expression similar to the canonical one in Table 12.1 but not Eq. (12.36). Only when $\nabla \times E = 0$ and $\nabla \cdot A_{\text{phys}} = 0$ (which is the definition in the Helmholtz decomposition) both hold, we can prove the above simplification (12.36).

For a uniform magnetic field, $A = \frac{B}{2}(-y, x, 0)$ in the symmetric gauge gives B along the z axis, and this already satisfies $\nabla \cdot A = 0$. Then, the explicit form of $(x \times A)_z$ is $\frac{B}{2}\rho^2$ with $\rho^2 = x^2 + y^2$. Since $\nabla \cdot E$ is nothing but the electric charge density, Eq. (12.36) under a uniform magnetic field eventually becomes

$$J_z^{\text{field}} = \frac{eB}{2}\langle \rho^2 \rangle = m\omega_L \langle \rho^2 \rangle \,. \tag{12.37}$$

This is precisely the potential angular momentum! There is a plain explanation of why J_z^{field} should appear to make the conserved angular momentum. Figure 12.1 is a corresponding illustration of a charged object placed in a uniform magnetic field. The red blob represents a charged particle distribution (i.e., charge density in classical physics and probability distribution in quantum mechanics). Such a charged object is a source resulting in Coulomb electric fields E, and $E \times B$ goes around the charged object. In this illustration, the charge is taken to be positive, but for an electron as we assumed in this section, the electric field should be directed oppositely and the Poynting vector goes in the other way around. Because of this circular structure of the Poynting vector, the electromagnetic fields have a nonzero angular momentum, which was found to be Eq. (12.37).

Still, the physical interpretation is quite non-trivial, we must say. Literally speaking, J_z^{field} is a purely electromagnetic contribution, and nevertheless, E extends from the charge source and in this sense we may well say that E is rather attributed to the matter property. If we are interested in the mechanical rotation as is the case in

Fig. 12.1 A charged object placed in a uniform magnetic field is surrounded by the Poynting vector $E \times B$ which carries an electromagnetic angular momentum contained in the conserved canonical angular momentum

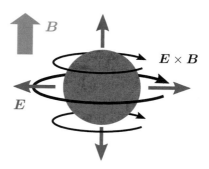

the Barnett and the Einstein–de Haas effects, however, we should count the kinetic angular momentum. Even in that case, this extra electromagnetic contribution could affect the kinetic angular momentum through the angular momentum conservation law.

12.5 Feynman's Angular Momentum Paradox and Possible Relevance to the Relativistic Nucleus–Nucleus Collision

Careful readers might have realized that the argument about J_z^{field} is essentially rooted in Feynman's angular momentum paradox in classical physics. The paradox is articulated in *The Feynman Lectures* and the original setup is composed of a conductor disk with a solenoid that controls the magnetic strength. For a detailed analysis of the original version of Feynman's angular momentum paradox, see Ref. [21], for example. Here, let us discuss a simplified version of Feynman's angular momentum paradox.

We suppose that a thin sphere is uniformly charged (whose total amount is denoted by Q), and a finite magnetic moment m is fixed at the center of the sphere (see Fig. 12.2). The electric (outside of the sphere) and the magnetic profiles are, respectively,

$$E = \frac{Q}{4\pi} \frac{x}{r^3}, \qquad B = \frac{1}{4\pi r^3}\left(\frac{3m \cdot xx}{r^2} - m\right). \qquad (12.38)$$

If m changes as a function of time, the magnetic field changes as well, which also results in an induction electric field due to Ampère's law. Then, the charged sphere feels a moment of force under this induced electric field, E_{ind}, and the sphere is accelerated for rotation. The space integrated moment of force is, after some patient calculations, found to take the form

$$N = \int dS \cdot \frac{x \times Q E_{\text{ind}}}{4\pi R^2} = -\frac{Q\dot{m}}{6\pi R}, \qquad (12.39)$$

where R denotes the radius of the sphere. Therefore, if m decreases, the sphere takes a positive moment of force to acquire a mechanical angular momentum. The question is: how can the angular momentum conservation law be satisfied? This phenomenon may sound similar to the Einstein–de Haas effect, but one should recall two important

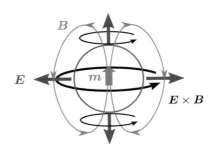

Fig. 12.2 A charged thin sphere (red circle) and a magnetic moment at the center of the sphere. The dipolar magnetic fields and the Coulomb electric fields make circulating Poynting vectors

differences. One is that the object should be charge neutral in the Einstein–de Haas effect, and another is that in this classical example there is no magnetization at all. There are many variants of Feynman's paradox, and they usually belong to classical physics (no spin effects).

Readers should be already aware of the resolution. As indicated in Fig. 12.2, the electromagnetic field generates circulating Poynting vectors. Actually, from explicit expressions of Eq. (12.38), we can obtain the angular momentum distribution as

$$x \times (E \times B) = \frac{Q}{(4\pi)^2 r^6}(r^2 m - xx \cdot m). \tag{12.40}$$

Therefore, the total angular momentum integrated in space outside of the sphere turns out to be

$$J^{\text{field}} = \frac{Qm}{6\pi R}. \tag{12.41}$$

It is obvious that the angular momentum in mechanical rotation originates from the loss in J^{field}, so that the total angular momentum is surely conserved. See Ref. [22] for related discussions on the Poynting vector contributions in classical electromagnetism. Interestingly, this result of Eq. (12.41) was extended to the one-loop QED level which turned out to be free from a short-distance cutoff [23].

In this classical example of Feynman's paradox, the essential point is that either E or B changes to make a finite difference in $x \times (E \times B)$ from which the mechanical rotation is induced. The novelty in the quantum mechanical example seen in the previous section is that quantum oscillations exhibit time dependence even for constant E and B. In both cases the important lesson is that as long as we prefer to use the Belinfante improved form for the EMT and the angular momenta, the covariant derivative in the matter sector makes all the expressions manifestly gauge invariant, and then we can access the kinetic angular momentum of the matter which is not necessarily conserved.

So far, we have been having general discussions not specifying any experimental realizations at all. Let us now consider some possible applications to the high-energy nucleus–nucleus collisions. It is known that the OAM in the non-central nucleus–nucleus collision can reach a gigantic value as large as $\sim 10^5 \hbar$ as evaluated in the AMPT model [24], supported by experimental data [25]. Here, we can make an order of magnitude estimate of extra angular momentum from the decay of the magnetic field using Eq. (12.37). Our following discussions may look different from Ref. [26] which addresses a possibility of the spin polarization by the induced electric fields. There are some discrepancies from spatial inhomogeneity as well as temporally decaying magnetic properties and also from hydrodynamic treatments, but we note that microscopically underlying physics is common.

The magnetic field created right after the collision is of order $eB \sim \text{GeV}^2$ at largest, and $\langle \rho^2 \rangle$ in the collision geometry is around $\sim 10 \, \text{fm}^2$. Therefore, if the magnetic field quickly decays whose time scale is $\sim 0.1 \, \text{fm/c}$, this field angular momentum, $J_z^{\text{field}} \sim 10 \, \text{GeV}^2 \cdot \text{fm}^2 \sim 100 \hbar$, is transferred to the angular momentum

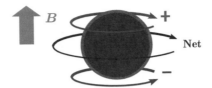

Fig. 12.3 A net induced angular momentum with faster rotating positively and negatively charged particles. There are more positively charged particles in a plasma because of protons in the participant particles

of a single particle. The net charge is $0.1Z \sim Z$ depending on the impact parameter and the baryon stopping, where $Z \sim 100$ is the atomic number of the heavy nucleus, and so the net angular momentum is of order $10^3 \sim 10^4 \hbar$. Here, we would emphasize that the time scale is irrelevant. This angular momentum arises as a consequence of the conservation law, and it is just there for any fast decaying B (except loss by polarized photon emissions). From this simple estimate, we can conclude that the net induced angular momentum is significantly smaller than the primarily produced angular momentum $\sim 10^5 \hbar$. This is, however, not yet the end of the story. In the reality of the nucleus–nucleus collision, a plasma state consists of positively and negatively charged particles and the net charge is only its small fraction. Then, we can anticipate at least an order of magnitude larger angular momenta for positively and negatively charged components in the opposite directions which mostly cancel to lead to the net angular momentum (see Fig. 12.3). If this two-component model is a good approximation (which is dictated by the interaction strength between two components), each charged sector could carry the induced angular momentum $\sim 10^4 \sim 10^5 \hbar$, comparable to the primarily produced angular momentum. Interestingly, such a two-component picture with opposite rotation has been confirmed in the numerical simulation for the Einstein–de Haas effect in cold atomic systems [27,28].

We have some more ideas [say, the global polarization should be also associated with the field angular momentum by Eq. (12.41) whose effect has never been studied] and have in mind applications to the local polarization measurements, but we shall stop our stories here. Such ideas as well as more detailed and quantitative calculations will be reported in a separate publication.

12.6   Epilogue

The interplay between the OAM and SAM is an old subject, but its entanglement with chirality in a relativistic framework is a quite new research field. The ultra-relativistic nucleus–nucleus collision experiments have been offering inspiring data, and high-energy nuclear physicists have become wiser and wiser over decades. Some people, especially researchers close to but not directly in our field, might have assumed that the physics of the relativistic nucleus–nucleus collision passed a peak. We must say, such an assumption is nothing but a hasty conclusion. The nucleus–nucleus collision still continues to provide us with surprises one after another.

Recent investigations on the OAM and SAM decomposition and their interactions are motivated by the Λ and $\bar{\Lambda}$ polarization measurements, but we should emphasize that this is not a hip excitement. Theoretically speaking, this is an extremely profound subject, and there are still many things that nobody has understood. One common criticism against such kind of theory problem would be what you call "profound" is just what I would call "academic", or give me any measurable observable? Indeed it is not easy to make a new proposal for the nucleus–nucleus collision. Nevertheless, we can export our ideas inspired by the nucleus–nucleus collision to other physics fields such as cold atomic systems and laser optics. Still, even if exported ideas are adapted in a different shape, we can proudly say that this is a tremendous achievement from the high-energy nuclear physics!

We also emphasize that the OAM/SAM decomposition and also the EMT measurements are of central interest to the future coming electron-ion collider (EIC) physics. At least three pretty independent communities, the heavy-ion collision, the proton spin, and the laser optics, have worked on very similar physics, and now is the time to put all our wisdom together toward the next generation breakthrough.

We would like to make acknowledgments. We thank Zebin Qiu for successful collaborations. K. F. is grateful to Kazuya Mameda for extremely useful discussions about ongoing projects on the Einstein–de Haas effect. K. F. also thanks Yoshi Hatta for interesting and critical (as always) conversations. K. F. also would like to acknowledge very useful and sometimes confusing (and so interesting) conversations with Francesco Becattini, Wojciech Florkowski, and Xu-Guang Huang.

References

1. Fukushima, K., Pu, S., Qiu, Z.: Eddy magnetization from the chiral Barnett effect. Phys. Rev. **A99**, 032105 (2019). https://doi.org/10.1103/PhysRevA.99.032105. arXiv:1808.08016 [hep-ph]
2. Barnett, S.J.: Gyromagnetic and electron-inertia effects. Rev. Mod. Phys. **7**, 129–166 (1935). https://doi.org/10.1103/RevModPhys.7.129
3. Barnett, S.M.: Relativistic electron vortices. Phys. Rev. Lett. **118**, 114802 (2017). https://doi.org/10.1103/PhysRevLett.118.114802
4. Tomonaga, S., Oka, T.: The Story of Spin. University of Chicago Press (1998)
5. Leader, E., Lorcé, C.: The angular momentum controversy: what's it all about and does it matter? Phys. Rept. **541**, 163–248 (2014). https://doi.org/10.1016/j.physrep.2014.02.010. arXiv:1309.4235 [hep-ph]
6. Wakamatsu, M.: Is gauge-invariant complete decomposition of the nucleon spin possible? Int. J. Mod. Phys. **A29**, 1430012 (2014). https://doi.org/10.1142/S0217751X14300129. arXiv:1402.4193 [hep-ph]
7. Florkowski, W., Ryblewski, R., Kumar, A.: Relativistic hydrodynamics for spin-polarized fluids. Prog. Part. Nucl. Phys. **108**, 103709 (2019). https://doi.org/10.1016/j.ppnp.2019.07.001. arXiv:1811.04409 [nucl-th]
8. Hattori, K., Hongo, M., Huang, X.-G., Matsuo, M., Taya, H.: Fate of spin polarization in a relativistic fluid: an entropy-current analysis. Phys. Lett. **B795**, 100–106 (2019). https://doi.org/10.1016/j.physletb.2019.05.040. arXiv:1901.06615 [hep-th]
9. Becattini, F., Tinti, L.: Thermodynamical inequivalence of quantum stress-energy and spin tensors. Phys. Rev. **D84**, 025013 (2011). https://doi.org/10.1103/PhysRevD.84.025013. arXiv:1101.5251 [hep-th]

10. Becattini, F., Tinti, L.: Nonequilibrium thermodynamical inequivalence of quantum stress-energy and spin tensors. Phys. Rev. **D87**(2), 025029 (2013). https://doi.org/10.1103/PhysRevD. 87.025029. arXiv:1209.6212 [hep-th]

11. Becattini, F., Florkowski, W., Speranza, E.: Spin tensor and its role in non-equilibrium thermodynamics. Phys. Lett. **B789**, 419–425 (2019). https://doi.org/10.1016/j.physletb.2018.12.016. arXiv:1807.10994 [hep-th]

12. Berry, M.V.: Optical currents. J. Opt. A: Pure Appl. Opt. **11**, 094001 (2009). https://doi.org/10. 1088/1464-4258/11/9/094001

13. Chen, X.-S., Lu, X.-F.,Sun, W.-M. , Wang, F., Goldman, T.: Spin and orbital angular momentum in gauge theories: Nucleon spin structure and multipole radiation revisited. Phys. Rev. Lett. **100**, 232002 (2008). https://doi.org/10.1103/PhysRevLett.100.232002. arXiv:0806.3166 [hep-ph]

14. Cameron, R.P., Barnett, S.M., Yao, A.M.: Optical helicity, optical spin and related quantities in electromagnetic theory. New J. Phys. **14**, 053050 (2012). https://doi.org/10.1088/1367-2630/ 14/5/053050

15. Wakamatsu, M.: On gauge-invariant decomposition of nucleon spin. Phys. Rev. **D81**, 114010 (2010). https://doi.org/10.1103/PhysRevD.81.114010. arXiv:1004.0268 [hep-ph]

16. Leader, E.: The photon angular momentum controversy: resolution of a conflict between laser optics and particle physics. Phys. Lett. **B756**, 303–308 (2016). https://doi.org/10.1016/j. physletb.2016.03.023. arXiv:1510.03293 [hep-ph]

17. Hatta, Y.: Gluon polarization in the nucleon demystified. Phys. Rev. **D84**, 041701 (2011). https:// doi.org/10.1103/PhysRevD.84.041701. arXiv:1101.5989 [hep-ph]

18. Lorcé, C.: Wilson lines and orbital angular momentum. Phys. Lett. **B719**, 185–190 (2013). https://doi.org/10.1016/j.physletb.2013.01.007. arXiv:1210.2581 [hep-ph]

19. Wakamatsu, M., Kitadono, Y., Zhang, P.-M.: The issue of gauge choice in the Landau problem and the physics of canonical and mechanical orbital angular momenta. Ann. Phys. **392**, 287–322 (2018). https://doi.org/10.1016/j.aop.2018.03.019. arXiv:1709.09766 [hep-ph]

20. Greenshields, C.R., Stamps, R.L., Franke-Arnold, S., Barnett, S.M.: Is the angular momentum of an electron conserved in a uniform magnetic field? Phys. Rev. Lett. **113**, 240404 (2014). https://doi.org/10.1103/PhysRevLett.113.240404

21. Lombardi, G.: Feynman's disk paradox. Am. J. Phys. **51**, 213–214 (1983). https://doi.org/10. 1119/1.13272

22. Higbie, J.: Angular momentum in the field of an electron. Am. J. Phys. **56**, 378–379 (1988). https://doi.org/10.1119/1.15597

23. Damski, B.: Electromagnetic angular momentum of the electron: one-loop studies. Nucl. Phys. **B949**, 114828 (2019). https://doi.org/10.1016/j.nuclphysb.2019.114828

24. Jiang, Y., Lin, Z.-W., Liao, J.: Rotating quark-gluon plasma in relativistic heavy ion collisions. Phys. Rev. **C94**, 044910 (2016). https://doi.org/10.1103/PhysRevC.94.044910, https://doi.org/10.1103/PhysRevC.95.049904. arXiv:1602.06580 [hep-ph]. [Erratum: Phys. Rev.C95,no.4,049904(2017)]

25. STAR Collaboration, Adamczyk, L., et al.: Global Λ hyperon polarization in nuclear collisions: evidence for the most vortical fluid. Nature **548**, 62–65 (2017). https://doi.org/10.1038/ nature23004. arXiv:1701.06657 [nucl-ex]

26. Guo, X., Liao, J., Wang, E.: Magnetic field in the charged subatomic swirl. arXiv:1904.04704 [hep-ph]

27. Kawaguchi, Y., Saito, H., Ueda, M.: Einstein–de Haas effect in dipolar Bose-Einstein condensates. Phys. Rev. Lett. **96**, 080405 (2006). https://doi.org/10.1103/PhysRevLett.96.080405

28. Ebling, U., Ueda, M.: Einstein-de Haas effect in a dipolar Fermi gas. arXiv:1701.05446 [cond-mat.quant-gas]

Printed in the United States
by Baker & Taylor Publisher Services